T0201718

Machine Learning Refined

With its intuitive yet rigorous approach to machine learning, this text provides students with the fundamental knowledge and practical tools needed to conduct research and build data-driven products. The authors prioritize geometric intuition and algorithmic thinking, and include detail on all the essential mathematical prerequisites, to offer a fresh and accessible way to learn. Practical applications are emphasized, with examples from disciplines including computer vision, natural language processing, economics, neuroscience, recommender systems, physics, and biology. Over 300 color illustrations are included and have been meticulously designed to enable an intuitive grasp of technical concepts, and over 100 in-depth coding exercises (in `Python`) provide a real understanding of crucial machine learning algorithms. A suite of online resources including sample code, data sets, interactive lecture slides, and a solutions manual are provided online, making this an ideal text both for graduate courses on machine learning and for individual reference and self-study.

Jeremy Watt received his PhD in Electrical Engineering from Northwestern University, and is now a machine learning consultant and educator. He teaches machine learning, deep learning, mathematical optimization, and reinforcement learning at Northwestern University.

Reza Borhani received his PhD in Electrical Engineering from Northwestern University, and is now a machine learning consultant and educator. He teaches a variety of courses in machine learning and deep learning at Northwestern University.

Aggelos K. Katsaggelos is the Joseph Cummings Professor at Northwestern University, where he heads the Image and Video Processing Laboratory. He is a Fellow of IEEE, SPIE, EURASIP, and OSA and the recipient of the IEEE Third Millennium Medal (2000).

Machine Learning Refined

Foundations, Algorithms, and Applications

JEREMY WATT

Northwestern University, Illinois

REZA BORHANI

Northwestern University, Illinois

AGGELOS K. KATSAGGELOS

Northwestern University, Illinois

CAMBRIDGE
UNIVERSITY PRESS

University Printing House, Cambridge CB2 8BS, United Kingdom

One Liberty Plaza, 20th Floor, New York, NY 10006, USA

477 Williamstown Road, Port Melbourne, VIC 3207, Australia

314–321, 3rd Floor, Plot 3, Splendor Forum, Jasola District Centre, New Delhi – 110025, India

79 Anson Road, #06–04/06, Singapore 079906

Cambridge University Press is part of the University of Cambridge.

It furthers the University's mission by disseminating knowledge in the pursuit of education, learning, and research at the highest international levels of excellence.

www.cambridge.org
Information on this title: www.cambridge.org/9781108480727
DOI: 10.1017/9781108690935

First published 2020

Printed and bound in Great Britain by Clays Ltd, Elcograf S.p.A.

A catalogue record for this publication is available from the British Library.

ISBN 978-1-108-48072-7 Hardback

Additional resources for this publication at www.cambridge.org/watt2

To our families:

Deb, Robert, and Terri

Soheila, Ali, and Maryam

Ειρηνη, Ζωη, Σοφια, and Ειρηνη

Contents

Preface

For eons we humans have sought out *rules* or *patterns* that accurately describe how important systems in the world around us work, whether these systems be agricultural, biological, physical, financial, etc. We do this because such rules allow us to understand a system better, accurately predict its future behavior and ultimately, control it. However, the process of finding the "right" rule that seems to govern a given system has historically been no easy task. For most of our history *data* (glimpses of a given system at work) has been an extremely scarce commodity. Moreover, our ability to *compute*, to try out various rules to see which most accurately represents a phenomenon, has been limited to what we could accomplish by hand. Both of these factors naturally limited the range of phenomena scientific pioneers of the past could investigate and inevitably forced them to use philosophical and/or visual approaches to rule-finding. Today, however, we live in a world awash in data, and have colossal computing power at our fingertips. Because of this, we lucky descendants of the great pioneers can tackle a much wider array of problems and take a much more empirical approach to rule-finding than our forbears could. Machine learning, the topic of this textbook, is a term used to describe a broad (and growing) collection of pattern-finding algorithms designed to properly identify system rules empirically and by leveraging our access to potentially enormous amounts of data and computing power.

In the past decade the user base of machine learning has grown dramatically. From a relatively small circle in computer science, engineering, and mathematics departments the users of machine learning now include students and researchers from every corner of the academic universe, as well as members of industry, data scientists, entrepreneurs, and machine learning enthusiasts. This textbook is the result of a complete tearing down of the standard curriculum of machine learning into its most fundamental components, and a curated reassembly of those pieces (painstakingly polished and organized) that we feel will most benefit this broadening audience of learners. It contains fresh and intuitive yet rigorous descriptions of the most fundamental concepts necessary to conduct research, build products, and tinker.

Book Overview

The second edition of this text is a complete revision of our first endeavor, with virtually every chapter of the original rewritten from the ground up and eight new chapters of material added, doubling the size of the first edition. Topics from the first edition, from expositions on gradient descent to those on One-versus-All classification and Principal Component Analysis have been reworked and polished. A swath of new topics have been added throughout the text, from derivative-free optimization to weighted supervised learning, feature selection, nonlinear feature engineering, boosting-based cross-validation, and more.

While heftier in size, the intent of our original attempt has remained unchanged: to explain machine learning, from first principles to practical implementation, in the simplest possible terms. A big-picture breakdown of the second edition text follows below.

Part I: Mathematical Optimization (Chapters 2–4)

Mathematical optimization is the workhorse of machine learning, powering not only the tuning of individual machine learning models (introduced in Part II) but also the framework by which we determine appropriate models themselves via cross-validation (discussed in Part III of the text).

In this first part of the text we provide a complete introduction to mathematical optimization, from basic zero-order (derivative-free) methods detailed in Chapter 2 to fundamental and advanced first-order and second-order methods in Chapters 3 and 4, respectively. More specifically this part of the text contains complete descriptions of local optimization, *random search* methodologies, *gradient descent*, and *Newton's method*.

Part II: Linear Learning (Chapters 5–9)

In this part of the text we describe the fundamental components of cost function based machine learning, with an emphasis on linear models.

This includes a complete description of *supervised learning* in Chapters 5–7 including linear regression, two-class, and multi-class classification. In each of these chapters we describe a range of perspectives and popular design choices made when building supervised learners.

In Chapter 8 we similarly describe *unsupervised learning*, and Chapter 9 contains an introduction to fundamental *feature engineering* practices including popular *histogram* features as well as various input normalization schemes, and *feature selection* paradigms.

Part III: Nonlinear Learning (Chapters 10–14)

In the final part of the text we extend the fundamental paradigms introduced in Part II to the general nonlinear setting.

We do this carefully beginning with a basic introduction to nonlinear supervised and unsupervised learning in Chapter 10, where we introduce the motivation, common terminology, and notation of nonlinear learning used throughout the remainder of the text.

In Chapter 11 we discuss how to *automate* the selection of appropriate nonlinear models, beginning with an introduction to *universal approximation*. This naturally leads to detailed descriptions of *cross-validation*, as well as *boosting*, *regularization*, *ensembling*, and *K-folds* cross-validation.

With these fundamental ideas in-hand, in Chapters 12–14 we then dedicate an individual chapter to each of the three popular universal approximators used in machine learning: *fixed-shape kernels*, *neural networks*, and *trees*, where we discuss the strengths, weaknesses, technical eccentricities, and usages of each popular universal approximator.

To get the most out of this part of the book we strongly recommend that Chapter 11 and the fundamental ideas therein are studied and understood before moving on to Chapters 12–14.

Part IV: Appendices

This shorter set of appendix chapters provides a complete treatment on advanced optimization techniques, as well as a thorough introduction to a range of subjects that the readers will need to understand in order to make full use of the text.

Appendix A continues our discussion from Chapters 3 and 4, and describes *advanced first- and second-order optimization techniques*. This includes a discussion of popular extensions of gradient descent, including *mini-batch optimization*, *momentum acceleration*, *gradient normalization*, and the result of combining these enhancements in various ways (producing e.g., the RMSProp and Adam first order algorithms) – and Newton's method – including *regularization* schemes and *Hessian-free* methods.

Appendix B contains a tour of *computational calculus* including an introduction to the derivative/gradient, higher-order derivatives, the Hessian matrix, numerical differentiation, forward and backward (backpropogation) automatic differentiation, and Taylor series approximations.

Appendix C provides a suitable background in *linear and matrix algebra*, including vector/matrix arithmetic, the notions of spanning sets and orthogonality, as well as eigenvalues and eigenvectors.

Readers: How To Use This Book

This textbook was written with first-time learners of the subject in mind, as well as for more knowledgeable readers who yearn for a more intuitive and serviceable treatment than what is currently available today. To make full use of the text one needs only a basic understanding of vector algebra (mathematical functions, vector arithmetic, etc.) and computer programming (for example, basic proficiency with a dynamically typed language like Python). We provide complete introductory treatments of other prerequisite topics including linear algebra, vector calculus, and automatic differentiation in the appendices of the text. Example "roadmaps," shown in Figures 0.1–0.4, provide suggested paths for navigating the text based on a variety of learning outcomes and university courses (ranging from a course on the essentials of machine learning to special topics – as described further under "Instructors: How to use this Book" below).

We believe that *intuitive leaps precede intellectual ones*, and to this end defer the use of probabilistic and statistical views of machine learning in favor of a fresh and consistent geometric perspective throughout the text. We believe that this perspective not only permits a more intuitive understanding of individual concepts in the text, but also that it helps establish revealing connections between ideas often regarded as fundamentally distinct (e.g., the logistic regression and Support Vector Machine classifiers, kernels and fully connected neural networks, etc.). We also highly emphasize the importance of *mathematical optimization* in our treatment of machine learning. As detailed in the "Book Overview" section above, optimization is the workhorse of machine learning and is fundamental at many levels – from the tuning of individual models to the general selection of appropriate nonlinearities via cross-validation. Because of this a strong understanding of mathematical optimization is requisite if one wishes to deeply understand machine learning, and if one wishes to be able to implement fundamental algorithms.

To this end, we place significant emphasis on the design and implementation of algorithms throughout the text with implementations of fundamental algorithms given in Python. These fundamental examples can then be used as building blocks for the reader to help complete the text's programming exercises, allowing them to "get their hands dirty" and "learn by doing," practicing the concepts introduced in the body of the text. While in principle any programming language can be used to complete the text's coding exercises, we highly recommend using Python for its ease of use and large support community. We also recommend using the open-source Python libraries NumPy, autograd, and matplotlib, as well as the Jupyter notebook editor to make implementing and testing code easier. A complete set of installation instructions, datasets, as well as starter notebooks for many exercises can be found at

https://github.com/jermwatt/machine_learning_refined

Instructors: How To Use This Book

Chapter slides associated with this textbook, datasets, along with a large array of instructional interactive Python widgets illustrating various concepts throughout the text, can be found on the github repository accompanying this textbook at

https://github.com/jermwatt/machine_learning_refined

This site also contains instructions for installing Python as well as a number of other free packages that students will find useful in completing the text's exercises.

This book has been used as a basis for a number of machine learning courses at Northwestern University, ranging from introductory courses suitable for undergraduate students to more advanced courses on special topics focusing on optimization and deep learning for graduate students. With its treatment of foundations, applications, and algorithms this text can be used as a primary resource or in fundamental component for courses such as the following.

Machine learning essentials treatment: an introduction to the essentials of machine learning is ideal for undergraduate students, especially those in quarter-based programs and universities where a deep dive into the entirety of the book is not feasible due to time constraints. Topics for such a course can include: gradient descent, logistic regression, Support Vector Machines, One-versus-All and multi-class logistic regression, Principal Component Analysis, K-means clustering, the essentials of feature engineering and selection, cross-validation, regularization, ensembling, bagging, kernel methods, fully connected neural networks, and trees. A recommended roadmap for such a course – including recommended chapters, sections, and corresponding topics – is shown in Figure 0.1.

Machine learning full treatment: a standard machine learning course based on this text expands on the essentials course outlined above both in terms of breadth and depth. In addition to the topics mentioned in the essentials course, instructors may choose to cover Newton's method, Least Absolute Deviations, multi-output regression, weighted regression, the Perceptron, the Categorical Cross Entropy cost, weighted two-class and multi-class classification, online learning, recommender systems, matrix factorization techniques, boosting-based feature selection, universal approximation, gradient boosting, random forests, as well as a more in-depth treatment of fully connected neural networks involving topics such as batch normalization and early-stopping-based regularization. A recommended roadmap for such a course – including recommended chapters, sections, and corresponding topics – is illustrated in Figure 0.2.

Mathematical optimization for machine learning and deep learning: such a course entails a comprehensive description of zero-, first-, and second-order optimization techniques from Part I of the text (as well as Appendix A) including: coordinate descent, gradient descent, Newton's method, quasi-Newton methods, stochastic optimization, momentum acceleration, fixed and adaptive steplength rules, as well as advanced normalized gradient descent schemes (e.g., Adam and RMSProp). These can be followed by an in-depth description of the feature engineering processes (especially standard normalization and PCA-sphering) that speed up (particularly first-order) optimization algorithms. All students in general, and those taking an optimization for machine learning course in particular, should appreciate the fundamental role optimization plays in identifying the "right" nonlinearity via the processes of boosting and regularization based cross-validation, the principles of which are covered in Chapter 11. Select topics from Chapter 13 and Appendix B – including backpropagation, batch normalization, and foward/backward mode of automatic differentiation – can also be covered. A recommended roadmap for such a course – including recommended chapters, sections, and corresponding topics – is given in Figure 0.3.

Introductory portion of a course on deep learning: such a course is best suitable for students who have had prior exposure to fundamental machine learning concepts, and can begin with a discussion of appropriate first order optimization techniques, with an emphasis on stochastic and mini-batch optimization, momentum acceleration, and normalized gradient schemes such as Adam and RMSProp. Depending on the audience, a brief review of fundamental elements of machine learning may be needed using selected portions of Part II of the text. A complete discussion of fully connected networks, including a discussion of backpropagation and forward/backward mode of automatic differentiation, as well as special topics like batch normalization and early-stopping-based cross-validation, can then be made using Chapters 11, 13, and Appendices A and B of the text. A recommended roadmap for such a course – including recommended chapters, sections, and corresponding topics – is shown in Figure 0.4. Additional recommended resources on topics to complete a standard course on deep learning – like convolutional and recurrent networks – can be found by visiting the text's github repository.

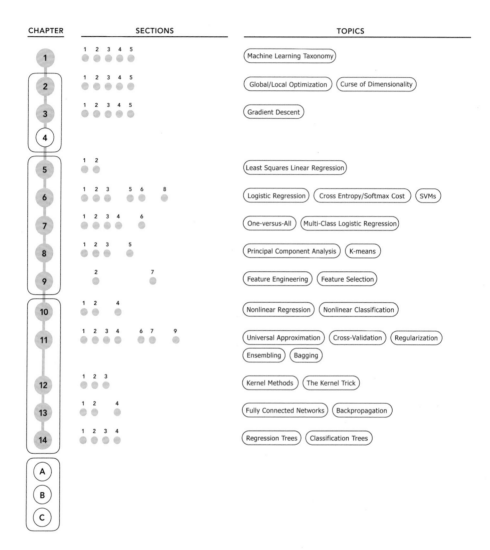

Figure 0.1 Recommended study roadmap for a course on the essentials of machine learning, including requisite chapters (left column), sections (middle column), and corresponding topics (right column). This essentials plan is suitable for time-constrained courses (in quarter-based programs and universities) or self-study, or where machine learning is not the sole focus but a key component of some broader course of study. Note that chapters are grouped together visually based on text layout detailed under "Book Overview" in the Preface. See the section titled "Instructors: How To Use This Book" in the Preface for further details.

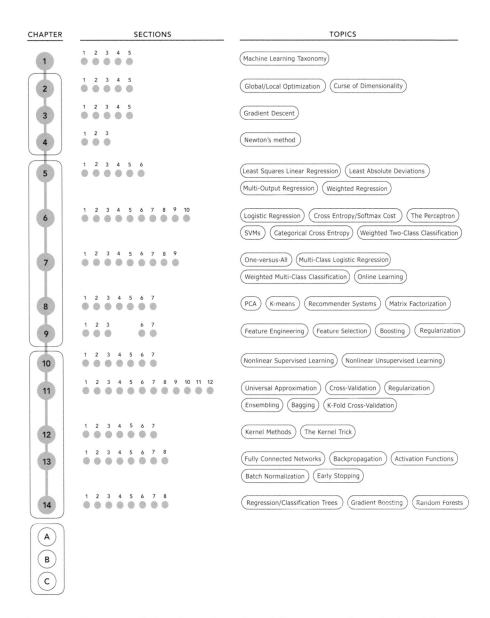

Figure 0.2 Recommended study roadmap for a full treatment of standard machine learning subjects, including chapters, sections, as well as corresponding topics to cover. This plan entails a more in-depth coverage of machine learning topics compared to the essentials roadmap given in Figure 0.1, and is best suited for senior undergraduate/early graduate students in semester-based programs and passionate independent readers. See the section titled "Instructors: How To Use This Book" in the Preface for further details.

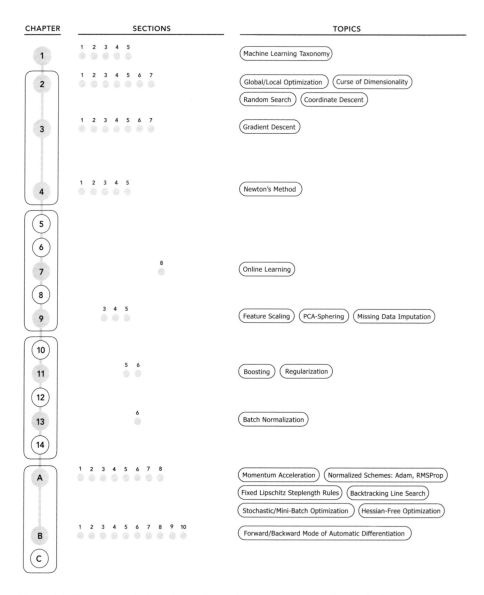

Figure 0.3 Recommended study roadmap for a course on mathematical optimization for machine learning and deep learning, including chapters, sections, as well as topics to cover. See the section titled "Instructors: How To Use This Book" in the Preface for further details.

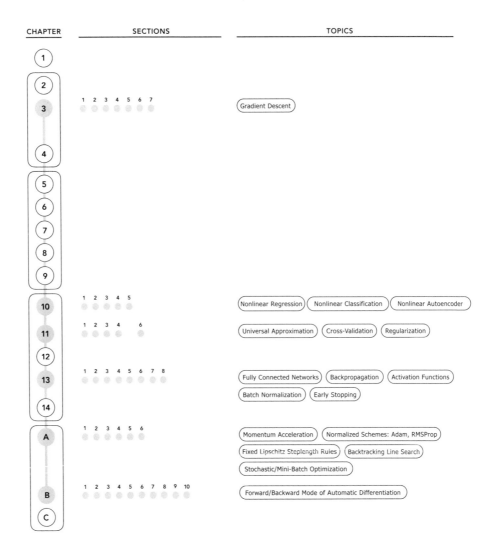

Figure 0.4 Recommended study roadmap for an introductory portion of a course on deep learning, including chapters, sections, as well as topics to cover. See the section titled "Instructors: How To Use This Book" in the Preface for further details.

Acknowledgements

This text could not have been written in anything close to its current form without the enormous work of countless genius-angels in the `Python` open-source community, particularly authors and contributers of `NumPy`, `Jupyter`, and `matplotlib`. We are especially grateful to the authors and contributors of `autograd` including Dougal Maclaurin, David Duvenaud, Matt Johnson, and Jamie Townsend, as `autograd` allowed us to experiment and iterate on a host of new ideas included in the second edition of this text that greatly improved it as well as, we hope, the learning experience for its readers.

We are also very grateful for the many students over the years that provided insightful feedback on the content of this text, with special thanks to Bowen Tian who provided copious amounts of insightful feedback on early drafts of the work.

Finally, a big thanks to Mark McNess Rosengren and the entire Standing Passengers crew for helping us stay caffeinated during the writing of this text.

1 Introduction to Machine Learning

1.1 Introduction

Machine learning is a unified algorithmic framework designed to identify computational models that accurately describe empirical data and the phenomena underlying it, with little or no human involvement. While still a young discipline with much more awaiting discovery than is currently known, today machine learning can be used to teach computers to perform a wide array of useful tasks including automatic detection of objects in images (a crucial component of driver-assisted and self-driving cars), speech recognition (which powers voice command technology), knowledge discovery in the medical sciences (used to improve our understanding of complex diseases), and predictive analytics (leveraged for sales and economic forecasting), to just name a few.

In this chapter we give a high-level introduction to the field of machine learning as well as the contents of this textbook.

1.2 Distinguishing Cats from Dogs: a Machine Learning Approach

To get a big-picture sense of how machine learning works, we begin by discussing a toy problem: teaching a computer how to distinguish between pictures of *cats* from those with *dogs*. This will allow us to informally describe the terminology and procedures involved in solving the typical machine learning problem.

Do you recall how you first learned about the difference between cats and dogs, and how they are different animals? The answer is probably no, as most humans learn to perform simple cognitive tasks like this very early on in the course of their lives. One thing is certain, however: young children do not need some kind of formal scientific training, or a zoological lecture on *felis catus* and *canis familiaris* species, in order to be able to tell cats and dogs apart. Instead, they learn by example. They are naturally presented with many images of what they are told by a *supervisor* (a parent, a caregiver, etc.) are either cats or dogs, until they fully grasp the two concepts. How do we know when a child can successfully distinguish between cats and dogs? Intuitively, when they encounter new (images of) cats and dogs, and can correctly identify each new example or, in other words, when they can *generalize* what they have learned to new, previously unseen, examples.

Like human beings, computers can be taught how to perform this sort of task in a similar manner. This kind of task where we aim to teach a computer to distinguish between different types or *classes* of things (here *cats* and *dogs*) is referred to as a *classification* problem in the jargon of machine learning, and is done through a series of steps which we detail below.

1. Data collection. Like human beings, a computer must be trained to recognize the difference between these two types of animals by learning from a batch of examples, typically referred to as a *training set* of data. Figure 1.1 shows such a training set consisting of a few images of different cats and dogs. Intuitively, the larger and more diverse the training set the better a computer (or human) can perform a learning task, since exposure to a wider breadth of examples gives the learner more experience.

Figure 1.1 A training set consisting of six images of cats (highlighted in blue) and six images of dogs (highlighted in red). This set is used to train a machine learning model that can distinguish between future images of cats and dogs. The images in this figure were taken from [1].

2. Feature design. Think for a moment about how we (humans) tell the difference between images containing cats from those containing dogs. We use color, size, the shape of the ears or nose, and/or some combination of these *features* in order to distinguish between the two. In other words, we do not just look at an image as simply a collection of many small square pixels. We pick out grosser details, or features, from images like these in order to identify what it is that we are looking at. This is true for computers as well. In order to successfully train a computer to perform this task (and any machine learning task more generally) we need to provide it with properly designed features or, ideally, have it find or *learn* such features itself.

Designing quality features is typically not a trivial task as it can be very application dependent. For instance, a feature like *color* would be less helpful in discriminating between cats and dogs (since many cats and dogs share similar hair colors) than it would be in telling grizzly bears and polar bears apart! Moreover, extracting the features from a training dataset can also be challenging. For example, if some of our training images were blurry or taken from a perspective where we could not see the animal properly, the features we designed might not be properly extracted.

However, for the sake of simplicity with our toy problem here, suppose we can easily extract the following two features from each image in the training set: *size of nose* relative to the size of the head, ranging from small to large, and *shape of ears*, ranging from round to pointy.

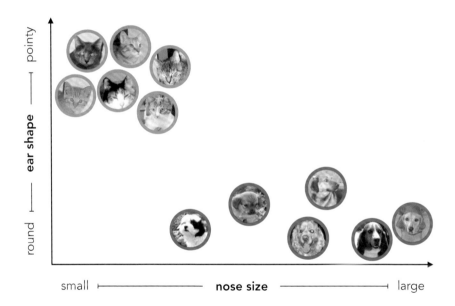

Figure 1.2 Feature space representation of the training set shown in Figure 1.1 where the horizontal and vertical axes represent the features *nose size* and *ear shape*, respectively. The fact that the cats and dogs from our training set lie in distinct regions of the feature space reflects a good choice of features.

Examining the training images shown in Figure 1.1, we can see that all cats have small noses and pointy ears, while dogs generally have large noses and round ears. Notice that with the current choice of features each image can now be represented by just two numbers: a number expressing the relative nose size, and another number capturing the pointiness or roundness of the ears. In other words, we can represent each image in our training set in a two-dimensional *feature space* where the features *nose size* and *ear shape* are the horizontal and vertical coordinate axes, respectively, as illustrated in Figure 1.2.

3. Model training. With our feature representation of the training data the machine learning problem of distinguishing between cats and dogs is now a simple geometric one: have the machine find a line or a curve that separates the cats from the dogs in our carefully designed feature space. Supposing for simplicity that we use a line, we must find the right values for its two parameters – a slope and vertical intercept – that define the line's orientation in the feature space. The process of determining proper parameters relies on a set of tools known as *mathematical optimization* detailed in Chapters 2 through 4 of this text, and the tuning of such a set of parameters to a training set is referred to as the training of a model.

Figure 1.3 shows a trained linear model (in black) which divides the feature space into cat and dog regions. This linear model provides a simple computational rule for distinguishing between cats and dogs: when the feature representation of a future image lies above the line (in the blue region) it will be considered a cat by the machine, and likewise any representation that falls below the line (in the red region) will be considered a dog.

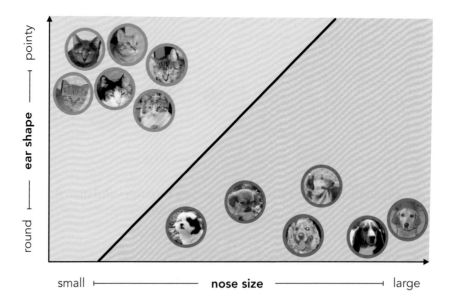

Figure 1.3 A trained linear model (shown in black) provides a computational rule for distinguishing between cats and dogs. Any new image received in the future will be classified as a cat if its feature representation lies above this line (in the blue region), and a dog if the feature representation lies below this line (in the red region).

Figure 1.4 A validation set of cat and dog images (also taken from [1]). Notice that the images in this set are not highlighted in red or blue (as was the case with the training set shown in Figure 1.1) indicating that the true identity of each image is not revealed to the learner. Notice that one of the dogs, the Boston terrier in the bottom right corner, has both a small nose and pointy ears. Because of our chosen feature representation the computer will think this is a cat!

4. Model validation. To validate the efficacy of our trained learner we now show the computer a batch of previously unseen images of cats and dogs, referred to generally as a *validation set* of data, and see how well it can identify the animal in each image. In Figure 1.4 we show a sample validation set for the problem at hand, consisting of three new cat and dog images. To do this, we take each new image, extract our designed features (i.e., nose size and ear shape), and simply check which side of our line (or *classifier*) the feature representation falls on. In this instance, as can be seen in Figure 1.5, all of the new cats and all but one dog from the validation set have been identified correctly by our trained model.

The misidentification of the single dog (a Boston terrier) is largely the result of our choice of features, which we designed based on the training set in Figure 1.1, and to some extent our decision to use a *linear* model (instead of a *nonlinear* one). This dog has been misidentified simply because its features, a small nose and pointy ears, match those of the cats from our training set. Therefore, while it first appeared that a combination of nose size and ear shape could indeed distinguish cats from dogs, we now see through validation that our training set was perhaps too small and not diverse enough for this choice of features to be completely effective in general.

We can take a number of steps to improve our learner. First and foremost we should collect more data, forming a larger and more diverse training set. Second, we can consider designing/including more discriminating features (perhaps eye color, tail shape, etc.) that further help distinguish cats from dogs using a linear model. Finally, we can also try out (i.e., train and validate) an array of *nonlinear* models with the hopes that a more complex rule might better distinguish between cats and dogs. Figure 1.6 compactly summarizes the four steps involved in solving our toy cat-versus-dog classification problem.

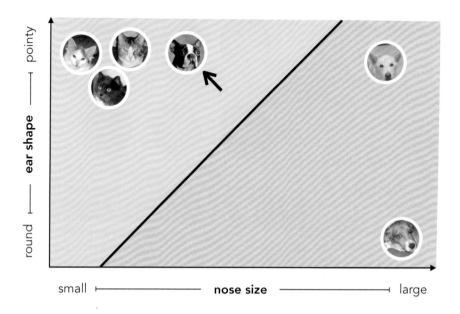

Figure 1.5 Identification of (the feature representation of) validation images using our trained linear model. The Boston terrier (pointed to by an arrow) is *misclassified* as a cat since it has pointy ears and a small nose, just like the cats in our training set.

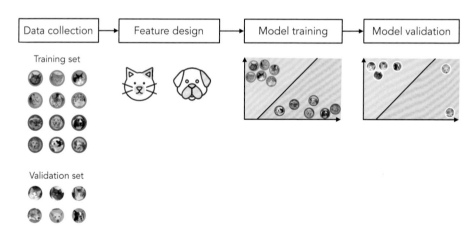

Figure 1.6 The schematic pipeline of our toy cat-versus-dog classification problem. The same general pipeline is used for essentially all machine learning problems.

1.3   The Basic Taxonomy of Machine Learning Problems

The sort of computational rules we can learn using machine learning generally fall into two main categories called *supervised* and *unsupervised* learning, which we discuss next.

1.3.1   Supervised learning

Supervised learning problems (like the prototypical problem outlined in Section 1.2) refer to the automatic learning of computational rules involving input/output relationships. Applicable to a wide array of situations and data types, this type of problem comes in two forms, called *regression* and *classification*, depending on the general numerical form of the output.

Regression

Suppose we wanted to predict the share price of a company that is about to go public. Following the pipeline discussed in Section 1.2, we first gather a training set of data consisting of a number of corporations (preferably active in the same domain) with known share prices. Next, we need to design feature(s) that are thought to be relevant to the task at hand. The company's revenue is one such potential feature, as we can expect that the higher the revenue the more expensive a share of stock should be. To connect the share price (output) to the revenue (input) we can train a simple linear model or *regression line* using our training data.

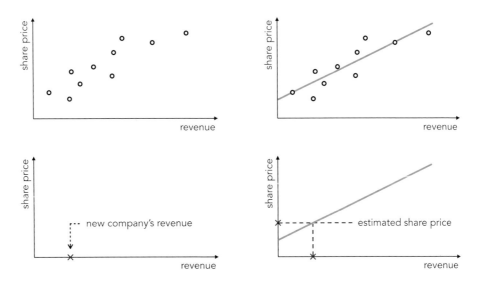

Figure 1.7 (top-left panel) A toy training dataset consisting of ten corporations' share price and revenue values. (top-right panel) A linear model is fit to the data. This trend line models the overall trajectory of the points and can be used for prediction in the future as shown in the bottom-left and bottom-right panels.

The top panels of Figure 1.7 show a toy dataset comprising share price versus revenue information for ten companies, as well as a linear model fit to this data. Once the model is trained, the share price of a new company can be predicted based on its revenue, as depicted in the bottom panels of this figure. Finally, comparing the predicted price to the actual price for a validation set of data we can test the performance of our linear regression model and apply changes as needed, for example, designing new features (e.g., total assets, total equity, number of employees, years active, etc.) and/or trying more complex nonlinear models.

This sort of task, i.e., fitting a model to a set of training data so that predictions about a *continuous-valued* output (here, share price) can be made, is referred to as regression. We begin our detailed discussion of regression in Chapter 5 with the linear case, and move to nonlinear models starting in Chapter 10 and throughout Chapters 11–14. Below we describe several additional examples of regression to help solidify this concept.

Example 1.1   The rise of student loan debt in the United States

Figure 1.8 (data taken from [2]) shows the total student loan debt (that is money borrowed by students to pay for college tuition, room and board, etc.) held by citizens of the United States from 2006 to 2014, measured quarterly. Over the eight-year period reflected in this plot the student debt has nearly tripled, totaling over one trillion dollars by the end of 2014. The regression line (in black) fits this dataset quite well and, with its sharp positive slope, emphasizes the point that student debt is rising dangerously fast. Moreover, if this trend continues, we can use the regression line to predict that total student debt will surpass two trillion dollars by the year 2026 (we revisit this problem later in Exercise 5.1).

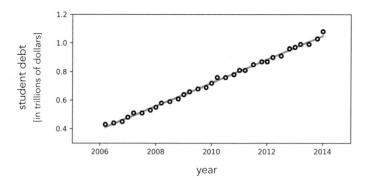

Figure 1.8 Figure associated with Example 1.1, illustrating total student loan debt in the United States measured quarterly from 2006 to 2014. The rapid increase rate of the debt, measured by the slope of the trend line fit to the data, confirms that student debt is growing very fast. See text for further details.

Example 1.2   Kleiber's law

Many natural laws in the sciences take the form of regression models. For example, after collecting a considerable amount of data comparing the body mass versus metabolic rate (a measure of at-rest energy expenditure) of a variety of animals, early twentieth-century biologist Max Kleiber found that the log of these two quantities are related *linearly*. This linear relationship can be seen visually by examining the dataset shown in Figure 1.9. Examining a similar dataset, Kleiber found the slope of the regression line to be around $\frac{3}{4}$ or, in other words, that metabolic rate $\propto$ mass$^{\frac{3}{4}}$.

This sublinear relationship means that compared to smaller-bodied species (like birds), larger-bodied species (like walruses) have lower metabolic rates, which is consistent with having lower heart rates and larger life spans (we revisit this problem later in Exercise 5.2).

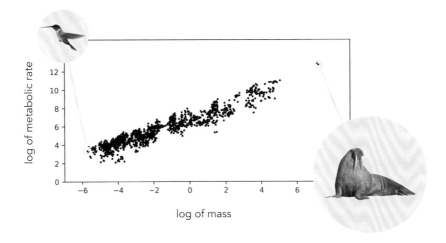

Figure 1.9 Figure associated with Example 1.2. A large set of body mass versus metabolic rate data points, transformed by taking the log of each value, for various animals over a wide range of different masses. See text for further details.

Example 1.3   Predicting box office success

In 1983 the Academy award winning screenwriter William Goldman coined the phrase "nobody knows anything" in his book *Adventures in the Screen Trade*, referring to his belief that at the time it was impossible for anyone to predict the success or failure of Hollywood movies. While this may be true – in the era of the internet – by leveraging data like the quantity of internet searches for a movie's trailer, as well as the amount of discussion about a movie on social networks (see, e.g., [3, 4]), machine learning can accurately predict opening box office revenue for certain films. Sales forecasting for a range of products/services including box office sales is often performed using regression since the output to be predicted is continuous (enough) in nature.

Example 1.4   Business and industrial applications

Examples of regression are plentiful in business and industry. For instance, using regression to accurately predict the *price* of consumer goods (from electronics, to automobiles, to houses) are hugely valuable enterprises unto themselves (explored further in Example 5.5). Regression is also commonly used in industiral applications to better understand a given system, e.g., how the configuration of an automobile affects its performance (see Example 5.6), so that such processes can be optimized.

Classification

The machine learning task of classification is similar in principle to that of regression, with the key difference between the two being that instead of predicting a continuous-valued output, with classification the output we aim at predicting takes on *discrete values* or *classes*. Classification problems arise in a host of forms. For example, object recognition where different objects from a set of images are distinguished from one another (e.g., handwritten digits for the automatic sorting of mail or street signs for driver-assisted and self-driving cars) is a very popular classification problem. The toy problem of distinguishing cats from dogs discussed in Section 1.2 falls into this bucket as well. Other common classification problems include speech recognition (recognizing different spoken words for voice recognition systems), determining the general sentiment of a social network like Twitter towards a particular product or service, as well as determining what kind of hand gesture someone is making from a finite set of possibilities (for use, for instance, in controlling a computer without a mouse).

Geometrically speaking, a common way of viewing the task of classification in two dimensions is one of finding a separating line (or, more generally, a separating curve) that accurately separates two kinds of data.[1] This is precisely the perspective on classification we took in describing the toy example in Section 1.2, where we used a line to separate (features extracted from) images of cats and dogs. New data from a validation set are then automatically classified by simply determining which side of the line the data lies on. Figure 1.10 illustrates the concept of a linear model or classifier used for performing classification on a two-dimensional toy dataset.

Many classification problems (e.g., handwritten digit recognition, discussed below) have naturally more than two classes. After describing linear two-class classification in Chapter 6 we detail linear multi-class classification in Chapter 7. The nonlinear extension of both problems is then described starting in Chapter 10 and throughout Chapters 11–14. Below we briefly describe several further examples of classification to help solidify this concept.

[1] In higher dimensions we likewise aim at determining a separating linear hyperplane (or, more generally, a nonlinear manifold).

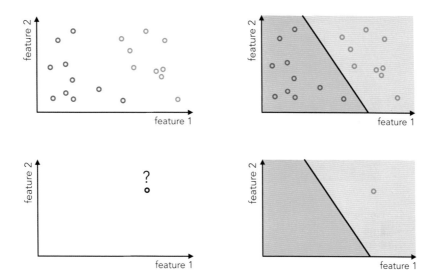

Figure 1.10 (top-left panel) A toy two-dimensional training set of data consisting of two distinct classes: *red* and *blue*. (top-right panel) A linear model is trained to separate the two classes. (bottom-left panel) A validation point whose class is unknown. (bottom-right panel) The validation point is classified as *blue* since it lies on the blue side of the trained linear classifier.

Example 1.5   Object detection

Object detection, a common classification problem (see, e.g., [5, 6, 7]), is the task of automatically identifying a specific object in a set of images or videos. Popular object detection applications include the detection of faces in images for organizational purposes and camera focusing, pedestrians for autonomous driving vehicles, and faulty components for automated quality control in electronics production. The same kind of machine learning framework, which we highlight here for the case of face detection, can be utilized for solving many such detection problems.

After training a (linear) classifier on a set of training data consisting of facial and nonfacial images, faces are sought after in a new validation image by sliding a (typically) square window over the entire image. At each location of the sliding window the image content inside is examined to see which side of the classifier it lies on. This is illustrated in Figure 1.11. If the (feature representation of the) content lies on the face side of the classifier the content is classified as a face, otherwise a nonface.

Example 1.6   Sentiment analysis

The rise of social media has significantly amplified the voice of consumers, providing them with an array of well-tended outlets on which to comment, discuss, and rate products and services (see, e.g., [8]). This has led many firms to seek out

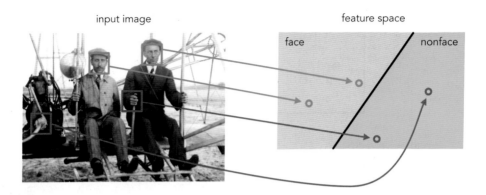

Figure 1.11 Figure associated with Example 1.5. To determine if any faces are present in an input image (in this instance an image of the Wright brothers, inventors of the airplane, sitting together in one of their first motorized flying machines in 1908) a small window is scanned across its entirety. The content inside the box at each instance is determined to be a face by checking which side of the learned classifier the feature representation of the content lies. In the figurative illustration shown here the area above and below the learned classifier (shown in blue and red) are the face and nonface sides of the classifier, respectively. See text for further details.

data intensive methods for gauging their customers' feelings towards recently released products, advertising campaigns, etc. Determining the aggregated feelings of a large base of customers, using text-based content like product reviews, tweets, and comments, is commonly referred to as *sentiment analysis*. Classification models are often used to perform sentiment analysis, learning to identify consumer data of either positive or negative feelings. We discuss this problem further in Example 9.1.

Example 1.7   Computer-aided diagnosis of medical conditions
A myriad of two-class classification problems naturally arise in medical settings when a healthcare professional looks to determine whether or not a patient suffers from a particular malady, and when medical researchers conduct research into what features distiniguish those suffering from such a malady from those who do not (in the hopes that a remedy can then be devised to ameliorate the malady based on these features). The modality of these sorts of experiments varies widely, from statistical measurements of affected areas (e.g., the shape and area of biopsied tumorous tissue; see Exercise 6.13), to biochemical markers, to information derived from radiological image data, to genes themselves (as explored further in Example 11.18).

   For instance, paired with a classification algorithm functional Magnetic Resonance Imaging (fMRI) of the brain is an increasingly useful method for diagnosing neurological disorders such as Autism, Alzheimer's, and Attention Deficit Hyperactivity Disorder (ADHD). To perform classification a dataset is acquired consisting of statistically based features extracted from fMRI brain scans of patients suffering from one such previously mentioned cognitive disorder, as well as individuals from a control group who are not afflicted. These fMRI brain scans capture neural activity patterns localized in different regions of the brain as patients perform simple activities such as tracking a small visual object. Figure 1.12, taken from [9], illustrates the result of applying a classification model to the problem of diagnosing patients with ADHD. This is discussed further in Example 11.19.

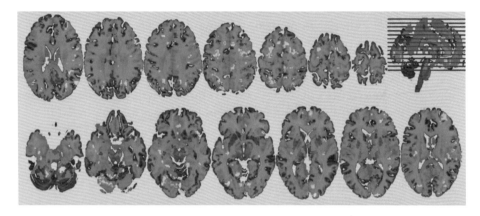

Figure 1.12 Figure associated with Example 1.7. See text for details.

Example 1.8 Spam detection
Spam detection is a standard text-based two-class classification problem. Implemented in most email systems, spam detection automatically identifies unwanted messages (e.g., advertisements), referred to as spam, from the emails users want to see (often referred to as ham). Once trained, a spam detector can remove unwanted messages without user input, greatly improving a user's email experience. This example is discussed in further detail in Examples 6.10 and 9.2.

Example 1.9 Financial applications
Two-class classification problems arise in all sorts of financial applications. They are often used in commercial lending to determine whether or not an individual should receive a commercial loan, credit card, etc., based on their historical financial information. This standard two-class classification problem, either "lend" or "do not lend," is explored in greater detail in Examples 6.11 and 9.7, with the latter example described in the context of *feature selection*.

Fraud detection is another hugely popular two-class classification problem in the financial space. The detection of fradulent financial (e.g., credit card) transactions is naturally framed as a two-class classification problem where the two classes consist of legitimate transactions and fraudulent ones. The challenge with such problems is typically that the record of fraudulent transactions is dwarfed many times over by valid ones, making such datasets highly *imbalanced*, an issue discussed further in Sections 6.8.4 and 6.9).

Example 1.10   Recognition problems

Recognition problems are a popular form of multi-class classification where the aim is to train a classifier to automatically distinguish between a collection of things, whether those things be human gestures (gesture recognition), various visual objects (object recognition), or spoken words (speech recognition).

For example, recognizing handwritten digits is a popular object recognition problem commonly built into the software of mobile banking applications, as well as more traditional Automated Teller Machines, to give users among others the ability to automatically deposit paper checks. In this application each class of data consists of (images of) several handwritten versions of a single digit in the range 0–9, resulting in a total of ten classes (see Figure 1.13).

Figure 1.13 Figure associated with Example 1.10. An illustration of various handwritten digits. See text for further details.

We discuss handwritten digit recognition in further detail later at several differnt points (for instance, in Example 7.10), and more general applications in object and speech recognition later in Sections 9.2.4 and 9.2.3, respectively.

1.3.2   **Unsupervised learning**

Unsupervised learning (unlike supervised learning problems outlined previously) deals with the automatic learning of computational rules that describe *input* data only. Often such rules are learned in order to simplify a dataset to allow for easier supervised learning, or for human analysis and interpretation. Two fundamental unsupervised problems, *dimension reduction* and *clustering*, allow for simplification of a dataset via two natural paths: by either reducing the ambient dimension of input data (in the former case) or by determining a small number of representatives that adequately describe the diversity of a larger set of data (in the latter case). Both subcategories of unsupervised learning are first introduced in Chapter 8 (where the linear version of each is detailed), and discussed further in Chapters 10–14 (where their nonlinear extensions are discussed).

Dimension reduction

The dimensionality of modern-day data such as images, videos, text documents, and genetic information, is often far too large for effective use of predictive modeling and analysis. For example, even a megapixel image, a medium-resolution image by today's standards, is a million-dimensional piece of data. This is true for a gray-scale image that has only one dimension for each of its one million pixels. A color megapixel image would have three million dimensions. Therefore, reducing the dimension of this sort of data is often crucial to effective application of many machine learning algorithms, making dimension reduction a common preprocessing step for prediction and analysis tasks.

Geometrically speaking, to reduce the dimension of a dataset means to squash or project it down onto a proper lower-dimensional line or curve (or more generally a linear hyperplane or nonlinear manifold), preferably one that retains as much of the original data's defining characteristics as possible.

This general idea is illustrated for two toy datasets in Figure 1.14, with the two-dimensional (left panel) and three-dimensional (right panel) data squashed (or projected) onto a proper one-dimensional line and two-dimensional hyperplane, respectively, reducing the ambient dimension of data by one in each case while retaining much of the shape of the original data. In practice, the reduction in dimension of modern-day large datasets can be much greater than achieved in this illustration.

Figure 1.14 Two toy datasets consisting of two-dimensional (left panel) and three-dimensional (right panel) input data, shown as hollow black circles. The data in each case is projected onto a lower-dimensional subspace, and is effectively lowered a dimension while retaining a good amount of the original data's structure.

Clustering

Clustering aims at identifying gross underlying structure in a set of input data by grouping together points that share some structural characteristic, e.g., proximity to one another in the feature space, which helps better organize or summarize

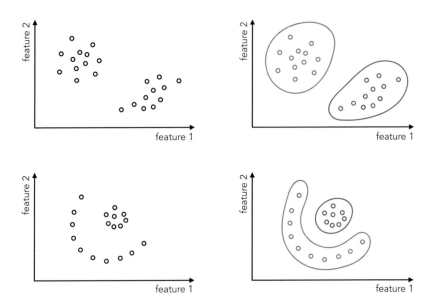

Figure 1.15 Two toy examples of the vast array in which input data can cluster. Clustering algorithms are designed to uncover these kinds of distinct structures. In each instance the distinct clusters in the original data (on the left) are colored for visualization purposes (on the right).

a set of training data for analysis by a human or machine interpreter. The structure of data can vary immensely, with data falling into globular clusters or along nonlinear manifolds as illustrated in Figure 1.15.

1.4   Mathematical Optimization

As we will see throughout the remainder of the book, we can formalize the search for parameters of a learning model via well-defined mathematical functions. These functions, commonly referred to as *cost* or *loss functions*, take in a specific set of model parameters and return a score indicating how well we would accomplish a given learning task using that choice of parameters. A high value indicates a choice of parameters that would give poor performance, while the opposite holds for a set of parameters providing a low value. For instance, recall the share price prediction example outlined in Figure 1.7 where we aimed at learning a regression line to predict a company's share price based on its revenue. This line is learned to the data by optimally tuning its two parameters: slope and vertical intercept. Geometrically, this corresponds to finding the set of parameters providing the smallest value (called a *minimum*) of a two-dimensional cost function, as shown pictorially in Figure 1.16. This concept plays a similarly fundamental role with classification (and indeed with all machine learning problems) as well. In Figure 1.10 we detailed how a general linear classifier is trained, with the ideal setting for its parameters again corresponding with the minimum of a cost function as illustrated pictorially in Figure 1.17.

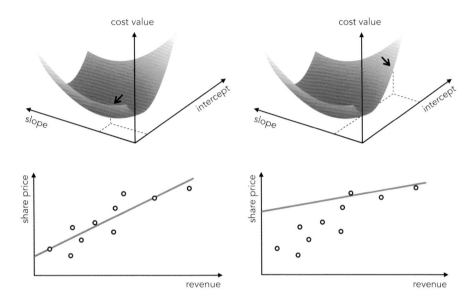

Figure 1.16 (top panels) A figurative drawing of the two-dimensional cost function associated with learning the slope and intercept parameters of a linear model for the share price regression problem discussed in the previous section and shown in Figure 1.7. Also shown here are two different sets of parameter values, one (left) at the minimum of the cost function and the other (right) at a point with larger cost function value. (bottom panels) The linear model corresponding to each set of parameters in the top panel. The set of parameters resulting in the best fit are found at the minimum of the cost surface.

Because a low value corresponds to a high-performing model in the case of both regression and classification (and, as we will see, for unsupervised learning problems as well) we will always look to *minimize* cost functions in order to find the ideal parameters of their associated learning models. As the study of computational methods for minimizing formal mathematical functions, the tools of *mathematical optimization* therefore play a fundamental role throughout the text. Additionally, as we will see later in the text starting in Chapter 11, optimization also plays a fundamental role in *cross-validation* or the learning of a proper *nonlinear* model automatically for any dataset. Because of these critical roles mathematical optimization plays in machine learning we begin this text with an exhaustive description of the fundamental tools of optimization in Chapters 2–4.

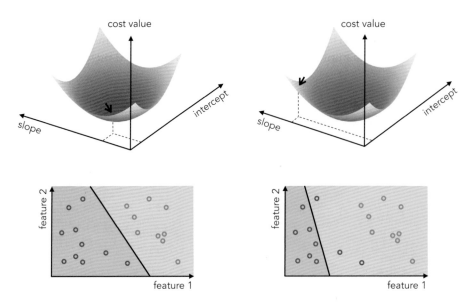

Figure 1.17 (top panels) A figurative drawing of the two-dimensional cost function associated with learning the slope and intercept parameters of a linear model separating two classes of the toy dataset first shown in Figure 1.10. Also shown here are two different sets of parameter values, one (left) corresponding to the minimum of the cost function and the other (right) corresponding to a point with larger cost function value. (bottom panels) The linear classifiers corresponding to each set of parameters in the top panels. The optimal set of parameters, i.e., those giving the minimum value of the associated cost function, allow for the best separation between the two classes.

1.5   Conclusion

In this chapter we have given a broad overview of machine learning, with an emphasis on critical concepts we will see repeatedly throughout the text. We began in Section 1.2 by describing a prototypical machine learning problem, as well as the steps typically taken to solve such a problem (summarized in Figure 1.6). In Section 1.3 we then introduced the fundamental families of machine learning problems – supervised and unsupervised learning – detailing a number of applications of both. Finally in Section 1.4 we motivated the need for mathematical optimization by the pursuit of ideal parameters for a machine learning model, which has direct correspondence to the geometric problem of finding the smallest value of an associated cost function (summarized pictorially in Figures 1.16 and 1.17).

Part I

Mathematical Optimization

2 Zero-Order Optimization Techniques

2.1 Introduction

The problem of determining the smallest (or largest) value a function can take, referred to as its *global minimum* (or *global maximum*), is a centuries-old pursuit that has numerous applications throughout the sciences and engineering. In this chapter we begin our investigation of mathematical optimization by describing the *zero-order optimization* techniques – also referred to as *derivative-free optimization* techniques. While not always the most powerful optimization tools at our disposal, zero-order techniques are conceptually the simplest tools available to us – requiring the least amount of intellectual machinery and jargon to describe. Because of this, discussing zero-order methods first allows us to lay bare, in a simple setting, a range of crucial concepts we will see throughout the chapters that follow in more complex settings – including the notions of *optimality*, *local optimization*, *descent directions*, *steplengths*, and more.

2.1.1 Visualizing minima and maxima

When a function takes in only one or two inputs we can attempt to visually identify its minima or maxima by plotting it over a large swath of its input space. While this idea certainly fails when a function takes in three or more inputs (since we can no longer visualize it properly), we begin nevertheless by first examining a number of low-dimensional examples to gain an intuitive feel for how we might effectively identify these desired minima or maxima in general.

Example 2.1 **Visual inspection of single-input functions for minima and maxima**

In the top-left panel of Figure 2.1 we plot the quadratic function

$$g(w) = w^2 \tag{2.1}$$

over a short region of its input space (centered around zero from $w = -3$ to $w = 3$). In this figure we also mark the evaluation of the function's global

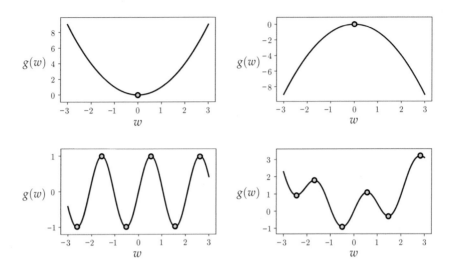

Figure 2.1 Figure associated with Example 2.1. Four example functions are shown with the evaluation of their minima and maxima highlighted by green dots. See text for further details.

minimum at $w = 0$ (that is, the point $(0, g(0))$ where $g(0) = 0$) with a green dot. Note that as we move farther away from the origin (in either the negative or positive direction) the evaluation of g becomes progressively larger, implying that its global maxima lies at $w = \pm\infty$.

In the top-right panel in Figure 2.1 we show the result of multiplying the previous quadratic function by -1, giving the new quadratic

$$g(w) = -w^2. \tag{2.2}$$

Doing so causes the function to flip upside down, with its global minima now lying at $w = \pm\infty$, and the input $w = 0$ that once provided the *global minimum* of g now returns its *global maximum*. The evaluation of this maximum is again marked with a green dot on the function.

In the bottom-left panel of Figure 2.1 we plot the sinusoidal function

$$g(w) = \sin(3w). \tag{2.3}$$

Here we can clearly see that (over the range we have plotted the function) there are three global minima and three global maxima (the evaluation of is each marked by a green dot on the function). Indeed if we drew this function out over a wider and wider swath of its input we would see that it has infinitely many such global minima and maxima (existing at every odd multiple of $\frac{\pi}{6}$).

In the bottom-right panel of Figure 2.1 we look at the sum of a sinusoidal and a quadratic function, which takes the algebraic form

$$g(w) = \sin(3w) + 0.3\,w^2. \tag{2.4}$$

Inspecting this function (over the range it is plotted) we can see that it has a global minimum around $w = -0.5$. The function also has other minima and maxima that are *locally optimal*, meaning values that are minimal or maximal only locally and with respect to just their neighbors (and not the function as a whole). For example, g has a local maximum near $w = 0.6$ and a local minimum near $w = 1.5$. The evaluation of both maxima and minima over the range of input shown for this function are marked by a green dot in the figure.

2.2   The Zero-Order Optimality Condition

With a number of simple examples illustrating minima and maxima we can now define them more formally. The task of determining a global minimum of a function g with N input variables $w_1, w_2, \ldots, w_N$ can formally be phrased as the following *minimization problem*

$$\underset{w_1, w_2, \ldots, w_N}{\text{minimize}}\ g\,(w_1, w_2, \ldots, w_N) \tag{2.5}$$

which can be rewritten much more compactly (by stacking all the inputs in an N-dimensional vector $\mathbf{w}$) as

$$\underset{\mathbf{w}}{\text{minimize}}\ g\,(\mathbf{w}). \tag{2.6}$$

By solving such minimization problem we aim to find a point $\mathbf{w}^\star$ such that

$$g\,(\mathbf{w}^\star) \le g\,(\mathbf{w})\ \text{ for all } \mathbf{w}. \tag{2.7}$$

This is the *zero-order* definition of a global minimum. In general, a function can have multiple or even infinitely many global minimum points (like the sinusoidal function in Equation (2.3)).

We can likewise describe mathematical points $\mathbf{w}^\star$ at which g has a global maximum. For such points we can write

$$g\,(\mathbf{w}^\star) \ge g\,(\mathbf{w})\ \text{ for all } \mathbf{w}. \tag{2.8}$$

This is the *zero-order* definition of a global maximum. To express our pursuit of a global maximum of a function we then write

$$\underset{\mathbf{w}}{\text{maximize}}\ g\,(\mathbf{w}). \tag{2.9}$$

Note that the concepts of minima and maxima of a function are always related to each other via multiplication by −1. That is, a global minimum of a function g is always a global maximum of the function $-g$, and vice versa. Therefore we can always express the maximization problem in Equation (2.9) in terms of a minimization problem, as

$$\underset{\mathbf{w}}{\text{minimize}} \quad - g\left(\mathbf{w}\right). \qquad (2.10)$$

Akin to zero-order definitions for global minima and maxima in Equations (2.7) and (2.8), there are zero-order definitions for local minima and maxima as well. For instance, we can say a function g has a *local* minimum at a point $\mathbf{w}^{\star}$ if

$$g\left(\mathbf{w}^{\star}\right) \leq g\left(\mathbf{w}\right) \quad \text{for all } \mathbf{w} \text{ near } \mathbf{w}^{\star}. \qquad (2.11)$$

The statement "for all $\mathbf{w}$ near $\mathbf{w}^{\star}$" is relative, simply describing the fact that a neighborhood (however small) around $\mathbf{w}^{\star}$ must exist such that, when evaluated at every point in this neighborhood, the function g attains its smallest value at $\mathbf{w}^{\star}$. The same formal zero-order definition can be made for local maxima as well, switching the $\leq$ sign to $\geq$.

Packaged together, these zero-order definitions for minima and maxima (collectively called optima) are often referred to as *the zero-order condition for optimality*. The phrase *zero-order* in this context refers to the fact that in each case, the optima of a function are defined in terms of the function itself (and nothing else). In further chapters we will see *higher-order* definitions of optimal points, specifically the *first-order* definitions that involve the first derivative of a function in Chapter 3, as well as *second-order* definitions involving a function's second derivative in Chapter 4.

2.3   Global Optimization Methods

In this section we describe the first approach one might take to approximately minimize an arbitrary function: evaluate the function using a large number of input points and treat the input that provides the lowest function value as the approximate global minimum of the function. This approach is called a *global* optimization method because it is capable of approximating the global optima of a function (provided a large enough number of evaluations are made).

The important question with this sort of optimization scheme is: how do we choose the inputs to try out with a generic function? We clearly cannot try them all since, even for a single-input continuous function, there are an infinite number of points to try.

We can take two approaches here to choosing our (finite) set of input points to test: either sample (i.e., guess) them uniformly over an evenly spaced grid (uniform sampling), or pick the same number of input points at random (random sampling). We illustrate both choices in Example 2.2.

Example 2.2 **Minimizing a quadratic function**
Here we illustrate two sampling methods for finding the global minimum of
the quadratic function

$$g(w) = w^2 + 0.2 \tag{2.12}$$

which has a global minimum at $w = 0$. For the sake of simplicity we limit the
range over which we search to $[-1, +1]$. In the top row of Figure 2.2 we show
the result of a uniform-versus-random sampling of four inputs, shown in blue
in each panel of the figure (with corresponding evaluations shown in green
on the function itself). We can see that by randomly sampling here we were
able to (by chance) achieve a slightly lower point when compared to sampling
the function evenly. However, using enough samples we can find an input very
close to the true global minimum of the function with either sampling approach.
In the bottom row we show the result of sampling 20 inputs uniformly versus
randomly, and we can see that, by increasing the number of samples, using
either approach, we are now able to approximate the global minimum with
much better precision.

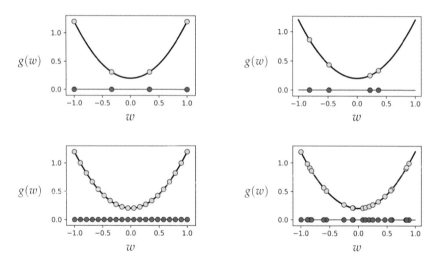

Figure 2.2 Figure associated with Example 2.2. Minimizing a simple function via
sampling – or "guessing." (top row) Sampling the input of a function four times evenly
(left panel) and at random (right panel). Here the inputs chosen are shown as blue dots,
and their evaluations by the function are shown as green dots. (bottom row) Sampling
uniformly (left panel) and randomly (right panel) 20 times. The more samples we take,
the more likely we are to find a point close to the global minimum using either
sampling approach. See text for further details.

Note that with both global optimization approaches discussed in Example 2.2 we are simply employing the zero-order optimality condition, since from a set of K chosen inputs $\left\{\mathbf{w}^k\right\}_{k=1}^K$ we are choosing the one input $\mathbf{w}^j$ with lowest evaluation on the cost function

$$g\left(\mathbf{w}^j\right) \le g\left(\mathbf{w}^k\right) \quad k = 1, 2, \ldots, K \tag{2.13}$$

which is indeed an approximation to the zero-order optimality condition discussed in the previous section.

While easy to implement and perfectly adequate for functions having low-dimensional input, as we see next, this naturally zero-order framework fails miserably when the input dimension of a function grows to even moderate size.

2.3.1   The curse of dimensionality

While this sort of global optimization based on zero-order evaluations of a function works fine for low-dimensional functions, it quickly fails as we try to tackle functions with larger number of inputs or, in other words, functions that take in N-dimensional input $\mathbf{w}$ where N is large. This makes such optimization methods essentially unusable in modern machine learning since the functions we often deal with have input dimensions ranging from the hundreds to the hundreds of thousands, or even millions.

To get a sense of why the global approach quickly becomes infeasible, imagine we use a uniform sampling of points across the input space of a single-input function, choosing (for the sake of argument) three points, each at a distance of d from the previous one, as illustrated in the top-left panel of Figure 2.3. Imagine now that the input space of the function increases by one, and that the range of each input is precisely that of the original single-input function, as illustrated in the top-middle panel of Figure 2.3. We still aim to cover the space evenly and with enough samples such that each input we evaluate is once again at a distance d from its closest neighbors in either direction. Notice, in order to do this in a now two-dimensional space we need to sample $3^2 = 9$ input points. Likewise if we increase the dimension of the input once again in the same fashion, in order to sample evenly across the input space so that each input is at a maximum distance of d from its neighbors in every input dimension we will need $3^3 = 27$ input points, as illustrated in the top-right panel of Figure 2.3. If we continue this thought experiment, for a general N-dimensional input we would need to sample 3^N points, a huge number even for moderate values of N. This is a simple example of the so-called *curse of dimensionality* which, generally speaking, describes the exponential difficulty one encounters when trying to deal with functions of increasing input dimension.

The curse of dimensionality remains an issue even if we decide to take samples randomly. To see why this is the case using the same hypothetical scenario, suppose now that, instead of fixing the distance d of each sample from its

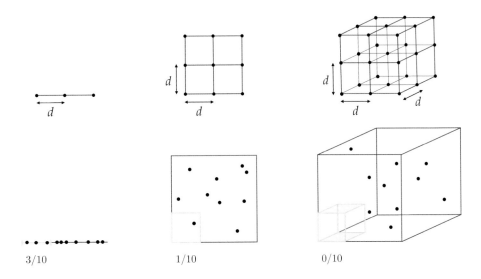

Figure 2.3 (top row) The number of input points we must sample uniformly if we wish each to be at a distance of d from its neighbors grows exponentially as the input dimension of a function increases. If three points are used to cover a single-input space in this way (left panel), $3^2 = 9$ points are required in two dimensions, and $3^3 = 27$ points in three dimensions (and this trend continues). Sampling randomly (bottom row) does not solve the problem either. See text for further details.

neighbors, we fix the total number of randomly chosen samples to a fixed value and look at how well they tend to distribute over an input space as we increase its dimension. From left to right in the bottom panels of Figure 2.3 we see one instance of how a total of ten points are randomly selected in $N = 1$, $N = 2$, and $N = 3$ dimensional space, respectively. Once again we are inhibited by the curse of dimensionality. As we increase the dimension of the input space the average number of samples per unit hypercube drops exponentially, leaving more and more regions of the space without a single sample or corresponding evaluation. In order to counteract this problem we would need to start sampling exponentially many points, leading to the same problem we encounter with the uniform sampling scheme.

2.4   Local Optimization Methods

As opposed to the global optimization techniques described in Section 2.3, where a large number of input points are sampled simultaneously with the lowest evaluation crowned the approximate global minimum, *local optimization methods* work by starting with a single input point and then by sequentially refining it, driving the point towards an approximate minimum point. Local optimization methods are by far the most popular mathematical optimization schemes used in machine learning today, and are the subject of the remainder of this part of the text. While there is substantial variation in the kinds of specific local optimization methods we will discuss going forward, nonetheless they all share a common overarching framework that we introduce in this section.

2.4.1   The big picture

Starting with a sample input, usually referred to as an *initial point* and denoted throughout the text by $\mathbf{w}^0$, local optimization methods refine this initial point sequentially, pulling it *downhill* towards points that are lower and lower on the function, eventually reaching a minimum as illustrated for a single-input function in Figure 2.4. More specifically, from $\mathbf{w}^0$ the point is pulled downhill to a new point $\mathbf{w}^1$ lower on the function, i.e., where $g\left(\mathbf{w}^0\right) > g\left(\mathbf{w}^1\right)$. The point $\mathbf{w}^1$ itself is then pulled downwards to a new point $\mathbf{w}^2$. Repeating this process K times yields a sequence of K points (excluding our starting initial point)

$$\mathbf{w}^0,\, \mathbf{w}^1,\, ...,\, \mathbf{w}^K \tag{2.14}$$

where each subsequent point is (generally speaking) on a lower and lower portion of the function, i.e.,

$$g\left(\mathbf{w}^0\right) > g\left(\mathbf{w}^1\right) > \cdots > g\left(\mathbf{w}^K\right). \tag{2.15}$$

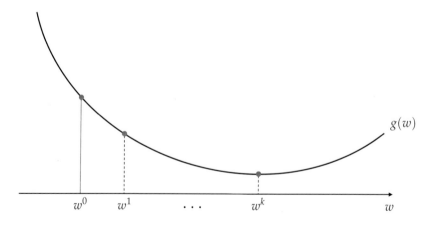

Figure 2.4 Local optimization methods work by minimizing a target function in a sequence of steps. Shown here is a generic local optimization method applied to minimize a single-input function. Starting with the initial point w^0, we move towards lower points on the cost function like a ball rolling downhill.

2.4.2   The general framework

In general the sequential refinement process enacted by a local optimization method works as follows. To take the first step from an initial point $\mathbf{w}^0$ to the very first update $\mathbf{w}^1$, what is called a *descent direction* at $\mathbf{w}^0$ is found. This is a direction vector $\mathbf{d}^0$ in the input space, beginning at $\mathbf{w}^0$ and pointing away from it towards a new point $\mathbf{w}^1$ with lower function evaluation. When such a direction is found the first update $\mathbf{w}^1$ is given by the sum

$$\mathbf{w}^1 = \mathbf{w}^0 + \mathbf{d}^0. \tag{2.16}$$

To refine the point $\mathbf{w}^1$ we look for a new descent direction $\mathbf{d}^1$, one that moves downhill stemming from the point $\mathbf{w}^1$. When we find such a direction the second update $\mathbf{w}^2$ is then formed as the sum

$$\mathbf{w}^2 = \mathbf{w}^1 + \mathbf{d}^1. \tag{2.17}$$

We repeat this process, producing a sequence of input points

$$
\begin{aligned}
\mathbf{w}^0 & \\
\mathbf{w}^1 &= \mathbf{w}^0 + \mathbf{d}^0 \\
\mathbf{w}^2 &= \mathbf{w}^1 + \mathbf{d}^1 \\
\mathbf{w}^3 &= \mathbf{w}^2 + \mathbf{d}^2 \\
&\vdots \\
\mathbf{w}^K &= \mathbf{w}^{K-1} + \mathbf{d}^{K-1}
\end{aligned}
\tag{2.18}
$$

where $\mathbf{d}^{k-1}$ is the descent direction determined at the kth step of the process, defining the kth step as $\mathbf{w}^k = \mathbf{w}^{k-1} + \mathbf{d}^{k-1}$ such that in the end the inequalities in Equation (2.15) are met. This is illustrated schematically with a generic function taking in two inputs in the top panel of Figure 2.5. The two-input function is illustrated here via a *contour plot*, a common visualization tool that allows us to project a function down onto its input space. Darker regions on the plot correspond to points with larger evaluations (higher on the function), while brighter regions correspond to points with smaller evaluations (lower on the function).

The descent directions in Equation (2.18) can be found in a multitude of ways. In the remaining sections of this chapter we discuss *zero-order* approaches for doing so, and in the following chapters we describe the so-called *first-* and *second-order* approaches, i.e., approaches that leverage the first- and/or second-order derivative of a function to determine descent directions. How the descent directions are determined is precisely what distinguishes major local optimization schemes from one another.

2.4.3   The steplength parameter

We can compute how far we travel at each step of a local optimization method by examining the general form of a local step. Making this measurement we can see that, at the kth step as defined in Equation (2.18), we move a distance equal to the length of the corresponding descent direction

$$\left\| \mathbf{w}^k - \mathbf{w}^{k-1} \right\|_2 = \left\| \mathbf{d}^{k-1} \right\|_2 . \tag{2.19}$$

This can mean that the *length* of descent vectors could be problematic even if they point in a descent direction, downhill. For example, if they are too long, as illustrated in the middle panel of Figure 2.5, then a local method can oscillate wildly at each update step, never reaching an approximate minimum. Likewise if they are too small in length, a local method will move so sluggishly slow, as illustrated in the bottom panel of Figure 2.5, that far too many steps would be required to reach an approximate minimum.

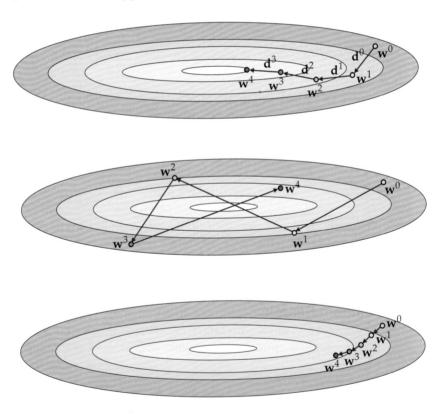

Figure 2.5 (top panel) Schematic illustration of a generic local optimization scheme applied to minimize a function taking in two inputs, with descent directions marked in black. See text for further details. (middle panel) Direction vectors are too large, causing a wild oscillatory behavior around the minimum point. (bottom panel) Direction vectors are too small, requiring a large number of steps be taken to reach the minimum.

Because of this potential problem many local optimization schemes come equipped with what is referred to as a *steplength parameter*, also called a *learning rate* in the jargon of machine learning. This parameter, which allows us to control the length of each update step (hence the name steplength parameter), is typically denoted by the Greek letter α. With a steplength parameter the generic kth update step is written as

$$\mathbf{w}^k = \mathbf{w}^{k-1} + \alpha \mathbf{d}^{k-1}. \tag{2.20}$$

The entire sequence of K steps is then similarly written as

$$\begin{aligned} \mathbf{w}^0 \\ \mathbf{w}^1 &= \mathbf{w}^0 + \alpha \mathbf{d}^0 \\ \mathbf{w}^2 &= \mathbf{w}^1 + \alpha \mathbf{d}^1 \\ \mathbf{w}^3 &= \mathbf{w}^2 + \alpha \mathbf{d}^2 \\ &\vdots \\ \mathbf{w}^K &= \mathbf{w}^{K-1} + \alpha \mathbf{d}^{K-1}. \end{aligned} \tag{2.21}$$

Note the only difference between this form for the kth step and the original is that now we scale the descent direction $\mathbf{d}^{k-1}$ by the steplength parameter $\alpha > 0$. With the addition of this parameter the distance traveled at the kth step of a generic local optimization scheme can be computed as

$$\left\| \mathbf{w}^k - \mathbf{w}^{k-1} \right\|_2 = \left\| \left(\mathbf{w}^{k-1} + \alpha \mathbf{d}^{k-1} \right) - \mathbf{w}^{k-1} \right\|_2 = \alpha \left\| \mathbf{d}^{k-1} \right\|_2. \tag{2.22}$$

In other words, the length of the kth step is now *proportional* to the length of the descent vector, and we can fine tune precisely how far we wish to travel in this direction by setting the value of α properly. A common choice is to set α to some *fixed* small value for each of the K steps. However (just like local optimization methods themselves), there are a number of ways of setting the steplength parameter which we will discuss later in the current and future chapters.

2.5 Random Search

In this section we describe our first local optimization algorithm: *random search*. With this instance of the general local optimization framework we seek out a descent direction at each step by examining a number of random directions stemming from our current point. This manner of determining a descent direction, much like the global optimization scheme described in Section 2.3, scales poorly with the dimension of input, which ultimately disqualifies random search for use with today's large-scale machine learning problems. However, this zero-order approach to local optimization is extremely useful as a simple example of the general framework introduced previously, allowing us to give a simple yet concrete algorithmic example of universally present ideas like *descent directions,* various choices for the *steplength parameter,* and issues of *convergence.*

2.5.1   The big picture

The defining characteristic of the *random search* (as is the case with every major local optimization scheme) lies in how the descent direction $\mathbf{d}^{k-1}$ at the kth local optimization update step $\mathbf{w}^k = \mathbf{w}^{k-1} + \mathbf{d}^{k-1}$ is chosen.

With random search we do (perhaps) the "laziest" possible thing one could think to do in order to find a descent direction: we sample a given number of random directions stemming from $\mathbf{w}^{k-1}$, evaluate each candidate update point, and choose the one that gives us the smallest evaluation (so long as it is indeed lower on the function than our current point). In other words, we look locally around the current point, in a certain number of random directions, for a point that has a lower evaluation, and if we find one we move to it.

To be more precise, at the kth step we generate P random directions $\{\mathbf{d}^p\}_{p=1}^P$ to try out, each stemming from the prior step $\mathbf{w}^{k-1}$ and leading to the candidate point $\mathbf{w}^{k-1} + \mathbf{d}^p$ for $p = 1, 2, ..., P$.

After evaluating all such P candidate points we pick the one that gives us the smallest evaluation, i.e., the one with the index given by

$$s = \underset{p=1,2,\dots,P}{\operatorname{argmin}}\ g\left(\mathbf{w}^{k-1} + \mathbf{d}^p\right). \tag{2.23}$$

Finally, if the best point found has a smaller evaluation than the current point, i.e., if $g\left(\mathbf{w}^{k-1} + \mathbf{d}^s\right) < g\left(\mathbf{w}^{k-1}\right)$, then we move to the new point $\mathbf{w}^k = \mathbf{w}^{k-1} + \mathbf{d}^s$, otherwise we either halt the method or try another batch of P random directions.

The random search method is illustrated in Figure 2.6 using a quadratic function where, for visualization purposes, the number of random directions to try is set relatively small to $P = 3$.

2.5.2   Controlling the length of each step

In order to better control the progress of random search we can *normalize* our randomly chosen directions to each have unit length, i.e., $\|\mathbf{d}\|_2 = 1$. This way we can adjust each step to have whatever length we desire by introducing a steplength parameter α (as discussed in Section 2.4.3). This more general step $\mathbf{w}^k = \mathbf{w}^{k-1} + \alpha \mathbf{d}$ now has length exactly equal to the steplength parameter α, as

$$\|\mathbf{w}^k - \mathbf{w}^{k-1}\|_2 = \|\alpha\, \mathbf{d}\|_2 = \alpha \|\mathbf{d}\|_2 = \alpha. \tag{2.24}$$

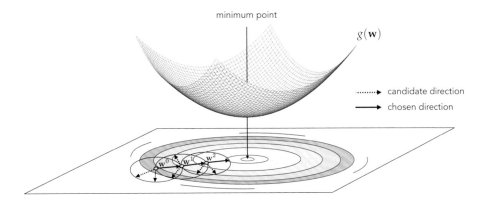

Figure 2.6 At each step the random search algorithm determines a descent direction by examining a number of random directions. The direction leading to the new point with the smallest evaluation is chosen as the descent direction, and the process repeated until a point near a local minimum is found. Here we show three prototypical steps performed by random search, where three random directions are examined at each step. At each step the best descent direction found is drawn as a solid black arrow while the other two inferior directions are shown as dashed black arrows.

Example 2.3   Minimizing a simple quadratic using random search

In this example we run random local search for $K = 5$ steps with $\alpha = 1$ for all steps, at each step searching for $P = 1000$ random directions to minimize the quadratic function

$$g(w_1, w_2) = w_1^2 + w_2^2 + 2. \tag{2.25}$$

Figure 2.7 shows the function in three dimensions on the top-left panel, along with the set of steps produced by the algorithm colored from green (at the start of the run where we initialize at $\mathbf{w}^0 = [3 \ 4]^T$) to red (when the algorithm halts). Directed arrows illustrate each descent direction chosen, connecting each step to its predecessor, and are shown to help illustrate the total path the algorithm takes. In the top-right panel we show the same function, but viewed from directly above as its *contour plot*.

Notice that if the dimension of the input N is greater than 2 we cannot make a plot like the ones shown in the figure to tell how well a particular run of any local optimization (here random search) is performed. A more general way to visualize the progress made by a local method is to plot the corresponding sequence of function evaluations against the step number, i.e., plotting the pairs $\left(k, g\left(\mathbf{w}^k\right)\right)$ for $k = 1, 2, ..., K$, as demonstrated in the bottom panel of Figure 2.7. This allows us to tell (regardless of the input dimension N of the function being minimized) how the algorithm performed, and whether we need to adjust any of its parameters (e.g., the maximum number of steps K or the value of α). This

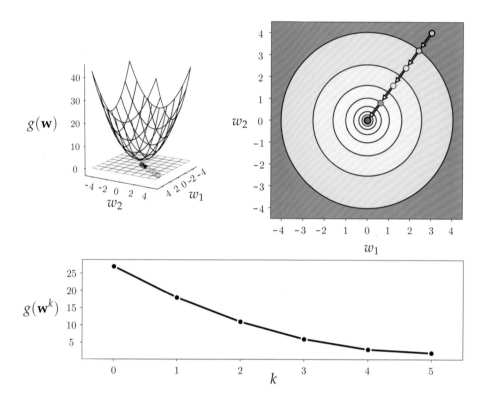

Figure 2.7 Figure associated with Example 2.3. See text for further details.

visualization is called a *cost function history plot*. An additional benefit of such a plot is that we can more easily tell the exact value of each function evaluation during the local optimization run.

Example 2.4   Minimizing a function with multiple local minima
In this example we show what one may need to do in order to find the global minimum of a function using a local optimization scheme like random search. For visualization purposes we use the single-input function

$$g(w) = \sin(3w) + 0.3w^2 \tag{2.26}$$

and initialize two runs, one starting at $w^0 = 4.5$ and another at $w^0 = -1.5$. For both runs we use a steplength of $\alpha = 0.1$ fixed for all $K = 10$ steps or iterations. As can be seen in Figure 2.8, depending on where we initialize, we may end up near a local (left panel) or global minimum (right panel). Here we illustrate the steps of each run as circles along the input axis with corresponding evaluations on the function itself as similarly colored x marks. The steps of each run are colored green near the start of the run to red when a run halts.

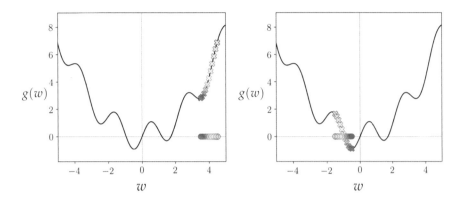

Figure 2.8 Figure associated with Example 2.4. Properly minimizing such a nonconvex function like the one shown here requires multiple runs of local optimization with different initializations. See text for further details.

2.5.3 Exploring fundamental steplength rules

In Examples 2.3 and 2.4 we set the steplength parameter α fixed for all steps of each run. Known as a *fixed steplength rule*, this is a very common choice of steplength parameter for local optimization methods in general. It is also possible to change the value of α from one step to another with what is often referred to as an *adjustable steplength rule*. Before exploring a very common adjustable steplength rule used in machine learning, called the *diminishing* steplength rule, we first show the importance of steplength tuning through a simple example.

Example 2.5 A failure to converge

In this example we use random search to minimize the quadratic function

$$g\left(w_1, w_2\right) = w_1^2 + w_2^2 + 2 \tag{2.27}$$

using the steplength $\alpha = 1$ (as we used in Example 2.3) but with a different initialization at $\mathbf{w}^0 = [1.5\ \ 2]^T$. However, with this initialization and as shown by examining the contour plot of this function in the left panel of Figure 2.9, the algorithm gets stuck at a non-optimal point (colored red) away from the global minimum point located at the origin, where here the contour plot is shown without color for better visualization. Also drawn in the same plot is a blue unit circle centered at the final red point, representing the location of all possible points the algorithm could take us to if we decided to take another step and move from where it halts at the red point. Notice how this blue circle encompasses one of the contours of the quadratic (in dashed red) on which the final red point lies. This means that every possible direction provides ascent, not descent, and the algorithm must therefore halt.

We need to be careful when choosing the steplength value with this simple

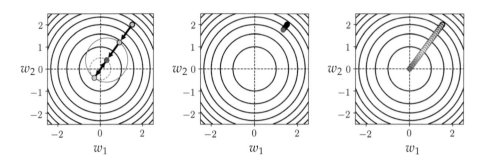

Figure 2.9 Figure associated with Example 2.5. Determining a proper steplength is crucial to optimal performance with random search – and by extension many local optimization algorithms. Here choosing the steplength too large leads to the method halting at a suboptimal point (left panel), setting it too small leads to very slow convergence towards the minimum of the function (middle panel), and setting it "just right" leads to ideal convergence to a point close to the function's minimum (right panel). See text for further details.

quadratic function, and by extension any general function. If, as shown in the middle panel of Figure 2.9, we repeat the same experiment but cut the steplength down to $\alpha = 0.01$, we do not reach a point anywhere near the global minimum within the same number of steps.

Setting the steplength parameter a little larger to $\alpha = 0.1$ for all steps, we make another run mirroring the previous one with results shown in the right panel of Figure 2.9. The algorithm now converges to a point much closer to the global minimum of the function at the origin.

In general, the combination of steplength and maximum number of iterations are best chosen together. The trade-off here is straightforward: a small steplength combined with a large number of steps can guarantee convergence to a local minimum, but can be computationally expensive. Conversely, a large steplength and small number of maximum iterations can be cheaper but less effective at finding the optimal. Often, in practice, these kinds of choices are made by making several runs of an algorithm and plotting their corresponding cost function histories to determine optimal parameter settings.

2.5.4   Diminishing steplength rules

A commonly used alternative to fixed steplength rules are the so-called *diminishing steplength rules* wherein we shrink or *diminish* the size of the steplength at each step of local optimization. One common way of producing a diminishing steplength rule is simply to set $\alpha = \frac{1}{k}$ at the kth step of the process. This provides the benefit of shrinking the distance between subsequent steps as we progress on a run, since with this choice of steplength and a unit-length descent direction vector we have

$$\|\mathbf{w}^k - \mathbf{w}^{k-1}\|_2 = \alpha \|\mathbf{d}^{k-1}\|_2 = \frac{1}{k}. \tag{2.28}$$

At the same time, if we sum up the total distance the algorithm travels in K steps (provided we indeed move at each step) we have

$$\sum_{k=1}^{K} \|\mathbf{w}^k - \mathbf{w}^{k-1}\|_2 = \sum_{k=1}^{K} \frac{1}{k}. \tag{2.29}$$

The beauty of this sort of diminishing steplength is that while the steplength $\alpha = \frac{1}{k}$ decreases to zero as k increases, the total distance traveled by the algorithm goes to infinity.[1] This means that a local algorithm employing this sort of diminishing steplength rule can – in theory – move around an infinite distance in search of a minimum all the while taking smaller and smaller steps, which allows it to work into any small nooks and crannies a function might have where any minimum lies.

2.5.5 Random search and the curse of dimensionality

As with the global optimization approach discussed in Section 2.3, the curse of dimensionality also poses a major obstacle to the practical application of the random search algorithm as the dimension of a function's input increases. In other words, for most functions it becomes *exponentially* more difficult to find a descent direction *at random* at a given point as its input dimension increases.

Take, for example, the single-input quadratic function $g(w) = w^2 + 2$ and suppose we take a single step using the random search algorithm beginning at $w^0 = 1$ with steplength $\alpha = 1$. As illustrated in the top panel of Figure 2.10, because the input dimension in this case is $N = 1$, to determine a descent direction we only have two directions to consider: the negative and positive directions from our initial point. One of these two directions will provide descent

---

[1] The sum $\sum_{k=1}^{\infty} \frac{1}{k}$ is often called the *harmonic series*, and one way to see that it diverges to infinity is by lumping together consecutive terms as

$$\sum_{k=1}^{\infty} \frac{1}{k} = 1 + \frac{1}{2} + \left(\frac{1}{3} + \frac{1}{4}\right) + \left(\frac{1}{5} + \frac{1}{6} + \frac{1}{7} + \frac{1}{8}\right) + \cdots$$

$$\geq 1 + \frac{1}{2} + 2\left(\frac{1}{4}\right) + 4\left(\frac{1}{8}\right) + \cdots \tag{2.30}$$

$$= 1 + \frac{1}{2} + \frac{1}{2} + \frac{1}{2} + \cdots .$$

In other words, the harmonic series is lower-bounded by an infinite sum of $\frac{1}{2}$ values, and thus diverges to infinity.

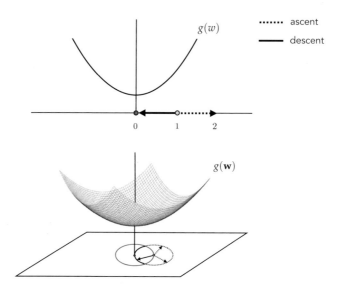

Figure 2.10 (top panel) When the input dimension is $N = 1$, there are only two unit directions we can move in, only one of which (the solid arrow) is a descent direction. (bottom panel) When the input dimension is $N = 2$, there are infinitely many unit directions to choose from, only a fraction of which whose endpoint lies inside the unit circle (points on the solid portion of the arc) are descent directions.

(here, the negative direction). In other words, we would have a 50 percent chance of finding a descent direction if were to choose one at random.

Now let us examine the same sort of quadratic function, this time one that takes in an $N = 2$ dimensional vector $\mathbf{w}$ as input: $g(\mathbf{w}) = \mathbf{w}^T\mathbf{w} + 2$, and imagine taking a single random search step starting at $\mathbf{w}^0 = [1 \ 0]^T$, a two-dimensional analog of the initialization we used for our one-dimensional quadratic. As illustrated in the bottom panel of Figure 2.10, we now have infinitely many unit directions to choose from, but only a fraction of which (less than 50 percent) provide descent. In other words, in two dimensions the chance of randomly selecting a descent direction drops with respect to its analogous value in one dimension. This decrease in the probability of randomly choosing a descent direction decreases *exponentially* as the input dimension N of this quadratic increases. Indeed one can compute that for a general N, the probability of choosing a descent direction at random starting at

$$\mathbf{w}^0 = \begin{bmatrix} 1 \\ 0 \\ \vdots \\ 0 \end{bmatrix}_{N \times 1} \tag{2.31}$$

for the quadratic function at hand is upper-bounded by $\frac{1}{2}\left(\frac{\sqrt{3}}{2}\right)^{N-1}$ (see Exercise 2.5). This means, for example, that when $N = 30$ the descent probability falls below 1 percent, making random search incredibly inefficient for minimizing even a simple quadratic function.

2.6   Coordinate Search and Descent

The coordinate search and descent algorithms are additional zero-order local methods that get around the inherent scaling issues of random local search by restricting the set of search directions to the coordinate axes of the input space. The concept is simple: random search was designed to minimize a function $g(w_1, w_2, ..., w_N)$ with respect to all of its parameters *simultaneously*. With coordinate-wise algorithms we attempt to minimize such a function with respect to one coordinate or parameter at a time (or, more generally, one subset of coordinates or parameters at a time) keeping all others fixed.

While this limits the diversity of descent directions that can be potentially discovered, and thus more steps are often required to determine approximate minima, these algorithms are far more scalable than random search.

2.6.1   Coordinate search

As illustrated in the left panel of Figure 2.11 for a prototypical $N = 2$ dimensional example, with *coordinate search* we seek out a descent direction by searching randomly among only the coordinate axes of the input space. This means in general that, for a function of input dimension N, we only look over the $2N$ directions from the set $\{\pm \mathbf{e}_n\}_{n=1}^{N}$, where $\mathbf{e}_n$ is a *standard basis vector* whose entries are set to zero except its nth entry which is set to 1.

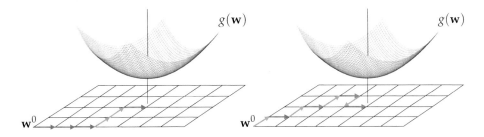

Figure 2.11 (left panel) With *coordinate search* we seek out descent directions only among the coordinates axes: at each step, colored alternately blue and red for better visualization, we try $2N = 4$ directions along the $N = 2$ coordinate axes, and pick the one resulting in the largest decrease in the function's evaluation. (right panel) With *coordinate descent* we (ideally) take a step immediately after examining the positive and negative directions along each coordinate.

It is this restricted set of directions we are searching over that distinguishes the coordinate search approach from the random search approach described in the previous section, where the set of directions at each step was made up of random directions. While the (lack of) diversity of the coordinate axes may limit the effectiveness of the possible descent directions it can encounter (and thus more steps are required to determine an approximate minimum), the restricted search makes coordinate search far more scalable than the random search method since at each step only $2N$ directions must be tested.

2.6.2   Coordinate descent

A slight twist on the coordinate search produces a much more effective algorithm at precisely the same computational cost, called *coordinate descent*. Instead of collecting each coordinate direction (along with its negative version), and then choosing a single best direction from this entire set, we can simply examine one coordinate direction (and its negative) at a time and step in this direction if it produces descent. This idea is illustrated in the right panel of Figure 2.11.

Whereas with coordinate search we evaluate the cost function $2N$ times (twice per coordinate) to produce a single step, this alternative approach takes the same number of function evaluations but potentially moves N steps in doing so. In other words, with coordinate descent we can, for the same cost as coordinate search, potentially minimize a function much faster. Indeed of all the zero-order methods detailed in this chapter, coordinate descent is by far the most practical.

Example 2.6   Coordinate search versus coordinate descent
In this example we compare the efficacy of coordinate search and the coordinate descent algorithms on the simple quadratic function

$$g(w_1, w_2) = 0.26\left(w_1^2 + w_2^2\right) - 0.48\, w_1 w_2. \tag{2.32}$$

In Figure 2.12 we compare 20 steps of coordinate search (left panel) and coordinate descent (right panel), using a diminishing steplength for both runs. Because coordinate descent effectively takes two steps for every single step taken by coordinate search, we get significantly closer to the function's minimum using the same number of function evaluations.

2.7   Conclusion

This chapter laid the groundwork for a wide range of fundamental ideas related to mathematical optimization (motivated in Section 1.4) that we will see repeat-

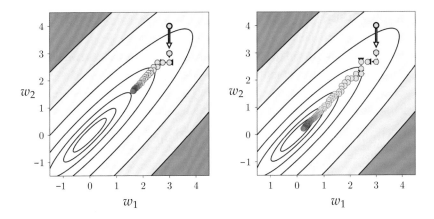

Figure 2.12 Figure associated with Example 2.6. A run of coordinate search (left panel) and coordinate descent (right panel) employed to minimize the same function. While both runs have the same computational cost, coordinate descent makes significantly greater progress. See text for further details.

edly not only in the next two chapters, but throughout the remainder of the text as well.

We began by introducing the concept of mathematical optimization, which is the mathematical/computational pursuit of a function's minima or maxima. Then in Section 2.2 we translated our informal definition of a function's minima and maxima into the language of mathematics, with this formal translation referred to as the *zero-order condition for optimality*. Leveraging this definition we then described global optimization methods (essentially the raw evaluation of a function over a fine grid of input points) in Section 2.3 which we saw – due to the curse of dimensionality – scales very poorly in terms of input dimension, and are thus not often very practical. Section 2.4 introduced the alternative to this limited framework – *local optimization* – which encompasses an enormous family of algorithms we discuss in the remainder of this chapter, as well as the two chapters that follow. Finally, in Sections 2.5 and 2.6 we described a number of examples of zero-order local optimization algorithms – including random search and coordinate search/descent. While the latter schemes can be very useful in particular applications, in general these zero-order local schemes are not as widely used in machine learning when compared to those we will see in the chapters to come, which leverage a function's first and/or second derivatives to more immediately formulate descent directions (instead of the raw search required by zero-order algorithms). However, the relative simplicity of zero-order schemes allowed us to flush out a range of critical concepts associated with local optimization – ideas that we will see echo repeatedly throughout the chapters that follow in a comparatively uncluttered setting – including the notions of descent directions, steplength/learning rates, diminishing steplength schemes, and cost function history plots.

2.8   Exercises

† The data required to complete the following exercises can be downloaded from the text's github repository at github.com/jermwatt/machine_learning_refined

2.1   Minimizing a quadratic function and the curse of dimensionality
Consider the simple quadratic function

$$g(\mathbf{w}) = \mathbf{w}^T \mathbf{w} \tag{2.33}$$

whose minimum is always at the origin regardless of the input dimension N.

(a) Create a range of these quadratics for input dimension $N = 1$ to $N = 100$, sample the input space of each $P = 100$ times uniformly on the hypercube $[-1, 1] \times [-1, 1] \times \cdots \times [-1, 1]$ (this hypercube has N sides), and plot the *minimum* value attained for each quadratic against the input dimension N.

(b) Repeat part (a) using $P = 100$, $P = 1000$, and $P = 10,000$ samples, and plot all three curves in the same figure. What sort of trend can you see in this plot as N and P increase?

(c) Repeat parts (a) and (b), this time replacing uniformly chosen samples with randomly selected ones.

2.2   Implementing random search in Python
Implement the random search algorithm in Python and repeat the experiment discussed in Example 2.4.

2.3   Using random search to minimize a nonconvex function
Use your implementation of random search in Exercise 2.2 to minimize the function

$$g(w_1, w_2) = \tanh(4w_1 + 4w_2) + \max\left(0.4w_1^2, 1\right) + 1. \tag{2.34}$$

Take a maximum of eight steps and search through $P = 1000$ random directions at each step, with a steplength $\alpha = 1$ and an initial point $\mathbf{w}^0 = [2\ 2]^T$.

2.4   Random search with diminishing steplength
In this exercise you will use random search and a diminishing steplength to minimize the *Rosenbrock function*

$$g(w_1, w_2) = 100\left(w_2 - w_1^2\right)^2 + (w_1 - 1)^2. \tag{2.35}$$

This function has a global minimum at the point $\mathbf{w}^\star = [1\ \ 1]^T$ located in a very narrow and curved valley.

Make two runs of random search using $P = 1000$, the initial point $\mathbf{w}^0 = [-2\ \ -2]^T$, and $K = 50$ steps. With the first run use a fixed steplength $\alpha = 1$, and – with the second run – a diminishing steplength as detailed in Section 2.5.4. Compare the two runs by either plotting the contour plot of the cost function (with each run plotted on top), or by constructing a cost function history plot.

2.5   Random descent probabilities

Consider the quadratic function $g(\mathbf{w}) = \mathbf{w}^T\mathbf{w} + 2$, which we aim to minimize using random search starting at $\mathbf{w}^0$ defined in Equation (2.31), with $\alpha = 1$ and $\left\|\mathbf{d}^0\right\|_2 = 1$.

(a) When $N = 2$, show that the *probability of descent* – i.e., the probability that $g(\mathbf{w}^0 + \alpha\mathbf{d}^0) < g(\mathbf{w}^0)$ for a randomly chosen unit direction $\mathbf{d}^0$ – is upper-bounded by $\frac{\sqrt{3}}{4}$. *Hint: see Figure 2.13.*

(b) Extend your argument in part (a) to find an upper-bound on the probability of descent for general N.

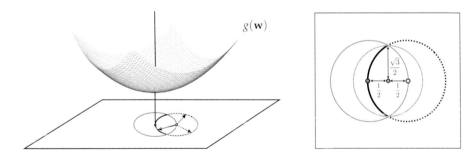

Figure 2.13 Figure associated with Exercise 2.5.

2.6   Revisiting the curse of dimensionality

In this exercise you will empirically confirm the curse of dimensionality problem described in Section 2.5.5 for the simple quadratic $g(\mathbf{w}) = \mathbf{w}^T\mathbf{w} + 2$.

Starting at the N-dimensional input point $\mathbf{w}^0$ defined in Equation (2.31), create P random unit directions $\{\mathbf{d}^p\}_{p=1}^P$ (where $\|\mathbf{d}^p\|_2 = 1$) and evaluate the point $\mathbf{w}^0 + \mathbf{d}^p$ via the quadratic for all p.

Next, produce a plot illustrating the portion of the sampled directions that provide a decrease in function evaluation (i.e., the number of *descent directions* divided by P) against N (for $N = 1, 2, ..., 25$) and for three values of P: $P = 100$, $P = 1000$, and $P = 10,000$. Describe the trend you observe.

2.7   Pseudo-code for the coordinate search algorithm
Devise pseudo-code for the coordinate search algorithm described in Section 2.6.1.

2.8   Coordinate search applied to minimize a simple quadratic
Compare five steps of the random search algorithm (with $P = 1000$ random directions tested at each step) to seven steps of coordinate search, using the same starting point $\mathbf{w}^0 = [3\ 4]^T$ and fixed steplength parameter value $\alpha = 1$ to minimize the quadratic function

$$g(w_1, w_2) = w_1^2 + w_2^2 + 2. \tag{2.36}$$

Plot this function along with the resulting runs of both algorithms, and describe any differences in behavior between the two runs.

2.9   Coordinate search with diminishing steplength
Implement the coordinate search algorithm detailed in Section 2.6.1 and use it to minimize the function

$$g(w_1, w_2) = 0.26\left(w_1^2 + w_2^2\right) - 0.48\,w_1 w_2 \tag{2.37}$$

using the diminishing steplength rule beginning at a random initial point. The global minimum of this function lies at the origin. Test your implementation by making sure it can reach a point significantly close (e.g., within 10^{-2}) to the origin from various random initializations.

2.10   Coordinate search versus coordinate descent
Implement the coordinate search and coordinate descent algorithms, and repeat the experiment discussed in Example 2.6.

3 First-Order Optimization Techniques

3.1   Introduction

In this chapter we describe fundamental optimization algorithms that leverage the *first derivative* or *gradient* of a function. These techniques, collectively called *first-order optimization methods*, are some of the most popular local optimization algorithms used to tackle machine learning problems today. We begin with a discussion of the *first-order optimality condition* which codifies how the first derivative(s) of a function characterizes its minima. We then discuss fundamental concepts related to hyperplanes, and in particular the first-order Taylor series approximation. As we will see, by exploiting a function's first-order information we can construct local optimization methods, foremost among them the extremely popular *gradient descent algorithm*, that naturally determine high-quality descent directions at a cost that is very often cheaper than even the coordinate-wise approaches described in the previous chapter.

3.2   The First-Order Optimality Condition

In Figure 3.1 we show two simple quadratic functions, one in two dimensions (left panel) and one in three dimensions (right panel), marking the global minimum on each with a green point. Also drawn in each panel is the line/hyperplane tangent to the function at its minimum point, also known as its first-order Taylor series approximation (see Appendix Section B.9 if you are not familiar with the notion of Taylor series approximation). Notice in both instances that the tangent line/hyperplane is perfectly flat. This sort of behavior is universal (for differentiable functions) regardless of the function one examines, and it holds regardless of the dimension of a function's input. That is, minimum values of a function are naturally located at *valley floors* where a tangent line or hyperplane is perfectly flat, and thus has zero-valued slope(s).

Because the *derivative* (see Appendix Sextion B.2) of a single-input function or the *gradient* (see Appendix Sextion B.4) of a multi-input function at a point gives precisely this slope information, the value of first-order derivatives provide a convenient way of characterizing minimum values of a function g. When $N = 1$, any point v of the single-input function $g(w)$ where

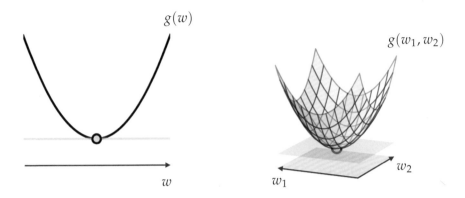

Figure 3.1 The first-order optimality condition characterizes points where the derivative or gradient of a function equals zero or, likewise, where the line or hyperplane tangent to a function has zero slope (i.e., it is completely horizontal and parallel to the input space). With a simple convex quadratic this condition identifies the global minimum of the function, illustrated here with a single-input (left panel) and multi-input (right panel) quadratic function.

$$\frac{\mathrm{d}}{\mathrm{d}w} g\left(v\right) = 0 \tag{3.1}$$

is a potential minimum. Analogously with multi-input functions, any N-dimensional point $\mathbf{v}$ where *every* partial derivative of g is zero, i.e.,

$$\begin{aligned}
\frac{\partial}{\partial w_1} g(\mathbf{v}) &= 0 \\
\frac{\partial}{\partial w_2} g(\mathbf{v}) &= 0 \\
&\vdots \\
\frac{\partial}{\partial w_N} g(\mathbf{v}) &= 0
\end{aligned} \tag{3.2}$$

is a potential minimum. This system of N equations is naturally referred to as the *first-order system of equations*. We can also write the first-order system more compactly using gradient notation as

$$\nabla g\left(\mathbf{v}\right) = \mathbf{0}_{N\times1}. \tag{3.3}$$

This very useful characterization of minimum points is the first-order analog to the zero-order condition for optimality discussed in Section 2.2, and is thus referred to as the *first-order optimality condition* (or the *first-order condition* for short). There are, however, two problems with the first-order characterization of minima.

Firstly, with few exceptions (including some interesting examples we detail in Section 3.2.1), it is virtually impossible to solve a general function's first-order systems of equations "by hand" (that is, to solve such equations algebraically for closed-form solutions). The other problem is that while the first-order condition defines only global minima for *convex* functions, like the quadratics shown in Figure 3.1, in general this condition captures not only the minima of a function but other points as well, including *maxima* and *saddle points* of *nonconvex* functions as we see in the example below. Collectively, minima, maxima, and saddle points are often referred to as *stationary* or *critical* points of a function.

Example 3.1   Visual inspection of single-input functions for stationary points
In the top row of Figure 3.2 we plot the functions

$$g(w) = \sin(2w)$$
$$g(w) = w^3 \qquad\qquad (3.4)$$
$$g(w) = \sin(3w) + 0.3w^2$$

along with their derivatives in the second row of the same figure. On each function we mark the points where its derivative is zero using a green dot (we likewise mark these points on each derivative itself), and show the tangent line corresponding to each such point in green as well.

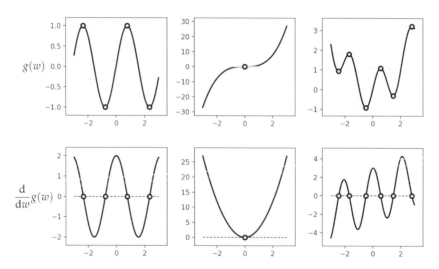

Figure 3.2 Figure associated with Example 3.1. From left to right in the top row, the functions $g(w) = \sin(2w)$, w^3, and $\sin(3w) + 0.3w^2$ are plotted along with their derivatives in the bottom row. See text for further details.

Examining these plots we can see that it is not only *global minima* that have zero derivatives, but a variety of other points as well, including (i) *local minima* or points that are the smallest with respect to their immediate neighbors, e.g., the one around the input value $w = -2.5$ in the top-right panel; (ii) *local* (and *global*) maxima or points that are the largest with respect to their immediate neighbors, e.g., the one around the input value $w = 0.5$ in the top-right panel; and (iii) *saddle points,* like the one shown in the top-middle panel, that are neither maximal nor minimal with respect to their immediate neighbors.

3.2.1   Special cases where the first-order system can be solved by hand

In principle, the benefit of using the first-order condition is that it allows us to transform the task of seeking out global minima to that of solving a system of equations, for which a wide range of algorithmic methods have been designed. The emphasis here on the word *algorithmic* is key, as solving a system of (potentially nonlinear) equations *by hand* is, generally speaking, very difficult (if not impossible).

   However, there are a handful of relatively simple but important examples where one can compute the solution to a first-order system by hand, or at least one can show algebraically that they reduce to a *linear* system of equations which can be easily solved numerically. By far the most important of these is the multi-input quadratic function (see Example 3.4) and the highly related *Rayleigh quotient* (see Exercise 3.3). These functions arise in many places in the study of machine learning, from fundamental models like linear regression, to second-order algorithm design, to the mathematical analysis of algorithms.

Example 3.2   Finding stationary points of single-input functions via the first-order condition

In this example we use the first-order condition for optimality to compute stationary points of the functions

$$
\begin{aligned}
g(w) &= w^3 \\
g(w) &= e^w \\
g(w) &= \sin(w) \\
g(w) &= a + bw + cw^2 \quad (c > 0).
\end{aligned}
\tag{3.5}
$$

- $g(w) = w^3$: the first-order condition gives $\frac{d}{dw}g(v) = 3v^2 = 0$, which we can visually identify as a saddle point at $v = 0$ (see top-middle panel of Figure 3.2).
- $g(w) = e^w$: the first-order condition gives $\frac{d}{dw}g(v) = e^v = 0$, which is only satisfied as v goes to $-\infty$, giving a minimum.
- $g(w) = \sin(w)$: the first-order condition gives stationary points wherever $\frac{d}{dw}g(v) = \cos(v) = 0$, which occurs at odd integer multiples of $\frac{\pi}{2}$, i.e., maxima at $v = \frac{(4k+1)\pi}{2}$ and minima at $v = \frac{(4k+3)\pi}{2}$ where k is any integer.

- $g(w) = a + bw + cw^2$: the first-order condition gives $\frac{d}{dw}g(v) = 2cv + b = 0$, with a minimum at $v = \frac{-b}{2c}$ (assuming $c > 0$).

Example 3.3   A simple-looking function

As mentioned previously, the vast majority of first-order systems cannot be solved by hand algebraically. To get a sense of this challenge here we show an example of a simple-enough looking function whose global minimum is not straightforward to compute by hand.

Consider the degree-four polynomial

$$g(w) = \frac{1}{50}\left(w^4 + w^2 + 10w\right) \tag{3.6}$$

which is plotted over a short range of inputs containing its global minimum in Figure 3.3.

Figure 3.3 Figure associated with Example 3.3. See text for details.

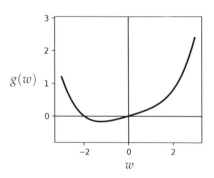

The first-order system here can be easily computed as

$$\frac{d}{dw}g(w) = \frac{1}{50}\left(4w^3 + 2w + 10\right) = 0 \tag{3.7}$$

which simplifies to

$$2w^3 + w + 5 = 0. \tag{3.8}$$

This has only one real solution

$$w = \frac{\sqrt[3]{\sqrt{2031} - 45}}{\sqrt[3]{36}} - \frac{1}{\sqrt[3]{6\left(\sqrt{2031} - 45\right)}} \tag{3.9}$$

which can be computed – after much toil – using centuries-old tricks developed for just such problems. In fact, had the polynomial in Equation (3.6) been of degree six or higher, we would not have been able to guarantee finding its stationary point(s) in closed form.

Example 3.4 **Stationary points of multi-input quadratic functions**

Take the general multi-input quadratic function

$$g(\mathbf{w}) = a + \mathbf{b}^T\mathbf{w} + \mathbf{w}^T\mathbf{C}\mathbf{w} \tag{3.10}$$

where $\mathbf{C}$ is an $N \times N$ symmetric matrix, $\mathbf{b}$ is an $N \times 1$ vector, and a is a scalar. Computing the gradient of g we have

$$\nabla g(\mathbf{w}) = 2\mathbf{C}\mathbf{w} + \mathbf{b}. \tag{3.11}$$

Setting this equal to zero gives a symmetric linear system of equations of the form

$$\mathbf{C}\mathbf{w} = -\frac{1}{2}\mathbf{b} \tag{3.12}$$

whose solutions are stationary points of the original function. Note here we have not explicitly solved for these stationary points, but have merely shown that the first-order system of equations in this particular case is in fact one of the easiest to solve numerically (see Example 3.6).

___

3.2.2   Coordinate descent and the first-order optimality condition

While solving the first-order system in Equation (3.2) *simultaneously* is often impossible, it is sometimes possible to solve such a system *sequentially*. In other words, in some (rather important) cases the first order system can be solved *one equation at a time*, the nth of which takes the form $\frac{\partial}{\partial w_n}g(\mathbf{v}) = 0$. This idea, which is a form of *coordinate descent*, is especially effective when each of these equations can be solved for in closed form (e.g., when the function being minimized is a quadratic).

To solve the first-order system sequentially, we first initialize at an input $\mathbf{w}^0$, and begin by updating the first coordinate by solving

$$\frac{\partial}{\partial w_1}g\left(\mathbf{w}^0\right) = 0 \tag{3.13}$$

for the optimal first weight $w_1^\star$. Note importantly in solving this equation for w_1 that all other weights are kept fixed at their initial values. We then update the first coordinate of the vector $\mathbf{w}^0$ with this solution $w_1^\star$, and call the updated set of weights $\mathbf{w}^1$. Continuing this pattern, to update the nth weight we solve

$$\frac{\partial}{\partial w_n}g\left(\mathbf{w}^{n-1}\right) = 0 \tag{3.14}$$

for $w_n^\star$. Again, when this equation is solved all other weights are kept fixed at their current values. We then update the nth weight using this value forming the updated set of weights $\mathbf{w}^n$.

After we sweep through all N weights a single time we can refine our solution by sweeping through the weights again (as with any other coordinate-wise method). At the kth such sweep we update the nth weight by solving the single equation

$$\frac{\partial}{\partial w_n} g\left(\mathbf{w}^{N(k-1)+n-1}\right) = 0 \tag{3.15}$$

to update the nth weight of $\mathbf{w}^{N(k-1)+n-1}$, and so on.

Example 3.5   Minimizing convex quadratic functions via coordinate descent
In this example we use coordinate descent to minimize the convex quadratic function

$$g(w_1, w_2) = w_1^2 + w_2^2 + 2 \tag{3.16}$$

whose minimum lies at the origin. We make the run initialized at $\mathbf{w}^0 = [3\ 4]^T$, where a single sweep through the coordinates (i.e., two steps) here perfectly minimizes the function. The path this run took is illustrated in the left panel of Figure 3.4 along with a contour plot of the function. One can easily check that each first-order equation in this case is linear and trivial to solve in closed form.

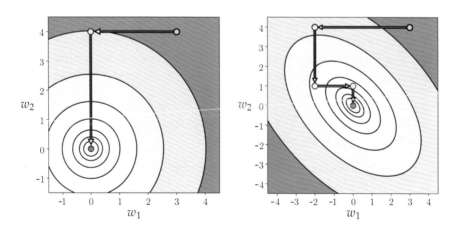

Figure 3.4 Figure associated with Example 3.5. See text for details.

We then apply coordinate descent to minimize the convex quadratic

$$g(w_1, w_2) = 2w_1^2 + 2w_2^2 + 2w_1 w_2 + 20 \tag{3.17}$$

whose contour plot is shown in the right panel of Figure 3.4. Here it takes two
full sweeps through the variables to find the global minimum, which again lies
at the origin.

Example 3.6   Solving symmetric linear systems of equations
In Example 3.4 we saw that the first-order system of a multi-input quadratic
function takes the form

$$\mathbf{C}\mathbf{w} = -\frac{1}{2}\mathbf{b} \tag{3.18}$$

where $\mathbf{C}$ is a symmetric matrix as described in that example. We can use the
coordinate descent algorithm to solve this system, thereby minimizing the corresponding quadratic function. Divorced from the concept of a quadratic we
can think of coordinate descent in a broader context as a method for solving
more general symmetric linear systems of equations, which is quite commonly
encountered in practice, e.g., at each and every step of Newton's method (as
detailed in Chapter 4).

3.3   The Geometry of First-Order Taylor Series

In this section we describe important characteristics of the *hyperplane* including the concept of the direction of *steepest ascent* and *steepest descent*. We then
study a special hyperplane: the first-order Taylor series approximation to a function, which defines the very essence of the extremely popular gradient descent
algorithm, introduced in Section 3.5.

3.3.1   The anatomy of hyperplanes

A general N-dimensional hyperplane can be characterized as

$$h(w_1, w_2, \ldots, w_N) = a + b_1 w_1 + b_2 w_2 + \cdots + b_N w_N \tag{3.19}$$

where a as well as b_1 through b_N are all scalar parameters. We can rewrite h more
compactly – using vector notation – as

$$h(\mathbf{w}) = a + \mathbf{b}^T \mathbf{w} \tag{3.20}$$

denoting

$$\mathbf{b} = \begin{bmatrix} b_1 \\ b_2 \\ \vdots \\ b_N \end{bmatrix} \quad \text{and} \quad \mathbf{w} = \begin{bmatrix} w_1 \\ w_2 \\ \vdots \\ w_N \end{bmatrix}. \tag{3.21}$$

When $N = 1$, Equation (3.20) simplifies to

$$h(w) = a + bw \tag{3.22}$$

which is the familiar formula for a (one-dimensional) line. Notice, $h(w) = a + bw$ is a *one*-dimensional thing living in a *two*-dimensional ambient space whose input space (characterized by w) is *one*-dimensional itself.

The same is true for general N. That is, $h(\mathbf{w}) = a + \mathbf{b}^T\mathbf{w}$ is an N-dimensional mathematical object living in an $(N+1)$-dimensional ambient space whose input space (characterized by $w_1, w_2, \ldots, w_N$) is N-dimensional.

3.3.2 Steepest ascent and descent directions

As we just saw, with a one-dimensional hyperplane the input space is also one-dimensional, implying that at any point w^0 in the input space there are only two directions to move in: to the left or right of w^0. This is illustrated in the left panel of Figure 3.5. Here, starting at w^0 and moving to the right (towards $+\infty$) *increases* the value of h, and hence it is an *ascent* direction. Conversely, moving to the left (towards $-\infty$) *decreases* the value of h, and hence it is a *descent* direction.

When $N > 1$, however, there are infinitely many directions to move in (as opposed to just two when $N = 1$) – some providing ascent, some providing descent, and some that preserve the value of h – as illustrated in the right panel of Figure 3.5 for $N = 2$. It is therefore logical to ask whether we can find the direction that produces the largest ascent (or descent), commonly referred to as the direction of *steepest ascent* (or *descent*).

Formalizing the search for the direction of steepest ascent at a given point $\mathbf{w}^0$, we aim to find the unit direction $\mathbf{d}$ such that the value of $h\left(\mathbf{w}^0 + \mathbf{d}\right)$ is maximal. In other words, we aim to solve

$$\underset{\mathbf{d}}{\text{maximize}} \ \ h\left(\mathbf{w}^0 + \mathbf{d}\right) \tag{3.23}$$

over all unit-length vectors $\mathbf{d}$. Note from Equation (3.20) that $h\left(\mathbf{w}^0 + \mathbf{d}\right)$ can be written as

$$a + \mathbf{b}^T\left(\mathbf{w}^0 + \mathbf{d}\right) = a + \mathbf{b}^T\mathbf{w}^0 + \mathbf{b}^T\mathbf{d} \tag{3.24}$$

where the first two terms on the right-hand side are constant with respect to $\mathbf{d}$.

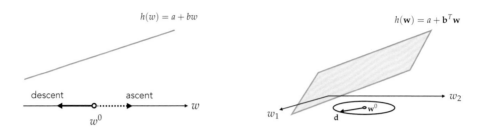

$$h(w) = a + bw \qquad\qquad h(\mathbf{w}) = a + \mathbf{b}^T\mathbf{w}$$

Figure 3.5 (left panel) At any given point w^0 in the input space of a one-dimensional hyperplane h, there are only two directions to travel in: one that increases the evaluation of h (or an *ascent* direction), and one that decreases it (or a *descent* direction). (right panel) In higher dimensions (here $N = 2$) there are infinitely many (unit) directions $\mathbf{d}$ to move into – starting at a given N-dimensional input point $\mathbf{w}^0$. As can be seen in this case, the endpoint of all such directions form a unit circle around and centered at $\mathbf{w}^0$.

Therefore maximizing the value of $h\left(\mathbf{w}^0 + \mathbf{d}\right)$ is equivalent to maximizing $\mathbf{b}^T\mathbf{d}$, which itself can be written, using the inner-product rule (see Appendix C), as

$$\mathbf{b}^T\mathbf{d} = \|\mathbf{b}\|_2\,\|\mathbf{d}\|_2 \cos(\theta). \tag{3.25}$$

Note, once again, that $\|\mathbf{b}\|_2$ (i.e., the length of $\mathbf{b}$) does not change with respect to $\mathbf{d}$, and that $\|\mathbf{d}\|_2 = 1$. Therefore the problem in Equation (3.23) reduces to

$$\underset{\theta}{\text{maximize}}\;\; \cos(\theta) \tag{3.26}$$

where θ is the angle between the vectors $\mathbf{b}$ and $\mathbf{d}$.

It is clear now, of all unit diretions, $\mathbf{d} = \frac{\mathbf{b}}{\|\mathbf{b}\|_2}$ provides the steepest ascent (where $\theta = 0$ and $\cos(\theta) = 1$). Similarly, we can show that the unit direction $\mathbf{d} = \frac{-\mathbf{b}}{\|\mathbf{b}\|_2}$ provides the steepest descent (where $\theta = \pi$ and $\cos(\theta) = -1$).

3.3.3   The gradient and the direction of steepest ascent/descent

A multi-input function $g(\mathbf{w})$ can be approximated locally around a given point $\mathbf{w}^0$ by a hyperplane $h(\mathbf{w})$

$$h(\mathbf{w}) = g(\mathbf{w}^0) + \nabla g(\mathbf{w}^0)^T\left(\mathbf{w} - \mathbf{w}^0\right) \tag{3.27}$$

which can be rewritten as $h(\mathbf{w}) = a + \mathbf{b}^T\mathbf{w}$ (to match our notation in the previous section) where

$$a = g(\mathbf{w}^0) - \nabla g(\mathbf{w}^0)^T\mathbf{w}^0 \quad \text{and} \quad \mathbf{b} = \nabla g(\mathbf{w}^0). \tag{3.28}$$

This hyperplane is the *first-order Taylor series approximation* of g at $\mathbf{w}^0$, and is tangent to g at this point (see Appendix B).

Because h is constructed to closely approximate g near the point $\mathbf{w}^0$, its steepest ascent and descent directions also tell us the direction to travel to increase or decrease the value of the underlying function g itself at/near the point $\mathbf{w}^0$.

3.4   Computing Gradients Efficiently

Think for a moment about how you perform basic arithmetic, e.g., multiplying two numbers. If the two numbers involved are relatively small, such as 35 times 21, you can likely do the multiplication in your head using a combination of *multiplication properties* and *simple multiplication results* you learned in elementary school. For example, you may choose to use the distributive property of multiplication to decompose 35×21 as

$$(30 + 5) \times (20 + 1) = (30 \times 20) + (30 \times 1) + (5 \times 20) + (5 \times 1) \qquad (3.29)$$

and then use the multiplication table that you have likely memorized to find 30×20, 30×1, and so on. We use this sort of strategy on a daily basis when making quick back-of-the-envelope calculations like computing interest on a loan or investment, computing how much to tip at a restaurant, etc.

However, even though the rules for multiplication are quite simple and work regardless of the two numbers being multiplied together, you would likely never compute the product of two arbitrarily large numbers, like 140283197 times 2241792341, using the same strategy. Instead you would likely use a *calculator* because it conveniently automates the process of multiplying two numbers of arbitrary size. A calculator allows you to compute with much greater efficiency and accuracy, and empowers you to use the fruits of arithmetic computation for more important tasks.

This is precisely how you can think about the computation of derivatives and gradients. Perhaps you can compute the derivative of a relatively simple mathematical function like $g(w) = \sin(w^2)$ easily, knowing a combination of *differentiation rules* as well as derivatives of certain *elementary functions* such as sinusoids and polynomials (see Appendix B for a review). In this particular case you can use the *chain rule* to write $\frac{d}{dw} g(w)$ as

$$\frac{d}{dw} g(w) = \left(\frac{d}{dw} w^2 \right) \cos(w^2) = 2w \cos(w^2). \qquad (3.30)$$

As with multiplication, even though the rules for differentiation are quite simple and work regardless of the function being differentiated, you would likely never compute the gradient of an arbitrarily complicated function, such as the one that follows, yourself and *by hand*

$$g(w_1, w_2) = 2^{\sin\left(w_1^2 + w_2^2\right)} \tanh\left(\cos\left(w_1 w_2\right)\right) \tanh\left(w_1 w_2^4 + \tanh\left(w_1 + w_2^2\right)\right) \qquad (3.31)$$

as it is extremely time consuming and easy to mess up (just like multiplication of two large numbers). Following the same logic a *gradient calculator* would allow for computing derivatives and gradients with much greater efficiency and accuracy, empowering us to use the fruits of gradient computation for more important tasks, e.g., for the popular first-order *local optimization* schemes that are the subject of this chapter and that are widely used in machine learning. Therefore throughout the remainder of the text the reader should feel comfortable using a *gradient calculator* as an alternative to hand computation.

Gradient calculators come in several varieties, from those that provide numerical approximations to those that literally automate the simple derivative rules for elementary functions and operations. We outline these various approaches in Appendix B. For `Python` users we strongly recommend using the open-source automatic differentiation library called `autograd` [10, 11, 12, 13] or `JAX` (an extension of `autograd` that runs on GPUs and TPUs). This is a high-quality and easy-to-use professional-grade gradient calculator that gives the user the power to easily compute the gradient for a wide variety of `Python` functions built using standard data structures and `autograd` operations. In Section B.10 we provide a brief tutorial on how to get started with `autograd`, as well as demonstrations of some of its core functionality.

3.5   Gradient Descent

In Section 3.3 we saw how the negative gradient $-\nabla g(\mathbf{w})$ of a function $g(\mathbf{w})$ computed at a particular point *always* defines a valid descent direction at that point. We could very naturally wonder about the efficacy of a local optimization method, that is one consisting of steps of the general form $\mathbf{w}^k = \mathbf{w}^{k-1} + \alpha \mathbf{d}^k$ (see Section 2.4), employing the negative gradient direction $\mathbf{d}^k = -\nabla g(\mathbf{w}^{k-1})$. Such a sequence of steps would then take the form

$$\mathbf{w}^k = \mathbf{w}^{k-1} - \alpha \nabla g(\mathbf{w}^{k-1}). \tag{3.32}$$

It seems intuitive, at least at the outset that, because each and every direction is guaranteed to be one of descent (provided we set α appropriately, as we must always do when using any local optimization method), taking such steps could lead us to a point near a local minimum of the target function g. The rather simple update step in Equation (3.32) is indeed an extremely popular local optimization method called the *gradient descent algorithm*, named so because it employs the (negative) *gradient* as the *descent* direction.

A prototypical path taken by gradient descent is illustrated in Figure 3.6 for a generic single-input function. At each step of this local optimization method we can think about drawing the first-order Taylor series approximation to the function, and taking the descent direction of this tangent hyperplane (i.e., the negative gradient of the function at this point) as our descent direction for the algorithm.

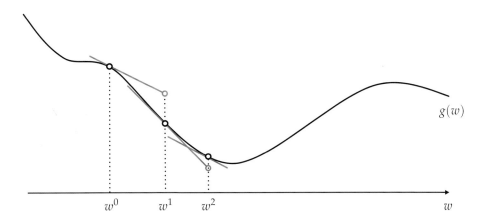

Figure 3.6 A figurative drawing of the gradient descent algorithm. Beginning at the initial point w^0 we make our first approximation to $g(w)$ at the point $(w^0, g(w^0))$ on the function (shown as a hollow black dot) with the first-order Taylor series approximation itself drawn in red. Moving in the negative gradient descent direction provided by this approximation we arrive at a point $w^1 = w^0 - \alpha \frac{d}{dw} g(w^0)$. We then repeat this process at w^1, moving in the negative gradient direction there, to $w^2 = w^1 - \alpha \frac{d}{dw} g(w^1)$, and so forth.

As we will see in this and many of our future chapters, the gradient descent algorithm is often a far better local optimization algorithm than the zero-order approaches discussed in the previous chapter. Indeed gradient descent, along with its extensions which we detail in Appendix A, is arguably the most popular local optimization algorithm used in machine learning today. This is largely because of the fact that the descent direction provided here (via the gradient) is almost always easier to compute (particularly as the dimension of the input increases) than seeking out a descent direction at random (as is done with the zero-order methods described in Sections 2.4 through 2.6). In other words, the fact that the negative gradient direction provides a descent direction for the function locally, combined with the fact that gradients are easy to compute (particularly when employing an automatic differentiator) makes gradient descent a superb local optimization method.

Example 3.7 Minimizing a nonconvex function using gradient descent
To find the global minimum of a general nonconvex function using gradient descent (or any local optimization method) one may need to run it several times with different initializations and/or steplength schemes. We showcase this fact using the nonconvex function

$$g(w) = \sin(3w) + 0.3w^2 \tag{3.33}$$

illustrated in the top panels of Figure 3.7. The same function was minimized in Example 2.4 using random search. Here we initialize two runs of gradient descent, one at $w^0 = 4.5$ (top-left panel) and another at $w^0 = -1.5$ (top-right panel), using a fixed steplength of $\alpha = 0.05$ for both runs. As can be seen by the results, depending on where we initialize we may end up near a local or global minimum of the function (we color the steps from green to red as the method proceeds from start to finish).

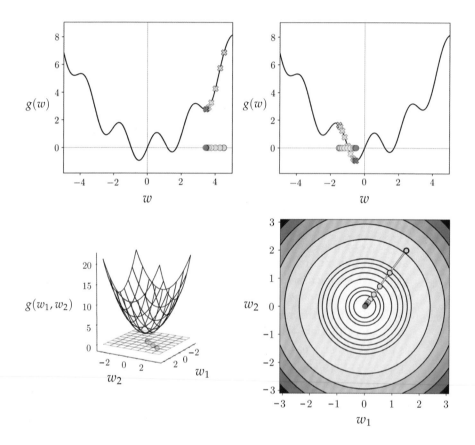

Figure 3.7 (top panels) Figure associated with Example 3.7. (bottom panels) Figure associated with Example 3.8. See text for details.

Example 3.8   Minimizing a convex multi-input function using gradient descent

In this example we run gradient descent on the convex multi-input quadratic function

$$g(w_1, w_2) = w_1^2 + w_2^2 + 2 \tag{3.34}$$

previously used in Example 2.3. We fix the steplength parameter at $\alpha = 0.1$ for all ten steps of the algorithm. In the bottom row of Figure 3.7 we illustrate the path taken by this run of gradient descent in the input space of the function, again coloring the steps from green to red as the method finishes. This is shown along with the three-dimensional surface of the function in the bottom-left panel, and "from above," showing the contours of the function on its input space in the bottom-right panel.

3.5.1 Basic steplength choices for gradient descent

As with all local methods, one needs to carefully choose the steplength or learning rate parameter α with gradient descent. While there are an array of available sophisticated methods for choosing α in the case of gradient descent, the most common choices employed in machine learning are those basic approaches first detailed in the simple context of zero-order methods in Section 2.5. These common choices include (i) using a fixed α value for each step of a gradient descent run, which for simplicity's sake commonly takes the form of 10^γ, where γ is an (often negative) integer, and (ii) using a diminishing steplength like $\alpha = \frac{1}{k}$ at the kth step of a run.

In both instances our aim in choosing a particular value for the steplength α at each step of gradient descent mirrors that of any other local optimization method: α should be chosen to induce the most rapid minimization possible. With the fixed steplength this often means choosing the *largest* possible value for α that leads to proper convergence.

Example 3.9   A fixed steplength selection for gradient descent

At each step of gradient descent we *always* have a descent direction – this is defined explicitly by the negative gradient itself. However, whether or not we actually descend in the function when taking a gradient descent step depends completely on how far we travel in the direction of the negative gradient, which we control via our steplength parameter. Set incorrectly, we can descend infinitely slowly, or even *ascend* in the function.

We illustrate this in Figure 3.8 using the simple single-input quadratic $g(w) = w^2$. Here we show the result of taking five gradient descent steps using three different fixed steplength values, all initialized at the same point $w^0 = -2.5$. The top row of this figure shows the function itself along with the evaluation at each step of a run (the value of α used in each run is shown at the top of each panel in this row), which are colored from green at the start of the run to red when the last step of the run is taken. From left to right each panel shows a different run with a slightly increased fixed steplength value α used for all steps. In the left panel the steplength is extremely small – so small that we do not descend very much at all. In the right panel, however, when we set the value of α too large the algorithm ascends in the function (ultimately *diverging*).

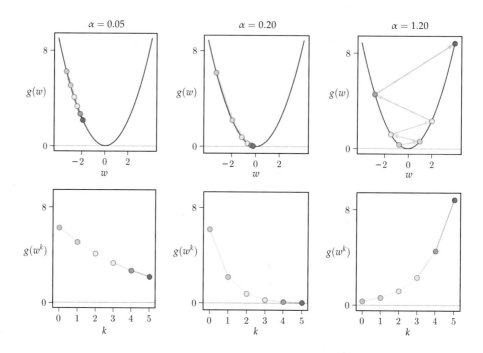

Figure 3.8 Figure associated with Example 3.9. See text for details.

In the bottom row we show the cost function history plot corresponding to each run of gradient descent in the top row of the figure. We also color these points from green (start of run) to red (end of run).

Example 3.10 Comparing fixed and diminishing steplengths

In Figure 3.9 we illustrate the comparison of a fixed steplength scheme and a diminishing one to minimize the function

$$g(w) = |w|. \tag{3.35}$$

Notice that this function has a single global minimum at $w = 0$. We make two runs of 20 steps of gradient descent, each initialized at the point $w^0 = 2$, the first with a fixed steplength rule of $\alpha = 0.5$ (shown in the top-left panel) for each and every step, and the second using the diminishing steplength rule $\alpha = \frac{1}{k}$ (shown in the top-right panel).

Here we can see that the fixed steplength run gets stuck, unable to descend towards the minimum of the function, while the diminishing steplength run settles down nicely to the global minimum (which is also reflected in the cost function history plot shown in the bottom panel of the figure). This is because the derivative of this function (defined everywhere but at $w = 0$) takes the form

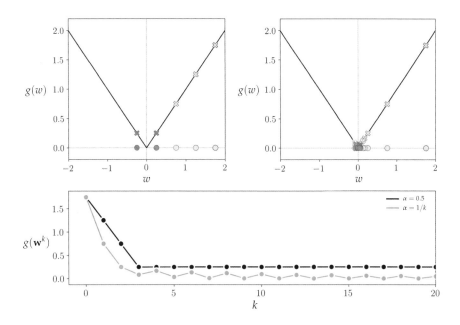

Figure 3.9 Figure associated with Example 3.10. See text for details.

$$\frac{\mathrm{d}}{\mathrm{d}w} g(w) = \begin{cases} +1 & \text{if } w > 0 \\ -1 & \text{if } w < 0 \end{cases} \tag{3.36}$$

which makes the use of any fixed steplength scheme problematic since the algorithm will always move at a fixed distance at each step. We face this potential issue with all local optimization methods (indeed we first saw it occur with a zero-order method in Example 2.5).

3.5.2   Oscillation in the cost function history plot: not always a bad thing

Remember that in practice – since we regularly deal with cost functions that take in far too many inputs – we use the *cost function history plot* (first described in Section 2.5.2) to tune our steplength parameter α, as well as debug implementations of the algorithm.

    Note that when employing the cost function history plot in choosing a proper steplength value, it is not ultimately important that the plot associated to a run of gradient descent (or any local optimization method) be *strictly decreasing* (that is showing that the algorithm *descended* at every single step). It is critical to find a value of α that allows gradient descent to find the lowest function value possible, even if it means that not every step *descends*. In other words, the *best* choice of α for a given minimization might cause gradient descent to "hop around" some, moving up and down, and not the one that shows descent in each and every step. Below we show an example illustrating this point.

Example 3.11   Oscillatory versus monotonically decreasing cost function history plots
In this example we show the result of three runs of gradient descent to minimize the function

$$g(\mathbf{w}) = w_1^2 + w_2^2 + 2\sin\left(1.5\left(w_1 + w_2\right)\right)^2 + 2 \tag{3.37}$$

whose contour plot is shown in Figure 3.10. You can see a *local minimum* around the point $[1.5 \ 1.5]^T$, and a global minimum near $[-0.5 \ -0.5]^T$. All three runs start at the same initial point $\mathbf{w}^0 = [3 \ 3]^T$ and take ten steps. The first run (shown in the top-left panel) uses a fixed steplength parameter $\alpha = 10^{-2}$, the second run (shown in the top-middle panel) uses $\alpha = 10^{-1}$, and the third run (shown in the top-right panel) uses $\alpha = 10^0$.

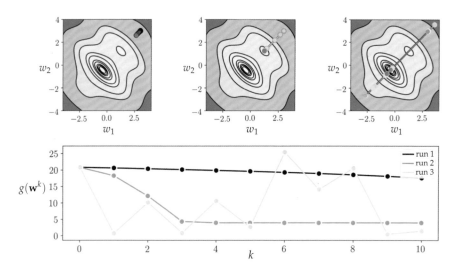

Figure 3.10 Figure associated with Example 3.11. See text for details.

In the bottom panel of the figure we plot the cost function history plot associated with each run of gradient descent, showing the first, second, and third run in black, pink, and blue, respectively. Here we can see that the value of our first choice was too small (as can also be seen in the top-left panel), the second choice leads to convergence to the *local* minimum (as can be seen in the top-middle panel), and final run while "hopping around" and not strictly decreasing at each step finds the lowest point out of all three runs. So while this run used the largest steplength $\alpha = 10^0$, clearly leading to oscillatory (and perhaps eventually divergent) behavior, it does indeed find the lowest point out of all three runs performed.

This function is rather pathological, i.e., it was designed specifically for the purposes of this example. However, in practice the moral expressed here, that it is just fine for the cost the function history associated with a run of gradient descent (or any local optimization algorithm more generally) to oscillate up and down (and not be perfectly smooth and monotonically decreasing), holds in general.

3.5.3 Convergence criteria

When does gradient descent stop? Technically, if the steplength is chosen wisely, the algorithm will halt near stationary points of a function, typically minima or saddle points. How do we know this? By the very form of the gradient descent step itself. If the step $\mathbf{w}^k = \mathbf{w}^{k-1} - \alpha \nabla g\left(\mathbf{w}^{k-1}\right)$ does not move from the prior point $\mathbf{w}^{k-1}$ significantly then this can mean only one thing: that the direction we are traveling in is *vanishing*, i.e., $-\nabla g\left(\mathbf{w}^{k-1}\right) \approx \mathbf{0}_{N \times 1}$. This is – by definition – a *stationary point* of the function (as detailed in Section 3.2).

In principle, then, we can wait for gradient descent to get sufficiently close to a stationary point by ensuring, for instance, that the magnitude of the gradient $\left\|\nabla g\left(\mathbf{w}^{k-1}\right)\right\|_2$ is sufficiently small. Other formal convergence criteria include (i) halting when steps no longer make sufficient progress (e.g., when $\frac{1}{N}\left\|\mathbf{w}^k - \mathbf{w}^{k-1}\right\|_2$ is smaller than some ϵ) or, (ii) when corresponding evaluations no longer differ substantially (e.g., when $\frac{1}{N}\left|g\left(\mathbf{w}^k\right) - g\left(\mathbf{w}^{k-1}\right)\right|$ is smaller than some ϵ). Finally, a practical way to halt gradient descent – as well as any other local optimization scheme – is to simply run the algorithm for a fixed number of maximum iterations. In machine learning applications this latter practical condition is very often employed, either alone or in combination with a formal stopping procedure.

How do we set the maximum iteration count? As with any local method this is typically set manually/heuristically, and is influenced by things like computing resources, knowledge of the particular function being minimized, and – very importantly – the choice of the steplength parameter α. Smaller choices for α, while more easily providing descent at each step, frequently require more steps for the algorithm to achieve significant progress. Conversely, if α is set too large gradient descent may bounce around erratically forever, never localizing in an adequate solution.

3.5.4 Python implementation

In this section we provide a basic Python implementation of the gradient descent algorithm. There are a number of variations one can use in practice including various halting conditions (described above), as well as various to compute the gradient itself. The inputs here include the function to minimize g, a steplength `alpha`, a maximum number of iterations `max_its` (our default stopping condition), and an initial point w that is typically chosen at random. The outputs include a history of weight updates and corresponding cost function value history (which one can use to produce a cost function history plot). The computation of the gradient function in line 16 employs by default the open-source automatic differentiation library `autograd` (detailed in Sections 3.4 and B.10) – although one can easily replace this with any other method for computing the gradient function.

```
1   # import automatic differentiator to compute gradient module
2   from autograd import grad
3
4   # gradient descent function
5   def gradient_descent(g, alpha, max_its, w):
6
7       # compute gradient module using autograd
8       gradient = grad(g)
9
10      # gradient descent loop
11      weight_history = [w]  # weight history container
12      cost_history = [g(w)] # cost function history container
13      for k in range(max_its):
14
15          # evaluate the gradient
16          grad_eval = gradient(w)
17
18          # take gradient descent step
19          w = w - alpha*grad_eval
20
21          # record weight and cost
22          weight_history.append(w)
23          cost_history.append(g(w))
24
25      return weight_history, cost_history
```

Given the input to g is N-dimensional, a general random initialization can be written as shown below where the NumPy function `random.randn` produces samples from a standard normal distribution with zero mean and unit standard deviation. It is also common to scale such initializations by small constants (e.g., 0.1 or smaller).

```
1   # a common random initialization scheme
2   import numpy as np
3   scale = 0.1
4   w = scale*np.random.randn(N,1)
```

3.6   Two Natural Weaknesses of Gradient Descent

As we saw in the previous section, gradient descent is a local optimization scheme that employs the negative gradient at each step. The fact that calculus provides us with a true descent direction in the form of the negative gradient direction, combined with the fact that gradients are often cheap to compute (whether or not one uses an automatic differentiator), means that we need not search for a reasonable descent direction at each step of the method as we needed to do with the zero-order methods detailed in the previous chapter. This is extremely advantageous, and is the fundamental reason why gradient descent is so popular in machine learning.

However, no basic local optimization scheme is without its shortcomings. In the previous chapter we saw how, for example, the natural shortcoming of random search limits its practical usage to functions of low-dimensional input. While gradient descent does not suffer from this particular limitation, the negative gradient has its own weaknesses as a descent direction, which we outline in this section.

3.6.1   Where do the weaknesses of the (negative) gradient direction originate?

Where do these weaknesses originate? Like any vector, the negative gradient always consists fundamentally of a *direction* and a *magnitude* (as illustrated in Figure 3.11). Depending on the function being minimized either one of these attributes – or both – can present challenges when using the negative gradient as a descent direction.

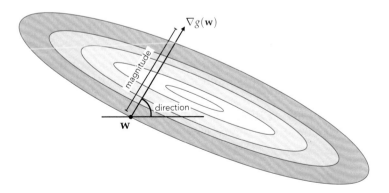

Figure 3.11 The gradient vector of any arbitrary function at any point consists of a *magnitude* and a *direction*.

The *direction* of the negative gradient can rapidly oscillate during a run of gradient descent, often producing *zig-zagging* steps that take considerable time to reach a minimum point. The *magnitude* of the negative gradient can vanish rapidly near stationary points, leading gradient descent to *slowly crawl* near minima and saddle points. This too can slow down gradient descent's progress near stationary points. These two problems, while not present when minimizing every single function, do present themselves in machine learning because many of the functions we aim to minimize have *long narrow valleys* – long flat areas where the contours of a function become increasingly parallel. Both of these issues – stemming from either the direction or magnitude of the negative gradient direction – are explored further below.

3.6.2   The (negative) gradient direction

A fundamental property of the (negative) gradient direction is that it always points perpendicular to the contours of a function. This statement is universally true, and holds for *any* (differentiable) function and at *all* of its inputs. That is, the gradient ascent/descent direction at an input $\mathbf{w}^0$ is always perpendicular to the contour $g(\mathbf{w}) = g(\mathbf{w}^0)$.

Example 3.12   The negative gradient direction
In Figure 3.12 we show the contour plot of (top-left panel) $g(\mathbf{w}) = w_1^2 + w_2^2 + 2$, (top-right panel) $g(\mathbf{w}) = w_1^2 + w_2^2 + 2\sin(1.5(w_1 + w_2))^2 + 2$, and (bottom panel) $g(\mathbf{w}) = \left(w_1^2 + w_2 - 11\right)^2 + \left(w_1 + w_2^2 - 6\right)^2$.

On each plot we also show the negative gradient direction defined at three random points. Each of the points we choose are highlighted in a unique color, with the contour on which they sit on the function colored in the same manner for visualization purposes. The *descent* direction defined by the gradient at each point is drawn as an arrow, and the tangent line to the contour at each input is also drawn (in both instances colored the same as their respective point).

In each instance we can see how the gradient descent direction is always *perpendicular* to the contour it lies on – in particular being perpendicular to the tangent line at each point on the contour (which is also shown). Because the gradient ascent directions will simply point in the opposite direction as the descent directions shown here, they too will be perpendicular to the contours.

3.6.3   The zig-zagging behavior of gradient descent

In practice, the fact that the negative gradient always points in a direction perpendicular to the contour of a function can – depending on the function being minimized – make the negative gradient direction oscillate rapidly or *zig-zag* during a run of gradient descent. This in turn can cause zig-zagging behavior in the gradient descent steps themselves and *too much* zig-zagging

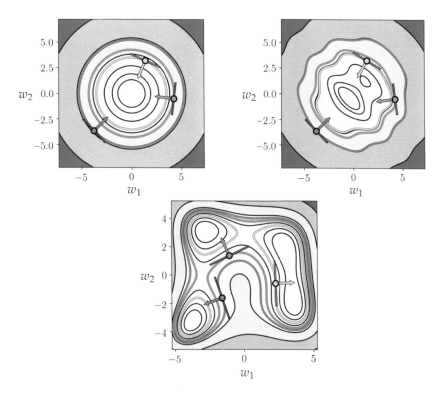

Figure 3.12 Figure associated with Example 3.12. Regardless of the function the negative gradient direction is always *perpendicular* to the function's contours. See text for further details.

slows minimization progress. When it occurs, many gradient descent steps are required to adequately minimize a function. We illustrate this phenomenon below using a set of simple examples.

The interested reader may note that we describe a popular solution to this zig-zagging behavior, called *momentum-accelerated* gradient descent, later in Appendix Section A.2.

Example 3.13   The zig-zagging behavior of gradient descent

In Figure 3.13 we illustrate the zig-zagging behavior of gradient descent with three $N = 2$ dimensional quadratic functions that take the general form $g(\mathbf{w}) = a + \mathbf{b}^T\mathbf{w} + \mathbf{w}^T\mathbf{C}\mathbf{w}$. In each case a and $\mathbf{b}$ are set to zero, and the matrix $\mathbf{C}$ is set so that each quadratic gets progressively narrower:

$$C = \begin{bmatrix} 0.50 & 0 \\ 0 & 12 \end{bmatrix} \qquad \text{(top panel of Figure 3.13)}$$

$$C = \begin{bmatrix} 0.10 & 0 \\ 0 & 12 \end{bmatrix} \qquad \text{(middle panel of Figure 3.13)} \qquad (3.38)$$

$$C = \begin{bmatrix} 0.01 & 0 \\ 0 & 12 \end{bmatrix} \qquad \text{(bottom panel of Figure 3.13)}$$

and hence the quadratic functions differ only in how we set the upper-left value of the matrix C. All three quadratics, whose contour plots are shown in the top, middle, and bottom panels of Figure 3.13 respectively, have the same global minimum at the origin. However, as we change this single value of C from quadratic to quadratic, we elongate the contours significantly along the horizontal axis, so much so that in the third case the contours seem almost completely parallel to each other near our initialization (an example of a *long narrow valley*).

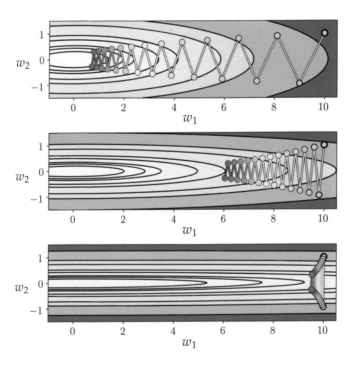

Figure 3.13 Figure associated with Example 3.13, illustrating the zig-zagging behavior of gradient descent. See text for further details.

We then make a run of 25 gradient descent steps to minimize each, using the same initialization at $\mathbf{w}^0 = [10 \ 1]^T$ and steplength value $\alpha = 10^{-1}$. In each case the weights found at each step are plotted on the contour plots and colored green (at the start of the run) to red (as we reach the maximum number of iterations). Examining the figure we can see – in each case, but increasingly from the first to third example – the zig-zagging behavior of gradient descent very clearly. Indeed not much progress is made with the third quadratic at all due to the large amount of zig-zagging.

We can also see the cause of this zig-zagging: the negative gradient direction constantly points in a perpendicular direction with respect to the contours of the function, and in very narrow functions these contours become almost parallel. While it is true that we can ameliorate this zig-zagging behavior by reducing the steplength value, this does not solve the underlying problem that zig-zagging produces – which is slow convergence.

3.6.4 The slow-crawling behavior of gradient descent

As we know from the first-order condition for optimality discussed in Section 3.2, the (negative) gradient vanishes at stationary points. That is, if $\mathbf{w}$ is a minimum, maximum, or saddle point then we know that $\nabla g(\mathbf{w}) = \mathbf{0}$. Notice that this also means that the magnitude of the gradient vanishes at stationary points, that is, $\|\nabla g(\mathbf{w})\|_2 = 0$. By extension, the (negative) gradient at points near a stationary point have non-zero direction but vanishing magnitude, i.e., $\|\nabla g(\mathbf{w})\|_2 \approx 0$.

The vanishing behavior of the magnitude of the negative gradient near stationary points has a natural consequence for gradient descent steps: they progress very slowly, or "crawl," near stationary points. This occurs because, *unlike* the zero-order methods discussed in Sections 2.5 and 2.6 (where we normalized the magnitude of each descent directions), the distance traveled during each step of gradient descent is not completely determined by the steplength value α. Indeed we can easily compute the general distance traveled by a gradient descent step as

$$\left\|\mathbf{w}^k - \mathbf{w}^{k-1}\right\|_2 = \left\|\left(\mathbf{w}^{k-1} - \alpha \nabla g\left(\mathbf{w}^{k-1}\right)\right) - \mathbf{w}^{k-1}\right\|_2 = \alpha \left\|\nabla g\left(\mathbf{w}^{k-1}\right)\right\|_2. \qquad (3.39)$$

In other words, the length of a general gradient descent step is equal to the value of the steplength parameter α times the magnitude of the descent direction.

The consequences of this are fairly easy to unravel. Since the magnitude of the gradient $\left\|\nabla g\left(\mathbf{w}^{k-1}\right)\right\|_2$ is large far away from stationary points, and because we often randomly initialize gradient descent in practice so that our initial points often lie far away from any stationary point of a function, the first few steps of a gradient descent run in general will be large, and make significant progress towards minimization. Conversely, when approaching a stationary point the magnitude of the gradient is small, and so the length traveled by a gradient descent step is also small. This means that gradient descent steps make little progress towards minimization when near a stationary point.

In short, the fact that the length of each step of gradient descent is proportional to the magnitude of the gradient means that gradient descent often starts off making significant progress but slows down significantly near minima and saddle points – a behavior we refer to as "slow-crawling." For particular functions this slow-crawling behavior can not only mean that many steps are required to achieve adequate minimization, but can also lead gradient descent to completely halt near saddle points of nonconvex functions.

The interested reader may note that we describe a popular solution to this slow-crawling behavior – called *normalized gradient descent* – later in Appendix Sections A.3 and A.4.

Example 3.14   The slow-crawling behavior of gradient descent

In the left panel of Figure 3.14 we plot the function

$$g(w) = w^4 + 0.1 \tag{3.40}$$

whose minimum is at the origin, which we will minimize using ten steps of gradient descent and a steplength parameter $\alpha = 10^{-1}$. We show the results of this run on the function itself (with steps colored from green at the start of the run to red at the final step). Here we can see that this run of gradient descent starts off taking large steps but crawls slowly as it approaches the minimum. Both of these behaviors are quite natural, since the magnitude of the gradient is large far from the global minimum and vanishes near it.

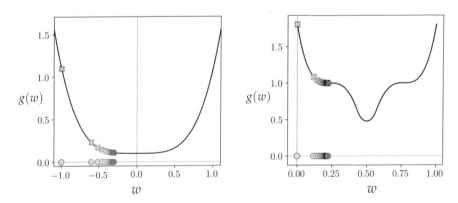

Figure 3.14 Figure associated with Example 3.14. See text for details.

In the right panel of Figure 3.14 we illustrate the crawling issue of gradient descent near saddle points using the nonconvex function

$$g(w) = \max^2(0, 1 + (3w - 2.3)^3) + \max^2(0, 1 + (-3w + 0.7)^3). \tag{3.41}$$

This function has a minimum at $w = \frac{1}{2}$ and saddle points at $w = \frac{7}{30}$ and $w = \frac{23}{30}$. We make a run of gradient descent on this function using 50 steps with $\alpha = 10^{-2}$, initialized at the origin.

Examining the right panel of the figure we can see how the gradient descent steps halt near the leftmost saddle point due to the settings (initialization and steplength parameter) chosen for this run. The fact that gradient descent crawls as it approaches this saddle point is quite natural (because the magnitude of the gradient vanishes here) but this prevents the algorithm from finding the global minimum.

3.7   Conclusion

In this chapter we described local optimization schemes that leverage a function's first derivative(s) to produce effective descent directions – otherwise known as *first-order methods*. Such methods constitute perhaps the most popular set of optimization tools used with machine learning problems.

We began in Section 3.2 by looking at how the first derivatives of a function provide a useful condition for characterizing its minima, maxima, and saddle points (together known as stationary points) via the first-order condition for optimality. In preparation for our discussion of first-order local methods we then described the *ascent* and *descent directions* of a hyperplane, as well as those provided by the tangent hyperplane associated with the first-order Taylor series approximation in Section 3.3. In Section 3.5 we saw how such descent directions, when employed in a local optimization framework, naturally lead to the construction of a popular local scheme known as *gradient descent*.

The gradient descent algorithm is widely used in machine learning, as the descent direction provided by the negative gradient is almost always readily available for use (and so a descent direction need not be sought out explicitly as with the zero-order methods described in Chapter 2). However, by its very nature the negative gradient direction has two inherent weaknesses when leveraged for local optimization – the *zig-zagging* and *slow-crawling* problems detailed in Section 3.6 – that reduce its effectiveness. These problems, and the corresponding solutions to each (collectively referred to as advanced first-order optimization methods), are discussed in significant detail in Appendix A of this text.

3.8   Exercises

† The data required to complete the following exercises can be downloaded from the text's github repository at `github.com/jermwatt/machine_learning_refined`

3.1   First-order condition for optimality

Use the first-order condition to find all stationary points of g (calculations should be done by hand). Then plot g and label the point(s) you find, and determine "by eye" whether each stationary point is a minimum, maximum, or saddle point. Note: stationary points can be at infinity!

(a) $g(w) = w \log(w) + (1 - w) \log(1 - w)$ where w lies between 0 and 1

(b) $g(w) = \log(1 + e^{w})$

(c) $g(w) = w \tanh(w)$

(d) $g(\mathbf{w}) = \frac{1}{2} \mathbf{w}^T \mathbf{C} \mathbf{w} + \mathbf{b}^T \mathbf{w}$ where $\mathbf{C} = \begin{bmatrix} 2 & 1 \\ 1 & 3 \end{bmatrix}$ and $\mathbf{b} = \begin{bmatrix} 1 \\ 1 \end{bmatrix}$

3.2   Stationary points of a simple quadratic function

A number of applications will find us employing a simple multi-input quadratic

$$g(\mathbf{w}) = a + \mathbf{b}^T \mathbf{w} + \mathbf{w}^T \mathbf{C} \mathbf{w} \tag{3.42}$$

where the matrix $\mathbf{C} = \frac{1}{\beta} \mathbf{I}$. Here $\mathbf{I}$ is the $N \times N$ identity matrix, and $\beta > 0$ a positive scalar. Find all stationary points of g.

3.3   Stationary points of the Rayleigh quotient

The Rayleigh quotient of an $N \times N$ matrix $\mathbf{C}$ is defined as the normalized quadratic function

$$g(\mathbf{w}) = \frac{\mathbf{w}^T \mathbf{C} \mathbf{w}}{\mathbf{w}^T \mathbf{w}} \tag{3.43}$$

where $\mathbf{w} \neq \mathbf{0}_{N \times 1}$. Compute the stationary points of this function.

3.4   First-order coordinate descent as a local optimization scheme

(a) Express the coordinate descent method described in Section 3.2.2 as a local optimization scheme, i.e., as a sequence of steps of the form $\mathbf{w}^k = \mathbf{w}^{k-1} + \alpha \, \mathbf{d}^k$.

(b) Code up the coordinate descent method for the particular case of a quadratic function and repeat the experiment described in Example 3.5.

3.5   Try out gradient descent

Run gradient descent to minimize the function

$$g(w) = \frac{1}{50}\left(w^4 + w^2 + 10w\right) \tag{3.44}$$

with an initial point $w^0 = 2$ and 1000 iterations. Make three separate runs using each of the steplength values $\alpha = 1$, $\alpha = 10^{-1}$, and $\alpha = 10^{-2}$. Compute the derivative of this function by hand, and implement it (as well as the function itself) in Python using NumPy.

Plot the resulting cost function history plot of each run in a single figure to compare their performance. Which steplength value works best for this particular function and initial point?

3.6   Compare fixed and diminishing steplength values for a simple example
Repeat the comparison experiment described in Example 3.10, producing the cost function history plot comparison shown in the bottom panel of Figure 3.9.

3.7   Oscillation in the cost function history plot
Repeat the experiment described in Example 3.11, producing cost function history plot shown in the bottom panel of Figure 3.10.

3.8   Tune fixed steplength for gradient descent
Take the cost function

$$g(\mathbf{w}) = \mathbf{w}^T \mathbf{w} \tag{3.45}$$

where $\mathbf{w}$ is an $N = 10$ dimensional input vector, and g is convex with a single global minimum at $\mathbf{w} = \mathbf{0}_{N\times 1}$. Code up gradient descent and run it for 100 steps using the initial point $\mathbf{w}^0 = 10 \cdot \mathbf{1}_{N\times 1}$, with three steplength values: $\alpha_1 = 0.001$, $\alpha_2 = 0.1$, and $\alpha_3 = 1$. Produce a cost function history plot to compare the three runs and determine which performs best.

3.9   Code up momentum-accelerated gradient descent
Code up the momentum-accelerated gradient descent scheme described in Section A.2.2 and use it to repeat the experiments detailed in Example A.1 using a cost function history plot to come to the same conclusions drawn by studying the contour plots shown in Figure A.3.

3.10   Slow-crawling behavior of gradient descent
In this exercise you will compare the standard and fully normalized gradient descent schemes in minimizing the function

$$g(w_1, w_2) = \tanh(4\,w_1 + 4\,w_2) + \max(1, 0.4\,w_1^2) + 1. \tag{3.46}$$

Using the initialization $\mathbf{w}^0 = [2\ 2]^T$ make a run of 1000 steps of the standard and fully normalized gradient descent schemes, using a steplength value of $\alpha = 10^{-1}$ in both instances. Use a cost function history plot to compare the two runs, noting the progress made with each approach.

3.11   Comparing normalized gradient descent schemes

Code up the full and component-wise normalized gradient descent schemes and repeat the experiment described in Example A.4 using a cost function history plot to come to the same conclusions drawn by studying the plots shown in Figure A.6.

3.12   Alternative formal definition of Lipschitz gradient

An alternative to defining the Lipschitz constant by Equation (A.49) for functions g with Lipschitz continuous gradient is given by

$$\left\| \nabla g\left(\mathbf{x}\right) - \nabla g\left(\mathbf{y}\right) \right\|_2 \le L \left\| \mathbf{x} - \mathbf{y} \right\|_2. \tag{3.47}$$

Using the limit definition of the derivative (see Section B.2.1) show that this definition is equivalent to the one given in Equation (A.49).

3.13   A composition of functions with Lipschitz gradient

Suppose f and g are two functions with Lipschitz gradient with constants L and K respectively. Using the definition of Lipschitz continuous gradient given in Exercise 3.12 show that the composition $f\left(g\right)$ also has Lipschitz continuous gradient. What is the corresponding Lipschitz constant of this composition?

4 Second-Order Optimization Techniques

In this chapter we describe fundamental optimization algorithms that leverage the first and second derivatives, or likewise the *gradient* and *Hessian* of a function. These techniques, collectively called *second-order optimization methods*, are popular in particular applications of machine learning today. In analogy to the previous chapter, here we begin with a discussion of the second-order optimality condition. We then discuss quadratic functions as well as the notion of curvature defined by second derivatives, and the second-order Taylor series expansion. By exploiting a function's first- and second-order derivative information we can construct powerful local optimization methods, including the popular *Newton's method* and its extensions (commonly referred to as *Hessian-free* optimizers).

4.1 The Second-Order Optimality Condition

When discussing convexity/concavity of general mathematical functions we often talk about convexity/concavity *at a point*. To determine whether a general single-input function $g(w)$ is *convex* or *concave* at a point v, we check its curvature or second derivative information at that point (assuming it is at least twice-differentiable there): if $\frac{d^2}{dw^2} g(v) \geq 0$ (or ≤ 0) then g is said to be convex (or concave) at v.

An analogous statement can be made for a function g with multi-dimensional input: if the Hessian matrix evaluated at a point $\mathbf{v}$, denoted by $\nabla^2 g(\mathbf{v})$, has all nonnegative (or nonpositive) *eigenvalues* then g is said to be convex (or concave) at $\mathbf{v}$, in which case the Hessian matrix itself is called positive (or negative) semi-definite.

Based on these point-wise convexity/concavity definitions, the function $g(w)$ is said to be convex *everywhere* if its second derivative $\frac{d^2}{dw^2} g(w)$ is always non-negative. Likewise $g(\mathbf{w})$ is convex *everywhere* if $\nabla^2 g(\mathbf{w})$ always has nonnegative eigenvalues. This is generally referred to as the *second-order definition of convexity*.

Example 4.1   Convexity of single-input functions
In this example we use the second-order definition of convexity to verify whether each of the functions shown in Figure 4.1 is convex or not.

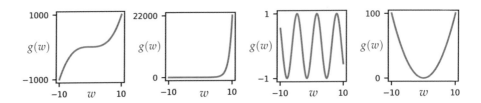

Figure 4.1 Figure associated with Example 4.1. From left to right, plots of functions $g(w) = w^3$, $g(w) = e^w$, $g(w) = \sin(w)$, and $g(w) = w^2$.

- $g(w) = w^3$ has second derivative $\frac{d^2}{dw^2} g(w) = 6w$, which is not always nonnegative, hence g is not convex.

- $g(w) = e^w$ has second derivative $\frac{d^2}{dw^2} g(w) = e^w$, which is positive for any choice of w, and g is therefore convex.

- $g(w) = \sin(w)$ has second derivative $\frac{d^2}{dw^2} g(w) = -\sin(w)$. Since this is not always nonnegative, g is nonconvex.

- $g(w) = w^2$ has second derivative $\frac{d^2}{dw^2} g(w) = 2$, and g is therefore convex.

Example 4.2   Convexity of multi-input quadratic functions

The multi-input quadratic function

$$g(\mathbf{w}) = a + \mathbf{b}^T \mathbf{w} + \mathbf{w}^T \mathbf{C} \mathbf{w} \tag{4.1}$$

has the Hessian matrix $\nabla^2 g(\mathbf{w}) = 2\mathbf{C}$ (assuming $\mathbf{C}$ is symmetric). Therefore its convexity is determined by studying the eigenvalues of $\mathbf{C}$.

By studying a few simple examples it is easy to come to some far-reaching conclusions about how the second derivative helps unveil the identity of stationary points. In Figure 4.2 we plot the three single-input functions we studied in Example 3.1 (defined in Equation (3.4)), along with their first- and second-order derivatives (shown in the top, middle, and bottom rows of the figure, respectively). In the top panels we mark the evaluation of all stationary points by the function in green (where we also show the tangent line in green). The corresponding evaluations by the first and second derivatives are marked in green as well in the middle and bottom panels of the figure, respectively.

By studying these simple examples in Figure 4.2 we can see consistent behavior of certain stationary points. In particular we can see consistency in how the value of a function's second derivative at a stationary point v helps us identify whether it is a local minimum, local maximum, or saddle point. A stationary point v is:

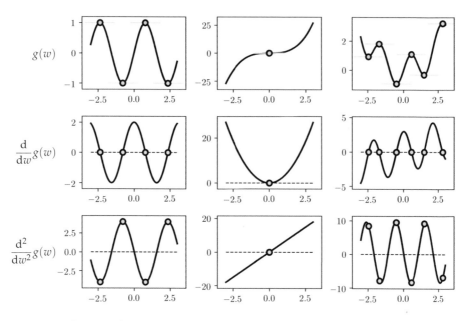

Figure 4.2 Three single-input functions along with their first- and second-order derivatives shown in the top, middle, and bottom panels, respectively. See text for further details.

- a local (or global) minimum if $\frac{d^2}{dw^2}g(v) > 0$ (since it occurs at *convex* portions of a function),
- a local (or global) maximum if $\frac{d^2}{dw^2}g(v) < 0$ (since it occurs at *concave* portions of a function),
- a saddle point if $\frac{d^2}{dw^2}g(v) = 0$ and $\frac{d^2}{dw^2}g(w)$ changes sign at $w = v$ (since it occurs at an *inflection point* of a function, i.e., where a function goes from concave to convex or vice versa).

These second-order characteristics hold more generally for any single-input function, and taken together form the *second-order condition for optimality* for single-input functions.

With multi-input functions the analogous second-order condition holds. As with all things having to do with convexity/concavity and the second-order derivative matrix (i.e., the Hessian), the second-order optimality condition for multi-input functions translates to the *eigenvalues* of the Hessian. More specifically, a stationary point $\mathbf{v}$ of a multi-input function $g(\mathbf{w})$ is:

- a local (or global) minimum if all eigenvalues of $\nabla^2 g(\mathbf{v})$ are positive (since it occurs at *convex* portions of a function),
- a local (or global) maximum if all eigenvalues of $\nabla^2 g(\mathbf{v})$ are negative (since it occurs at *concave* portions of a function),
- a saddle point if the eigenvalues of $\nabla^2 g(\mathbf{v})$ are of mixed values, i.e., some negative and some positive (since it occurs at an *inflection point* of a function).

Notice when the input dimension N equals 1, these rules reduce to those stated for single-input functions as the Hessian matrix collapses into a single second-order derivative.

4.2   The Geometry of Second-Order Taylor Series

As we will see throughout this chapter, quadratic functions naturally arise when studying second-order optimization methods. In this section we first discuss quadratic functions with an emphasis on how we determine their overall shape, and whether they are *convex*, *concave*, or have a more complicated geometry. We then study quadratic functions generated by the second-order Taylor series approximation (see Appendix Section B.9 for a review of this concept), and in particular how these fundamental quadratics inherently describe the *local* curvature of a twice-differentiable function.

4.2.1   The general shape of single-input quadratic functions

The basic formula for a quadratic function with a single input takes the familiar form

$$g(w) = a + bw + cw^2 \tag{4.2}$$

where a, b, and c are all constant values controlling the shape of the function. In particular, the constant c controls the *convexity* or *concavity* of the function or, in other words, whether the quadratic faces upwards or downwards. When the value of c is *nonnegative* the quadratic function is *convex* and points *upwards*, regardless of how the other parameters are set. Conversely, when the value of c is *nonpositive* the quadratic is *concave* and points *downwards*. When $c = 0$ the quadratic reduces to a linear function (which can be considered both convex and concave).

In the left column of Figure 4.3 we plot two simple quadratics: the convex quadratic $g(w) = 6w^2$ (top-left panel) and the concave quadratic $g(w) = -w^2$ (bottom-left panel) to illustrate how the value of c controls the shape and convexity of the general quadratic function.

4.2.2   The general shape of multi-input quadratic functions

The multi-input quadratic function takes a form that is completely generalized from the single-input case, which we write as

$$g(\mathbf{w}) = a + \mathbf{b}^T \mathbf{w} + \mathbf{w}^T \mathbf{C} \mathbf{w} \tag{4.3}$$

where the input $\mathbf{w}$ is N-dimensional, a remains a constant, $\mathbf{b}$ is an $N \times 1$ vector, and C an $N \times N$ matrix (which we assume is *symmetric* for our purposes). Because this quadratic is defined along many dimensions it can take on a wider variety of shapes than its single-input analog. For example, it can be convex along certain input dimensions and concave along others.

The generalization of the single-input test for convexity/concavity is no longer whether or not the *values* of C are positive or negative, but whether its *eigenvalues* are (see Appendix Section C.4.3). If the eigenvalues of the matrix are *all nonnegative* the quadratic is *convex*, if *all nonpositive* it is *concave*, if all equal zero it reduces to a linear function that is both convex and concave, and otherwise (i.e., if some of its eigenvalues are positive and others negative) it is neither convex nor concave.

In the middle and right columns of Figure 4.3 we show several examples of multi-input quadratic functions with $N = 2$ inputs. In all examples we have set a and b to zero and simply change the values of C. For simplicity in all four cases C is chosen to be a diagonal matrix so that its eigenvalues are conveniently the entries on its diagonal.

$$C = \begin{bmatrix} 1 & 0 \\ 0 & 0 \end{bmatrix} \qquad \text{(top-middle panel of Figure 4.3)}$$

$$C = \begin{bmatrix} 1 & 0 \\ 0 & 1 \end{bmatrix} \qquad \text{(top-right panel of Figure 4.3)}$$

$$C = \begin{bmatrix} 0 & 0 \\ 0 & -1 \end{bmatrix} \qquad \text{(bottom-middle panel of Figure 4.3)}$$

$$C = \begin{bmatrix} 1 & 0 \\ 0 & -1 \end{bmatrix} \qquad \text{(bottom-right panel of Figure 4.3)}$$

4.2.3 Local curvature and the second-order Taylor series

Another way to think about the local convexity or concavity of a function g at a point v is via its second-order Taylor series approximation at that point (see Section B.9). This fundamental approximation taking the form

$$h(w) = g(v) + \left(\frac{d}{dw}g(v)\right)(w - v) + \frac{1}{2}\left(\frac{d^2}{dw^2}g(v)\right)(w - v)^2 \qquad (4.4)$$

is a true quadratic built using the (first and) second derivative of the function. Not only does the second-order approximation match the curvature of the underlying function at each point v in the function's domain, but if the function is convex at that point (due to its second derivative being nonnegative) then the second-order Taylor series is *convex everywhere*. Likewise if the function is concave at v, this approximating quadratic is *concave everywhere*.

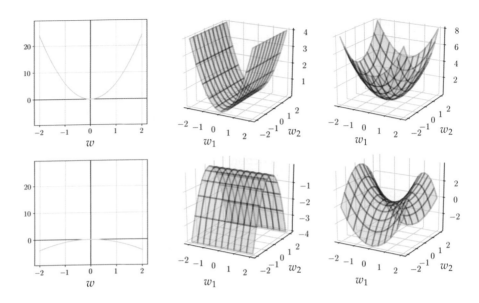

Figure 4.3 (top-left panel) A convex single-input quadratic. (bottom-left panel) a concave single-input quadratic. (top-middle and top-right panels) Two convex two-input quadratics. (bottom-middle panel) A concave two-input quadratic. (bottom-right panel) A two-input quadratic that is neither convex nor concave.

This concept holds analogously for multi-input functions as well. The second-order Taylor series approximation to a function taking in N-dimensional input at a point $\mathbf{v}$ is given by

$$h(\mathbf{w}) = g(\mathbf{v}) + \nabla g(\mathbf{v})^T (\mathbf{w} - \mathbf{v}) + \frac{1}{2}(\mathbf{w} - \mathbf{v})^T \nabla^2 g(\mathbf{v})(\mathbf{w} - \mathbf{v}). \tag{4.5}$$

Again, when the function g is convex at the point $\mathbf{v}$ the corresponding quadratic function is convex everywhere. Similar statements can be made when the function g is concave at the point $\mathbf{v}$, or neither convex or concave there.

Example 4.3   Local convexity/concavity and the second-order Taylor series
In Figure 4.4 we show the function

$$g(w) = \sin(3w) + 0.1w^2 \tag{4.6}$$

drawn in black, along with the second-order Taylor series quadratic approximation shown in turquoise at three example points (one per panel). We can see in this figure that the local convexity/concavity of the function is perfectly reflected in the shape of the associated quadratic approximation. That is, at points of local convexity (as in the first and third panel of the figure) the associated quadratic approximation is convex everywhere. Conversely, at points of local concavity

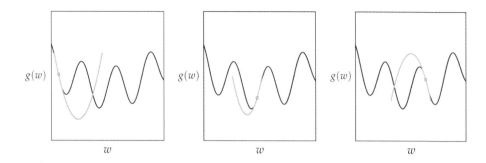

Figure 4.4 Figure associated with Example 4.3. See text for details.

(as in the second panel of the figure) the associated quadratic approximation is universally concave.

4.3   Newton's Method

Since the first-order Taylor series approximation to a function leads to the local optimization framework of gradient descent (see Section 3.5), it seems intuitive that higher-order Taylor series approximations might similarly yield descent-based algorithms as well. In this section we introduce a local optimization scheme based on the second-order Taylor series approximation, called *Newton's method* (named after its creator, Isaac Newton). Because it uses second derivative information, Newton's method has natural strengths and weaknesses when compared to gradient descent. In summary we will see that the cumulative effect of these trade-offs is, in general, that Newton's method is especially useful for minimizing *convex* functions of a moderate number of inputs.

4.3.1   The descent direction

We saw in our discussion of gradient descent that the first-order Taylor series approximation, being a hyperplane itself, conveniently provides us with a descent direction (see Section 3.3. By comparison a quadratic function has *stationary points* that are global minima when the quadratic is convex, and global maxima when it is concave. We can compute the stationary point(s) of a quadratic function fairly easily using the first-order condition for optimality (see Section 3.2).

   For the single-input case, the second-order Taylor series approximation centered at a point v is shown in Equation (4.4). Using the first-order condition to solve for the stationary point $w^\star$ of this quadratic (see Example 3.2) by setting its derivative to zero and solving, we find that

$$w^\star = v - \frac{\frac{\mathrm{d}}{\mathrm{d}w}g(v)}{\frac{\mathrm{d}^2}{\mathrm{d}w^2}g(v)}. \tag{4.7}$$

Equation (4.7) says that in order to get to the point $w^\star$ we move from v in the *direction* given by $-\frac{\frac{\mathrm{d}}{\mathrm{d}w}g(v)}{\frac{\mathrm{d}^2}{\mathrm{d}w^2}g(v)}$.

The same kind of calculation can be made in the case of multi-input second-order Taylor series approximation shown in Equation (4.5). Setting the gradient of the quadratic approximation to zero (as shown in Example 3.4) and solving gives the stationary point

$$\mathbf{w}^\star = \mathbf{v} - \left(\nabla^2 g(\mathbf{v})\right)^{-1} \nabla g(\mathbf{v}). \tag{4.8}$$

This is the direct analog of the single-input solution in Equation (4.7), and indeed reduces to it when $N = 1$. It likewise says that in order to get to the stationary point $\mathbf{w}^\star$ we move from $\mathbf{v}$ in the direction given by $-\left(\nabla^2 g(\mathbf{v})\right)^{-1} \nabla g(\mathbf{v})$. When might this direction be a descent direction? Let us examine a simple example first to build up our intuition.

Example 4.4   Stationary points of approximating quadratics

In the top row of Figure 4.5 we show the convex function

$$g(w) = \frac{1}{50}\left(w^4 + w^2\right) + 0.5 \tag{4.9}$$

drawn in black, along with three second-order Taylor series approximations shown in light blue (one per panel), each centered at a distinct input point. In each panel the point of expansion is shown as a red circle and its evaluation by the function as a red x, the stationary point $w^\star$ of the second-order Taylor series as a green circle, and the evaluations of both the quadratic approximation and the function itself at $w^\star$ are denoted by a blue and green x, respectively.

Since the function g itself is convex everywhere, the quadratic approximation not only matches the curvature at each point but is always convex and facing upwards. Therefore its stationary point is always a global minimum. Notice importantly that the minimum of the quadratic approximation $w^\star$ always leads to a lower point on the function than the evaluation of the function at v, i.e., $g(w^\star) < g(v)$.

In the bottom row of Figure 4.5 we show similar panels as those described above, only this time for the nonconvex function

$$g(w) = \sin(3w) + 0.1w^2 + 1.5. \tag{4.10}$$

However, the situation is now clearly different, with nonconvexity being the

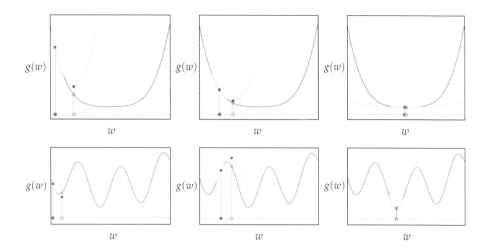

Figure 4.5 Figure associated with Example 4.4. See text for details.

culprit. In particular at concave portions of the function (like the one shown in the middle panel) since the quadratic is also concave, the stationary point $w^\star$ of the quadratic approximation is a global maximum of the approximator, and tends to lead towards points that *increase* the value of the function (not *decrease* it).

From our cursory investigation of these two simple examples we can intuit an idea for a local optimization scheme: repeatedly traveling to points defined by the stationary point of the second-order Taylor series approximation. For convex functions, where each quadratic approximation's stationary point seems to lower the original function's initial evaluation, this idea could provide an efficient algorithm to minimize a cost function. This is indeed the case, and the resulting algorithm is called the *Newton's method*.

4.3.2 The algorithm

Newton's method is a local optimization algorithm produced by repeatedly taking steps to stationary points of the second-order Taylor series approximations of a function. At the kth step of this process for a single-input function, we make a second-order Taylor series approximation centered at the point w^{k-1}

$$h(w) = g(w^{k-1}) + \left(\frac{\mathrm{d}}{\mathrm{d}w}g(w^{k-1})\right)(w - w^{k-1}) + \frac{1}{2}\left(\frac{\mathrm{d}^2}{\mathrm{d}w^2}g(w^{k-1})\right)(w - w^{k-1})^2 \quad (4.11)$$

and solve for its stationary point to create the update w^k as

$$w^k = w^{k-1} - \frac{\frac{d}{dw}g(w^{k-1})}{\frac{d^2}{dw^2}g(w^{k-1})}. \tag{4.12}$$

More generally with multi-input functions taking in N-dimensional input, at the kth step we form the second-order quadratic approximation

$$h(\mathbf{w}) = g(\mathbf{w}^{k-1}) + \nabla g(\mathbf{w}^{k-1})^T(\mathbf{w} - \mathbf{w}^{k-1}) + \frac{1}{2}(\mathbf{w} - \mathbf{w}^{k-1})^T \nabla^2 g\left(\mathbf{w}^{k-1}\right)(\mathbf{w} - \mathbf{w}^{k-1}) \tag{4.13}$$

and solve for a stationary point of this approximator, giving the update $\mathbf{w}^k$ as[1]

$$\mathbf{w}^k = \mathbf{w}^{k-1} - \left(\nabla^2 g(\mathbf{w}^{k-1})\right)^{-1} \nabla g(\mathbf{w}^{k-1}). \tag{4.15}$$

This is a local optimization scheme that fits right in with the general form we have seen in the previous two chapters, i.e.,

$$\mathbf{w}^k = \mathbf{w}^{k-1} + \alpha \mathbf{d}^k \tag{4.16}$$

where, in the case of Newton's method, $\mathbf{d}^k = -\left(\nabla^2 g(\mathbf{w}^{k-1})\right)^{-1} \nabla g(\mathbf{w}^{k-1})$ and $\alpha = 1$. The fact that the steplength parameter α is implicitly set to 1 here follows naturally from the derivation we have seen.

Notice, the Newton's update formula in Equation (4.15) requires that we invert an $N \times N$ Hessian matrix (where N is the input dimension). However, in practice, $\mathbf{w}^k$ is typically found via solving[2] the equivalent symmetric system of equations

$$\nabla^2 g\left(\mathbf{w}^{k-1}\right)\mathbf{w} = \nabla^2 g\left(\mathbf{w}^{k-1}\right)\mathbf{w}^{k-1} - \nabla g(\mathbf{w}^{k-1}) \tag{4.17}$$

which can be done more cost-effectively compared to finding its closed form solution via Equation (4.15).

---

[1] From the perspective of first-order optimization the kth Newton's method step in Equation (4.12) applied to a single-input function can also be considered a gradient descent step with self-adjusting steplength parameter

$$\alpha = \frac{1}{\frac{d^2}{dw^2}g\left(w^{k-1}\right)}, \tag{4.14}$$

which adjusts the length traveled based on the underlying curvature of the function, akin to the self-adjusting steplength perspective of normalized gradient steps discussed in Appendix Section A.3. Although this interpretation does not generalize directly to the multi-input case in Equation (4.15), by discarding the off-diagonal entries of the Hessian matrix one can form a generalization of this concept for the multi-input case. See Appendix Section A.8.1 for further details.

[2] One can solve this system using coordinate descent as outlined in Section 3.2.2. When more than one solution exists the smallest possible solution (e.g., in the ℓ_2 sense) is typically taken. This is also referred to as the *pseudo-inverse* of $\nabla^2 g(\mathbf{w})$.

As illustrated in the top panel of Figure 4.6 for a single-input function, starting at an initial point w^0 Newton's method produces a sequence of points
$w^1, w^2, \ldots$, etc., that minimize g by repeatedly creating the second-order Taylor
series quadratic approximation to the function, and traveling to a stationary
point of this quadratic. Because Newton's method uses quadratic as opposed
to linear approximations at each step, with a quadratic more closely mimicking
the associated function, it is often much more effective than gradient descent
in the sense that it requires far fewer steps for convergence [14, 15]. However,
this reliance on quadratic information also makes Newton's method naturally
more difficult to use with nonconvex functions since at concave portions of
such a function the algorithm can climb to a local maximum, as illustrated in
the bottom panel of Figure 4.6, or oscillate out of control.

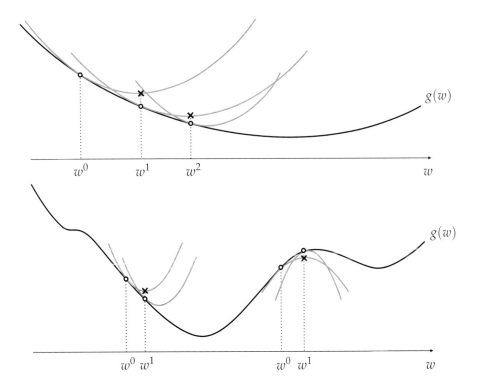

Figure 4.6 Newton's method illustrated. To find a minimum of g, Newton's method
hops down the stationary points of quadratic approximations generated by its
second-order Taylor series. (top panel) For convex functions these quadratic
approximations are themselves always convex (whose only stationary points are
minima), and the sequence leads to a minimum of the original function. (bottom panel)
For nonconvex functions quadratic approximations can be concave or convex
depending on where they are constructed, leading the algorithm to possibly converge
to a maximum.

Example 4.5   Minimization of a convex function using Newton's method
In Figure 4.7 we show the process of performing Newton's method to minimize the function

$$g(w) = \frac{1}{50}\left(w^4 + w^2 + 10w\right) + 0.5 \tag{4.18}$$

beginning at the point $w^0 = 2.5$, marked as a green dot in the top-left panel and corresponding evaluation of the function marked as a green **x**. The top-right panel of the figure shows the first Newton step, with the corresponding quadratic approximation shown in green and its minimum shown as a magenta circle along with the evaluation of this minimum on the quadratic shown as a blue **x**. The remaining panels show the next iterations of Newton's method.

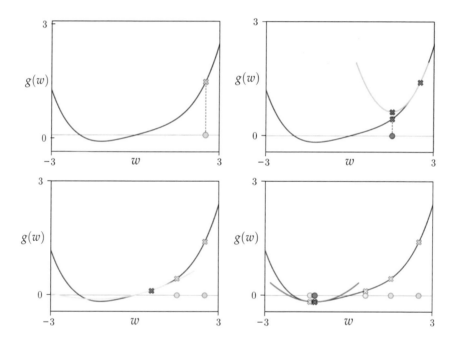

Figure 4.7 Figure associated with Example 4.5, animating a run of Newton's method applied to the function in Equation (4.18). See text for further details.

Example 4.6   Comparison to gradient descent
As illustrated in the right panel of Figure 4.8, a single Newton step is all that is required to completely minimize the convex quadratic function

$$g(w_1, w_2) = 0.26\left(w_1^2 + w_2^2\right) - 0.48\,w_1 w_2. \tag{4.19}$$

This can be done with a single step because the second-order Taylor series approximation to a quadratic function is simply the quadratic function itself. Thus Newton's method reduces to solving the linear first-order system of a quadratic function. We compare the result of this single Newton step (shown in the right panel of Figure 4.8) to a corresponding run of 100 steps of gradient descent in the left panel of the figure.

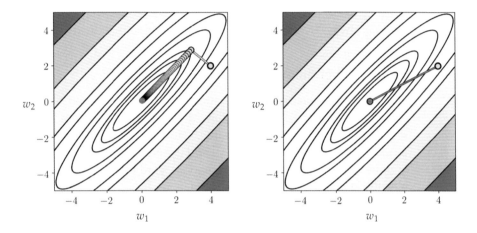

Figure 4.8 Figure associated with Example 4.6. See text for details.

4.3.3   Ensuring numerical stability

Near flat portions of a function the numerator $\frac{d}{dw}g(w^{k-1})$ and denominator $\frac{d^2}{dw^2}g(w^{k-1})$ of the single-input Newton update in Equation (4.12) both have small, near-zero values. This can cause serious numerical problems once each (but especially the denominator) shrinks below *machine precision*, i.e., the smallest value a computer can interpret as being nonzero.

One simple and common way to avoid this potential *division-by-zero* problem is to add a small positive value ϵ to the denominator, either when it shrinks below a certain value or for all iterations. This *regularized* Newton's step then takes the form

$$w^k = w^{k-1} - \frac{\frac{d}{dw}g(w^{k-1})}{\frac{d^2}{dw^2}g(w^{k-1}) + \epsilon}. \tag{4.20}$$

The value of the *regularization parameter* ϵ is typically set to a small positive[3] value (e.g., 10^{-7}).

---

[3] This adjustment is made when the function being minimized is known to be convex, since in this case $\frac{d^2}{dw^2}g(w) \geq 0$ for all w.

The analogous adjustment for the general multi-input Newtons's update is to add $\epsilon \mathbf{I}_{N \times N}$ (an $N \times N$ identity matrix scaled by a small positive ϵ value) to the Hessian matrix in Equation (4.15), giving[4]

$$\mathbf{w}^k = \mathbf{w}^{k-1} - \left(\nabla^2 g(\mathbf{w}^{k-1}) + \epsilon \mathbf{I}_{N \times N} \right)^{-1} \nabla g(\mathbf{w}^{k-1}). \qquad (4.21)$$

Adding this additional term to the Hessian guarantees that the matrix $\nabla^2 g(\mathbf{w}^{k-1}) + \epsilon \mathbf{I}_{N \times N}$ is always invertible, provided a large enough value for ϵ is used.

4.3.4   Steplength choices

While we have seen in the derivation of Newton's method that (being a local optimization approach) it does have a steplength parameter α, it is implicitly set to $\alpha = 1$ and so appears "invisible." However, in principle, one can explicitly introduce a steplength parameter α and use adjustable methods (e.g., backtracking line search as introduced in Section A.4) to tune it. An explicitly weighted Newton step then takes the form

$$\mathbf{w}^k = \mathbf{w}^{k-1} - \alpha \left(\nabla^2 g(\mathbf{w}^{k-1}) \right)^{-1} \nabla g(\mathbf{w}^{k-1}) \qquad (4.22)$$

with the standard Newton step falling out when $\alpha = 1$.

4.3.5   Newton's method as a zero-finding algorithm

Newton's method was first invented not as a local optimization algorithm, but as a *zero-finding* algorithm. In other words, Newton's method was first invented to find zeros of a function f, i.e., where $f(\mathbf{w}) = \mathbf{0}_{N \times 1}$. Traditionally the function f examined was some sort of polynomial function. In the context of local optimization we can think of Newton's method as an approach for *iteratively* solving the first-order system (see Section 3.2)

$$\nabla g(\mathbf{v}) = \mathbf{0}_{N \times 1}. \qquad (4.23)$$

Take the case where our input dimension $N = 1$. Generally speaking, finding zeros of an arbitrary function is not a trivial affair. Instead of trying to solve the first-order equation $\frac{d}{dw} g(v) = 0$ directly let us try to set up an iterative procedure where we find an approximate a solution to this equation by solving a related sequence of simpler problems. Following the same sort of logic we used in deriving gradient descent and Newton's method previously, instead of trying to find a zero of the function itself, let us try to find a zero of the tangent line provided by our function's *first-order Taylor series* approximation. Finding

---

[4] As with the original Newton step in Equation (4.15), it is virtually always more numerically efficient to compute this update by solving the associated linear system $\left(\nabla^2 g(\mathbf{w}^{k-1}) + \epsilon \mathbf{I}_{N \times N} \right) \mathbf{w} = \left(\nabla^2 g(\mathbf{w}^{k-1}) + \epsilon \mathbf{I}_{N \times N} \right) \mathbf{w}^{k-1} - \nabla g(\mathbf{w}^{k-1})$ for $\mathbf{w}$.

the point(s) at which a line or – more generally – a hyperplane equals zero is a comparatively trivial affair.

To write out the first step of this scheme, remember first and foremost that we are thinking of this as an iterative method applied to the derivative function $\frac{d}{dw} g(w)$. This means that, beginning at a point w^0, our linear first-order Taylor series approximation to the derivative function

$$h(w) = \frac{d}{dw} g\left(w^0\right) + \frac{d^2}{dw^2} g\left(w^0\right)\left(w - w^0\right) \tag{4.24}$$

naturally involves the second derivative of the function g (it is, after all, the first-order approximation of this function's derivative). We can easily compute where this line crosses the input axis by setting the equation above equal to zero and solving. Doing this, and calling the solution w^1, we have

$$w^1 = w^0 - \frac{\frac{d}{dw} g\left(w^0\right)}{\frac{d^2}{dw^2} g\left(w^0\right)}. \tag{4.25}$$

Examined closely we can see that this is indeed a Newton step. Since we have only found the zero of a linear approximation to $\frac{d}{dw} g(w)$ and not to this function itself, it is natural to repeat this procedure to refine our approximation. At the kth such step our update takes the form

$$w^k = w^{k-1} - \frac{\frac{d}{dw} g\left(w^{k-1}\right)}{\frac{d^2}{dw^2} g\left(w^{k-1}\right)} \tag{4.26}$$

which is exactly the form of Newton step in Equation (4.12). Precisely the analogous reasoning applied to multi-input functions (where $N > 1$), starting with the desire to iteratively solve the first-order system, leads to deriving the multi-input Newton's step shown in Equation (4.15).

4.3.6   Python implementation

In this section we provide a simple implementation of Newton's method in Python, leveraging the excellent `autograd` automatic differentiation and `NumPy` libraries (see Sections 3.4 and B.10). In particular we employ the `grad` and `hessian` modules from `autograd` to compute the first and second derivatives of a general input function automatically.

```
1   # import autograd's automatic differentiator
2   from autograd import grad
3   from autograd import hessian
4
5   # import NumPy library
6   import numpy as np
7
```

```
8   # Newton's method
9   def newtons_method(g, max_its, w):
10
11      # compute gradient/Hessian using autograd
12      gradient = grad(g)
13      hess = hessian(g)
14
15      # set numerical stability parameter
16      epsilon = 10**(-7)
17      if 'epsilon' in kwargs:
18          epsilon = kwargs['epsilon']
19
20      # run the Newton's method loop
21      weight_history = [w]  # container for weight history
22      cost_history = [g(w)] # container for cost function history
23      for k in range(max_its):
24
25          # evaluate the gradient and hessian
26          grad_eval = gradient(w)
27          hess_eval = hess(w)
28
29          # reshape hessian to square matrix
30          hess_eval.shape = (int((np.size(hess_eval))**(0.5)),int((np.
                size(hess_eval))**(0.5)))
31
32          # solve second-order system for weight update
33          A = hess_eval + epsilon*np.eye(w.size)
34          b = grad_eval
35          w = np.linalg.solve(A, np.dot(A,w)-b)
36
37          # record weight and cost
38          weight_history.append(w)
39          cost_history.append(g(w))
40
41      return weight_history,cost_history
```

Notice, while we used a maximum iterations convergence criterion the potentially high computational cost of each Newton step often incentivizes the use of
more formal convergence criteria (e.g., halting when the norm of the gradient
falls below a pre-defined threshold). This also often incentivizes the inclusion of
checkpoints that measure and/or adjust the progress of a Newton's method run
in order to avoid problems near flat areas of a function. Additionally, one can
use the same kind of initialization for this implementation of Newton's method
as described for gradient descent in Section 3.5.4.

4.4   Two Natural Weaknesses of Newton's Method

Newton's method is a powerful algorithm that makes enormous progress towards finding a function's minimum at each step, compared to zero- and first-
order methods that can require a large number of steps to make equivalent

progress. Since both first- and second-order (i.e., curvature) information are employed, Newton's method does not suffer from the problems inherent to first-order methods (e.g., the zig-zagging problem we saw in Section 3.6.3). However, Newton's method suffers from its own unique weaknesses – primarily in dealing with *nonconvexity*, as well as *scaling* with input dimension – which we briefly discuss here. While these weaknesses do not prevent Newton's method (as we have described it) from being widely used in machine learning, they are (at least) worth being aware of.

4.4.1 Minimizing nonconvex functions

As discussed in the previous section, Newton's method can behave very badly when applied to minimizing nonconvex functions. Since each step is based on the second-order approximation to a function, initiated at *concave* point/region Newton's method will naturally take a step *uphill*. This fact is illustrated for a prototypical nonconvex function in the bottom row of Figure 4.6. The interested reader can see Section A.7, where we describe a simple and common approach to adjusting Newton's method to address this issue.

4.4.2 Scaling limitations

Since the quadratic approximation used by Newton's method matches a function very well *locally*, the method can converge to a global minimum in far fewer steps (than first-order methods) particularly when close to a minimum. However, a Newton's method step is computationally far more expensive than a first-order step, requiring the storage and computation of not just a gradient but an entire $N \times N$ Hessian matrix of second derivative information as well. Simply storing the Hessian for a single step of Newton's method, with its N^2 entries, can quickly become challenging for even moderately sized input. For example, if the input to a function has dimension $N = 10,000$ the corresponding Hessian matrix has $100,000,000$ entries. The kind of functions used in machine learning applications can easily have tens of thousands to hundreds of thousands or even millions of inputs, making the complete storage of an associated Hessian impossible.

Later in Section A.8 we discuss basic ways of ameliorating this problem, which involve adjusting the basic Newton's method step by replacing the Hessian with some sort of approximation that does not suffer from this inherent scaling issue.

4.5 Conclusion

In this chapter we cap off our discussion of mathematical optimization in this part of the text by describing second-order optimization techniques, i.e., those that leverage both the first and second derivative(s) of a function in forming descent directions.

We began in Section 4.2 with a review of the second-order condition for optimality. We then briefly touched on function curvature as defined by its second derivative(s) in Section 4.1 before immediately applying this concept in detailing the keystone second-order local optimization method – *Newton's method* – in Section 4.3.

Afterwards in Section 4.4 we touched on two natural problems with the Newton's method scheme – its application to the minimization of nonconvex functions and to functions with high-dimensional input. The interested reader should note that in Appendix Sections A.7 and A.8 we detail common adjustments to the standard Newton's scheme for ameliorating these problems, with the latter set of adjustments referred to as *Hessian-free* optimization.

4.6   Exercises

† The data required to complete the following exercises can be downloaded from the text's github repository at `github.com/jermwatt/machine_learning_refined`

4.1   Determining the eigenvalues of a symmetric matrix
In this exercise we investigate an alternative approach to checking that the eigenvalues of an $N \times N$ symmetric matrix $\mathbf{C}$ (e.g., a Hessian matrix) are all nonnegative. This approach does not involve explicitly computing the eigenvalues themselves, and is significantly easier to employ in practice.

(a) Let $\mathbf{C}$ be an $N \times N$ symmetric matrix. Show that if $\mathbf{C}$ has all nonnegative eigenvalues then the quantity $\mathbf{z}^T \mathbf{C} \mathbf{z} \geq 0$ for all $\mathbf{z}$. *Hint: use the eigenvalue decomposition of $\mathbf{C}$ (see Appendix Section C.4).*

(b) Show the converse. That is, if an $N \times N$ symmetric matrix $\mathbf{C}$ satisfies $\mathbf{z}^T \mathbf{C} \mathbf{z} \geq 0$ for all $\mathbf{z}$ then it must have all nonnegative eigenvalues.

(c) Use this method to verify that the second-order definition of convexity holds for the quadratic function

$$g(\mathbf{w}) = a + \mathbf{b}^T \mathbf{w} + \mathbf{w}^T \mathbf{C} \mathbf{w} \tag{4.27}$$

where $a = 1$, $\mathbf{b} = [1\ 1]^T$, and $\mathbf{C} = \begin{bmatrix} 1 & 1 \\ 1 & 1 \end{bmatrix}$.

(d) Show that the eigenvalues of $\mathbf{C} + \lambda \mathbf{I}_{N \times N}$ can all be made to be positive by setting λ large enough. What is the smallest value of λ that will make this happen?

4.2   Outer-product matrices have all nonnegative eigenvalues
(a) Use the method described in Exercise 4.1 to verify that for any $N \times 1$ vector $\mathbf{x}$, the $N \times N$ outer-product matrix $\mathbf{xx}^T$ has all nonnegative eigenvalues.

(b) Similarly show that for any set of P vectors $\mathbf{x}_1, \mathbf{x}_2, ..., \mathbf{x}_P$ of length N that the sum of outer-product matrices $\sum_{p=1}^{P} \delta_p \mathbf{x}_p \mathbf{x}_p^T$ has all nonnegative eigenvalues if each $\delta_p \geq 0$.

(c) Show that the matrix $\sum_{p=1}^{P} \delta_p \mathbf{x}_p \mathbf{x}_p^T + \lambda \mathbf{I}_{N \times N}$ where each $\delta_p \geq 0$ and $\lambda > 0$ has all positive eigenvalues.

4.3   An alternative way to check the second-order definition of convexity
Recall that the second-order definition of convexity for a multi-input function $g(\mathbf{w})$ requires that we verify whether or not the eigenvalues of $\nabla^2 g(\mathbf{w})$ are nonnegative for each input $\mathbf{w}$. However, to explicitly compute the eigenvalues of the Hessian in order to check this is a cumbersome or even impossible task for all but the nicest of functions. Here we use the result of Exercise 4.1 to express the second-order definition of convexity in a way that is often much easier to employ in practice.

(a) Use the result of Exercise 4.1 to conclude that nonnegativity of the eigenvalues of $\nabla^2 g(\mathbf{w})$ are nonnegative at every $\mathbf{w}$ is equivalently stated as the inequality $\mathbf{z}^T \left(\nabla^2 g(\mathbf{w}) \right) \mathbf{z} \geq 0$ holding at each $\mathbf{w}$ for all $\mathbf{z}$.

(b) Use this manner of expressing the second-order definition of convexity to verify that the general quadratic function $g(\mathbf{w}) = a + \mathbf{b}^T \mathbf{w} + \mathbf{w}^T \mathbf{C} \mathbf{w}$ where $\mathbf{C}$ is symmetric and known to have all nonnegative eigenvalues, always defines a convex function.

(c) Verify that $g(\mathbf{w}) = -\cos\left(2\pi \mathbf{w}^T \mathbf{w}\right) + \mathbf{w}^T \mathbf{w}$ is nonconvex by showing that it does *not* satisfy the second-order definition of convexity.

4.4   Newton's method I
Repeat the experiment described in Example 4.5. Instead of plotting the re-

sulting path taken by Newton's method (as shown in Figure 4.7), create a cost function history plot to ensure your algorithm properly converges to a point near the global minimum of the function. You may employ the implementation of Newton's method described in Section 4.3.6 as a base for this exercise.

4.5   Newton's method II

(a) Use the first-order optimality condition (see Section 3.2) to determine the unique stationary point of the function $g(\mathbf{w}) = \log\left(1 + e^{\mathbf{w}^T\mathbf{w}}\right)$ where $\mathbf{w}$ is two-dimensional (i.e., $N = 2$).

(b) Use the second-order definition of convexity to verify that $g(\mathbf{w})$ is convex, implying that the stationary point found in part (a) is a global minimum. *Hint: to check the second-order definition use Exercise 4.2.*

(c) Perform Newton's method to find the minimum of the function $g(\mathbf{w})$ determined in part (a). Initialize your algorithm at $\mathbf{w}^0 = \mathbf{1}_{N\times 1}$ and make a plot of the cost function history for ten iterations of Newton's method in order to verify that your algorithm works properly and is converging. You may use the implementation given in Section 4.3.6 as a base for this part of the exercise.

(d) Now run your Newton's method code from part (c) again, this time initializing at the point $\mathbf{w}^0 = 4 \cdot \mathbf{1}_{N\times 1}$. While this initialization is further away from the unique minimum of $g(\mathbf{w})$ than the one used in part (c) your Newton's method algorithm should converge *faster* starting at this point. At first glance this result seems very counterintuitive, as we (rightfully) expect that an initial point closer to a minimum will provoke more rapid convergence of Newton's method! Explain why this result actually makes sense for the particular function $g(\mathbf{w})$ we are minimizing here.

4.6   Finding square roots

Use Newton's method to compute the square root of 999. Briefly explain how you set up the relevant cost function that was minimized to obtain this square root. Explain how you use zero- or first-order optimization methods (detailed in Chapters 2 and 3) to do this as well.

4.7   Nonconvex minimization using Newton's method

Use (regularized) Newton's method to minimize the function

$$g(w) = \cos(w) \tag{4.28}$$

beginning at $w = 0.1$. In particular make sure you achieve decrease in function value at *every* step of Newton's method.

4.8   **Newtonian descent**
(a) Show that when $g(\mathbf{w})$ is convex the Newton step in Equation (4.15) does indeed decrease the evaluation of g, i.e., $g(\mathbf{w}^k) \le g(\mathbf{w}^{k-1})$.

(b) Show, regardless of the function g being minimized, that ϵ in Equation (4.21) can be set large enough so that a corresponding Newton step can lead to a lower portion of the function, i.e., $g(\mathbf{w}^k) \le g(\mathbf{w}^{k-1})$.

4.9   **Newton's method as a self-adjusting gradient descent method**
Implement the subsampled Newton's step outlined in Appendix Section A.8.1 and given in Equation (A.78) formed by ignoring all off-diagonal elements of the Hessian, and compare it to gradient descent using the test function

$$g(\mathbf{w}) = a + \mathbf{b}^T\mathbf{w} + \mathbf{w}^T\mathbf{C}\mathbf{w} \qquad (4.29)$$

where $a = 0$, $\mathbf{b} = \begin{bmatrix} 0 \\ 0 \end{bmatrix}$, and $\mathbf{C} = \begin{bmatrix} 0.5 & 2 \\ 1 & 9.75 \end{bmatrix}$.

Make a run of each local method for 25 steps beginning at the initial point $\mathbf{w}^0 = [10\ 1]^T$, using the largest fixed steplength value of the form 10^γ (where γ is an integer) for gradient descent. Make a contour plot of the test function and plot the steps from each run on top of it to visualize how each algorithm performs.

4.10   **The Broyden–Fletcher–Goldfarb–Shanno (BFGS) method**
Start with the same assumption as in Example A.12 (i.e., a recursion based on a rank-2 difference between $\mathbf{S}^k$ and its predecessor) and employ the secant condition to derive a recursive update for $\mathbf{S}^k$ in terms of $\mathbf{S}^{k-1}$, $\mathbf{a}^k$, and $\mathbf{b}^k$. Next, use the Sherman–Morrison identity to rewrite your update in terms of $\mathbf{F}^k$, the inverse of $\mathbf{S}^k$.

Part II

Linear Learning

5 Linear Regression

5.1 Introduction

In this chapter we formally describe the supervised learning problem called *linear regression*, or the fitting of a representative line (or, in higher dimensions, a hyperplane) to a set of input/output data points as first broadly detailed in Section 1.3.1. Regression in general may be performed for a variety of reasons: to produce a so-called trend line (or more generally – as we will see later – a curve) that can be used to help visually summarize, drive home a particular point about the data under study, or to learn a model so that precise predictions can be made regarding output values in the future. In the description given here we will describe various ways of designing appropriate cost functions for regression, including a discussion of the *Least Squares* and *Least Absolute Deviations* costs, as well as appropriate metrics for measuring the quality of a trained regressor (that is one whose parameters have been fully optimized), and various common extensions of the basic regression concept (including weighted and multi-output regression).

5.2 Least Squares Linear Regression

In this section we formally introduce the problem of *linear regression*, or the fitting of a representative line (or hyperplane in higher dimensions) to a set of input/output data points. We also walk through the design of the popular *Least Squares* cost function commonly employed to appropriately tune the parameters of a regressor in general.

5.2.1 Notation and modeling

Data for regression problems generally come in the form of a set of P input/output observation pairs

$$(\mathbf{x}_1, y_1), (\mathbf{x}_2, y_2), ..., (\mathbf{x}_P, y_P) \tag{5.1}$$

or $\left\{ \left(\mathbf{x}_p, y_p \right) \right\}_{p=1}^{P}$ for short, where $\mathbf{x}_p$ and y_p denote the input and output of the pth observation, respectively. Each input $\mathbf{x}_p$ is in general a column vector of length N

$$\mathbf{x}_p = \begin{bmatrix} x_{1,p} \\ x_{2,p} \\ \vdots \\ x_{N,p} \end{bmatrix} \tag{5.2}$$

and each output y_p is scalar-valued (we consider vector-valued outputs in Section 5.6). Geometrically speaking, the linear regression problem is then one of fitting a *hyperplane* to a scatter of points in $(N+1)$-dimensional space.

In the simplest instance where the inputs are also scalar-valued (i.e., $N = 1$), linear regression simplifies to fitting a *line* to the associated scatter of data points in two-dimensional space. A line in two dimensions is determined by two parameters: a vertical intercept w_0 and slope w_1. We must set the values of these parameters in a way that the following approximate linear relationship holds between the input/output data

$$w_0 + x_p w_1 \approx y_p, \quad p = 1, ..., P. \tag{5.3}$$

Notice that we have used the approximately equal sign in Equation (5.3) because we can never be absolutely sure that all data lies completely on a single line. More generally, when dealing with N-dimensional input we have a bias weight and N associated slope weights to tune properly in order to fit a hyperplane, with the analogous linear relationship written as

$$w_0 + x_{1,p} w_1 + x_{2,p} w_2 + \cdots + x_{N,p} w_N \approx y_p, \quad p = 1, ..., P. \tag{5.4}$$

Both the single-input and general multi-input cases of linear regression are illustrated figuratively in Figure 5.1. Each dimension of the input is referred to as a *feature* or *input feature* in machine learning parlance. Hence we will often refer to the parameters $w_1, w_2, ..., w_N$ as the *feature-touching weights*, the only weight not touching a feature is the bias w_0.

The linear relationship in Equation (5.4) can be written more compactly, using the notation $\mathring{\mathbf{x}}$ to denote an input $\mathbf{x}$ with a 1 placed on top of it. This means that we stack a 1 on top of each of our input points $\mathbf{x}_p$

$$\mathring{\mathbf{x}}_p = \begin{bmatrix} 1 \\ x_{1,p} \\ x_{2,p} \\ \vdots \\ x_{N,p} \end{bmatrix} \tag{5.5}$$

for all $p = 1, ..., P$. Now placing all parameters into a single column vector $\mathbf{w}$

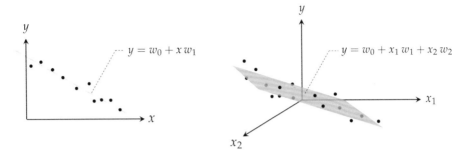

Figure 5.1 (left panel) A simulated dataset in two dimensions along with a well-fitting line. A line in two dimensions is defined as $w_0 + xw_1 = y$, where w_0 is referred to as the bias and w_1 the slope, and a point (x_p, y_p) lies close to it if $w_0 + x_pw_1 \approx y_p$. (right panel) A simulated three-dimensional dataset along with a well-fitting hyperplane. A hyperplane in general is defined as $w_0 + x_1w_1 + x_2w_2 + \cdots + x_Nw_N = y$, where again w_0 is the bias and $w_1, w_2, \ldots, w_N$ the hyperplane's coordinate-wise slopes, and a point (x_p, y_p) lies close to it if $w_0 + x_{1,p}w_1 + x_{2,p}w_2 + \cdots + x_{N,p}w_N \approx y_p$. Here $N = 2$.

$$\mathbf{w} = \begin{bmatrix} w_0 \\ w_1 \\ w_2 \\ \vdots \\ w_N \end{bmatrix} \tag{5.6}$$

we may write the general desired linear relationships in Equation (5.4) more compactly as

$$\mathring{\mathbf{x}}_p^T \mathbf{w} \approx y_p \quad p = 1, ..., P. \tag{5.7}$$

5.2.2 The Least Squares cost function

To find the parameters of the hyperplane that best fits a regression dataset we must first form a cost function that measures how well a linear model with a particular choice of parameters fits the regression data. One of the more popular choices for doing this is called the *Least Squares* cost function. For a given set of parameters in the vector $\mathbf{w}$ this cost function computes the total *squared error* between the associated hyperplane and the data, as illustrated pictorially in the Figure 5.2. Naturally then the best fitting hyperplane is the one whose parameters *minimize* this error.

Focusing on just the pth approximation in Equation (5.7) we ideally want $\mathring{\mathbf{x}}_p^T \mathbf{w}$ to be as close as possible to the pth output y_p, or equivalently, the error or difference between the two, i.e., $\mathring{\mathbf{x}}_p^T \mathbf{w} - y_p$, to be as small as possible. By squaring

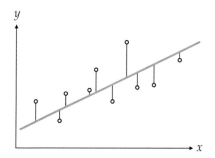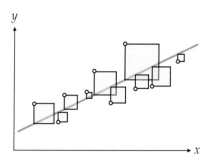

Figure 5.2 (left panel) A prototypical two-dimensional dataset along with a line fit to the data using the Least Squares framework, which aims at recovering the linear model that minimizes the total squared length of the solid error bars. (right panel) The Least Squares error can be thought of as the total area of the blue squares, having solid black error bars as sides. The cost function is called *Least Squares* because it allows us to determine a set of parameters whose corresponding line minimizes the sum of these square errors.

this quantity (so that both negative and positive errors of the same magnitude are treated equally) we can define

$$g_p(\mathbf{w}) = \left(\mathring{\mathbf{x}}_p^T \mathbf{w} - y_p\right)^2 \tag{5.8}$$

as a *point-wise cost function* that measures the error of a model (here a linear one) on the individual point $\left(\mathbf{x}_p, y_p\right)$. Now since we want all P such values to be small simultaneously we can take their average over the entire dataset, forming the *Least Squares cost function*[1] for linear regression

$$g(\mathbf{w}) = \frac{1}{P}\sum_{p=1}^{P} g_p(\mathbf{w}) = \frac{1}{P}\sum_{p=1}^{P} \left(\mathring{\mathbf{x}}_p^T \mathbf{w} - y_p\right)^2. \tag{5.9}$$

Notice that the larger the Least Squares cost becomes the larger the squared error between the corresponding linear model and the data, and hence the poorer we represent the given dataset using a linear model. Therefore we want to find the optimal parameter vector $\mathbf{w}$ that *minimizes* $g(\mathbf{w})$, or written formally we want to solve the unconstrained optimization problem

---

[1] Technically speaking, the Least Squares cost function $g(\mathbf{w})$ is a function of both the the weights $\mathbf{w}$ as well as the data. However, for notational simplicity we often choose not to show the dependency on data explicitly. Otherwise we would have to write the cost function as

$$g\left(\mathbf{w}, \left\{\left(\mathring{\mathbf{x}}_p, y_p\right)\right\}_{p=1}^{P}\right)$$

and things start to get too messy. Moreover, for a given dataset the weights $\mathbf{w}$ are the important input to the function as they are what we need to tune in order to produce a good fit. From an optimization perspective the dataset itself is considered *fixed*. We will make this sort of notational simplification for virtually all future machine learning cost functions we study as well.

$$\underset{\mathbf{w}}{\text{minimize}} \ \frac{1}{P}\sum_{p=1}^{P}\left(\mathring{\mathbf{x}}_p^T\mathbf{w} - y_p\right)^2. \tag{5.10}$$

using the tools of local optimization detailed in Chapters 2, 3, and 4.

5.2.3 Minimization of the Least Squares cost function

The Least Squares cost function for linear regression in Equation (5.9) can be proven to be *convex* for any dataset (see Section 5.9). In Example 5.1 we showcase this fact using a toy linear regression dataset.

Example 5.1 Verifying convexity by visual inspection
The top panel of Figure 5.3 shows a toy linear regression dataset, consisting of $P = 50$ input/output pairs randomly selected off of the line $y = x$, with a small amount of random noise added to each output. In the bottom-left panel of this figure we plot the three-dimensional surface of the Least Squares cost function associated with this dataset, with its contour plot shown in two dimensions in the bottom-right panel. We can see, by the *upward bending* shape of the cost function's surface on the left or by the elliptical shape of its contour lines on the right, that the Least Squares cost function is indeed convex for this particular dataset.

Because of its convexity, and because the Least Squares cost function is *infinitely differentiable*, we can apply virtually any local optimization scheme to minimize it properly. However, the generic practical considerations associated with each local optimization method still apply: that is, the zero- and second-order methods do not scale gracefully, and with gradient descent we must choose a fixed steplength value, a diminishing steplength scheme, or an adjustable method like backtracking line search (as detailed in Chapter 3). Because the Least Squares cost is a convex quadratic function, a *single* Newton step can perfectly minimize it. This is sometimes referred to as minimizing the Least Squares via solving its *normal equations* (see Exercise 5.3 for further discussion).

Example 5.2 Using gradient descent
In Figure 5.4 we show the result of minimizing the Least Squares cost using the toy dataset presented in Example 5.1. We use gradient descent and employ a fixed steplength value $\alpha = 0.5$ for all 75 steps until approximately reaching the minimum of the function.

The figure shows progression of the gradient descent process (from left to

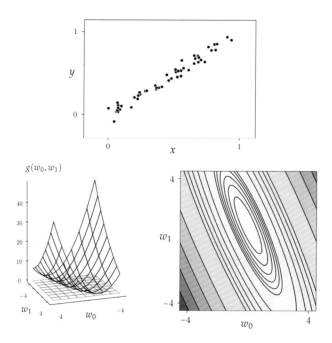

Figure 5.3 Figure associated with Example 5.1. See text for details.

right) both in terms of the Least Squares cost minimization (top row) and line provided by the corresponding weights (bottom row). The optimization steps in the top row are colored from green at the start of the run (leftmost panel) to red at its finale (rightmost panel). The linear model is colored to match the step of gradient descent (green near the beginning and red towards the end). Examining the figure, as gradient descent approaches the minimum of the cost function the corresponding parameters provide a better and better fit to the data, with the best fit occurring at the end of the run at the point closest to the Least Squares cost's minimizer.

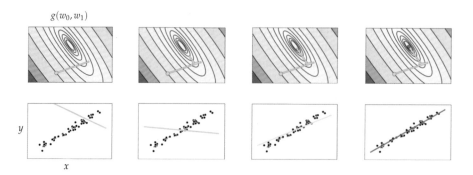

Figure 5.4 Figure associated with Example 5.2. See text for details.

Whenever we use a local optimization method like gradient descent we must properly tune the steplength parameter α (as described previously, e.g., in Section 3.5). In Figure 5.5 we show the cost function history plot for two steplength values: $\alpha = 0.01$ (in purple), and $\alpha = 0.5$ (in black), which we ended up using for the run shown in Figure 5.4. This illustrates why (in machine learning contexts) the steplength parameter is often referred to as the *learning rate*, since this value does indeed determine how quickly the proper parameters of our linear regression model (or any machine learning model in general) are learned.

Figure 5.5 Figure associated with Example 5.2. See text for details.

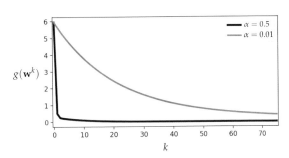

5.2.4 Python **implementation**

When implementing a cost function like Least Squares it is helpful to think in a modular fashion, with the aim of lightening the amount of mental "bookkeeping" required, by breaking down the cost into a few distinct components. Here we break the cost function down into two main parts: the *model* that is a linear combination of input and weights, and the *cost* itself (i.e., squared error).

We express our (linear) model as a function worthy enough of its own notation, as

$$\text{model}\left(\mathbf{x}_p, \mathbf{w}\right) = \mathring{\mathbf{x}}_p^T \mathbf{w}. \tag{5.11}$$

If we were to go back and use this new modeling notation we could re-write our ideal settings of the weights in Equation (5.7), as

$$\text{model}\left(\mathbf{x}_p, \mathbf{w}\right) \approx y_p \tag{5.12}$$

and likewise our Least Squares cost function in Equation (5.9), as

$$g\left(\mathbf{w}\right) = \frac{1}{P} \sum_{p=1}^{P} \left(\text{model}\left(\mathbf{x}_p, \mathbf{w}\right) - y_p\right)^2. \tag{5.13}$$

This kind of simple deconstruction of the Least Squares cost lends itself to an organized, modular, and easily extendable implementation. Starting with the model, note that while it is more compact and convenient *mathematically* to write the linear combination $\mathring{x}_p^T \mathbf{w}$ by tacking a 1 on top of the raw input $\mathbf{x}_p$, in *implementing* this we can more easily compute the linear combination by exposing the bias and feature-touching weights *separately* as

$$\mathring{x}_p^T \mathbf{w} = w_0 + \mathbf{x}_p^T \omega \tag{5.14}$$

where vector ω contains all the feature-touching weights

$$\omega = \begin{bmatrix} w_1 \\ w_2 \\ \vdots \\ w_N \end{bmatrix}. \tag{5.15}$$

Remember that w_0 is called the *bias* since it controls where our linear model pierces the y axis, and $w_1, w_2, ..., w_N$ are called *feature-touching weights* because they touch each individual dimension of the input (which in the jargon of machine learning are called *features*).

Using the efficient NumPy's np.dot operation[2] we can implement the linear model as

```
1  a = w[0] + np.dot(x_p.T,w[1:])
```

which matches the right-hand side of Equation (5.14), where w[0] denotes the bias w_0 and w[1:] denotes the remaining N feature-touching weights in ω. Wrapping this into a Python function we have our linear model implemented as

```
1  # compute linear combination of input point
2  def model(x_p,w):
3      # compute linear combination and return
4      a = w[0] + np.dot(x_p.T,w[1:])
5      return a.T
```

which we can then use to form the associated Least Squares cost function.

```
1  # a least squares function for linear regression
2  def least_squares(w,x,y):
3      # loop over points and compute cost contribution from each input/
           output pair
4      cost = 0
5      for p in range(y.size):
```

---

[2] As a general rule whenever vectorized implementations are available, one must refrain from implementing algebraic expressions in Python entry-wise using for instance, explicit for loops.

```
6          # get pth input/output pair
7          x_p = x[:,p][:,np.newaxis]
8          y_p = y[p]
9
10         ## add to current cost
11         cost += (model(x_p,w)  - y_p)**2
12
13      # return average least squares error
14      return cost/float(y.size)
```

Notice here we explicitly show *all* of the inputs to the cost function, not just the $(N + 1) \times 1$ weights **w**, whose Python variable is denoted by w. The Least Squares cost also takes in all inputs (with ones stacked on top of each point) $\mathring{x}_p$, which together we denote by the $(N + 1) \times P$ Python variable x as well as the entire set of corresponding outputs which we denote as the $1 \times P$ variable y.

Notice that this really is a direct implementation of the algebraic form of the cost in Equation (5.13), where we think of the cost as the sum of squared errors of a linear model of input against its corresponding output. However, explicit for loops (including list comprehensions) written in Python are rather slow due to the very nature of the language.

It is easy to get around most of this inefficiency by replacing explicit for loops with numerically equivalent operations performed using NumPy operations. The NumPy package is an API for some very efficient vector/matrix manipulation libraries written in C. Broadly speaking, when writing a Pythonic function like this one with heavy use of NumPy functionality one tries to package each step of computation – which previously was being formed sequentially at each data point – together for the entire dataset simultaneously. This means we do away with the explicit for loop over each of our P points and make the same computations (numerically speaking) for every point simultaneously. Below we provide one such NumPy-heavy version of the Least Squares implementation shown previously that is far more efficient.

Note that in using these functions the input variable x (containing the entire set of P inputs) is of size $N \times P$, and its corresponding output y is of size $1 \times P$. Here we have written this code – and in particular the model function – to mirror its respective formula as close as possible.

```
1  # compute linear combination of input points
2  def model(x,w):
3      a = w[0] + np.dot(x.T,w[1:])
4      return a.T
5
6  # an implementation of the least squares cost function for linear
       regression
7  def least_squares(w):
8      # compute the least squares cost
9      cost = np.sum((model(x,w)  - y)**2)
10     return cost/float(y.size)
```

Notice too that for simplicity we write the Pythonic Least Squares cost function `least_squares(w)` instead of `least_squares(w,x,y)`, where in the latter case we explicitly list its other two arguments: the input `x` and output `y` data. This is done for notational simplicity – we do this with our math notation as well denoting our Least Squares cost $g(\mathbf{w})$ instead of $g(\mathbf{w}, \mathbf{x}, \mathbf{y})$ – and either format is perfectly fine practically speaking as `autograd` (see Section B.10) will correctly differentiate both forms (since by default it computes the gradient of a `Python` function with respect to its first input only). We will use this kind of simplified Pythonic notation when introducing future machine learning cost functions as well.

While we recommend most users employ the automatic differentiator library `autograd` (see Section 3.4) to perform both gradient descent and Newton's method on our machine learning cost functions, here one can (since this cost function is simple enough to) "hard code" the gradient formally by writing it out "by hand" (using the derivative rules detailed in Section B.3). Doing so one can compute the gradient of the Least Squares cost in closed form as

$$\nabla g(\mathbf{w}) = \frac{2}{P} \sum_{p=1}^{P} \mathring{\mathbf{x}}_p \left(\mathring{\mathbf{x}}_p^T \mathbf{w} - y_p \right). \tag{5.16}$$

Furthermore, in performing Newton's method one can also compute the Hessian of the Least Squares cost by hand. Moreover, since the cost is a convex quadratic *only a single Newton step can completely minimize it*. This single Newton step solution is often referred to as minimizing the Least Squares cost via its *normal equations*. The system of equations solved in taking this single Newton step is equivalent to the *first-order system* (see Section 3.2) for the Least Squares cost function

$$\left(\sum_{p=1}^{P} \mathring{\mathbf{x}}_p \mathring{\mathbf{x}}_p^T \right) \mathbf{w} = \sum_{p=1}^{P} \mathring{\mathbf{x}}_p y_p. \tag{5.17}$$

5.3   Least Absolute Deviations

In this section we discuss a slight twist on the derivation of the Least Squares cost function that leads to an alternative cost for linear regression called *Least Absolute Deviations*. This alternative cost function is much more robust to outliers in a dataset than the original Least Squares.

5.3.1   Susceptibility of Least Squares to outliers

One downside of using the squared error in the Least Squares cost (as a measure that we then minimize to recover optimal linear regression parameters) is that *squaring the error increases the importance of larger errors.* In particular, squaring errors of length greater than 1 makes these values considerably larger. This forces weights learned via the Least Squares cost to produce a linear fit that is especially focused on trying to minimize these large errors, sometimes due to *outliers* in a dataset. In other words, the Least Squares cost produces linear models that tend to *overfit* to outliers in a dataset. We illustrate this fact via a simple dataset in Example 5.3.

Example 5.3   Least Squares overfits to outliers
In this example we use the dataset plotted in Figure 5.6, which can largely be represented by a proper linear model with the exception of a single outlier, to show how the Least Squares cost function for linear regression tends to create linear models that overfit to outliers. We tune the parameters of a linear regressor to this dataset by minimizing the Least Squares cost via gradient descent (see Section 3.5), and plot the associated linear model on top of the data. This fit (shown in black) does not fit the majority of the data points well, bending upward clearly with the aim of minimizing the large squared error on the singleton outlier point.

Figure 5.6 Figure associated with Example 5.3. See text for details.

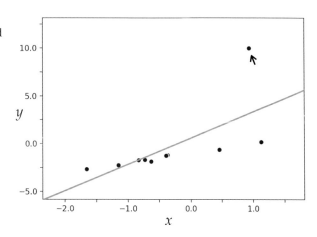

5.3.2   **Replacing squared error with absolute error**

Our original derivation of the Least Squares cost function in Section 5.2 aimed at learning a set of ideal weights so that we have

$$\mathring{\mathbf{x}}_p^T \mathbf{w} \approx y_p \quad p = 1, ..., P \tag{5.18}$$

for a dataset of P points $\left\{\left(\mathbf{x}_p, y_p\right)\right\}_{p=1}^{P}$. We then squared the difference between both sides of each desired approximation

$$g_p(\mathbf{w}) = \left(\mathring{\mathbf{x}}_p^T \mathbf{w} - y_p\right)^2 \quad p = 1, ..., P \tag{5.19}$$

and took the average of these P squared error terms to form the full Least Squares cost function.

As an alternative to using a *squared error* for our point-wise cost in Equation (5.19) we can instead measure the *absolute error* for each desired approximation

$$g_p(\mathbf{w}) = \left|\mathring{\mathbf{x}}_p^T \mathbf{w} - y_p\right| \quad p = 1, ..., P. \tag{5.20}$$

By using absolute error instead of the squared version we still treat negative and positive errors equally, but we do not exaggerate the importance of large errors greater than 1. Taking the average of these absolute error point-wise costs gives us the cousin of Least Squares, the so-called *Least Absolute Deviations* cost function

$$g(\mathbf{w}) = \frac{1}{P}\sum_{p=1}^{P} g_p(\mathbf{w}) = \frac{1}{P}\sum_{p=1}^{P}\left|\mathring{\mathbf{x}}_p^T \mathbf{w} - y_p\right|. \tag{5.21}$$

The only price we pay in employing the absolute error instead of the squared error is a technical one: while this cost function is also always convex regardless of the input dataset, since its second derivative is zero (almost everywhere) we can use only zero- and first-order methods to properly minimize it (and not second-order methods).

Example 5.4   Least Squares versus Least Absolute Deviations

In Figure 5.7 we compare the result of tuning a linear model by minimizing the Least Squares versus the Least Absolute Deviation cost functions employing the dataset in Example 5.3. In both cases we run gradient descent for the same number of steps and using the same choice of steplength parameter. We show the cost function histories for both runs in the right panel of Figure 5.7, where the runs using the Least Squares and Least Absolute Deviations are shown in black and magenta respectively. Examining the cost function histories we can see that the cost function value of the Least Absolute Deviations cost is considerably lower than that of Least Squares. This alone provides evidence that the former will provide a considerably better fit than Least Squares.

This advantage can also be seen in the left panel of Figure 5.7 where we plot and compare the best fit line provided by each run. The Least Squares fit is shown in black, while the Least Absolute Deviation fit is shown in magenta. The latter fits considerably better, since it does not exaggerate the large error produced by the single outlier.

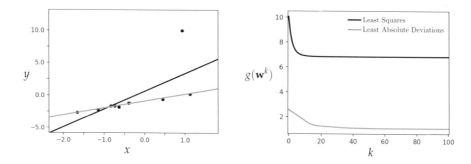

Figure 5.7 Figure associated with Example 5.4. See text for details.

5.4 Regression Quality Metrics

In this brief section we describe how to make predictions using a trained regression model followed by simple metrics for judging the quality of such a model.

5.4.1 Making predictions using a trained model

If we denote the optimal set of weights found by minimizing a regression cost function by $\mathbf{w}^{\star}$ then our fully trained linear model can be written as

$$\text{model}\,(\mathbf{x}, \mathbf{w}^{\star}) = \mathring{\mathbf{x}}^{T}\mathbf{w}^{\star} = w_0^{\star} + x_1 w_1^{\star} + x_2 w_2^{\star} + \cdots + x_N w_N^{\star}. \qquad (5.22)$$

Regardless of how we determine optimal parameters $\mathbf{w}^{\star}$, by minimizing a regression cost like the Least Squares or Least Absolute Deviations, we make predictions employing our linear model in the same way. That is, given an input $\mathbf{x}_0$ (whether it is from our training dataset or a brand new input) we predict its output y_0 by simply passing it along with our trained weights into our model as

$$\text{model}\,(\mathbf{x}_0, \mathbf{w}^{\star}) = y_0. \qquad (5.23)$$

This is illustrated pictorially on a prototypical linear regression dataset in Figure 5.8.

5.4.2 Judging the quality of a trained model

Once we have successfully minimized a linear regression cost function it is an easy matter to determine the quality of our regression model: we simply evaluate

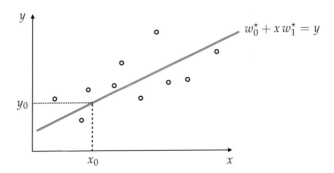

Figure 5.8 Once optimal parameters $w_0^\star$ and $w_1^\star$ of a regression line are found via minimizing an appropriate cost function they may be used to predict the output value of any input x_0 by substituting it into Equation (5.22). Here $N = 1$.

a cost function using our optimal weights. For example, we can evaluate the quality of a trained model using the Least Squares cost, which is especially natural to use when we employ this cost in training. To do this we plug in our learned model parameters along with the data into the Least Squares cost, giving the so-called Mean Squared Error (or MSE for short)

$$\text{MSE} = \frac{1}{P} \sum_{p=1}^{P} \left(\text{model} \left(\mathbf{x}_p, \mathbf{w}^\star \right) - y_p \right)^2. \tag{5.24}$$

The name for this regression quality metric describes precisely what the Least Squares cost computes, i.e., the average (or mean) squared error. To reduce the affect of outliers and other large values, often the *square root* of this value is taken as a regression quality metric. This is referred to as the *Root Mean Squared Error* or RMSE for short.

In the same way we can employ the Least Absolute Deviations cost to determine the quality of our trained model. Plugging in our learned model parameters along with the data into this cost computes the Mean Absolute Deviations (or MAD for short), which is precisely what this cost function computes

$$\text{MAD} = \frac{1}{P} \sum_{p=1}^{P} \left| \text{model} \left(\mathbf{x}_p, \mathbf{w}^\star \right) - y_p \right|. \tag{5.25}$$

These two metrics differ in precisely the ways we have seen their respective cost functions differ (e.g., the MSE measure is far more sensitive to outliers). In general, the *lower* one can make these quality metrics (by proper tuning of the model weights) the *better* the quality of the corresponding trained model, and vice versa. However, the threshold for what one considers "good" or "great"

performance can depend on personal preference, an occupational or institution-
ally set benchmark, or some other problem-dependent concern.

Example 5.5   Predicting house prices

As described in Example 1.4, linear regression has a myriad of business applications. Predicting the price of a given commodity is a particularly popular
application, and the so-called Boston Housing dataset [16] is a readily available
example of such a problem. This dataset consists of a set of basic statistics (features) and corresponding prices (in dollars) for $P = 506$ houses in the city of
Boston in the United States. The $N = 13$-dimensional input features include: per
capita crime rate by town (feature 1), average number of rooms per dwelling (feature 6), weighted distances to five Boston employment centers (feature 8), and
the percentage of the local population deemed "lower class," denoted LSTAT
(feature 13). One can easily attain reasonable RMSE and MAD metrics of around
4500 and 3000, respectively (see Exercise 5.9). Later on we drill deeper into this
dataset, examining its critical features in a process known as *feature selection* in
Examples 9.6 and 9.11.

Example 5.6   Predicting Automobile Miles-per-Gallon

As detailed earlier in Example 1.4, linear regression has a host of industrial
applications involved in accurately predicting the behavior of particular systems. The Autombile Miles-per-Gallon (or Auto-MPG for short) dataset [17] is
a popular dataset used for such purposes, and consists of a set of basic data on
$P = 398$ automobiles. The $N = 6$ input features are to be used to predict the
MPG of each car. The input features of this dataset include: number of cylinders
in the car's engine block (feature 1), the total engine displacement (feature 2),
the horsepower of the car's motor (feature 3), the weight of the car (feature
4), the car's acceleration ability measured in seconds taken to accelerate to a
benchmark speed from standstill (feature 5), and the year the car was produced
(feature 6).[3] One can attain reasonable RMSE and MAD metrics of around 3.3
and 2.5 MPG, respectively (see Exercise 5.9). Later on we drill deeper into this
data, examining its critical features in a process known as *feature selection* in
Exercise 9.10.

5.5   Weighted Regression

Because regression cost functions can be decomposed over individual data
points, we see in this section that it is possible to weight these points in or-

---

[3] The final feature of the original dataset (called "origin") was removed as no meaningful
description of it could be found.

der to emphasize or de-emphasize their importance to a regression model. This practice is called *weighted regression*.

5.5.1   Dealing with duplicates

Imagine we have a linear regression dataset that contains multiple copies of the same point, generated not by error but for example by necessary quantization (or binning) of input features in order to make human-in-the-loop analysis or modeling of the data easier. Needless to say, "duplicate" data points should not be thrown away in a situation like this.

Example 5.7   Quantization of input features can create duplicate points
In Figure 5.9 we show a raw set of data from a modern reenactment of Galileo's famous ramp experiment where, in order to quantify the effect of gravity, he repeatedly rolled a ball down a ramp to determine the relationship between distance and amount of time it takes an object to fall to the ground. This dataset consists of multiple trials of the same experiment, where each output's numerical value has been rounded to two decimal places. Performing this natural numerical rounding (sometimes referred to as *quantizing*) produces multiple duplicate data points, which we denote visually by scaling the dot representing each point in the image. The larger the dot's radius, the more duplicate points it represents.

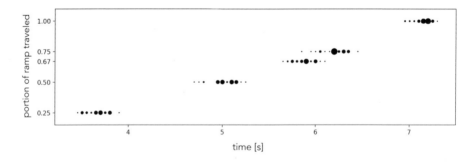

Figure 5.9 Figure associated with Example 5.7. See text for details.

Let us now examine what happens to a regression cost function (e.g., Least Squares) when a dataset contains duplicate data points. Specifically we assume there are β_p versions of the input/output pair $\left(\mathbf{x}_p,\, y_p\right)$ in our data. For regression datasets we have seen so far (excluding the one shown in Figure 5.9) we have always had $\beta_p = 1$ for all $p = 1,\ldots,P$. Using our `model` notation to represent our linear model (see, e.g., Section 5.4.1) we can write the sum of all point-wise squared errors as

$$\underbrace{\left(\text{model}\left(\mathbf{x}_1, \mathbf{w}\right) - y_1\right)^2 + \cdots + \left(\text{model}\left(\mathbf{x}_1, \mathbf{w}\right) - y_1\right)^2}_{\beta_1}$$

$$+ \underbrace{\left(\text{model}\left(\mathbf{x}_2, \mathbf{w}\right) - y_2\right)^2 + \cdots + \left(\text{model}\left(\mathbf{x}_2, \mathbf{w}\right) - y_2\right)^2}_{\beta_2}$$

$$\vdots$$

$$+ \underbrace{\left(\text{model}\left(\mathbf{x}_P, \mathbf{w}\right) - y_P\right)^2 + \cdots + \left(\text{model}\left(\mathbf{x}_P, \mathbf{w}\right) - y_P\right)^2}_{\beta_P}.$$

(5.26)

The natural grouping in Equation (5.26) then helps us write the overall Least Squares cost function as

$$g\left(\mathbf{w}\right) = \frac{1}{\beta_1 + \beta_2 + \cdots + \beta_P} \sum_{p=1}^{P} \beta_p \left(\text{model}\left(x_p, \mathbf{w}\right) - y_p\right)^2. \tag{5.27}$$

As we can see here the Least Squares cost function naturally *collapses* into a weighted version of itself in the sense that we can combine summands so that a repeated point in the dataset is represented in the cost function by a single weighted summand. Since the weights $\beta_1, \beta_2, ..., \beta_P$ are fixed for any given dataset, we can minimize a weighted regression cost precisely as we would any other (by tuning $\mathbf{w}$ alone). Finally notice that setting $\beta_p = 1$ (for all p) in Equation (5.27) gives us back the original (unweighted) Least Squares cost function in Equation (5.13).

5.5.2   Weighting points by confidence

Weighted regression can also be employed when we wish to weight each point based on our confidence in the *trustworthiness* of each data point. For example if our dataset came in two batches, one batch from a trustworthy source and another from a less trustworthy source (where some data points could be noisy or fallacious), we would want to weight data points from the trustworthy source more in our final regression. This can be done very easily using precisely the weighted regression paradigm introduced previously, only now we set the weights $\beta_1, \beta_2, ..., \beta_P$ ourselves based on our confidence of each point. If we believe that a point is very trustworthy we can set its corresponding weight β_p high, and vice versa. Notice in the extreme case, a weight value of $\beta_p = 0$ effectively removes its corresponding data point from the cost function, implying we do not trust that point at all.

Example 5.8   Adjusting a single data point's weight to reflect confidence
In Figure 5.10 we show how adjusting the weight associated with a single data point affects the final learned model in a weighted linear regression setting. The toy regression dataset shown in this figure includes a red data point whose diameter changes in proportion to its weight. As the weight (which can be interpret as "confidence") is increased from left to right, the regressor focuses more and more on representing the red point. If we increase its weight enough the fully trained regression model naturally starts fitting to this single data point alone while disregarding all other points, as illustrated in the rightmost panel of Figure 5.10.

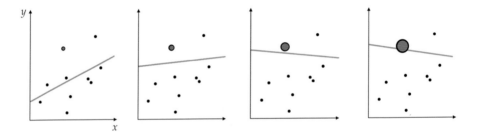

Figure 5.10 Figure associated with Example 5.8. See text for details.

5.6   Multi-Output Regression

Thus far we have assumed that data points for linear regression consist of *vector-valued* inputs and *scalar-valued* outputs. In other words, a prototypical regression data point takes the form of an input/output pair $\left(\mathbf{x}_p,\, y_p\right)$ where the input $\mathbf{x}_p$ is an N-dimensional vector, and the output y_p a scalar. While this configuration covers the vast majority of regression cases one may encounter in practice, it is possible to perform (linear) regression where *both* input and output are vector-valued. This is often called *multi-output regression*, which we now discuss.

5.6.1   Notation and modeling

Suppose our regression dataset consists of P input/output pairs

$$\left(\mathbf{x}_1, \mathbf{y}_1\right), \left(\mathbf{x}_2, \mathbf{y}_2\right), ..., \left(\mathbf{x}_P, \mathbf{y}_P\right) \tag{5.28}$$

where each input $\mathbf{x}_p$ is N-dimensional and each output $\mathbf{y}_p$ is C-dimensional. While in principle we can treat $\mathbf{y}_p$ as a $C \times 1$ column vector, in order to keep the formulae that follows looking similar to what we have already seen in the scalar case we will treat the input as an $N \times 1$ *column* vector and the output as a $1 \times C$ *row* vector, as[4]

$$\mathbf{x}_p = \begin{bmatrix} x_{1,p} \\ x_{2,p} \\ \vdots \\ x_{N,p} \end{bmatrix} \qquad \mathbf{y}_p = \begin{bmatrix} y_{0,p} & y_{1,p} & \cdots & y_{C-1,p} \end{bmatrix}. \qquad (5.29)$$

If we assume that a *linear* relationship holds between the input $\mathbf{x}_p$ and *just* the cth dimension of the output $y_{c,p}$, we are back to precisely the sort of regression framework we have seen thus far, and we can write

$$\mathring{\mathbf{x}}_p^T \mathbf{w}_c \approx y_{c,p} \qquad p = 1, ..., P \qquad (5.30)$$

where $\mathbf{w}_c$ is a set of weights

$$\mathbf{w}_c = \begin{bmatrix} w_{0,c} \\ w_{1,c} \\ w_{2,c} \\ \vdots \\ w_{N,c} \end{bmatrix} \qquad (5.31)$$

and $\mathring{\mathbf{x}}_p$ is the vector formed by stacking 1 on top of $\mathbf{x}_p$. If we then further assume that a linear relationship holds between the input and *all* C entries of the output, we can place each weight vector $\mathbf{w}_c$ into the cth column of an $(N+1) \times C$ weight *matrix* $\mathbf{W}$ as

$$\mathbf{W} = \begin{bmatrix} w_{0,0} & w_{0,1} & \cdots & w_{0,c} & \cdots & w_{0,C-1} \\ w_{1,0} & w_{1,1} & \cdots & w_{1,c} & \cdots & w_{1,C-1} \\ w_{2,0} & w_{2,1} & \cdots & w_{2,c} & \cdots & w_{2,C-1} \\ \vdots & \vdots & \cdots & \vdots & \cdots & \vdots \\ w_{N,0} & w_{N,1} & \cdots & w_{N,c} & \cdots & w_{N,C-1} \end{bmatrix} \qquad (5.32)$$

and write the entire set of C linear models via a vector-matrix product

$$\mathring{\mathbf{x}}_p^T \mathbf{W} = \begin{bmatrix} \mathring{\mathbf{x}}_p^T \mathbf{w}_0 & \mathring{\mathbf{x}}_p^T \mathbf{w}_1 & \cdots & \mathring{\mathbf{x}}_p^T \mathbf{w}_{C-1} \end{bmatrix}. \qquad (5.33)$$

This allows us to write the entire set of C linear relationships very compactly as

---

[4] Notice that unlike the input we index the output starting from 0. We do this because eventually we will stack a 1 on top of each input $\mathbf{x}_p$ (as we did with standard regression in Section 5.2) and this entry will have the zeroth index of our input.

$$\mathring{\mathbf{x}}_p^T \mathbf{W} \approx \mathbf{y}_p \qquad p = 1, ..., P. \qquad (5.34)$$

5.6.2   Cost functions

The thought process involved in deriving a regression cost function for the case of multi-output regression mirrors almost exactly the scalar-output case discussed in Sections 5.2 and 5.3. For example, to derive a Least Squares cost function we begin (in the same way we did in Section 5.2) by taking the difference of both sides of Equation (5.34). However, the error associated with the pth point, written as $\mathring{\mathbf{x}}_p^T \mathbf{W} - \mathbf{y}_p$, is now a vector of C values. To square this error we must therefore employ the *squared ℓ_2 vector norm* (see Section C.5 if not familiar with this vector norm). The Least Squares cost function in this case is then the average squared ℓ_2 norm of each point's error, written as

$$g(\mathbf{W}) = \frac{1}{P} \sum_{p=1}^{P} \left\| \mathring{\mathbf{x}}_p^T \mathbf{W} - \mathbf{y}_p \right\|_2^2 = \frac{1}{P} \sum_{p=1}^{P} \sum_{c=0}^{C-1} \left(\mathring{\mathbf{x}}_p^T \mathbf{w}_c - y_{c,p} \right)^2. \qquad (5.35)$$

Notice that when $C = 1$, this reduces to the original Least Squares cost we saw in Section 5.2.

Likewise, the Least Absolute Deviations cost (which measures the absolute value of each error as opposed to its square) for our present case takes the analogous form

$$g(\mathbf{W}) = \frac{1}{P} \sum_{p=1}^{P} \left\| \mathring{\mathbf{x}}_p^T \mathbf{W} - \mathbf{y}_p \right\|_1 = \frac{1}{P} \sum_{p=1}^{P} \sum_{c=0}^{C-1} \left| \mathring{\mathbf{x}}_p^T \mathbf{w}_c - y_{c,p} \right| \qquad (5.36)$$

where $\|\cdot\|_1$ is the ℓ_1 vector norm, the generalization of the absolute value function for vectors (see Section C.5.1 if not familiar with this vector norm).

Just like their scalar-valued versions, these cost functions are always convex regardless of the dataset used. They also decompose over the weights $\mathbf{w}_c$ associated with each output dimension. For example, we can rewrite the right-hand side of the Least Absolute Deviations cost in Equation (5.36) by swapping the summands over P and C, giving

$$g(\mathbf{W}) = \sum_{c=0}^{C-1} \left(\frac{1}{P} \sum_{p=1}^{P} \left| \mathring{\mathbf{x}}_p^T \mathbf{w}_c - y_{c,p} \right| \right) = \sum_{c=0}^{C-1} g_c(\mathbf{w}_c) \qquad (5.37)$$

where we have denoted $g_c(\mathbf{w}_c) = \frac{1}{P} \sum_{p=1}^{P} \left| \mathring{\mathbf{x}}_p^T \mathbf{w}_c - y_{c,p} \right|$. Since the weights from each of the C subproblems do not interact we can, if desired, minimize each g_c for an optimal setting of $\mathbf{w}_c$ independently, and then take their sum to form the full cost function g.

Example 5.9   Fitting a linear model to a multi-output regression dataset

In Figure 5.11 we show an example of multi-output linear regression using a toy dataset with input dimension $N = 2$ and output dimension $C = 2$, where we have plotted the input and *one* output value in each of the two panels of the figure.

   We tune the parameters of an appropriate linear model via minimizing the Least Squares cost using gradient descent, and illustrate the fully trained model (shown in green in each panel) by evaluating a fine mesh of points in the input region of the dataset.

Figure 5.11
Figure associated with Example 5.9. See text for details.

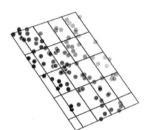

5.6.3   Python implementation

Because `Python` and `NumPy` have such flexible syntax, we can implement the linear model

$$\text{model}(\mathbf{x}, \mathbf{W}) = \mathring{\mathbf{x}}^T \mathbf{W} \tag{5.38}$$

precisely as we did in the scalar-output case in Section 5.2.4. In *implementing* this linear combination we need not form the adjusted input $\mathring{\mathbf{x}}_p$ (by tacking a 1 on top of the raw input $\mathbf{x}_p$) and can more easily compute the linear combination by exposing the biases as

$$\mathring{\mathbf{x}}_p^T \mathbf{W} = \mathbf{b} + \mathbf{x}_p^T \mathcal{W} \tag{5.39}$$

where we denote the bias $\mathbf{b}$ and the feature-touching weights $\mathcal{W}$ as

$$\mathbf{b} = \begin{bmatrix} w_{0,0} \\ w_{0,1} \\ w_{0,2} \\ \vdots \\ w_{0,C-1} \end{bmatrix} \qquad \mathcal{W} = \begin{bmatrix} w_{0,1} & w_{0,2} & \cdots & w_{0,C-1} \\ w_{1,1} & w_{1,2} & \cdots & w_{1,C-1} \\ w_{2,1} & w_{2,2} & \cdots & w_{2,C-1} \\ \vdots & \vdots & \vdots & \vdots \\ w_{N,1} & w_{N,2} & \cdots & w_{N,C-1} \end{bmatrix}. \tag{5.40}$$

This notation is used to match the Pythonic slicing operation (as shown in the implementation given below), which we implement in Python analogously as follows.

$$a = w[0] + np.dot(x\_p.T,w[1:])$$

That is, $\mathbf{b} = w[0]$ denotes the bias and $\mathcal{W} = w[1:]$ denotes the remaining feature-touching weights. Another reason to implement in this way is that the particular linear combination $\mathbf{x}_p^T\mathcal{W}$ – implemented the model using np.dot as np.dot(x_p.T,w[1:]) below – is an especially efficient since NumPy's np.dot operation is far more efficient than constructing a linear combination in Python via an explicit for loop.

Multi-output regression cost functions can also be implemented in Python precisely as we have seen previously. For example, our linear model and Least Squares cost can be written as shown below.

```
1   # linear model
2   def model(x,w):
3       a = w[0] + np.dot(x.T,w[1:])
4       return a.T
5
6   # least squares cost
7   def least_squares(w):
8       cost = np.sum((model(x,w) - y)**2)
9       return cost/float(np.size(y))
```

Note that since any cost for multi-output regression takes in a *matrix* of parameters, when using autograd as part of your optimization process it can be very convenient to first *flatten* your chosen cost – as explained in Section B.10.3 – prior to minimization. Doing this avoids the need to explicitly loop over weights in your local optimization routine, allowing you to directly employ the basic Python implementations of, e.g., gradient descent (see Section 3.5.4) and Newton's method (see Section 4.3.6) without modification.

5.7   Conclusion

We began Part II of the text in this chapter by describing linear regression, the simplest of our *supervised learning* problems.

More specifically, we began in Section 5.2 by introducing important notation, the formal linear model, as well as the *Least Squares* cost function for regression. In Section 5.3 we then introduced the *Least Absolute Deviations* cost, which is far less sensitive to outliers, but at the cost of not being twice differentiable (thus second-order methods are not directly applicable to its minimization). Having already described methods of mathematical optimization in Part I of the text we quickly deal with the minimization of these cost functions for linear regression. Next in Section 5.5 we described *weighted regression*, a twist on the standard scheme that allows for complete control over how each point is emphasized during regression. Finally in Section 5.6 we discussed various metrics for quantifying the quality of a trained regression model.

5.8   Exercises

† The data required to complete the following exercises can be downloaded from the text's github repository at `github.com/jermwatt/machine_learning_refined`

5.1   Fitting a regression line to the student debt data

Fit a linear model to the US student load debt dataset shown in Figure 1.8 by minimizing the associated linear regression Least Squares problem using a single Newton step (also known as solving the normal equations). If this linear trend continues what will the total student debt be in 2030?

5.2   Kleiber's law and linear regression

After collecting and plotting a considerable amount of data comparing the body mass versus metabolic rate (a measure of at rest energy expenditure) of a variety of animals, early twentieth-century biologist Max Kleiber noted an interesting relationship between the two values. Denoting by x_p and y_p the body mass (in kg) and metabolic rate (in kJ/day) of a given animal respectively, treating the body mass as the input feature Kleiber noted (by visual inspection) that the natural log of these two values were linearly related. That is

$$w_0 + \log\left(x_p\right)w_1 \approx \log\left(y_p\right). \tag{5.41}$$

In Figure 1.9 we show a large collection of transformed data points

$$\left\{\left(\log\left(x_p\right), \log\left(y_p\right)\right)\right\}_{p=1}^{P} \tag{5.42}$$

each representing an animal ranging from a small black-chinned hummingbird in the bottom-left corner to a large walrus in the top-right corner.

(a) Fit a linear model to the data shown in Figure 1.9.

(b) Use the optimal parameters you found in part (a) along with the properties of the log function to write the nonlinear relationship between the body mass x and the metabolic rate y.

(c) Use your fitted line to determine how many calories an animal weighing 10 kg requires (note each calorie is equivalent to 4.18 J).

5.3   The Least Squares cost function and a single Newton step

As mentioned in Section 5.2.3, a single Newton step can perfectly minimize the Least Squares cost for linear regression. Use a single Newton step to perfectly minimize the Least Squares cost over the dataset shown in Figure 5.12. This dataset roughly lies on a hyperplane, thus the fit provided by a perfectly minimized Least Squares cost should fit very well. Use a cost function history plot to check that you have tuned the parameters of your linear model properly.

Figure 5.12 Figure associated with Exercise 5.3. See text for further details.

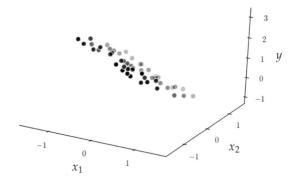

5.4   Solving the normal equations

As discussed in Section 5.2.4, the Least Squares cost function for linear regression can be perfectly minimized using a single Newton's method step (see Section 4.3) – with the corresponding system of equations shown in Equation (5.17) also referred to as the *normal equations*. In what circumstancs do you think this is *not* an excellent solution method for minimizing the Least Squares cost? Why? *Hint: see Section 4.4.2.*

5.5   Lipschitz constant for the Least Squares cost

Compute the Lipschitz constant (see Section A.6.4) of the Least Squares cost function.

5.6   Compare the Least Squares and Least Absolute Deviation costs

Repeat the experiment outlined in Example 5.4. You will need to implement the Least Absolute Deviations cost, which can be done similarly to the Least Squares implementation in Section 5.2.4.

5.7   Empirically confirm convexity for a toy dataset

Empirically confirm that the Least Absolute Deviations cost function is convex for the toy dataset shown in Section 5.3.

5.8   **The Least Absolute Deviations cost is convex**
Prove that the Least Absolute Deviations cost is convex using the zero-order definition of convexity given below.

An unconstrained function g is convex if and only if any line segment connecting two points on the graph of g lies *above* its graph. Figure 5.13 illustrates this definition of a convex function.

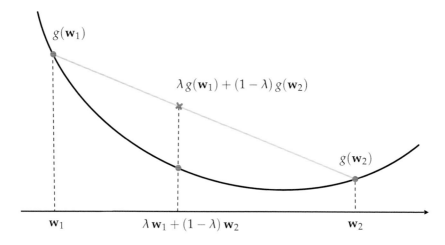

Figure 5.13 Figure associated with Exercise 5.8.

Stating this geometric fact algebraically, g is convex if and only if for all $\mathbf{w}_1$ and $\mathbf{w}_2$ in the domain of g and all $\lambda \in [0, 1]$, we have

$$g \left(\lambda \mathbf{w}_1 + (1 - \lambda) \mathbf{w}_2 \right) \leq \lambda g \left(\mathbf{w}_1 \right) + (1 - \lambda) g \left(\mathbf{w}_2 \right). \tag{5.43}$$

5.9   **Housing price and Automobile Miles-per-Gallon prediction**
Verify the quality metrics given in Examples 5.5 and 5.6 for the Boston Housing and Automobile Miles-per-Gallon datasets. Because of the large variations in the input values of these datasets you should *standard normalize* the input features of each – as detailed in Section 9.3 – prior to optimization.

5.10   **Improper tuning and weighted regression**
Suppose someone proposed tuning the point-wise weights $\beta_1, ..., \beta_P$ by minimizing the weighted Least Squares cost in Equation (5.27) with respect to both $\beta_1, ..., \beta_P$ and $\mathbf{w}$, the weights of the linear model, while keeping $\beta_1, ..., \beta_P$ all non-negative (thus the smallest value taken by the cost function is zero). Suppose you are able to optimize to completion – what would go wrong in terms of the efficacy of the final model on a general dataset? Draw a picture to explain your ideas.

5.11   **Multi-output regression**
Repeat the experiment outlined in Example 5.9. You can use the implementation described in Section 5.6.3 as a base for your implementation.

5.9   Endnotes

5.9.1   Proof that the Least Squares cost function is always convex

Here we show that the Least Squares cost function for linear regression is always a convex quadratic. Examining just the pth summand of the Least Squares cost, we have

$$\left(\mathring{\mathbf{x}}_p^T \mathbf{w} - y_p\right)^2 = \left(\mathring{\mathbf{x}}_p^T \mathbf{w} - y_p\right)\left(\mathring{\mathbf{x}}_p^T \mathbf{w} - y_p\right) = y_p^2 - 2\mathring{\mathbf{x}}_p^T \mathbf{w} y_p + \mathring{\mathbf{x}}_p^T \mathbf{w} \, \mathring{\mathbf{x}}_p^T \mathbf{w} \tag{5.44}$$

where we have arranged the terms in increasing order of degree.

Now, since $\mathring{\mathbf{x}}_p^T \mathbf{w} = \mathbf{w}^T \mathring{\mathbf{x}}_p$ we can switch around the first inner-product in the third term on the right, giving equivalently

$$y_p^2 - 2\mathring{\mathbf{x}}_p^T \mathbf{w} y_p + \mathbf{w}^T \mathring{\mathbf{x}}_p \mathring{\mathbf{x}}_p^T \mathbf{w}. \tag{5.45}$$

This is only the pth summand. Summing over all the points gives analogously

$$\begin{aligned} g(\mathbf{w}) &= \frac{1}{P} \sum_{p=1}^{P} \left(y_p^2 - 2\mathring{\mathbf{x}}_p^T \mathbf{w} y_p + \mathbf{w}^T \mathring{\mathbf{x}}_p \mathring{\mathbf{x}}_p^T \mathbf{w}\right) \\ &= \frac{1}{P} \sum_{p=1}^{P} y_p^2 - \frac{2}{P} \sum_{p=1}^{P} y_p \mathring{\mathbf{x}}_p^T \mathbf{w} + \frac{1}{P} \sum_{p=1}^{P} \mathbf{w}^T \mathring{\mathbf{x}}_p \mathring{\mathbf{x}}_p^T \mathbf{w}. \end{aligned} \tag{5.46}$$

From here we can spot that indeed the Least Squares cost function is a quadratic, since denoting

$$\begin{aligned} a &= \frac{1}{P} \sum_{p=1}^{P} y_p^2 \\ \mathbf{b} &= -\frac{2}{P} \sum_{p=1}^{P} \mathring{\mathbf{x}}_p y_p \\ \mathbf{C} &= \frac{1}{P} \sum_{p=1}^{P} \mathring{\mathbf{x}}_p \mathring{\mathbf{x}}_p^T \end{aligned} \tag{5.47}$$

we can write the Least Squares cost equivalently as

$$g(\mathbf{w}) = a + \mathbf{b}^T \mathbf{w} + \mathbf{w}^T \mathbf{C} \mathbf{w}. \tag{5.48}$$

Furthermore, because the matrix $\mathbf{C}$ is constructed from a sum of *outer-product* matrices it is also convex, since the eigenvalues of such a matrix are always nonnegative (see Sections 4.1 and 4.2 for further details about convex quadratic functions of this form).

6 Linear Two-Class Classification

6.1 Introduction

In this chapter we discuss *linear two-class classification*, another kind of *supervised learning* problem (as introduced in Section 1.3.1). At the outset the difference between classification and regression (detailed in the previous chapter) is quite subtle: two-class (or binary) classification is the name we give to a supervised learning problem when the output of a dataset takes on only two *discrete* values, often referred to as two *classes*. Many popular machine learning problems fall into this category, including face detection (the two classes here include face and nonface objects) and object detection in general, text-based sentiment analysis (the two classes here consist of written product reviews ascribing a positive or negative opinion), automatic diagnosis of medical conditions (the two classes in this case refer to medical data corresponding to patients who either do or do not have a specific malady), and more.

This subtle difference is important, and spurs the development of new cost functions that are better-suited to deal with such data. These new cost functions are formulated based on a wide array of motivating perspectives including *logistic regression*, *Perceptron*, and *Support Vector Machines*, which we discuss in detail. While these perspectives widely differ on the surface they all (as we will see) reduce to virtually the same essential principle for two-class classification. As in the prior chapter, here we will also look at metrics that help us determine the quality of a trained classifier, as well as extensions on the basic concept (including categorical and weighted two-class classification).

6.2 Logistic Regression and the Cross Entropy Cost

In this section we describe a fundamental framework for linear two-class classification referred to as *logistic regression* employing the so-called *Cross Entropy* cost function.

6.2.1   **Notation and modeling**

Two-class classification is a particular instance of regression wherein the data still comes in the form of P input/output pairs $\left\{\left(\mathbf{x}_p, y_p\right)\right\}_{p=1}^{P}$, and each input $\mathbf{x}_p$ is an N-dimensional vector. However, the corresponding output y_p is no longer continuous but takes on only two discrete numbers. While the actual value of these numbers is in principle arbitrary, particular value pairs are more helpful than others for derivation purposes. In this section we suppose that the output of our data takes on either the value 0 or +1, i.e., $y_p \in \{0, +1\}$. Often in the context of classification the output values y_p are called *labels*, and all points sharing the same label value are referred to as a *class* of data. Hence a dataset containing points with label values $y_p \in \{0, +1\}$ is said to be a dataset consisting of two classes.

The simplest way such a dataset can be distributed is on a set of adjacent *steps* as illustrated in the top two panels of Figure 6.1 for $N = 1$ (on the left) and $N = 2$ (on the right). Here the *bottom step* is the region of the space containing class 0, i.e., all of the points that have label value $y_p = 0$. Likewise the *top step* contains class 1, i.e., all of the points having label value $y_p = +1$. From this perspective, the problem of two-class classification can be naturally viewed as a case of *nonlinear* regression where our goal is to regress (or fit) a nonlinear step function to the data. We call this the *regression perspective* on classification.

Alternatively we can change perspective and view the dataset directly from "above" where we can imagine looking down on the data from a point high up on the y axis. In other words, we look at the data as if it is projected onto the $y = 0$ plane. From this perspective, which is illustrated in the bottom panels of Figure 6.1, we remove the vertical y dimension of the data and visually represent the dataset using its input only, displaying the output values of each point by coloring the points one of two unique colors: we use blue for points with label $y_p = 0$, and red for those having label $y_p = +1$. From this second perspective which we call the *perceptron perspective* and illustrate in the bottom two panels of Figure 6.1, the edge separating the two steps (and consequently the data points on them) when projected onto the input space, takes the form of a single point when $N = 1$ (as illustrated in the bottom-left panel) and a *line* when $N = 2$ (as illustrated in the bottom-right panel). For general N what separates the two classes of data will be a *hyperplane*,[1] which is also called a *decision boundary* in the context of classification.

In the current section and the one that follows we focus solely on the regression perspective on two-class classification. We come back to the Perceptron perspective again in Section 6.4.

---

[1] A point and a line are special low-dimensional instances of a hyperplane.

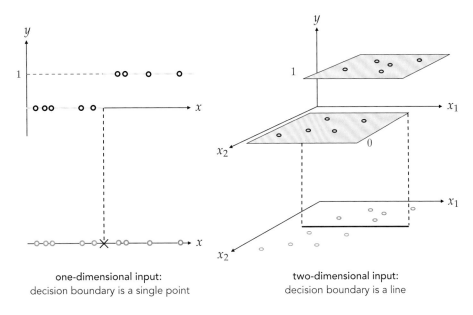

one-dimensional input:
decision boundary is a single point

two-dimensional input:
decision boundary is a line

Figure 6.1 Two perspectives on classification illustrated using single-input (left column) and two-input (right column) toy datasets. The regression perspective shown in the top panels is equivalent to the perceptron perspective shown in the bottom panels, where we look at each respective dataset from "above." In the Perceptron perspective we also mark the decision boundary. This is where the step function (colored in yellow in the top panels) transitions from its bottom to top step. See text for further details.

6.2.2 Fitting a discontinuous step function

Adopting the regression perspective on two-class classification, we might be tempted at first to simply apply the linear regression framework described in Chapter 5 to fit such data. We do exactly this in Example 6.1.

Example 6.1   Fitting a linear regressor to classification data

In Figure 6.2 we show a simple two-class dataset where we have fit a line to this dataset via linear regression (shown in green). The line itself provides a poor representation of this data since its output takes on just two discrete values. Even when we pass this fully tuned linear regressor through a discrete step function by assigning the label +1 to all output values greater than 0.5 and the label 0 to all output values less than 0.5, the resulting step function (shown in dashed red in the figure) still provides a less than ideal representation for the data. This is because the parameters of the (green) line were tuned first (before passing the resulting model through the step) causing the final step model to fail to properly identify two of the points on the top step. In the parlance of classification these types of points are referred to as *misclassified points* or *misclassifications* for short.

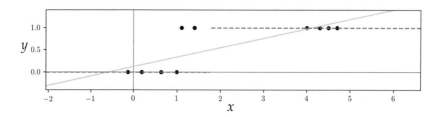

Figure 6.2 Figure associated with Example 6.1. See text for details.

Example 6.1 hints at the fact that using linear regression outright to represent classification data is a poor option. Even after passing the resulting linear model through a step function the result still did not capture the true step function on which the data of that example lay. Instead of tuning the parameters of a linear model (performing linear regression) and then passing the result through a step function, we can in principle do a better job if we tune the parameters of the linear model *after* passing it through a step function.

To describe this sort of regression more formally, first recall our notation (introduced in Section 5.2) for denoting a linear model of N-dimensional input

$$\mathring{x}^T \mathbf{w} = w_0 + x_1 w_1 + x_2 w_2 + \cdots + x_N w_N \tag{6.1}$$

where

$$\mathbf{w} = \begin{bmatrix} w_0 \\ w_1 \\ w_2 \\ \vdots \\ w_N \end{bmatrix} \quad \text{and} \quad \mathring{x} = \begin{bmatrix} 1 \\ x_1 \\ x_2 \\ \vdots \\ x_N \end{bmatrix}. \tag{6.2}$$

Next, let us denote a *step function* algebraically as[2]

$$\text{step}(t) = \begin{cases} 1 & \text{if } t > 0 \\ 0 & \text{if } t < 0 \end{cases} \tag{6.3}$$

Shoving our linear model in Equation (6.1) through this gives us a step function[3]

$$\text{step}\left(\mathring{x}^T \mathbf{w}\right) \tag{6.4}$$

---

[2] What happens with step (0) is, for our purposes, arbitrary. It can be set to any fixed value or left undefined as we have done here.

[3] Technically, step $\left(\mathring{x}^T \mathbf{w} - 0.5\right)$ is the function that maps output values $\mathring{x}^T \mathbf{w}$ greater (smaller) than 0.5 to 1 (0). However, we can fuse the constant -0.5 into the bias weight w_0 by rewriting it as $w_0 \longleftarrow w_0 - 0.5$ (after all it is a parameter that must be learned) so that the step function can be expressed more compactly, as is done in Equation (6.4).

with a *linear decision boundary* between its lower and upper steps, defined by all points $\mathring{x}$ where $\mathring{x}^T \mathbf{w} = 0$. Any input lying *exactly* on the decision boundary can be assigned a label at random.

To tune the weight vector $\mathbf{w}$ properly we can (once again, as with linear regression in Chapter 5) set up a Least Squares cost function by reflecting on the sort of ideal relationship we want to find between the input and output of our dataset. Ideally, we want the point $(\mathbf{x}_p, y_p)$ to lie on the correct side of the optimal decision boundary or, in other words, the output y_p to lie on the proper step. Expressed algebraically, this ideal desire can be written as

$$\text{step}\left(\mathring{\mathbf{x}}_p^T \mathbf{w}\right) = y_p \qquad p = 1, ..., P. \tag{6.5}$$

To find weights that satisfy this set of P equalities we form a Least Squares cost function by squaring the difference between both sides of each equality, and taking their average

$$g(\mathbf{w}) = \frac{1}{P} \sum_{p=1}^{P} \left(\text{step}\left(\mathring{\mathbf{x}}^T \mathbf{w}\right) - y_p\right)^2. \tag{6.6}$$

Our ideal weights then correspond to a minimizer of this cost function.

Unfortunately it is very difficult (if not impossible) to properly minimize this Least Squares cost using local optimization, as at virtually every point the function is *completely flat* locally (see Example 6.2). This problem, which is inherited from our use of the step function, renders both gradient descent and Newton's method ineffective, since both methods immediately halt at flat areas of a cost function.

Example 6.2   Visual inspection of classification cost functions
In the left panel of Figure 6.3 we plot the Least Squares cost function in Equation (6.6) for the dataset shown previously in Figure 6.2, over a wide range of values of its two parameters w_0 and w_1. This Least Squares surface consists of discrete steps at many different levels, and each step is *completely flat*. Because of this local optimization methods (like those detailed in Chapters 2–4) cannot not be used to effectively minimize it.

In the middle and right panels of Figure 6.3 we plot the surfaces of two related cost functions over the same dataset. In the following we introduce the cost function shown in the middle panel in Figure 6.3, and the cost in the right panel in Figure 6.3. Both are far superior, in terms of our ability to properly minimize them, than the step-based Least Squares cost function shown on the left.

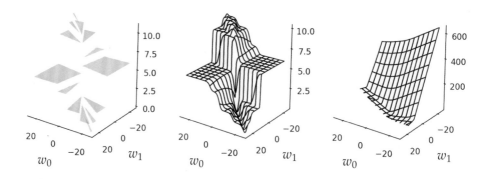

Figure 6.3 Figure associated with Example 6.2. See text for details.

6.2.3   The logistic sigmoid function

To make the minimization of the Least Squares cost possible we can replace the step function in Equation (6.6) with a *continuous* approximation that matches it closely. The *logistic sigmoid function*

$$\sigma(x) = \frac{1}{1 + e^{-x}} \tag{6.7}$$

is such an approximation. In Figure 6.4 we plot this function (left panel) as well as several internally weighted versions of it (right panel). As we can see for the correct setting of internal weights the logistic sigmoid can be made to look arbitrarily similar to the step function.

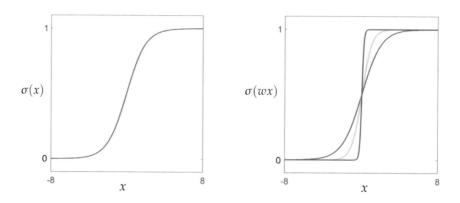

Figure 6.4 (left panel) Plot of the sigmoid function $\sigma(x)$. (right panel) By increasing the weight w in $\sigma(wx)$ from $w = 1$ (shown in red) to $w = 2$ (shown in green) and finally to $w = 10$ (shown in blue), the internally weighted version of the sigmoid function becomes an increasingly good approximator of the step function.

6.2.4   Logistic regression using the Least Squares cost

Swapping out the step function with its sigmoid approximation in Equation (6.5) gives the related set of *approximate* equalities we desire to hold

$$\sigma\left(\mathring{\mathbf{x}}_p^T\mathbf{w}\right)\approx y_p \qquad p=1,...,P \qquad (6.8)$$

as well as the corresponding Least Squares cost function

$$g\left(\mathbf{w}\right)=\frac{1}{P}\sum_{p=1}^{P}\left(\sigma\left(\mathring{\mathbf{x}}^T\mathbf{w}\right)-y_p\right)^2. \qquad (6.9)$$

Fitting a logistic sigmoid to classification data by minimizing this cost function is often referred to as performing *logistic regression*.[4] While the resulting cost function is generally nonconvex, it can be properly minimized nonetheless using a host of local optimization techniques.

Example 6.3   Using normalized gradient descent
In Figure 6.5 we show how normalized gradient descent (a slight variation on the standard gradient descent algorithm outlined in Chapter 3, see Appendix Section A.3 for further details) can be used to minimize the Least Squares cost in Equation (6.9) over the dataset first shown in Figure 6.2. Here a of run normalized gradient descent was performed, initialized at the point $w_0=-w_1=20$. Plotted in the left panel is the sigmoidal fit provided by properly minimizing the Least Squares cost. In the right panel a contour plot of the cost function is shown with the (normalized) gradient descent path colored green to red as the run progresses towards the cost function's minimizer. A surface plot of this cost function (in three dimensions) is shown in the middle panel of Figure 6.3. Although this cost function is very flat in many places, normalized gradient descent is designed specifically to deal with costs like this (see Section A.3).

6.2.5   Logistic regression using the Cross Entropy cost

The *squared error* point-wise cost $g_p\left(\mathbf{w}\right)=\left(\sigma\left(\mathring{\mathbf{x}}_p^T\mathbf{w}\right)-y_p\right)^2$ that we average over all P points in the data to form the Least Squares cost in Equation (6.9) is universally defined, regardless of the values taken by the output by y_p. However, because we *know* that the output we deal with in a two-class classification setting is limited to the *discrete* values $y_p\in\{0,1\}$, it is reasonable to wonder if we can create a more appropriate cost customized to deal with just such values.

One such point-wise cost, which we refer to as the *log error*, is defined as follows

[4] Because we are essentially performing *regression* using a *logistic* function.

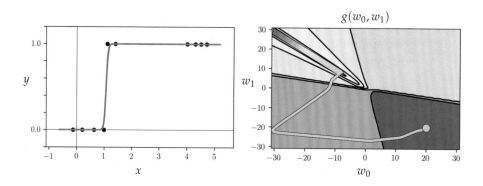

Figure 6.5 Figure associated with Example 6.3. See text for details.

$$g_p(\mathbf{w}) = \begin{cases} -\log\left(\sigma\left(\mathring{\mathbf{x}}_p^T \mathbf{w}\right)\right) & \text{if } y_p = 1 \\ -\log\left(1 - \sigma\left(\mathring{\mathbf{x}}_p^T \mathbf{w}\right)\right) & \text{if } y_p = 0. \end{cases} \tag{6.10}$$

First, notice that this point-wise cost is always nonnegative (regardless of the input and weight values) with a minimum[5] value of 0.

Secondly, notice how this *log error* cost penalizes violations of our desired (approximate) equalities in Equation (6.8) much more harshly than a *squared error* does, as can be seen in Figure 6.6 where both are plotted for comparison.

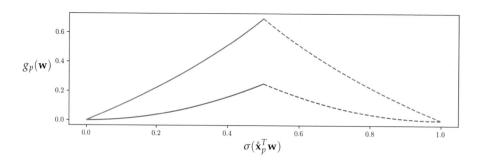

Figure 6.6 Visual comparison of the squared error (in blue) and the log error (in red) for two cases: $y_p = 0$ (solid curves) and $y_p = 1$ (dashed curves). In both cases the log error penalizes deviation from the true label value to a greater extent than the squared error.

Finally, notice that since our label values $y_p \in \{0, 1\}$ we can write the *log error* in Equation (6.10) equivalently in a single line as

$$g_p(\mathbf{w}) = -y_p \log\left(\sigma\left(\mathring{\mathbf{x}}_p^T \mathbf{w}\right)\right) - \left(1 - y_p\right) \log\left(1 - \sigma\left(\mathring{\mathbf{x}}_p^T \mathbf{w}\right)\right). \tag{6.11}$$

---

[5] Technically, an infimum.

This equivalent form allows us to write the overall cost function (formed by taking the average of the point-wise costs over all P data points) as

$$g(\mathbf{w}) = -\frac{1}{P} \sum_{p=1}^{P} y_p \log \left(\sigma \left(\mathring{\mathbf{x}}_p^T \mathbf{w} \right) \right) + \left(1 - y_p \right) \log \left(1 - \sigma \left(\mathring{\mathbf{x}}_p^T \mathbf{w} \right) \right). \tag{6.12}$$

This is referred to as the *Cross Entropy cost* for logistic regression.

6.2.6 Minimizing the Cross Entropy cost

The right panel of Figure 6.3 shows the surface of the Cross Entropy cost taken over the dataset shown in Figure 6.2. That the plotted surface looks convex is not accidental. Indeed (unlike Least Squares) the Cross Entropy cost is always convex regardless of the dataset used (see chapter's exercises). This means that a wider variety of local optimization schemes can be used to properly minimize it (compared to the generally nonconvex sigmoid-based Least Squares cost) including standard gradient descent schemes (see Section 3.5) and second-order Newton's methods (see Section 4.3). For this reason the Cross Entropy cost is very often used in practice to perform logistic regression.

Example 6.4   Minimizing the Cross Entropy logistic regression
In this example we repeat the experiments of Example 6.3 using (instead) the Cross Entropy cost and standard gradient descent, initialized at the point $w_0 = 3$ and $w_1 = 3$. The left panel of Figure 6.7 shows the sigmoidal fit provided by properly minimizing the Cross Entropy cost. In the right panel a contour plot of the cost function is shown with the gradient descent path colored green to red as the run progresses towards the cost's minimizer. A surface plot of this cost function is shown in the right panel of Figure 6.3.

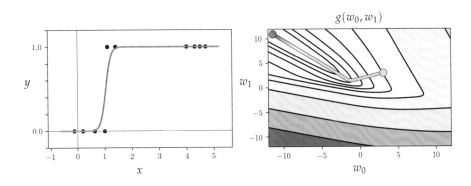

Figure 6.7 Figure associated with Example 6.4. See text for details.

6.2.7   **Python implementation**

We can implement the Cross Entropy costs very similarly to the way we did the Least Squares cost for linear regression (see Section 5.2.4) breaking down our implementation into a linear model and the error function itself. Our linear model takes in both an appended input point $\mathring{x}_p$ and a set of weights $\mathbf{w}$

$$\text{model}\left(\mathbf{x}_p, \mathbf{w}\right) = \mathring{\mathbf{x}}_p^T \mathbf{w} \tag{6.13}$$

which we can still implement as shown below.

```
1  # compute linear combination of input point
2  def model(x,w):
3      a = w[0] + np.dot(x.T,w[1:])
4      return a.T
```

We can then implement the Cross Entropy cost using the log error in Equation (6.10) as shown below.

```
1   # define sigmoid function
2   def sigmoid(t):
3       return 1/(1 + np.exp(-t))
4
5   # the convex cross-entropy cost function
6   def cross_entropy(w):
7       # compute sigmoid of model
8       a = sigmoid(model(x,w))
9
10      # compute cost of label 0 points
11      ind = np.argwhere(y == 0)[:,1]
12      cost = -np.sum(np.log(1 - a[:,ind]))
13
14      # add cost on label 1 points
15      ind = np.argwhere(y==1)[:,1]
16      cost -= np.sum(np.log(a[:,ind]))
17
18      # compute cross-entropy
19      return cost/y.size
```

To minimize this cost we can use virtually any local optimization method detailed in Chapters 2–4. For first- and second-order methods (e.g., gradient descent and Newton's method schemes) an autograd (see Section 3.4) can be used to automatically compute its gradient and Hessian.

Alternatively, one can indeed compute the gradient and Hessian of the Cross Entropy cost in closed form, and implement them directly. Using the simple derivative rules outlined in Section B.3 the gradient can be computed as

$$\nabla g\left(\mathbf{w}\right) = -\frac{1}{P}\sum_{p=1}^{P}\left(y_p - \sigma\left(\mathring{\mathbf{x}}_p^T \mathbf{w}\right)\right)\mathring{\mathbf{x}}_p. \tag{6.14}$$

In addition to employing Newton's method "by hand" one can hand compute the Hessian of the Cross Entropy function as

$$\nabla^2 g\left(\mathbf{w}\right) = \frac{1}{P}\sum_{p=1}^{P}\sigma\left(\mathring{\mathbf{x}}_p^T \mathbf{w}\right)\left(1 - \sigma\left(\mathring{\mathbf{x}}_p^T \mathbf{w}\right)\right)\mathring{\mathbf{x}}_p\mathring{\mathbf{x}}_p^T. \tag{6.15}$$

6.3 Logistic Regression and the Softmax Cost

In the previous section we saw how to derive logistic regression when employing label values $y_p \in \{0, 1\}$. However, as mentioned earlier, these label values are arbitrary, and one can derive logistic regression using a different set of label values, e.g., $y_p \in \{-1, +1\}$. In this brief section we do just this, tracing out entirely similar steps to what we have seen previously, resulting in new cost function called the *Softmax cost* for logistic regression. While the Softmax differs in form algebraically, it is in fact equivalent to the Cross Entropy cost. However, conceptually speaking the Softmax cost is considerably valuable, since it allows us to sew together the many different perspectives on two-class classification into a single coherent idea, as we will see in the sections that follow.

6.3.1 Different labels, same story

If we change the label values from $y_p \in \{0, 1\}$ to $y_p \in \{-1, +1\}$ much of the story we saw unfold previously unfolds here as well. That is, instead of our data ideally sitting on a step function with lower and upper steps taking on the values 0 and 1 respectively, they take on values −1 and +1 as shown in Figure 6.8 for prototypical cases where $N = 1$ (left panel) and $N = 2$ (right panel).

This particular step function is called a *sign function*, since it returns the numerical *sign* of its input

$$\text{sign}(x) = \begin{cases} +1 & \text{if } x > 0 \\ -1 & \text{if } x < 0 \end{cases}. \tag{6.16}$$

Shoving a *linear* model through the sign function gives us a step function

$$\text{sign}\left(\mathring{\mathbf{x}}^T \mathbf{w}\right) \tag{6.17}$$

with a *linear* decision boundary between its two steps defined by all points $\mathring{\mathbf{x}}$ where $\mathring{\mathbf{x}}^T \mathbf{w} = 0$. Any input lying *exactly* on the decision boundary can be

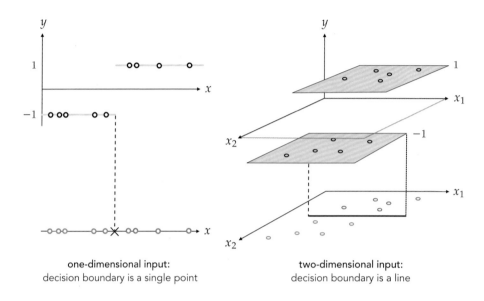

one-dimensional input:
decision boundary is a single point

two-dimensional input:
decision boundary is a line

Figure 6.8 The analogous setup to Figure 6.1, only here we use label values $y_p \in \{-1, +1\}$.

assigned a label at random. A point is classified correctly when its true label is predicted correctly, that is when $\text{sign}\left(\mathring{\mathbf{x}}_p^T \mathbf{w}\right) = y_p$. Otherwise, it is said to have been *mislcassified*.

As when using label values $y_p \in \{0, 1\}$, we can again attempt to form a Least Squares cost using the sign function. However, just as with the original step-based Least Squares in Equation (6.6) this too would be completely flat almost everywhere, and therefore extremely difficult to minimize properly.

Akin to what we did previously, we can look to replace the discontinuous $\text{sign}(\cdot)$ function with a *smooth approximation*: a slightly adjusted version of the *logistic sigmoid* function so that its values range between -1 and 1 (instead of 0 and 1). This scaled version of the sigmoid, called the *hyperbolic tangent function*, is written as

$$\tanh(x) = 2\,\sigma\,(x) - 1 = \frac{2}{1 + e^{-x}} - 1. \tag{6.18}$$

Given that the sigmoid function $\sigma\,(\cdot)$ ranges smoothly between 0 and 1, it is easy to see why $\tanh(\cdot)$ ranges smoothly between -1 and $+1$.

Analogous to the Least Squares cost in Equation (6.9), we can form a Least Squares cost for recovering optimal model weights using the $\tanh(\cdot)$ function

$$g(\mathbf{w}) = \frac{1}{P} \sum_{p=1}^{P} \left(\tanh\left(\mathring{\mathbf{x}}_p^T \mathbf{w}\right) - y_p\right)^2 \tag{6.19}$$

which is likewise nonconvex with undesirable flat regions, requiring specialized local methods for its proper minimization (see Example 6.3).

As with 0/1 labels here too we can employ the point-wise *log error* cost

$$
g_p(\mathbf{w}) = \begin{cases} -\log\left(\sigma\left(\mathring{\mathbf{x}}_p^T \mathbf{w}\right)\right) & \text{if } y_p = +1 \\ -\log\left(1 - \sigma\left(\mathring{\mathbf{x}}_p^T \mathbf{w}\right)\right) & \text{if } y_p = -1 \end{cases} \tag{6.20}
$$

which we can then use to form the so-called *Softmax cost* for logistic regression

$$
g(\mathbf{w}) = \frac{1}{P} \sum_{p=1}^{P} g_p(\mathbf{w}). \tag{6.21}
$$

As with the Cross Entropy cost it is far more common to express the Softmax cost differently by re-writing the log error in a equivalent way as follows. First, notice that because $1 - \sigma(x) = \sigma(-x)$ the point-wise cost in Equation (6.20) can be rewritten equivalently as $-\log\left(\sigma\left(-\mathring{\mathbf{x}}^T \mathbf{w}\right)\right)$ and so the point-wise cost function can be written as

$$
g_p(\mathbf{w}) = \begin{cases} -\log\left(\sigma\left(\mathring{\mathbf{x}}_p^T \mathbf{w}\right)\right) & \text{if } y_p = +1 \\ -\log\left(\sigma\left(-\mathring{\mathbf{x}}_p^T \mathbf{w}\right)\right) & \text{if } y_p = -1. \end{cases} \tag{6.22}
$$

Now notice, because of the particular choice of label values we are using here, i.e., $y_p \in \{-1, +1\}$, that we can move the label value in each case *inside* the inner most parentheses, and write both cases in a single line as

$$
g_p(\mathbf{w}) = -\log\left(\sigma\left(y_p \mathring{\mathbf{x}}_p^T \mathbf{w}\right)\right). \tag{6.23}
$$

Finally, since $-\log(x) = \log\left(\frac{1}{x}\right)$, we can write the point-wise cost in Equation (6.23) equivalently (using the definition of the sigmoid) as

$$
g_p(\mathbf{w}) = \log\left(1 + e^{-y_p \mathring{\mathbf{x}}_p^T \mathbf{w}}\right). \tag{6.24}
$$

Substituting this form of the point-wise log error function into Equation (6.21) we have a more common appearance of the *Softmax cost* for logistic regression

$$
g(\mathbf{w}) = \frac{1}{P} \sum_{p=1}^{P} \log\left(1 + e^{-y_p \mathring{\mathbf{x}}_p^T \mathbf{w}}\right). \tag{6.25}
$$

This cost function, like the Cross Entropy cost detailed in the previous Section, is always convex regardless of the dataset used (see chapter's exercises). Moreover, as we can see here by its derivation, the Softmax and Cross Entropy cost functions are completely equivalent (upon change of label value $y_p = -1$ to $y_p = 0$ and vice versa), having been built using the same point-wise log error cost function.

Example 6.5 **Minimizing Softmax logistic regression using standard gradient descent**

In this example we repeat the experiments of Example 6.4, swapping out labels $y_p = 0$ with $y_p = -1$, to form the Softmax cost and use gradient descent (with the same initial point, steplength parameter, and number of iterations) for its minimization. The results are shown in Figure 6.9.

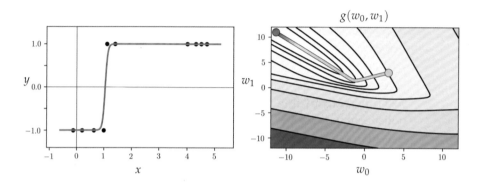

Figure 6.9 Figure associated with Example 6.5. See text for details.

6.3.2   Python **implementation**

If we express the Softmax cost using the log error as in Equation (6.21), then we can implement it almost entirely the same way we did with the Cross Entropy cost as shown in Section 6.2.7.

To implement the Softmax cost as shown in Equation (6.25) we first implement the linear model, which takes in both an appended input point $\mathring{x}_p$ and a set of weights $\mathbf{w}$ as

$$\text{model}\left(\mathbf{x}_p, \mathbf{w}\right) = \mathring{\mathbf{x}}_p^T \mathbf{w} . \tag{6.26}$$

With this notation for our model, the corresponding Softmax cost can be written as

$$g(\mathbf{w}) = \frac{1}{P} \sum_{p=1}^{P} \log\left(1 + e^{-y_p \text{model}\left(\mathbf{x}_p, \mathbf{w}\right)}\right).$$

We can then implement the cost in chunks – first the model function below precisely as we did with linear regression (see Section 5.2.4).

```
1  # compute linear combination of input point
2  def model(x,w):
3      a = w[0] + np.dot(x.T,w[1:])
4      return a.T
```

We can then implement the Softmax cost as shown below.

```
1  # the convex softmax cost function
2  def softmax(w):
3      cost = np.sum(np.log(1 + np.exp(-y*model(x,w))))
4      return cost/float(np.size(y))
```

As alternative to using an automatic differentiator (which we use by default – employing autograd (see Section 3.4), one can perform gradient descent and Newton's method here by hand computing the gradient and Hessian of the Softmax cost function. Using the simple derivative rules outlined in Section B.3 gradient can be computed as [6]

$$\nabla g(\mathbf{w}) = -\frac{1}{P}\sum_{p=1}^{P}\frac{e^{-y_p\mathring{\mathbf{x}}_p^T\mathbf{w}}}{1+e^{-y_p\mathring{\mathbf{x}}_p^T\mathbf{w}}}y_p\mathring{\mathbf{x}}_p. \tag{6.27}$$

In addition to employ Newton's method "by hand" one can hand compute the Hessian of the Softmax function as

$$\nabla^2 g(\mathbf{w}) = \frac{1}{P}\sum_{p=1}^{P}\left(\frac{1}{1+e^{y_p\mathring{\mathbf{x}}_p^T\mathbf{w}}}\right)\left(1-\frac{1}{1+e^{y_p\mathring{\mathbf{x}}_p^T\mathbf{w}}}\right)\mathring{\mathbf{x}}_p\mathring{\mathbf{x}}_p^T. \tag{6.28}$$

6.3.3   Noisy classification datasets

The discrete nature of the output in classification makes the concepts of *noise* and *noisy* data different in linear classification than we saw previously with linear regression in Chapter 5. As described there, in the case of linear regression, noise causes the data to *not* fall precisely on a single line (or hyperplane in higher dimensions). With linear two-class classification, noise manifests itself in our *inability* to find a single line (or hyperplane in higher dimensions) to separate the two classes of data. Figure 6.10 shows such a noisy classification dataset consisting of $P = 100$ points, whose two classes can be separated by a

---

[6] One can express the gradient algebraically in several ways. However, writing the gradient in this way helps avoid numerical problems associated with using the exponential function on a modern computer. This is due to the exponential "overflowing" with large exponents, e.g., e^{1000}, as these numbers are too large to store explicitly on the computer and so are represented symbolically as ∞. This becomes a problem when evaluating $\frac{e^{1000}}{1+e^{1000}}$ which, although basically equal to the value 1, is thought of by the computer to be a NaN (not a number).

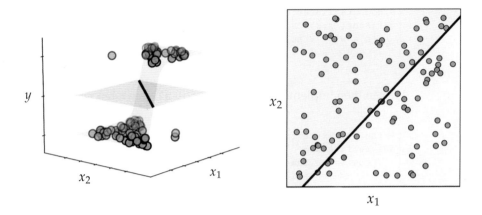

Figure 6.10 A two-class classification dataset viewed from the regression perspective on the left and from the perceptron perspective on the right, with three *noisy* data points pointed to by small arrows. See text for further details.

line but not perfectly. The left panel of this figure shows the data in three dimensions (viewed from a regression perspective) along with the trained classifier: a three-dimensional hyperbolic tangent function. The right panel shows the same dataset in two dimensions (viewed from a perceptron perspective) along with the learned linear decision boundary. Here the two half-spaces created by the decision boundary are also colored (light blue and light red) according to the class confined within each. As you can see, there are three points in this case (two blue points and one red point) that look like they are on the *wrong side* of our classifier. Such *noisy* points are often misclassified by a trained classifier, meaning that their true label value will not be correctly predicted. Two-class classification datasets typically have noise of this kind, and thus are not often perfectly separable by a linear decision boundary.

6.4   The Perceptron

As we have seen with logistic regression in the previous section, we treat classification as a particular form of nonlinear regression (employing – with the choice of label values $y_p \in \{-1, +1\}$ – a tanh nonlinearity). This results in the learning of a proper nonlinear regressor, and a corresponding *linear decision boundary*

$$\mathring{x}^T w = 0. \tag{6.29}$$

Instead of learning this decision boundary as a result of a nonlinear regression, the *Perceptron* derivation described in this section aims at determining this ideal linear decision boundary directly. While we will see how this direct approach leads back to the *Softmax cost function*, and that practically speaking

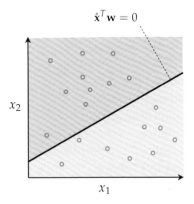
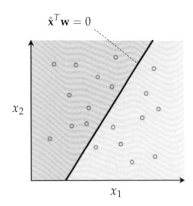

Figure 6.11 With the Perceptron we aim to directly learn the linear decision boundary $\mathring{\mathbf{x}}^T \mathbf{w} = 0$ (shown here in black) to separate two classes of data, colored red (class +1) and blue (class −1), by dividing the input space into a red half-space where $\mathring{\mathbf{x}}^T \mathbf{w} > 0$, and a blue half-space where $\mathring{\mathbf{x}}^T \mathbf{w} < 0$. (left panel) A linearly separable dataset where it is possible to learn a hyperplane to perfectly separate the two classes. (right panel) A dataset with two overlapping classes. Although the distribution of data does not allow for perfect linear separation, the Perceptron still aims to find a hyperplane that minimizes the number of misclassified points that end up in the wrong half-space.

the Perceptron and logistic regression often result in learning *the same linear decision boundary*, the Perceptron's focus on learning the decision boundary directly provides a valuable new perspective on the process of two-class classification. In particular – as we will see here – the Perceptron provides a simple geometric context for introducing the important concept of *regularization* (an idea we will see arise in various forms throughout the remainder of the text).

6.4.1 The Perceptron cost function

As we saw in the previous section with our discussion of logistic regression (where our output/label values $y_p \in \{-1, +1\}$), in the simplest instance our two classes of data are largely separated by a *linear decision boundary* given by the collection of input $\mathbf{x}$ where $\mathring{\mathbf{x}}^T \mathbf{w} = 0$, with each class (mostly) lying on either side. This decision boundary is a *point* when the dimension of the input is $N = 1$, a *line* when $N = 2$, and is more generally for arbitrary N a *hyperplane* defined in the input space of a dataset.

This classification scenario can be best visualized in the case $N = 2$, where we view the problem of classification "from above" – showing the input of a dataset colored to denote class membership. The default coloring scheme we use here – matching the scheme used in the previous section – is to color points with label $y_p = -1$ blue and $y_p = +1$ red. The linear decision boundary is here a line that best separates points from the $y_p = -1$ class from those of the $y_p = +1$ class, as shown for a prototypical dataset in Figure 6.11.

The linear decision boundary cuts the input space into two *half-spaces*, one lying "above" the hyperplane where $\mathring{x}^T\mathbf{w} > 0$, and one lying "below" it where $\mathring{x}^T\mathbf{w} < 0$. Notice then, as depicted visually in Figure 6.11, that a proper set of weights $\mathbf{w}$ define a linear decision boundary that separates a two-class dataset as well as possible with as many members of one class as possible lying above it, and likewise as many members as possible of the other class lying below it. Because we can always flip the orientation of an ideal hyperplane by multiplying it by -1 (or likewise because we can always swap our two label values) we can say in general that when the weights of a hyperplane are tuned properly members of the class $y_p = +1$ lie (mostly) "above" it, while members of the $y_p = -1$ class lie (mostly) "below" it. In other words, our *desired* set of weights define a hyperplane where as often as possible we have that

$$\begin{aligned} \mathring{x}_p^T\mathbf{w} > 0 \qquad &\text{if} \quad y_p = +1 \\ \mathring{x}_p^T\mathbf{w} < 0 \qquad &\text{if} \quad y_p = -1. \end{aligned} \qquad (6.30)$$

Because of our choice of label values we can consolidate the ideal conditions above into the single equation below

$$-y_p\mathring{x}_p^T\mathbf{w} < 0. \qquad (6.31)$$

Again we can do so specifically because we chose the label values $y_p \in \{-1, +1\}$. Likewise by taking the maximum of this quantity and zero we can then write this ideal condition, which states that a hyperplane correctly classifies the point x_p, equivalently forming a *point-wise cost*

$$g_p(\mathbf{w}) = \max\left(0, -y_p\mathring{x}_p^T\mathbf{w}\right) = 0. \qquad (6.32)$$

Note that the expression $\max\left(0, -y_p\mathring{x}_p^T\mathbf{w}\right)$ is always nonnegative, since it returns zero if x_p is classified correctly, and returns a *positive value* if the point is classified incorrectly. The functional form of this point-wise cost $\max(0, \cdot)$ is often called a *rectified linear unit* for historical reasons (see Section 13.3). Because these point-wise costs are nonnegative and equal to *zero* when our weights are tuned correctly, we can take their average over the entire dataset to form a proper cost function as

$$g(\mathbf{w}) = \frac{1}{P}\sum_{p=1}^{P}\max\left(0, -y_p\mathring{x}_p^T\mathbf{w}\right). \qquad (6.33)$$

When minimized appropriately this cost function can be used to recover the ideal weights satisfying the desired equations above as often as possible.

6.4.2   Minimizing the Perceptron cost

This cost function goes by many names such as the *Perceptron* cost, the *rectified linear unit* cost (or ReLU cost for short), and the *hinge cost* (since when plotted a ReLU function looks like a hinge – see Figure 6.12). This cost function is *always convex* but only has a single (discontinuous) derivative in each input dimension. This implies that we can only use zero- and first-order local optimization schemes (but not Newton's method). Note that the Perceptron cost *always* has a trivial solution at $\mathbf{w} = \mathbf{0}$, since indeed $g(\mathbf{0}) = 0$, thus one may need to take care in practice to avoid finding it (or a point too close to it) accidentally.

6.4.3   The Softmax approximation to the Perceptron

Here we describe a common approach to ameliorating the optimization issues detailed above concerning the Perceptron cost. Somewhat akin to our replacement of the discrete step function with a smooth approximating sigmoid function in previous sections, here we replace the max function portion of the Perceptron cost with a smooth (or at least twice differentiable) alternative that closely matches it everywhere. We do this via the *Softmax* function defined as

$$\text{soft}(s_0, s_1, ..., s_{C-1}) = \log(e^{s_0} + e^{s_1} + \cdots + e^{s_{C-1}}) \tag{6.34}$$

where $s_0, s_1, ..., s_{C-1}$ are any C scalar values – which is a generic smooth approximation to the *max* function, i.e.,

$$\text{soft}(s_0, s_1, ..., s_{C-1}) \approx \max(s_0, s_1, ..., s_{C-1}). \tag{6.35}$$

To see why the Softmax approximates the max function let us look at the simple case when $C = 2$. Suppose momentarily that $s_0 \leq s_1$, so that $\max(s_0, s_1) = s_1$. Therefore $\max(s_0, s_1)$ can be written as $\max(s_0, s_1) = s_0 + (s_1 - s_0)$, or equivalently as $\max(s_0, s_1) = \log(e^{s_0}) + \log(e^{s_1 - s_0})$. Written in this way we can see that $\log(e^{s_0}) + \log(1 + e^{s_1 - s_0}) = \log(e^{s_0} + e^{s_1}) = \text{soft}(s_0, s_1)$ is always larger than $\max(s_0, s_1)$ but not by much, especially when $e^{s_1 - s_0} \gg 1$. Since the same argument can be made if $s_0 \geq s_1$ we can say generally that $\text{soft}(s_0, s_1) \approx \max(s_0, s_1)$. The more general case follows similarly as well.

Returning to the Perceptron cost function in Equation (6.33), we replace the pth summand with its Softmax approximation making our point-wise cost

$$g_p(\mathbf{w}) = \text{soft}\left(0, -y_p \mathring{\mathbf{x}}_p^T \mathbf{w}\right) = \log\left(e^0 + e^{-y_p \mathring{\mathbf{x}}_p^T \mathbf{w}}\right) = \log\left(1 + e^{-y_p \mathring{\mathbf{x}}_p^T \mathbf{w}}\right) \tag{6.36}$$

giving the overall cost function as

$$g(\mathbf{w}) = \sum_{p=1}^{P} \log\left(1 + e^{-y_p \mathring{\mathbf{x}}_p^T \mathbf{w}}\right) \tag{6.37}$$

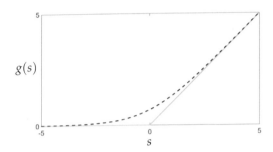

Figure 6.12 Plots of the Perceptron $g(s) = \max(0, s)$ (shown in green) as well as its smooth Softmax approximation $g(s) = \text{soft}(0, s) = \log(1 + e^s)$ (shown in dashed black).

which is the *Softmax cost* we saw previously derived from the logistic regression perspective on two-class classification. This is why the cost function is called *Softmax*, since it derives from the general Softmax approximation to the max function.

Note that *like* the Perceptron cost – as we already know – the Softmax cost is convex. However, *unlike* the Perceptron cost, the Softmax cost has infinitely many derivatives and Newton's method can therefore be used to minimize it. Moreover, it does not have a trivial solution at zero like the Perceptron cost does. Nonetheless, the fact that the Softmax cost so closely approximates the Perceptron shows just how closely aligned – in the end – both logistic regression and the Perceptron perspectives on classification truly are. Practically speaking their differences lie in how well – for a particular dataset – one can optimize either cost function, along with (what is very often slight) differences in the quality of each cost function's learned decision boundary. Of course when the Softmax is employed from the Perceptron perspective there is no qualitative difference between the Perceptron and logistic regression at all.

6.4.4 The Softmax cost and linearly separable datasets

Imagine that we have a dataset whose two classes can be perfectly separated by a hyperplane, and that we have chosen an appropriate cost function to minimize it in order to determine proper weights for our model. Imagine further that we are *extremely lucky* and our *initialization* $\mathbf{w}^0$ produces a linear decision boundary $\mathring{\mathbf{x}}^T \mathbf{w}^0 = 0$ with *perfect separation*. This means, according to Equation (6.31), that for each of our P points we have that $-y_p \mathring{\mathbf{x}}_p^T \mathbf{w}^0 < 0$ and likewise that the point-wise Perceptron cost in Equation (6.33) is zero for every point, i.e.,

$$g_p\left(\mathbf{w}^0\right) = \max\left(0, -y_p \mathring{\mathbf{x}}_p^T \mathbf{w}^0\right) = 0$$ and so the Perceptron cost in Equation 6.33 is *exactly equal to zero*.

Since the Perceptron cost value is already zero, its lowest value, any local optimization algorithm will halt immediately (that is, we would never take a single optimization step). However, this will *not* be the case if we used the same initialization but employed the Softmax cost instead of the Perceptron.

Since we always have that $e^{-y_p \mathring{x}_p^T w^0} > 0$, the Softmax point-wise cost is always nonnegative $g_p(w^0) = \log\left(1 + e^{-y_p \mathring{x}_p^T w^0}\right) > 0$ and hence too the Softmax cost. This means that in applying any local optimization scheme like, e.g., gradient descent we will indeed take steps away from the initialization w^0 in order to drive the value of the Softmax cost lower and lower towards its minimum at zero. In fact – with data that is indeed linearly separable – the Softmax cost achieves this lower bound *only when the magnitude of the weights grows to infinity*. This is clear from the fact that each individual term $\log\left(1 + e^{-C}\right) = 0$ only as $C \longrightarrow \infty$. Indeed if we multiply our initialization w^0 by any constant $C > 1$ we can *decrease* the value of any negative exponential involving one of our data points since $e^{-C} < 1$ and so $e^{-y_p \mathring{x}_p^T C w^0} = e^{-C} e^{-y_p \mathring{x}_p^T w^0} < e^{-y_p \mathring{x}_p^T w^0}$.

This likewise decreases the Softmax cost as well with the minimum achieved only as $C \longrightarrow \infty$. However, importantly, regardless of the scalar $C > 1$ value involved the decision boundary defined by the initial weights $\mathring{x}^T w^0 = 0$ *does not change location*, since we still have that $C \mathring{x}^T w^0 = 0$ (indeed this is true for any nonzero scalar C). So even though the location of the separating hyperplane need not change, with the Softmax cost we still take more and more steps in minimization since (in the case of perfectly linearly separable data) its minimum lies off at infinity. This fact can cause severe numerical instability issues with local optimization schemes that make *large progress* at each step – particularly Newton's method (see Section 4.3) – since they will tend to rapidly diverge to infinity.[7]

Example 6.6   Perfectly separable data and the Softmax cost
In applying Newton's method to minimize the Softmax cost over perfectly linearly separable data, it is easy to run into numerical instability issues as the global minimum of the cost technically lies at infinity. Here we examine a simple instance of this behavior using the single input dataset shown Figure 6.9. In the top row of Figure 6.13 we illustrate the progress of five Newton steps in beginning at the point $w = \begin{bmatrix} 1 \\ 1 \end{bmatrix}$. Within five steps we have reached a point providing a very good fit to the data (in the top-left panel of the figure we plot the tanh $(\cdot)$ fit using the logistic regression perspective on the Softmax cost), and one that is already quite large in magnitude (as can be seen in the top-right panel of the figure, where the contour plot of the cost function is shown). We can see here by the trajectory of the steps in the right panel, which are traveling linearly towards the minimum out at $\begin{bmatrix} -\infty \\ \infty \end{bmatrix}$, that the location of

---

[7] Notice: because the Softmax and Cross Entropy costs are equivalent (as discussed in the previous section), this issue equally presents itself when using the Cross Entropy cost as well.

the linear decision boundary (here a point) is not changing after the first step or two. In other words, after the first few steps each subsequent step is simply multiplying its predecessor by a scalar value $C > 1$.

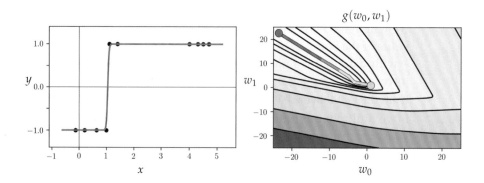

Figure 6.13 (top row) Figure associated with Example 6.6. See text for details.

Notice that if we simply flip one of the labels – making this dataset not perfectly linearly separable – the corresponding cost function does not have a global minimum out at infinity, as illustrated in the contour plot shown in the bottom row of Figure 6.13.

6.4.5   **Normalizing feature-touching weights**

How can we prevent this potential problem – of the weights learned via local optimization (and Newton's method in particular) shooting off to infinity – when employing the Softmax/Cross Entropy cost with perfectly linearly separable data? One simple approach would be simply to employ our local optimization schemes more carefully – by taking fewer steps and/or halting optimization if the magnitude of the weights grows larger than a large predefined constant. Another approach is to control the magnitude of the weights during the optimization procedure itself. Both approaches are generally referred to in the jargon of machine learning as *regularization strategies* (described in great detail in Section 11.6). The former strategy (a form of what is called *early stopping*) is straightforward, requiring slight adjustments to the way we have typically employed local optimization, but the latter approach requires some further explanation which we now provide.

To control the magnitude of **w** means that we want to control the size of the $N + 1$ individual weights it contains

$$\mathbf{w} = \begin{bmatrix} w_0 \\ w_1 \\ \vdots \\ w_N \end{bmatrix}. \tag{6.38}$$

We can do this by *directly* controlling the size of just N of these weights, and it is particularly convenient to do so using the final N feature-touching weights $w_1, w_2, ..., w_N$ because these define the *normal vector* to the linear decision boundary $\mathring{\mathbf{x}}^T \mathbf{w} = 0$. To more easily introduce the geometric concepts that follow we will use our bias/feature weight notation for $\mathbf{w}$ first introduced in Section 5.2.4. This provides us with individual notation for the bias and feature-touching weights as

$$\text{(bias): } b = w_0 \quad \text{(feature-touching weights): } \quad \omega = \begin{bmatrix} w_1 \\ w_2 \\ \vdots \\ w_N \end{bmatrix}. \tag{6.39}$$

With this notation we can express a linear decision boundary as

$$\mathring{\mathbf{x}}^T \mathbf{w} = b + \mathbf{x}^T \omega = 0. \tag{6.40}$$

To see why this notation is useful, first note how, geometrically speaking, the feature-touching weights ω define the *normal vector of the linear decision boundary*. The *normal vector* to a hyperplane (like our decision boundary) is always *perpendicular* to it, as illustrated in Figure 6.14. We can always compute the error – also called the signed distance – of a point $\mathbf{x}_p$ to a linear decision boundary in terms of the normal vector ω.

To see how this is possible, imagine we have a point $\mathbf{x}_p$ lying "above" the linear decision boundary on a translate of the decision boundary where $b + \mathbf{x}^T \omega = \beta > 0$, as illustrated in the Figure 6.14 (the same simple argument that follows can be made if $\mathbf{x}_p$ lies "below" it as well). To compute the distance of $\mathbf{x}_p$ to the decision boundary imagine we know the location of its *vertical projection* onto the decision boundary, which will call $\mathbf{x}'_p$. To compute our desired error we want to compute the signed distance between $\mathbf{x}_p$ and its vertical projection, i.e., the length of the vector $\mathbf{x}'_p - \mathbf{x}_p$ times the sign of β, which here is $+1$ since we assume the point lies above the decision boundary, i.e., $d = \left\| \mathbf{x}'_p - \mathbf{x}_p \right\|_2 \text{sign}(\beta) = \left\| \mathbf{x}'_p - \mathbf{x}_p \right\|_2$. Now, because *this vector is also perpendicular* to the decision boundary (and so is *parallel* to the normal vector ω) the *inner-product rule* (see Section C.2) gives

$$\left(\mathbf{x}'_p - \mathbf{x}_p \right)^T \omega = \left\| \mathbf{x}'_p - \mathbf{x}_p \right\|_2 \left\| \omega \right\|_2 = d \left\| \omega \right\|_2. \tag{6.41}$$

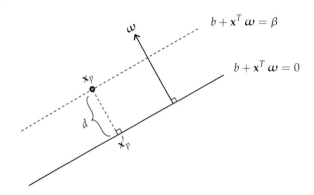

Figure 6.14 A linear decision boundary written as $b + \mathbf{x}^T \boldsymbol{\omega} = 0$ has a normal vector $\boldsymbol{\omega}$ defined by its feature-touching weights. To compute the signed distance of a point $\mathbf{x}_p$ to the boundary we mark the translation of the decision boundary passing through this point as $b + \mathbf{x}^T \boldsymbol{\omega} = \beta$, and the projection of the point onto the decision boundary as $\mathbf{x}'_p$.

Now if we take the difference between our decision boundary and its translation evaluated at $\mathbf{x}'_p$ and $\mathbf{x}_p$ respectively, we have simplifying

$$\beta - 0 = \left(b + \left(\mathbf{x}'_p\right)^T \boldsymbol{\omega}\right) - \left(b + \mathbf{x}_p^T \boldsymbol{\omega}\right) = \left(\mathbf{x}'_p - \mathbf{x}_p\right)^T \boldsymbol{\omega}. \tag{6.42}$$

Since both formulae are equal to $\left(\mathbf{x}'_p - \mathbf{x}_p\right)^T \boldsymbol{\omega}$ we can set them equal to each other, which gives

$$d \left\| \boldsymbol{\omega} \right\|_2 = \beta \tag{6.43}$$

or in other words that the signed distance d of $\mathbf{x}_p$ to the decision boundary is

$$d = \frac{\beta}{\left\| \boldsymbol{\omega} \right\|_2} = \frac{b + \mathbf{x}_p^T \boldsymbol{\omega}}{\left\| \boldsymbol{\omega} \right\|_2}. \tag{6.44}$$

Note that we need not worry dividing by zero here since if the feature-touching weights $\boldsymbol{\omega}$ were all zero, this would imply that the bias $b = 0$ as well and we have no decision boundary at all. Also notice, this analysis implies that if the feature-touching weights have *unit length* as $\|\boldsymbol{\omega}\|_2 = 1$ then the signed distance d of a point $\mathbf{x}_p$ to the decision boundary is given *simply by its evaluation* $b + \mathbf{x}_p^T \boldsymbol{\omega}$. Finally note that if $\mathbf{x}_p$ were to lie below the decision boundary and $\beta < 0$ nothing about the final formulae derived above will change.

We mark this point-to-decision-boundary distance on points in Figure 6.15; here the input dimension is $N = 3$ and the decision boundary is a true hyperplane.

Remember, as detailed above, we can scale any linear decision boundary by

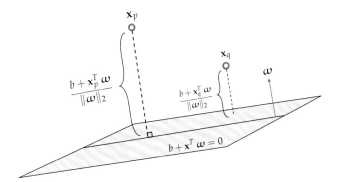

Figure 6.15 Visual representation of the distance to the hyperplane $b + \mathbf{x}^T \omega$, of two points $\mathbf{x}_p$ and $\mathbf{x}_q$ lying above it.

a nonzero scalar C and it still defines the same hyperplane. So if – in particular – we multiply by $C = \frac{1}{\|\omega\|_2}$ we have

$$\frac{b + \mathbf{x}^T \omega}{\|\omega\|_2} = \frac{b}{\|\omega\|_2} + \mathbf{x}^T \frac{\omega}{\|\omega\|_2} = 0. \tag{6.45}$$

We do not change the nature of our decision boundary and now our feature-touching weights have unit length as $\left\| \frac{\omega}{\|\omega\|_2} \right\|_2 = 1$. In other words, regardless of how large our weights $\mathbf{w}$ were to begin with we can always normalize them in a consistent way by dividing them by the magnitude of ω.

6.4.6   Regularizing two-class classification

The normalization scheme described above is particularly useful in the context of the technical issue with the Softmax/Cross Entropy highlighted above because clearly a decision boundary that perfectly separates two classes of data *can be feature-weight normalized* to prevent its weights from growing too large (and diverging to infinity). Of course we do not want to wait to perform this normalization until *after* we run our local optimization, since this will not prevent the magnitude of the weights from potentially diverging, but *during* optimization. We can achieve this by *constraining* the Softmax/Cross Entropy cost so that feature-touching weights always have length one, i.e., $\|\omega\|_2 = 1$. Formally this minimization problem (employing the Softmax cost) can be phrased as follows

$$\underset{b,\,\omega}{\text{minimize}} \quad \frac{1}{P} \sum_{p=1}^{P} \log\left(1 + e^{-y_p\left(b + \mathbf{x}_p^T \omega\right)}\right)$$

$$\text{subject to} \quad \|\omega\|_2^2 = 1. \tag{6.46}$$

By solving this *constrained* version of the Softmax cost we can still learn a decision boundary that perfectly separates two classes of data, but we avoid divergence in the magnitude of the weights by keeping their magnitude *feature-weight normalized*. This formulation can indeed be solved by simple extensions of the local optimization methods detailed in Chapters 2–4. However, a more popular approach in the machine learning community is to "relax" this constrained formulation and instead solve the highly related unconstrained *regularized* version of the original Softmax cost. This relaxed form of the problem consists in minimizing a cost function that is a linear combination of our original Softmax cost and the magnitude of the feature weights

$$g\left(b, \omega\right) = \frac{1}{P} \sum_{p=1}^{P} \log\left(1 + e^{-y_p\left(b + \mathbf{x}_p^T \omega\right)}\right) + \lambda \left\|\omega\right\|_2^2 \tag{6.47}$$

which we can minimize using any of our familiar local optimization schemes. Here the term $\left\|\omega\right\|_2^2$ is referred to as a *regularizer*, and the parameter $\lambda \geq 0$ is called a *regularization parameter*. The parameter λ is used to balance how strongly we pressure one term or the other in minimizing their sum. In minimizing the first term, our Softmax cost, we are still looking to learn an excellent linear decision boundary. In also minimizing the second term, the magnitude of the feature-touching weights, we incentivize the learning of *small* weights. This prevents the divergence of their magnitude since if their size does start to grow our entire cost function "suffers" because of it, and becomes large. Because of this the value of λ is typically chosen to be small (and positive) in practice, although some fine-tuning can be useful.

Example 6.7   The regularized Softmax cost

Here we repeat the experiment of described in Example 6.6, but add a regularizer with $\lambda = 10^{-3}$ to the Softmax cost as in Equation (6.47). In the right panel of Figure 6.16 we show the contour plot of the regularized cost, and we can see that its global minimum no longer lies at infinity. However, we still learn a perfect decision boundary as illustrated in the left panel by a tightly fitting $\tanh(\cdot)$ function.

6.5   Support Vector Machines

In this section we describe *Support Vector Machines* [18], or SVMs for short. This approach provides interesting theoretical insight into the two-class classification process – particularly under the assumption that the data is perfectly linearly separable. However, we will see that in the more realistic scenario when data is

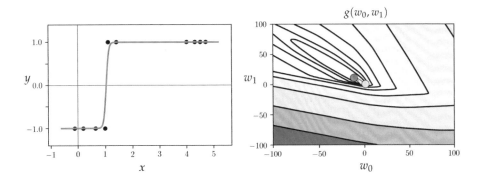

Figure 6.16 Figure associated with Example 6.7. See text for details.

not perfectly separable the Support Vector Machines approach does not provide a learned decision boundary that substantially differs from those provided by logistic regression or the Perceptron.

6.5.1 The Margin-Perceptron

Suppose once again that we have a two-class classification training dataset of P points $\left\{\left(\mathbf{x}_p, y_p\right)\right\}_{p=1}^{P}$ with the labels $y_p \in \{-1, +1\}$. Also suppose for the time being that we are dealing with a two-class dataset that is perfectly linearly separable with a known linear decision boundary $\mathring{\mathbf{x}}^T \mathbf{w} = 0$ like the one illustrated in Figure 6.17.

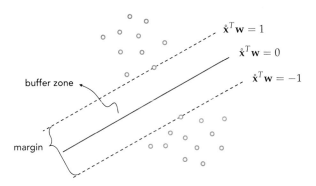

Figure 6.17 For linearly separable data the width of the buffer zone (in gray) confined between two evenly spaced translates of a separating hyperplane that just touch each respective class, defines the margin of that separating hyperplane.

This separating hyperplane creates a buffer between the two classes confined between two evenly shifted (equidistant) versions of itself: one version that lies on the positive side of the separator and just touches the class having labels

$y_p = +1$ (colored red) taking the form $\mathring{x}^T w = +1$, and one lying on the negative side of it just touching the class with labels $y_p = -1$ (colored blue) taking the form $\mathring{x}^T w = -1$. The translations above and below the separating hyperplane are more generally defined as $\mathring{x}^T w = +\beta$ and $\mathring{x}^T w = -\beta$ respectively, where $\beta > 0$. However, by dividing off β in both equations and reassigning the variables as $w \longleftarrow \frac{w}{\beta}$ we can leave out the redundant parameter β and have the two translations as stated $\mathring{x}^T w = \pm 1$.

The fact that all points in the $+1$ class lie exactly on or on the positive side of $\mathring{x}^T w = +1$, and all points in the -1 class lie exactly on or on the negative side of $\mathring{x}^T w = -1$ can be written formally as the following conditions

$$\begin{aligned} \mathring{x}^T w &\geq 1 && \text{if } y_p = +1 \\ \mathring{x}^T w &\leq -1 && \text{if } y_p = -1 \end{aligned} \tag{6.48}$$

which is a generalization of the conditions which led to the Perceptron cost in Equation (6.30).

We can combine these conditions into a single statement by multiplying each by their respective label values, giving the single inequality $y_p \mathring{x}^T w \geq 1$ which can be equivalently written as a point-wise cost

$$g_p(w) = \max\left(0, 1 - y_p \mathring{x}^T w\right) = 0. \tag{6.49}$$

Again, this value is always nonnegative. Summing up all P equations of the form above gives the *Margin-Perceptron* cost

$$g(w) = \sum_{p=1}^{P} \max\left(0, 1 - y_p \mathring{x}^T w\right). \tag{6.50}$$

Notice the striking similarity between the original Perceptron cost from the previous section and the Margin-Perceptron cost above: naively we have just "added a 1" to the nonzero input of the max function in each summand. However, this additional 1 prevents the issue of a trivial zero solution with the original Perceptron discussed previously, which simply does not arise here.

If the data is indeed perfectly linearly separable any hyperplane passing between the two classes will have parameters w where $g(w) = 0$. However, the Margin-Perceptron is still a valid cost function even if the data is not linearly separable. The only difference is that with such a dataset we can not make the criteria above hold for all points in the dataset. Thus a violation for the pth point adds the positive value of $1 - y_p \mathring{x}^T w$ to the cost function.

6.5.2   Relation to the Softmax cost

As with the Perceptron, we can smooth out the Margin-Perceptron by replacing the max operator with Softmax (see Section 6.4.3). Doing so in one summand of the Margin-Perceptron gives the related summand

$$\text{soft}\left(0, 1 - y_p \, \mathring{x}^T \mathbf{w}\right) = \log\left(1 + e^{1 - y_p \, \mathring{x}^T \mathbf{w}}\right). \tag{6.51}$$

Right away, if we were to sum over all P we could form a Softmax-like cost function that closely matches the Margin-Perceptron. But note how in the derivation of the margin Perceptron above the "1" used in the $1 - y_p\left(\mathring{x}^T \mathbf{w}\right)$ component of the cost could have been chosen to be any number we desire. Indeed we chose the value "1" out of convenience. Instead we could have chosen any value $\epsilon > 0$ in which case the set of P conditions stated in Equation (6.48) would be equivalently stated as

$$\max\left(0, \, \epsilon - y_p \, \mathring{x}^T \mathbf{w}\right) = 0 \tag{6.52}$$

for all p and the Margin-Perceptron equivalently stated as

$$g\left(\mathbf{w}\right) = \sum_{p=1}^{P} \max\left(0, \, \epsilon - y_p \, \mathring{x}^T \mathbf{w}\right) \tag{6.53}$$

and, finally, the softmax version of one summand here being

$$\text{soft}\left(0, \epsilon - y_p \, \mathring{x}^T \mathbf{w}\right) = \log\left(1 + e^{\epsilon - y_p \, \mathring{x}^T \mathbf{w}}\right). \tag{6.54}$$

When ϵ is quite small we of course have that $\log\left(1 + e^{\epsilon - y_p \, \mathring{x}^T \mathbf{w}}\right) \approx \log\left(1 + e^{-y_p \, \mathring{x}^T \mathbf{w}}\right)$, the same summand used in the Softmax cost. Thus we can, roughly speaking, interpret Softmax cost function as a smoothed version of our Margin-Perceptron cost as well.

6.5.3   Maximum margin decision boundaries

When two classes of data are perfectly linearly separable infinitely many hyperplanes perfectly divide up the data. In Figure 6.18 we display two such hyperplanes for a prototypical perfectly separable dataset. Given that both classifiers (as well as any other decision boundary perfectly separating the data) would perfectly classify this dataset, is there one that we can say is the "best" of all possible separating hyperplanes?

One reasonable standard for judging the quality of these hyperplanes is via their margin lengths, that is the distance between the evenly spaced translates that just touch each class. The larger this distance is the intuitively better the associated hyperplane separates the entire space given the particular distribution of the data. This idea is illustrated pictorially in the figure. In this illustration two separators are shown along with their respective margins. While both perfectly distinguish between the two classes the green separator (with smaller margin) divides up the space in a rather awkward fashion given how the data is distributed, and will therefore tend to more easily misclassify future data points.

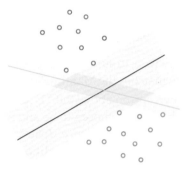

Figure 6.18 Infinitely many linear decision boundaries can perfectly separate a dataset like the one shown here, where two linear decision boundaries are shown in green and black. The decision boundary with the maximum margin – here the one shown in black – is intuitively the best choice. See text for further details.

On the other hand, the black separator (having a larger margin) divides up the space more evenly with respect to the given data, and will tend to classify future points more accurately.

In our venture to recover the maximum margin separating decision boundary, it will be convenient to use our individual notation for the bias and feature-touching weights (used in, e.g., Section 6.4.5)

$$(\text{bias}): b = w_0 \quad (\text{feature-touching weights}): \quad \omega = \begin{bmatrix} w_1 \\ w_2 \\ \vdots \\ w_N \end{bmatrix}. \qquad (6.55)$$

With this notation we can express a linear decision boundary as

$$\mathring{x}^T \mathbf{w} = b + x^T \omega = 0. \qquad (6.56)$$

To find the separating hyperplane with maximum margin we aim to find a set of parameters so that the region defined by $b + x^T \omega = \pm 1$, with each translate just touching one of the two classes, has the largest possible margin. As depicted in Figure 6.19 the margin can be determined by calculating the distance between any two points (one from each translated hyperplane) both lying on the normal vector ω. Denoting by x_1 and x_2 the points on vector ω belonging to the *positively* and *negatively* translated hyperplanes, respectively, the margin is computed simply as the length of the line segment connecting x_1 and x_2, i.e., $\|x_1 - x_2\|_2$.

The margin can be written much more conveniently by taking the difference of the two translates evaluated at x_1 and x_2 respectively, as

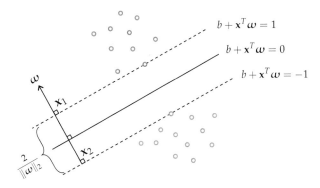

Figure 6.19 The margin of a separating hyperplane can be calculated by measuring the distance between the two points of intersection of the normal vector ω and the two equidistant translations of the hyperplane. This distance can be shown to have the value of $\frac{2}{\|\omega\|_2}$ (see text for further details).

$$\left(w_0 + \mathbf{x}_1^T\mathbf{w}\right) - \left(w_0 + \mathbf{x}_2^T\mathbf{w}\right) = \left(\mathbf{x}_1 - \mathbf{x}_2\right)^T \omega = 2. \tag{6.57}$$

Using the inner-product rule,[8] and the fact that the two vectors $\mathbf{x}_1 - \mathbf{x}_2$ and ω are parallel to each other, we can solve for the margin directly in terms of ω, as

$$\|\mathbf{x}_1 - \mathbf{x}_2\|_2 = \frac{2}{\|\omega\|_2}. \tag{6.58}$$

Therefore finding the separating hyperplane with maximum margin is equivalent to finding the one with the *smallest* possible normal vector ω.

6.5.4 The hard-margin and soft-margin SVM problems

In order to find a separating hyperplane for the data with minimum length normal vector we can simply combine our desire to minimize $\|\omega\|_2^2$ subject to the constraint (defined by the Margin-Perceptron) that the hyperplane perfectly separates the data (given by the margin criterion described above). This results in the so-called *hard-margin* support vector machine problem

$$\underset{b,\,\omega}{\text{minimize}} \ \|\omega\|_2^2$$
$$\text{subject to} \ \max\left(0,\, 1 - y_p\left(b + \mathbf{x}_p^T\omega\right)\right) = 0, \quad p = 1, ..., P. \tag{6.59}$$

The constraints here guarantee that the hyperplane we recover separates the data perfectly. While there are *constrained optimization* algorithms that are

---

[8] Using the inner-product rule (see Section C.2) we have from Equation (6.57) that
$2 = \left(\mathbf{x}_1 - \mathbf{x}_2\right)^T \omega = \|\mathbf{x}_1 - \mathbf{x}_2\|_2 \|\omega\|_2$. Rearranging gives the expression shown in Equation (6.58).

designed to solve problems like this as stated (see e.g., [14, 15, 19]), we can also solve the hard-margin problem by *relaxing* the constraints and forming an unconstrained formulation of the problem (to which we can apply familiar algorithmic methods to minimize). This is precisely the *regularization* approach detailed previously in Section 6.4.6. To do this we merely bring the constraints up, forming a single cost function

$$g(b, \omega) = \sum_{p=1}^{P} \max\left(0,\ 1 - y_p\left(b + \mathbf{x}_p^T \omega\right)\right) + \lambda \left\|\omega\right\|_2^2 \tag{6.60}$$

to be minimized. Here the parameter $\lambda \geq 0$ is called a penalty or regularization parameter (as we saw previously in Section 6.4.6). When λ is set to a small positive value we put more "pressure" on the cost function to make sure the constraints indeed hold, and (in theory) when λ is made very small the formulation above matches the original constrained form. Because of this, λ is often set to be quite small in practice (the optimal value being the one that results in the original constraints of Equation (6.59) being satisfied).

This *regularized* form of the Margin-Perceptron cost function is referred to as the *soft-margin Support Vector Machine cost.*[9]

Example 6.8   The SVM decision boundary
In the left panel of Figure 6.20 we show three boundaries learned via minimizing the Margin-Perceptron cost (shown in Equation (6.50)) three separate times with different random initializations. In the right panel of this figure we show the decision boundary provided by properly minimizing the soft-margin SVM cost in Equation (6.60) with regularization parameter $\lambda = 10^{-3}$.

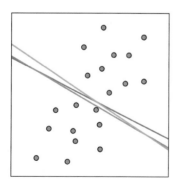
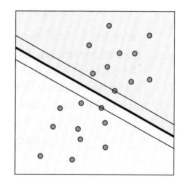

Figure 6.20 Figure associated with Example 6.8. See text for details.

---

[9] There are other relaxations of the hard-margin SVM problem used in practice (see, e.g., [20]); however, they have no theoretical or practical advantage over the one presented here [21, 22].

Each of the boundaries shown in the left panel perfectly separates the two classes, but the SVM decision boundary in the right panel provides the maximum margin. Note how in the right panel the translates of the decision boundary pass through points from both classes – equidistant from the SVM linear decision boundary. These points are called *support vectors*, hence the name Support Vector Machines.

6.5.5 SVMs and noisy data

A very big practical benefit of the soft-margin SVM problem in Equation (6.60) is that it allows us it to deal with noisy imperfectly (linearly) separable data – which arise far more commonly in practice than datasets that are perfectly linearly separable. Whereas "noise" makes at least one of the constraints in the hard-margin problem in Equation (6.59) impossible to satisfy (and thus the problem is technically impossible to solve), the soft-margin relaxation can always be properly minimized and is therefore much more frequently used in practice.

6.6 Which Approach Produces the Best Results?

Once we forgo the strong (and unrealistic) assumption of perfectly (linear) separability the added value of a "maximum margin hyperplane" provided by the SVM solution disappears since we *no longer have a margin to begin with*. Thus with many datasets in practice the soft-margin problem does *not* provide a solution remarkably different than the Perceptron or logistic regression. Indeed – with datasets that are not linearly separable – it often returns *exactly* the same solution provided by the Perceptron or logistic regression.

Furthermore, let us consider the soft-margin SVM problem in Equation (6.60), and *smooth* the Margin-Perceptron portion of the cost using the *Softmax* function as detailed in Section 6.5.2. This results in a smoothed out soft-margin SVM cost function of the form

$$g\left(b,\omega\right) = \sum_{p=1}^{P}\log\left(1 + e^{-y_p\left(b + \mathbf{x}_p^T\omega\right)}\right) + \lambda\left\|\omega\right\|_2^2. \tag{6.61}$$

While this is interpreted through the lens of SVMs, it can also be immediately identified as a regularized Softmax cost (i.e., as a regularized Perceptron or logistic regression[10]). Therefore we can see that all three methods of linear two-class classification we have seen – logistic regression, the Perceptron, and SVMs – are very deeply connected, and why they tend to provide similar results on realistic (not linearly separable) datasets.

[10] Indeed this soft-margin SVM cost even at times referred to as the *log-loss SVM* (see, e.g., [21])

6.7   The Categorical Cross Entropy Cost

In Sections 6.2 and 6.3 we saw how two different choices for label values, $y_p \in \{0, 1\}$ or $y_p \in \{-1, +1\}$, result in precisely the same two-class classification via Cross Entropy/Softmax cost function minimization. In each instance we formed a *log error* cost per data point, and averaging these over all P data points provided a proper and convex cost function. In other words, the numerical values of the label pairs themselves were largely used just to *simplify* the expression of these cost functions. Given the pattern for deriving convex cost functions for logistic regression, given any two *numerical* label values $y_p \in \{a, b\}$ it would be a straightforward affair to derive an appropriate convex cost function based on a log error like point-wise cost.

However, the true range of label value choices is even broader than this – than two *numerical* values. We can indeed use *any two distinct objects as labels* as well, i.e., two unordered values. However, regardless of how we define our labels we still end up building the same sort of two-class classifier we have seen previously – tuning its weights by minimization of a familiar cost function like, for example, the Cross Entropy/Softmax cost.

To drive home this point, in this brief section we show how to derive the same Cross Entropy cost function seen in Section 6.2 employing *categorical labels* instead of numerical ones. This leads to the derivation of the so-called *Categorical Cross Entropy cost* function, which – as we will see – is equivalent to the Cross Entropy cost.

6.7.1   One-hot encoded categorical labels

Suppose we begin with a two-class classification dataset $\left\{ \left(\mathbf{x}_p, y_p \right) \right\}_{p=1}^{P}$ with N-dimensional input and transform our original numerical label values $y_p \in \{0, 1\}$ with *one-hot encoded vectors* of the form

$$y_p = 0 \longleftarrow \mathbf{y}_p = \begin{bmatrix} 1 \\ 0 \end{bmatrix} \quad y_p = 1 \longleftarrow \mathbf{y}_p = \begin{bmatrix} 0 \\ 1 \end{bmatrix}. \tag{6.62}$$

Each vector representation uniquely identifies its corresponding label value, but now our label values are no longer *ordered numerical values*, and our dataset now takes the form $\left\{ \left(\mathbf{x}_p, \mathbf{y}_p \right) \right\}_{p=1}^{P}$, where $\mathbf{y}_p$ is defined as above. However, our goal here will remain the same: to properly tune a set of $N + 1$ weights $\mathbf{w}$ so as to regress the input to the output of our dataset.

6.7.2   Choosing a nonlinearity

With these new *categorical labels* our classification task – when viewed as a *regression problem* – is a special case of *multi-output regression* (as detailed in Section 5.6) where we aim to regress N-dimensional input against two dimensional categorical labels using a nonlinear function of the linear combination $\mathring{\mathbf{x}}_p^T \mathbf{w}$. Because our categorical labels have length two we need to use a nonlinear function of this linear combination that produces *two* outputs as well. Since the labels are one-hot encoded and we are familiar with the sigmoid function (see Section 6.2.3), it then makes sense to use the following nonlinear function of each input point $\mathbf{x}_p$

$$\sigma_p = \begin{bmatrix} \sigma\left(\mathring{\mathbf{x}}_p^T \mathbf{w}\right) \\ 1 - \sigma\left(\mathring{\mathbf{x}}_p^T \mathbf{w}\right) \end{bmatrix}. \tag{6.63}$$

Why? Because suppose for a particular point that $\mathbf{y}_p = \begin{bmatrix} 1 \\ 0 \end{bmatrix}$ and $\mathbf{w}$ is tuned so that $\sigma\left(\mathring{\mathbf{x}}_p^T \mathbf{w}\right) \approx 1$. By the definition of the sigmoid this implies that $1 - \sigma\left(\mathring{\mathbf{x}}_p^T \mathbf{w}\right) \approx 0$ and so that – for this point – $\sigma_p \approx \begin{bmatrix} 1 \\ 0 \end{bmatrix} = \mathbf{y}_p$ which is indeed our desire. And, of course, this same idea holds if $\mathbf{y}_p = \begin{bmatrix} 0 \\ 1 \end{bmatrix}$ as well.

Thus with this nonlinear transformation an ideal setting of our weights $\mathbf{w}$ will force

$$\sigma_p \approx \mathbf{y}_p \tag{6.64}$$

to hold for as many points as possible.

6.7.3 Choosing a cost function

As was the case with numerical labels, here we could also very well propose a standard point-wise cost taken from our experience with regression such as, for example, the Least Squares

$$g_p(\mathbf{w}) = \left\| \sigma_p - \mathbf{y}_p \right\|_2^2 \tag{6.65}$$

and minimize the average of these over all P points to tune $\mathbf{w}$. However, as was the case with numerical labels, here because our *categorical labels* take a very precise binary form we are better off employing a *log error* (see Section 6.2) to better incentivize learning (producing a convex cost function). Denoting $\log \sigma_p$ the vector formed by taking the $\log(\cdot)$ of each entry of σ_p, here the log error takes the form

$$g_p(\mathbf{w}) = -\mathbf{y}_p^T \log \sigma_p = -y_{p,1} \log\left(\sigma\left(\mathring{\mathbf{x}}_p^T \mathbf{w}\right)\right) - y_{p,2} \log\left(1 - \sigma\left(\mathring{\mathbf{x}}_p^T \mathbf{w}\right)\right) \tag{6.66}$$

where note here that $\mathbf{y}_p = \begin{bmatrix} y_{p,1} \\ y_{p,2} \end{bmatrix}$. Taking the average of these P point-wise costs gives the so-called *Categorical Cross Entropy* cost function for two-class classification. Here the "categorical" part of this name refers to the fact that our labels are one-hot encoded categorical (i.e., unordered) vectors.

However, it is easy to see that cost function is precisely the log error we found in Section 6.2. In other words, written in terms of our original *numerical* labels, the pointwise cost above is precisely

$$g_p(\mathbf{w}) = -y_p \log\left(\sigma\left(\mathring{\mathbf{x}}_p^T \mathbf{w}\right)\right) - \left(1 - y_p\right)\log\left(1 - \sigma\left(\mathring{\mathbf{x}}_p^T \mathbf{w}\right)\right). \quad (6.67)$$

Therefore, even though we employed categorical versions of our original numerical label values, the cost function we minimize in the end to properly tune **w** is precisely the same we have seen previously – the Cross Entropy/Softmax cost where numerical labels were employed.

6.8   Classification Quality Metrics

In this section we describe simple metrics for judging the quality of a trained two-class classification model, as well as how to make predictions using one.

6.8.1   Making predictions using a trained model

If we denote by $\mathbf{w}^\star$ the optimal set of weights found by minimizing a classification cost function, employing by default label values $y_p \in \{-1, +1\}$, then note we can write our fully tuned linear model as

$$\text{model}(\mathbf{x}, \mathbf{w}^\star) = \mathring{\mathbf{x}}^T \mathbf{w}^\star = w_0^\star + x_1 w_1^\star + x_2 w_2^\star + \cdots + x_N w_N^\star. \quad (6.68)$$

This fully trained model defines an optimal decision boundary for the training dataset which takes the form

$$\text{model}(\mathbf{x}, \mathbf{w}^\star) = 0. \quad (6.69)$$

To predict the label y' of an input $\mathbf{x}'$ we then process this model through an appropriate step. Since by default we use label values $y_p \in \{-1, +1\}$, this step function is conveniently defined by the sign $(\cdot)$ function (as detailed in Section 6.3), and the predicted label for $\mathbf{x}'$ is given as

$$\text{sign}\left(\text{model}(\mathbf{x}', \mathbf{w}^\star)\right) = y'. \quad (6.70)$$

This evaluation, which will always take on values ± 1 if $\mathbf{x}'$ does not lie *exactly* on the decision boundary (in which case we assign a random value from ± 1), simply computes which side of the decision boundary the input $\mathbf{x}'$ lies on. If it lies "above" the decision boundary then $y' = +1$, and if "below" then $y' = -1$. This is illustrated for a prototypical dataset in Figure 6.21.

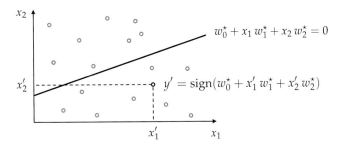

Figure 6.21 Once a decision boundary has been learned for the training dataset with optimal parameters $w_0^\star$ and $\mathbf{w}^\star$, the label y of a new point $\mathbf{x}$ can be predicted by simply checking which side of the boundary it lies on. In the illustration shown here $\mathbf{x}$ lies below the learned hyperplane, and as a result is given the label $\mathrm{sign}(\mathring{\mathbf{x}}'^T\mathbf{w}^\star) = -1$.

6.8.2 Confidence scoring

Once a proper decision boundary is learned, we can judge its *confidence* in any point based on *the point's distance to the decision boundary*. We say that our classifier has *zero* confidence in points lying along the decision boundary itself, because the boundary cannot tell us accurately which class such points belong to (which is why they are randomly assigned a label value if we ever need to make a prediction there). Likewise we say that *near* the decision boundary we have *little confidence* in the classifier's predictions. Why is this? Imagine we apply a small perturbation to the decision boundary, changing its location ever so slightly. Some points *very close to the original boundary* will end up on the *opposite* side of the new boundary, and will consequently have *different predicted labels*. Conversely, this is why we have *high confidence* in the predicted labels of points *far* from the decision boundary. These predicted labels will not change if we make a small change to the location of the decision boundary.

The notion of "confidence" can be made precise and normalized to be universally applicable by running the point's distance to the boundary through the *sigmoid function* (see Section 6.2.3). This gives the confidence that a point belongs to class +1.

The signed distance d from a point to the decision boundary provided by our trained model can be computed (see Section 6.4.5) as

$$d = \frac{b^\star + \mathbf{x}_p^T \omega^\star}{\left\| \omega^\star \right\|_2} \tag{6.71}$$

where we denote

$$\text{(bias): } b^\star = w_0^\star \quad \text{(feature-touching weights): } \quad \omega^\star = \begin{bmatrix} w_1^\star \\ w_2^\star \\ \vdots \\ w_N^\star \end{bmatrix}. \tag{6.72}$$

By evaluating d using the *sigmoid function* we squash it smoothly onto the interval $[0, 1]$, giving a "confidence" score

$$\text{confidence in the predicted label of a point } \mathbf{x} = \sigma(d). \tag{6.73}$$

When this value equals 0.5 the point lies on the boundary itself. If the value is greater than 0.5 the point lies on the positive side of the decision boundary and so we have larger confidence in its predicted label being +1. When the value is less than 0.5 the point lies on the *negative* side of the classifier, and so we have less confidence that it truly has label value +1. Because normalization employing the sigmoid squashes $(-\infty, +\infty)$ down to the interval $[0, 1]$ this confidence value is often interpreted as a *probability*.

6.8.3   Judging the quality of a trained model using *accuracy*

Once we have successfully minimized a cost function for linear two-class classification it can be a delicate matter to determine our trained model's quality. The simplest metric for judging the quality of a fully trained model is to *count the number of misclassifications* it forms over our training dataset. This is a raw count of the number of training data points $\mathbf{x}_p$ whose true label y_p is predicted *incorrectly* by our trained model.

To compare the point $\mathbf{x}_p$'s predicted label $\hat{y}_p = \text{sign} \left(\text{model} \left(\mathbf{x}_p, \mathbf{w}^\star \right) \right)$ and true label y_p we can use an identity function $\mathcal{I}(\cdot)$ and compute

$$\mathcal{I}\left(\hat{y}_p, y_p\right) = \begin{cases} 0 & \text{if } \hat{y}_p = y_p \\ 1 & \text{if } \hat{y}_p \neq y_p. \end{cases} \tag{6.74}$$

Summing all P points gives the total number of misclassifications of our trained model

$$\text{number of misclassifications} = \sum_{p=1}^{P} \mathcal{I}\left(\hat{y}_p, y_p\right). \tag{6.75}$$

Using this we can also compute the *accuracy*, denoted by $\mathcal{A}$, of a trained model. This is simply the percentage of training dataset whose labels are *correctly* predicted by the model, that is

$$\mathcal{A} = 1 - \frac{1}{P} \sum_{p=1}^{P} \mathcal{I}\left(\hat{y}_p, y_p\right). \qquad (6.76)$$

The accuracy ranges from 0 (no points are classified correctly) to 1 (all points are classified correctly).

Example 6.9   Comparing cost function and misclassification histories

Our classification cost functions are – in the end – based on smooth approximations to a discrete step function (as detailed in Sections 6.2 and 6.3). This is the function we truly wish to use, i.e., the function through which we truly want to tune the parameters of our model. However, since we cannot optimize this parameterized step function directly we settle for a smooth approximation. The consequences of this practical choice are seen when we compare the cost function history from a run of gradient descent to the corresponding misclassification count measured at each step of the run. In Figure 6.22 we show such a comparison using the dataset shown in Figure 6.10. In fact we show such results of three independent runs of gradient descent, with the history of misclassifications shown in the left panel and corresponding Softmax cost histories shown in the right panel. Each run is color-coded to distinguish it from the other runs.

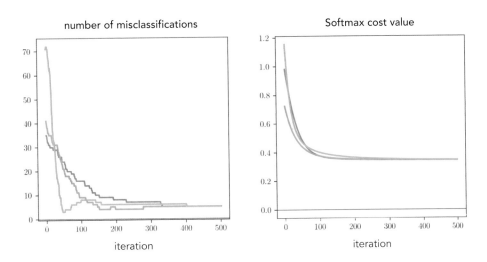

Figure 6.22 Figure associated with Example 6.9. See text for details.

Comparing the left and right panels of the figure we can see that the number of misclassifications and Softmax evaluations at each step of a gradient descent run do not perfectly track one another. That is, it is not the case that just because the cost function value is decreasing that so too is the number of misclassifications.

Again, this occurs because our Softmax cost is only an approximation of the true quantity we would like to minimize.

This simple example has an extremely practical implication: after a running a local optimization to minimize a two-class classification cost function the best step (the one that provides the best classification result), and corresponding weights, are associated with the lowest *number of misclassifications* (or likewise the *highest accuracy*), **not** the lowest cost function value.

6.8.4   Judging the quality of a trained model using *balanced accuracy*

Classification accuracy is an excellent basic measurement of a trained classifier's performance. However, in certain scenarios using the accuracy metric can paint an incomplete picture of how well we have really solved a classification problem. For example, when a dataset consists of *highly imbalanced classes* – that is, when a dataset has far more examples of one class than the other – the "accuracy" of a trained model loses its value as a quality metric. This is because when one class greatly outnumbers the other in a dataset an accuracy value close to 1 can be misleading. For example, if one class makes up 95 percent of all data points, a naive classifier that blindly assigns the label of the majority class to *every training point* achieves an accuracy of 95 percent. But here misclassifying 5 percent amounts to *completely misclassifying an entire class of data*.

This idea of "sacrificing" members of the smaller class by misclassifying them (instead of members from the majority class) is – depending on the application – very undesirable. For example, in distinguishing between healthy individuals and sufferers of a rare disease (the minority class), one might rather misclassify a healthy individual (a member of the majority class) and give them further testing than miss out on correctly detecting someone who truly has the disease. As another example, with financial fraud detection (introduced in Example 1.9) one is more accepting of misclassified valid transactions than undetected fraudulent ones, as the former can typically be easily dealt with by further human inspection (e.g., by having a customer review possibly fraudulent transactions).

These sorts of scenarios point to a problem with the use of *accuracy* as a proper metric for diagnosing classifier performance on datasets with highly imbalanced classes: because it weights *misclassifications from both classes equally* it fails to convey how well a trained model performs on each class of the data individually. This results in the potential for strong performance on a very large class of data masking poor performance on a very small one. The simplest way to improve the accuracy metric is to take this potential problem into account, and instead of computing accuracy over *both classes together* to compute accuracy on *each class individually and average the results*.

If we denote the *indices* of those points with labels $y_p = +1$ and $y_p = -1$ as Ω_{+1} and Ω_{-1} respectively, then we can compute the number of misclassifications on each class individually (employing the notation and indicator function introduced above) as

$$
\begin{aligned}
\text{number of misclassifications on +1 class} &= \sum_{p \in \Omega_{+1}} \mathcal{I}\left(\hat{y}_p, y_p\right) \\
\text{number of misclassifications on −1 class} &= \sum_{p \in \Omega_{-1}} \mathcal{I}\left(\hat{y}_p, y_p\right).
\end{aligned}
\tag{6.77}
$$

The accuracy on each class individually can then be likewise computed as (denoting the accuracy on class +1 and −1 individually as $\mathcal{A}_{+1}$ and $\mathcal{A}_{-1}$ respectively)

$$
\begin{aligned}
\mathcal{A}_{+1} &= 1 - \frac{1}{|\Omega_{+1}|} \sum_{p \in \Omega_{+1}} \mathcal{I}\left(\hat{y}_p, y_p\right) \\
\mathcal{A}_{-1} &= 1 - \frac{1}{|\Omega_{-1}|} \sum_{p \in \Omega_{-1}} \mathcal{I}\left(\hat{y}_p, y_p\right).
\end{aligned}
\tag{6.78}
$$

Note here the $|\Omega_{+1}|$ and $|\Omega_{-1}|$ denote the number of points belonging to the +1 and −1 class respectively. We can then combine these two metrics into a single value by *taking their average*. This combined metric is called *balanced accuracy* (which we denote as $\mathcal{A}_{\text{balanced}}$)

$$
\mathcal{A}_{\text{balanced}} = \frac{\mathcal{A}_{+1} + \mathcal{A}_{-1}}{2}.
\tag{6.79}
$$

Notice if both classes have equal representation then balanced accuracy reduces to the overall accuracy value $\mathcal{A}$.

The balanced accuracy metric ranges from 0 to 1. When equal to 0 no point is classified correctly, and when both classes are classified perfectly $\mathcal{A}_{\text{balanced}} = 1$. Values of the metric in between 0 and 1 measure how well – on average – each class is classified individually. If, for example, one class of data is classified completely correct and the other completely incorrect (as in our imaginary scenario where we have an imbalanced dataset with 95 percent membership in one class and 5 percent in the other, and where have simply classified the entire space as the majority class) then $\mathcal{A}_{\text{balanced}} = 0.5$.

Thus balanced accuracy is a simple metric for helping us understand whether our learned model has "behaved poorly" on highly imbalanced datasets (see e.g., Example 6.12). In order to *improve the behavior* of our learned model in such instances we have to adjust the way we perform two class classification. One popular way of doing this – called weighted classification – is discussed in Section 6.9.

6.8.5   The confusion matrix and additional quality metrics

Additional metrics for judging the quality of a trained model for two-class classification can be formed using the *confusion matrix*, shown in the Figure 6.23. A confusion matrix is a simple look-up table where classification results are broken down by actual (across rows) and predicted (across columns) class membership. Here we denote by A the number of data points whose actual label, +1, is identical to the label assigned to them by the trained classifier. The

Figure 6.23 The *confusion matrix* can be used to produce additional quality metrics for two-class classification.

predicted label

$$
\begin{array}{c}
\text{actual label} \\
\begin{array}{cc} +1 & -1 \end{array}
\end{array}
\quad
\begin{array}{c}
+1 \\ -1
\end{array}
\left[
\begin{array}{cc}
A & B \\
C & D
\end{array}
\right]
$$

other diagonal entry D is similarly defined as the number of data points whose predicted class, -1, is equal to their actual class. The off-diagonal entries denoted by B and C represent the two types of classification errors wherein the actual and predicted labels do not match one another. In practice we want these two values to be as small as possible.

Our *accuracy* metric can be expressed in terms of the confusion matrix quantities shown in the figure as

$$
\mathcal{A} = \frac{A + D}{A + B + C + D} \tag{6.80}
$$

and our accuracy on each individual class likewise as

$$
\begin{aligned}
\mathcal{A}_{+1} &= \frac{A}{A+C} \\
\mathcal{A}_{-1} &= \frac{D}{B+D}.
\end{aligned} \tag{6.81}
$$

In the jargon of machine learning these individual accuracy metrics are often called *precision* ($\mathcal{A}_{+1}$) and *specificity* ($\mathcal{A}_{-1}$) respectively. The *balanced accuracy* metric can likewise be expressed as

$$
\mathcal{A}_{\text{balanced}} = \frac{1}{2}\frac{A}{A+C} + \frac{1}{2}\frac{D}{B+D}. \tag{6.82}
$$

Example 6.10   Spam detection

In this example we perform two-class classification on a popular spam detection (see Example 1.9) dataset taken from [23]. This dataset consists of $P = 4601$ data points, 1813 spam and 2788 ham emails, with each data point consisting of various input features for an email (described in further detail in Example 9.2), as well as a binary label indicating whether or not the email is spam. Properly minimizing the Softmax cost we can achieve a 93 percent accuracy over the entire dataset, along with the following confusion matrix.

		Predicted	
		ham	spam
Actual	ham	2664	124
	spam	191	1622

Example 6.11   Credit check

In this example we examine a two-class classification dataset consisting of $P = 1000$ samples, each a set of statistics extracted from loan application to a German bank (taken from [24]). Each input has an associated label: either a *good* (700 examples) or *bad* (300 examples) credit risk as determined by financial professionals. In learning a classifier for this dataset we create an automatic credit risk assessment tool that can help decide whether or not future applicants are good candidates for loans.

The $N = 20$ dimensional input features in this dataset include: the individual's current account balance with the bank (feature 1), the duration (in months) of previous credit with the bank (feature 2), the payment status of any prior credit taken out with the bank (feature 3), and the current value of their savings/stocks (feature 6). Properly minimizing the Perceptron cost we can achieve a 75 percent accuracy over the entire dataset, along with the following confusion matrix.

		Predicted	
		bad	good
Actual	bad	285	15
	good	234	466

6.9   Weighted Two-Class Classification

Because our two-class classification cost functions are *summable over individual points* we can – as we did with regression in Section 5.5 – weight individual points in order to emphasize or deemphasize their importance to a classification model. This is called *weighted classification*. This idea is often implemented when dealing with *highly imbalanced* two class datasets (see Section 6.8.4).

6.9.1   Weighted two-class classification

Just as we saw with regression in Section 5.5, weighted classification cost functions naturally arise due to repeated points in a dataset. For example, with metadata (for example, census data) datasets it is not uncommon to receive duplicate data points due to multiple individuals reporting the same answers to a survey.

In Figure 6.24 we take a standard census dataset and plot a subset of it along a single input feature. With only one feature taken into account we end up with multiple entries of the same data point, which we show visually via the radius of each point (the more times a given data point appears in the dataset the larger we make the radius). These data points should not be thrown away – they did not arise due to some error in data collecting – they represent the true dataset.

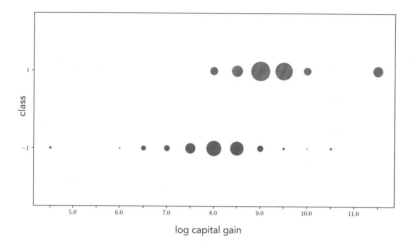

Figure 6.24 An example of duplicate entries in a metadata dataset. Here a single input feature of this dataset is plotted along with the labels for each point. There are many duplicates in this slice of the data, which are visually depicted by the radius of each point (the larger the radius, the more duplicates of that point exist in the dataset).

Just as with a regression cost, if we examine any two-class classification cost it will "collapse," with summands containing identical points naturally combining into weighted terms. One can see this by performing the same kind of simple exercise used in Section 6.24 to illustrate this fact for regression. This leads to the notion of *weighting* two-class cost functions, like for example the weighted Softmax cost, which we write below using the generic `model` notation used used in Section 7.6 to represent our linear model

$$g\left(\mathbf{w}\right) = \sum_{p=1}^{P} \beta_p \log\left(1 + e^{-y_p \text{model}\left(x_p, \mathbf{w}\right)}\right). \tag{6.83}$$

Here the values $\beta_1, \beta_2, ..., \beta_P$ are fixed *point-wise* weights. That is, a unique point $\left(x_p, y_p\right)$ in the dataset has weight $\beta_p = 1$, whereas if this point is repeated R times in the dataset then one instance of it will have weight $\beta_p = R$ while the others have weight $\beta_p = 0$. Since these weights are fixed (i.e., they are *not* parameters that need to be tuned, like $\mathbf{w}$) we can minimize a weighted classification cost precisely as we would any other via a local optimization scheme like gradient descent or Newton's method.

6.9.2   Weighting points by confidence

Just as with regression (see Section 5.5), we can also think of *assigning* the fixed weight values in Equation (6.83) based on our "confidence" of the legitimacy of a data point. If we believe that a point is very trustworthy we can set its corresponding weight β_p closer to 1, and the more untrustworthy we find a point the smaller we set β_p in the range $0 \le \beta_p \le 1$, where $\beta_p = 0$ implies we do not trust the point at all. In making these weight selections we of course determine how important each data point is in the training of the model.

6.9.3 Dealing with class imbalances

Weighted classification – in the manner detailed above – is often used to deal with imbalanced datasets. These are datasets that contain far more examples of one class of data than the other. With such datasets it is often easy to achieve a high accuracy by misclassifying points from the smaller class (as described in Section 6.8.4).

One way of ameliorating this issue is to use a weighted classification cost to alter the behavior of the learned classifier so that it weights points in the smaller class more, and points in the larger class less. In order to produce this outcome it is common to assign such weights *inversely proportional to the number of members of each class*. This weights members of the majority and minority classes so that – overall – each provides an equal contribution to the weighted classification.

That is if we denote by Ω_{+1} and Ω_{-1} the index sets for the points in classes $+1$ and -1, respectively, then first note that $P = |\Omega_{+1}| + |\Omega_{-1}|$. Then denoting by β_{+1} and β_{-1} the weight for each member of class $+1$ and -1 respectively we can set these class-wise weights inversely proportional to the number of points in each class as

$$\begin{aligned} \beta_{+1} &\propto \tfrac{1}{|\Omega_{+1}|} \\ \beta_{-1} &\propto \tfrac{1}{|\Omega_{-1}|}. \end{aligned} \qquad (6.84)$$

Example 6.12   Class imbalance and weighted classification

In the left panel of Figure 6.25 we show a toy dataset with severe class imbalance. Here we also show the linear decision boundary resulting from minimizing the Softmax cost over this dataset using five steps of Newton's method, and color each region of the space based on how this trained classifier labels points. There are only three (of a total of 55) points in total misclassified here (one blue and two red – giving an accuracy close to 95 percent); however, those that are misclassified constitute almost half of the minority (red) class. While this is not reflected in a gross misclassification or accuracy metric, it is reflected in a balanced accuracy (see Section 6.8.4) which is significantly lower, at around 79 percent.

In the middle and right panels we show the result of increasing the weights of each member of the minority class from $\beta = 1$ to $\beta = 5$ (middle panel) and $\beta = 10$ (right panel). These weights are denoted visually in the figure by increasing the radius of the points in proportion to the value of β used (thus their radius increases from left to right). Also shown in the middle and right panels is the result of properly minimizing the weighted Softmax cost in Equation (6.83) using the same optimization procedure (i.e., five steps of Newton's method). As the value of β is increased on the minority class, we encourage fewer misclassifications of its members (at the expense here of additional misclassifications of the majority class). In the right panel of the figure – where $\beta = 10$ – we have one more misclassification than in the original run with an accuracy of 93 percent. However, with the assumption that misclassifying minority class members is far more perilous than misclassifying members of the majority class, here the trade-off is well worth it as no members of the minority class are misclassified. Moreover, we achieve a significantly improved *balanced accuracy score* of 96 percent over the 79 percent achieved with the original (unweighted) run.

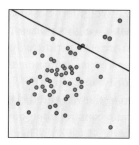 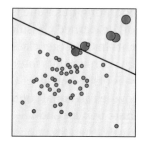 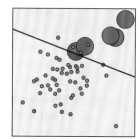

Figure 6.25 Figure associated with Example 6.12. See text for details.

6.10   Conclusion

In this chapter we detailed the problem of linear two-class classification, where we look to automatically distinguish between different types of distinct things using a linear decision boundary. We saw a wide range of perspectives for framing classification, with each shedding unique light on the process itself.

We began by discussing the regression perspective on classification in Sections 6.2 and 6.3 – where logistic regression was described – and followed this by detailing ways of viewing the problem "from above" with both the *Perceptron* and *Support Vector Machines* perspectives described in Sections 6.4 and 6.5, respectively. While these disparate perspectives differ in the how they couch the origins of two-class classification, we saw how they all naturally lead to the minimization of the same sort of cost function (e.g., the *Cross Entropy* or *Softmax* cost) using the same linear model. This unifying realization – discussed further in Section 7.4 – helps explain why (on average) these various perspectives on two-class classification result in similar performance in practice (based on the metrics introduced in Section 6.8). Practically speaking this makes the various perspectives, cost functions, and labeling schemes for two-class classification (including categorical labeling schemes, as detailed in Section 6.8) essentially interchangeable, allowing one great flexibility in conceptualizing and constructing two-class classifiers. In future chapters we will rely heavily on this idea when layering on new concepts, introducing them in the context of a single perspective (e.g., the perceptron) for the sake of simplicity, while appreciating implicitly that such new concepts automatically apply to all other perspectives on two-class classification as well.

6.11   Exercises

† The data required to complete the following exercises can be downloaded from the text's github repository at `github.com/jermwatt/machine_learning_refined`

6.1   Implementing sigmoidal Least Squares cost

Repeat the experiment described in Example 6.3 by coding up the Least Squares cost function shown in Equation (6.9) and the normalized gradient descent algorithm detailed in Section A.3. You need not reproduce the contour plot shown in the right panel of Figure 6.5; however, you can verify that your implementation is working properly by reproducing the final fit shown in the left panel of that figure. Alternatively show that your final result produces zero misclassifications (see Section 7.6.3).

6.2   Show the equivalence of the log error and Cross Entropy point-wise cost

Show that – with label values $y_p \in \{0, 1\}$ – the log error in Equation (6.10) is equivalent to the Cross Entropy point-wise cost in Equation (6.11).

6.3   Implementing the Cross Entropy cost

Repeat the experiment described in Example 6.4 by coding up the Cross Entropy cost function shown in Equation (6.12) as detailed in Section 6.2.7. You need not reproduce the contour plot shown in the right panel of Figure 6.7; however, you can verify that your implementation is working properly by reproducing the final fit shown in the left panel of that figure. Alternatively show that your final result produces zero misclassifications (see Section 7.6.3).

6.4   Compute the Lipschitz constant of the Cross Entropy cost

Compute the Lipschitz constant (Section A.6.4) of the Cross Entropy cost shown in Equation (6.12).

6.5   **Confirm gradient and Hessian calculations**
Confirm that the gradient and Hessian of the Cross Entropy cost are as shown
in Section 6.2.7.

6.6   **Show the equivalence of the log error and Softmax point-wise cost**
Show that – with label values $y_p \in \{-1, +1\}$ – the log error in Equation (6.22) is
equivalent to the Softmax point-wise cost in Equation (6.24).

6.7   **Implementing the Softmax cost**
Repeat the experiment described in Example 6.5 by coding up the Softmax
cost function shown in Equation (6.25). You need not reproduce the contour
plot shown in the right panel of Figure 6.9; however, you can verify that your
implementation is working properly by reproducing the final fit shown in the
left panel of that figure. Alternatively show that your final result produces zero
misclassifications (see Section 7.6.3).

6.8   **Implementing the Log Error version of Softmax**
Repeat the experiment described in Section 6.3.3 and shown in Figure 6.10
using the log error based Softmax cost shown in Equation (6.21) and any local
optimization scheme you wish. You need not reproduce the plots shown in the
figure to confirm your implementation works properly, but should be able to
achieve a minimum of five misclassifications (see Section 7.6.3).

6.9   **Using gradient descent to minimize the Perceptron cost**
Use the standard gradient descent scheme to minimize the Perceptron cost
function in Equation (6.33) over the dataset shown in Figure 6.10. Make two
runs of gradient descent using fixed steplength values $\alpha = 10^{-1}$ and 10^{-2}, with
50 steps each (and random initializations). Produce a cost function history plot
and history of misclassification (see Section 7.6.3) at each step of the run. Which
run achieves perfect classification first?

6.10   **The perceptron cost is convex**
Show that the Perceptron cost given in Equation (6.33) is convex using the
zero-order definition of convexity described in Exercise 5.8.

6.11   **The Softmax cost is convex**
Show that the Softmax cost function given in Equation (6.25) is convex by verifying that it satisfies the second-order definition of convexity. *Hint: the Hessian,
already given in Equation (6.3.2), is a weighted outer-product matrix like the one described in Exercise 4.2.*

6.12   **The regularized Softmax**

Repeat the experiment described in Example 6.7 and shown in Figure 6.16. You need not reproduce the plots shown in the figure to confirm your implementation works properly, but should be able to achieve a result that has five or fewer misclassifications.

6.13   **Compare the efficacy of two-class cost functions I**

Compare the efficacy of the Softmax and the Perceptron cost functions in terms of the minimal number of misclassifications each can achieve by proper minimization via gradient descent on a breast cancer dataset. This dataset consists of $P = 699$ data points, each point consisting of $N = 9$ input of attributes of a single individual and output label indicating whether or not the individual does or does not have breast cancer. You should be able to achieve around 20 misclassifications with each method.

6.14   **Compare the efficacy of two-class cost functions II**

Compare the efficacy of the Softmax and the Perceptron cost functions in terms of the minimal number of misclassifications each can achieve by proper minimization via gradient descent on the spam detection dataset introduced in Example 6.10. You should be able to achieve at least 90 percent accuracy with both methods. Because of the large variations in the input values of this dataset you should *standard normalize* the input features of the dataset – as detailed in Section 9.3 – prior to optimization.

6.15   **Credit check**

Repeat the experiment described in Example 6.11. Using an optimizer of your choice, try to achieve something close to the results reported. Make sure you *standard normalize* the input features of the dataset – as detailed in Section 9.3 – prior to optimization.

6.16   **Weighted classification and balanced accuracy**

Repeat the experiment described in Example 6.12 and shown in Figure 6.25. You need not reproduce the plots shown in the figure to confirm your implementation works properly, but should be able to achieve similar results to those reported there.

7 Linear Multi-Class Classification

7.1   Introduction

In practice many classification problems (e.g., face recognition, hand gesture recognition, recognition of spoken phrases or words, etc.) have more than just two classes we wish to distinguish. However, a single linear decision boundary naturally divides the input space only in two subspaces, and therefore is fundamentally insufficient as a mechanism for differentiating between more than two classes of data. In this chapter we describe how to generalize what we have seen in the previous chapter to deal with this multi-class setting – referred to as linear *multi-class classification*. As with prior chapters, we describe several perspectives on multi-class classification including *One-versus-All* and the *multi-class Perceptron*, quality metrics, and extensions of basic principles (including categorical and weighted multi-class classification). Here we also describe mini-batch optimization – first detailed in Section A.5 in the context of first order optimization – through the lens of machine learning, summarizing its use in optimizing both supervised and unsupervised learners.

7.2   One-versus-All Multi-Class Classification

In this section we explain the fundamental multi-classification scheme called *One-versus-All*, step by step using a single toy dataset with three classes.

7.2.1   Notation and modeling

A multi-class classification dataset $\left\{\left(\mathbf{x}_p,\, y_p\right)\right\}_{p=1}^{P}$ consists of C distinct classes of data. As with the two-class case we can in theory use *any* set of C distinct label values to distinguish between the classes. For convenience, here we use label values $y_p \in \{0, 1, \ldots, C-1\}$. In what follows we employ the prototypical toy dataset shown in Figure 7.1 to help us derive our fundamental multi-class classification scheme.

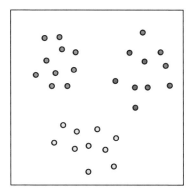

Figure 7.1 A prototypical classification dataset with $C = 3$ classes. Here points with label value $y_p = 0$, $y_p = 1$, and $y_p = 2$ are colored blue, red, and green, respectively.

7.2.2 Training C One-versus-All classifiers

A good first step in the development of multi-class classification is to simplify the problem to something we are already familiar with: *two-class classification*. We already know how to distinguish each class of our data from the other $C - 1$ classes. This sort of two-class classification problem (discussed in the previous chapter) is indeed simpler, and is adjacent to the actual problem we want to solve, i.e., learn a classifier that can distinguish between all C classes simultaneously. To solve this adjacent problem we learn C two-class classifiers over the entire dataset, with the cth classifier trained to distinguish the cth class from the remainder of the data, and hence called a *One-versus-Rest* or *One-versus-All* classifier.

To solve the cth two-class classification subproblem we simply assign temporary labels $\tilde{y}_p$ to the entire dataset, giving $+1$ labels to the cth class and -1 labels to the remainder of the dataset

$$\tilde{y}_p = \begin{cases} +1 & \text{if } y_p = c \\ -1 & \text{if } y_p \neq c \end{cases} \tag{7.1}$$

where y_p is the original label for the pth point of the multi-class dataset. We then run a two-class classification scheme of our choice (by minimizing any classification cost detailed in the previous chapter). Denoting the optimal weights learned for the cth classifier as $\mathbf{w}_c$ where

$$\mathbf{w}_c = \begin{bmatrix} w_{0,c} \\ w_{1,c} \\ w_{2,c} \\ \vdots \\ w_{N,c} \end{bmatrix} \qquad (7.2)$$

we can then express the corresponding decision boundary associated with the cth two-class classification (distinguishing class c from all other data points) simply as

$$\mathring{\mathbf{x}}^T \mathbf{w}_c = 0. \qquad (7.3)$$

In Figure 7.2 we show the result of solving this set of subproblems on the prototypical dataset shown first in Figure 7.1. In this case we learn $C = 3$ distinct linear two-class classifiers, which are shown in the top row of the figure. Each learned decision boundary is illustrated in the color of the single class being distinguished from the rest of the data, with the remainder of the data colored gray in each instance. In the bottom row of the figure the dataset is shown once again along with all three learned two-class decision boundaries.

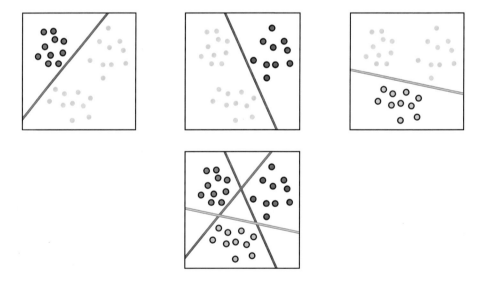

Figure 7.2 (top row) Three one-versus-all linear two-class classifiers learned to the dataset shown in Figure 7.1. (bottom row) All three classifiers shown on top of the original dataset. See text for further details.

Having solved the adjacent problem by learning C One-versus-All classifiers we can now ask: is it possible to somehow *combine* these two-class classifiers to solve the original multi-class problem? As we will see, it most certainly is.

But first it is helpful to break up the problem into three pieces for (i) points that lie on the positive side of a *single* two-class classifier, (ii) points that lie on the positive side of *more than one* classifier, and (iii) points that lie on the positive side of *none* of the classifiers. Notice, these three cases exhaust all the possible ways a point can lie in the input space in relation to the C two-class classifiers.

7.2.3   Case 1. Labeling points on the positive side of a single classifier

Geometrically speaking, a point $\mathbf{x}$ that lies on the positive side of the cth classifier but on the negative side of all the rest, satisfies the following inequalities: $\mathring{\mathbf{x}}^T \mathbf{w}_c > 0$, and $\mathring{\mathbf{x}}^T \mathbf{w}_j < 0$ for all $j \neq c$. First, notice that since all classifier evaluations of $\mathbf{x}$ are negative except for the cth one (which is positive), we can write

$$\mathring{\mathbf{x}}^T \mathbf{w}_c = \max_{j=0,\dots,C-1} \mathring{\mathbf{x}}^T \mathbf{w}_j. \tag{7.4}$$

Next, that the cth classifier evaluates $\mathbf{x}$ positively means that (from its perspective) $\mathbf{x}$ belongs to class c. Similarly, because every other classifier evaluates $\mathbf{x}$ negatively (from their perspective) $\mathbf{x}$ does *not* belong to any class $j \neq c$. Altogether we have all C individual classifiers *in agreement* that $\mathbf{x}$ should receive the label $y = c$. Therefore, using Equation (7.4) we can write the label y as

$$y = \operatorname*{argmax}_{j=0,\dots,C-1} \mathring{\mathbf{x}}^T \mathbf{w}_j. \tag{7.5}$$

In Figure 7.3 we show the result of classifying all points in the space of our toy dataset that lie on the positive side of a *single* classifier. These points are colored to match their respective classifier. Notice there still are regions left uncolored in Figure 7.3. These include regions where points are either on the positive side of more than one classifier or on the positive side of none of the classifiers. We discuss these cases next.

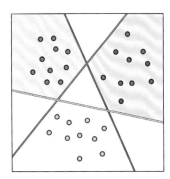

Figure 7.3 One-versus-All classification of points lying on the positive side of a single classifier. See text for details.

7.2.4   Case 2. Labeling points on the positive side of more than one classifier

When a point **x** falls on the positive side of more than one classifier it means that, unlike the previous case, more than one classifier will claim **x** as one of their own. In a situation like this and, as we saw in Section 6.8.2, we can consider the *signed distance* of **x** to a two-class decision boundary as a measure of our *confidence* in how the point should be labeled. The *farther* a point is on the *positive* side of a classifier, the *more* confidence we have that this point should be labeled +1. Intuitively, this is a simple geometric concept: the larger a point's distance to the boundary the deeper into one region of a classifier's half-space it lies, and thus we can be much more confident in its class identity than a point closer to the boundary. Another way to think about is to imagine what would happen if we slightly perturbed the decision boundary? Those points originally close to its boundary might end up on the other side of the perturbed hyperplane, changing classes, whereas those points farther from the boundary are less likely to be so affected and hence we can be more confident in their class identities to begin with.

In the present context we can intuitively extend this idea. If a point lies on the positive side of multiple two-class classifiers, we should assign it the label corresponding to the decision boundary it is *farthest* from.

Let us now see how this simple idea plays out when used on our toy dataset. In the left and middle panels of Figure 7.4 we show two example points (drawn in black) in a region of our toy dataset that lie on the *positive* side of more than one two-class classifier. In each case we highlight the distance from the new point to both decision boundaries in dashed black, with the projection of the point onto each decision boundary shown as an **x** in the same color as its respective classifier.

Beginning with the point shown in the left panel, notice that it lies on the positive side of both the red and blue classifiers. However, we can be more confident that the point should belong to the blue class since the point lies a greater distance from the blue decision boundary than the red (as we can tell by examining the length of the dashed lines emanating from the point to each boundary). By the same logic the point shown in the middle panel is best assigned to the red class, being at a greater distance from the red classifier than the blue. If we repeat this logic for every point in this region (as well as those points in the other two triangular regions where two or more classifiers are positive) and color each point the color of its respective class, we will end up shading each such region as shown in the right panel of Figure 7.4. The line segments separating each pair of colored regions in this panel (e.g., the green and the red areas of the bottom-right triangular region) consist of points lying equidistant to (and on the positive sides of) two of the One-versus-All classifiers. In other words, these are parts of our multi-class decision boundary.

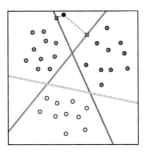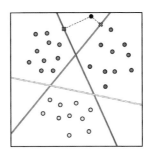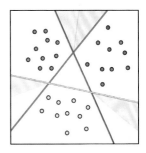

Figure 7.4 One-versus-All classification of points lying on the positive side of more than one classifier. Here a point (shown in black) lies on the positive side of both the red and blue classifiers with its distance to each decision boundary noted by dashed lines connecting it to each classifier. See text for further details.

Decomposing the weights $\mathbf{w}_j$ of our jth classifier by separating the bias weight from feature-touching weights as

$$\text{(bias): } b_j = w_{0,j} \quad \text{(feature-touching weights): } \boldsymbol{\omega}_j = \begin{bmatrix} w_{1,j} \\ w_{2,j} \\ \vdots \\ w_{N,j} \end{bmatrix} \qquad (7.6)$$

allows us to write the *signed distance* from a point $\mathbf{x}$ to its decision boundary as

$$\text{signed distance of } \mathbf{x} \text{ to } j\text{th boundary} = \frac{\mathring{\mathbf{x}}^T \mathbf{w}_j}{\left\| \boldsymbol{\omega}_j \right\|_2}. \qquad (7.7)$$

Now, if we *normalize* the weights of each linear classifier by the length of its normal vector (containing all feature-touching weights) as

$$\mathbf{w}_j \longleftarrow \frac{\mathbf{w}_j}{\left\| \boldsymbol{\omega}_j \right\|_2} \qquad (7.8)$$

which we will assume to do from here on, then this distance is simply written as the raw evaluation of the point via the decision boundary

$$\text{signed distance of } \mathbf{x} \text{ to } j\text{th boundary} = \mathring{\mathbf{x}}^T \mathbf{w}_j. \qquad (7.9)$$

To assign a point in one of our current regions we seek out the classifier which *maximizes* this quantity. Expressed algebraically, the label y assigned to this point is given (after weight-normalization) as

$$y = \operatorname*{argmax}_{j=0,\ldots,C-1} \mathring{\mathbf{x}}^T \mathbf{w}_j. \qquad (7.10)$$

This is precisely the same rule we found in Equation (7.5) for regions of the space where only a single classifier is positive. Note that normalizing the weights of each classifier (by the magnitude of their feature-touching weights) does not affect the validity of this labeling scheme for points lying on the positive side of just a single classifier as dividing a set of raw evaluations by a positive number does not change the mathematical *sign* of those raw evaluations: the sole classifier that positively evaluates a point will remain the sole classifier whose signed distance to the point in question is positive.

7.2.5   Case 3. Labeling points on the positive side of no classifier

When a point $\mathbf{x}$ lies on the positive side of *none* of our C One-versus-All classifiers (or in other words, on the *negative* side of all of them), it means that each of our classifiers designates it as *not* in their respective class. Thus we cannot argue (as we did in the previous case) that one classifier is more confident in its class identity. What we can do instead is ask which classifier is the *least* confident about $\mathbf{x}$ *not* belonging to their class? The answer is *not* the decision boundary it is farthest from as was the case previously, but the one it is *closest* to. Here again the notion of *signed distance to decision boundary* comes in handy, noticing that assigning $\mathbf{x}$ to the decision boundary that it is closest to means assigning it to the boundary that it has the largest signed distance to (since now this signed distance, as well as the other signed distances, are all negative). Formally, again assuming that the weights of each classifier have been normalized, we can write

$$y = \underset{j=0,\dots,C-1}{\operatorname{argmax}} \; \mathring{\mathbf{x}}^T \mathbf{w}_j. \tag{7.11}$$

In the left and middle panels of Figure 7.5 we show two example points in a region of our toy dataset that lie on the *negative* side of all three of our two-class classifiers. This is the white triangular region in the middle of the data. Again we highlight the distance from each point to all three decision boundaries in dashed black, with the projection of the point onto each decision boundary shown as an x in the same color as its respective classifier.

Starting with the point in the left panel, since it lies on the negative side of all three classifiers the best we can do is assign it to the class that it is closest to (or "least offends"), here the blue class. Likewise the point shown in the middle panel can be assigned to the green class as it lies closest to its boundary. If we repeat this logic for every point in the region and color each point the color of its respective class, we will end up shading this central region as shown in the right panel of Figure 7.5. The line segments separating the colored regions in the middle of this panel consists of points lying equidistant to (and on the negative sides of) all the One-versus-All classifiers. In other words, these are parts of our multi-class decision boundary.

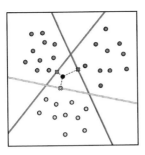 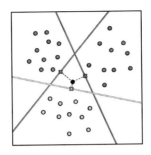 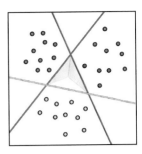

Figure 7.5 One-versus-All classification of points lying on the negative side of all classifiers. See text for further details.

7.2.6 Putting it all together

We have now deduced, by breaking down the problem of how to combine our C two-class classifiers to perform multi-class classification into three exhaustive cases, that a *single* rule can be used to assign a label y to any point $\mathbf{x}$ as

$$y = \underset{j=0,\dots,C-1}{\operatorname{argmax}} \; \mathring{\mathbf{x}}^T \mathbf{w}_j \qquad (7.12)$$

assuming that the (feature-touching) weights of each classifier have been normalized as in Equation (7.8). We call this the *fusion rule* since we can think of it as a rule that fuses our C two-class classifiers together to determine the label that must be assigned to a point based on the notion of signed distance to decision boundary.

In Figure 7.6 we show the result of applying the fusion rule in Equation (7.12) to the entire input space of our toy dataset we have used throughout this section. This includes the portions of the space where points lie (i) on the positive side of a single classifier (as shown in Figure 7.3), (ii) on the positive side of more than one classifier (as shown in the right panel of Figure 7.4), and (iii) on the positive side of none of the classifiers (as shown in the right panel of Figure 7.5).

In the left panel of Figure 7.6 we show the entire space colored appropriately according to the fusion rule, along with the three original two-class classifiers. In the right panel we highlight (in black) the line segments that define borders between the regions of the space occupied by each class. This is indeed our *multi-class classifier* or *decision boundary* provided by the fusion rule. In general when using linear two-class classifiers the boundary resulting from the fusion rule will always be *piece-wise* linear (as with our simple example here). While the fusion rule explicitly defines this piece-wise linear boundary, it does not provide us with a nice, closed form formula for it (as with two-class frameworks like logistic regression or SVMs).

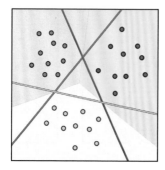 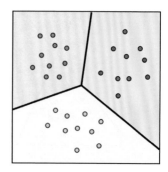

Figure 7.6 The result of applying the fusion rule in Equation (7.12) to the input space of our toy dataset, with regions colored according to their predicted label along with the original two-class decision boundaries (left panel) and fused multi-class decision boundary in black (right panel). See text for further details.

7.2.7 The One-versus-All (OvA) algorithm

When put together, the two-step process we have now seen (i.e., learn C One-versus-All two-class classifiers and then combine them using the fusion rule) is generally referred to as the *One-versus-All multi-class classification algorithm* or OvA algorithm for short. In practice it is common to see implementations of the OvA algorithm that skip the normalization step shown in Equation (7.8). This can theoretically lead to poor classification accuracy due to differently sized normal vectors, creating out of scale distance-to-classifier measurements. However, because often each classifier is trained using the same local optimization scheme the resulting magnitude of each trained normal vector can end up being around the same magnitude, hence reducing the difference between normalized and unnormalized evaluation of input points by each classifier.

Example 7.1   Classification of a dataset with $C = 4$ classes using OvA

In this example we apply the OvA algorithm to classify a toy dataset with $C = 4$ classes shown in the left panel of Figure 7.7. Here blue, red, green, and yellow points have label values 0, 1, 2, and 3, respectively. In the middle panel of Figure 7.7 we show the input space colored according to the fusion rule, along with the $C = 4$ individually learned two-class classifiers. Notice that none of these two-class classifiers perfectly separates its respective class from the rest of the data. Nevertheless, the final multi-class decision boundary shown in the right panel of Figure 7.7 does a fine job of distinguishing the four classes.

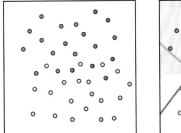 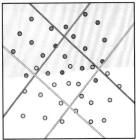 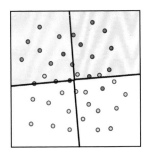

Figure 7.7 Figure associated with Example 7.1. See text for details.

Example 7.2   Regression view of the fusion rule

In deriving the fusion rule in Equation (7.12) we viewed the problem of multi-class classification from the perceptron perspective (first discussed in the two-class case in Section 6.4), meaning that we viewed our data in the input feature space alone, coloring the value of each label instead of treating the labels as an output dimension (visually speaking) and plotting the data in three dimensions. However, if we do this, viewing our multi-class data "from the side" in what we can think of as the regression perspective, we can indeed view the fusion rule as a *multi-level step function*. This is shown in Figure 7.8 for the primary toy dataset used throughout this section. The left panel shows the data and fusion rule "from above," while in the right panel the same setup shown "from the side" with the fusion rule displayed as a discrete step function.

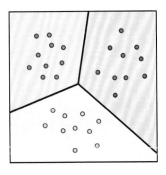   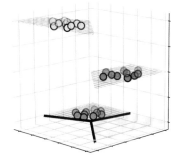

Figure 7.8 Fusion rule for a toy classification dataset with $C = 3$ classes, shown from the perceptron perspective (left panel) and from the regression perspective (right panel). Note that the jagged edges on some of the steps in the right panel are merely an artifact of the plotting mechanism used to generate the three dimensional plot. In reality the edges of each step are smooth like the fused decision boundary shown in the input space. See Example 7.2 for further details.

7.3   Multi-Class Classification and the Perceptron

In this section we discuss a natural alternative to One-versus-All (OvA) multi-class classification detailed in the previous section. Instead of training C two class classifiers *first* and *then* fusing them into a single decision boundary (via the *fusion rule*), we train all C classifiers *simultaneously* to directly satisfy the fusion rule. In particular we derive the *multi-class Perceptron* cost for achieving this feat, which can be thought of as a direct generalization of the two-class Perceptron described in Section 6.4.

7.3.1   The multi-class Perceptron cost function

In the previous section on OvA multi-class classification we saw how the fusion rule in Equation (7.12) defined class ownership for *every* point $\mathbf{x}$ in the input space of the problem. This, of course, includes all (input) points $\mathbf{x}_p$ in our training dataset $\left\{ \left(\mathbf{x}_p, y_p \right) \right\}_{p=1}^{P}$. Ideally, with all two-class classifiers properly tuned, we would like the fusion rule to hold true for as many of these points as possible

$$y_p = \operatorname*{argmax}_{j=0,...,C-1} \mathring{\mathbf{x}}_p^T \mathbf{w}_j. \tag{7.13}$$

Instead of tuning our C two-class classifiers one-by-one and then combining them in this way, we can learn the weights of all C classifiers *simultaneously* so as to satisfy this ideal condition as often as possible.

To get started in constructing a proper cost function whose minimizers satisfy this ideal, first note that *if* Equation (7.13) is to hold for our pth point then we can say that the following must be true as well

$$\mathring{\mathbf{x}}_p^T \mathbf{w}_{y_p} = \max_{j=0,...,C-1} \mathring{\mathbf{x}}_p^T \mathbf{w}_j. \tag{7.14}$$

In words, Equation (7.14) simply says that the (signed) distance from the point $\mathbf{x}_p$ to its class decision boundary is *greater* than (or equal to) its distance to every other two-class decision boundary. This is what we ideally want for all of our training data points.

Subtracting $\mathring{\mathbf{x}}_p^T \mathbf{w}_{y_p}$ from the right-hand side of Equation (7.14) then gives us a good candidate for a point-wise cost function that is always nonnegative and minimal at zero, defined as

$$g_p\left(\mathbf{w}_0, ..., \mathbf{w}_{C-1} \right) = \left(\max_{j=0,...,C-1} \mathring{\mathbf{x}}_p^T \mathbf{w}_j \right) - \mathring{\mathbf{x}}_p^T \mathbf{w}_{y_p}. \tag{7.15}$$

Notice that if our weights $\mathbf{w}_0, ..., \mathbf{w}_{C-1}$ are set ideally, $g_p\left(\mathbf{w}_0, ..., \mathbf{w}_{C-1} \right)$ should be zero for as many points as possible. With this in mind, we can then form a cost function by taking the average of the point-wise cost in Equation (7.15) over the entire dataset, as

$$g\left(\mathbf{w}_0, ..., \mathbf{w}_{C-1}\right) = \frac{1}{P} \sum_{p=1}^{P} \left[\left(\max_{j=0,...,C-1} \mathring{\mathbf{x}}_p^T \mathbf{w}_j\right) - \mathring{\mathbf{x}}_p^T \mathbf{w}_{y_p}\right]. \tag{7.16}$$

This cost function, which we refer to hereafter as the *multi-class Perceptron cost*, provides a way to tune all classifier weights *simultaneously* in order to recover weights that satisfy the fusion rule as well as possible.

7.3.2 Minimizing the multi-class Perceptron cost

Like its two-class analog discussed in Section 6.4, the multi-class Perceptron is always convex regardless of the dataset employed (see chapter's exercises). It also has a trivial solution at zero. That is, when $\mathbf{w}_j = \mathbf{0}$ for all $j = 0, ..., C - 1$ the cost is minimal. This undesirable behavior can often be avoided by initializing any local optimization scheme used to minimize it away from the origin. Note also that we are restricted to using zero- and first-order optimization methods to minimize the multi-class Perceptron cost as its second derivative is zero (wherever it is defined).

Example 7.3   Minimizing the multi-class Perceptron
In this example we minimize the multi-class Perceptron in Equation (7.16) over the toy multi-class dataset originally shown in Figure 7.1, minimizing the cost via the standard gradient descent procedure (see Section 3.5).

In the left panel of Figure 7.9 we plot the dataset, the final multi-class classification over the entire space, and each individual One-versus-All decision boundary (that is, where $\mathring{\mathbf{x}}^T \mathbf{w}_c = 0$ for $c = 0, 1$, and 2). In the right panel we show the fused multi-class decision boundary formed by combining these individual One-versus-All boundaries via the fusion rule. Note in the left panel that because we did *not* train each individual two-class classifier in a One-versus-All fashion (as was done in the previous section), each individual learned two-class classifier performs quite poorly in separating its class from rest of the data. This is perfectly fine, as it is the fusion of these linear classifiers (via the fusion rule) that provides the final multi-class decision boundary shown in the right panel of the figure, which achieves perfect classification.

7.3.3 Alternative formulations of the multi-class Perceptron cost

The multi-class Perceptron cost in Equation (7.16) can also be derived as a *direct generalization* of its two-class version introduced in Section 6.4. Using the following simple property of the max function

Figure 7.9 Figure associated with Example 7.3.

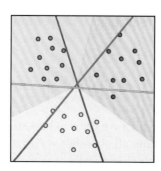 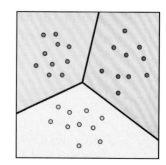

$$\max\left(s_0, s_1, ..., s_{C-1}\right) - z = \max\left(s_0 - z, s_1 - z, ..., s_{C-1} - z\right), \qquad (7.17)$$

where $s_0, s_1, ..., s_{C-1}$ and z are scalar values, we can write each summand on the right hand side of Equation (7.16) as

$$\max_{j=0,...,C-1} \mathring{\mathbf{x}}_p^T\left(\mathbf{w}_j - \mathbf{w}_{y_p}\right). \qquad (7.18)$$

Notice that for $j = y_p$ we have $\mathring{\mathbf{x}}_p^T\left(\mathbf{w}_j - \mathbf{w}_{y_p}\right) = 0$. This allows us to rewrite the quantity in Equation (7.18) equivalently as

$$\max_{\substack{j=0,...,C-1 \\ j\neq y_p}} \left(0, \mathring{\mathbf{x}}_p^T\left(\mathbf{w}_j - \mathbf{w}_{y_p}\right)\right) \qquad (7.19)$$

and hence the entire multi-class Perceptron cost as

$$g\left(\mathbf{w}_0, ..., \mathbf{w}_{C-1}\right) = \frac{1}{P}\sum_{p=1}^{P} \max_{\substack{j=0,...,C-1 \\ j\neq y_p}} \left(0, \mathring{\mathbf{x}}_p^T\left(\mathbf{w}_j - \mathbf{w}_{y_p}\right)\right). \qquad (7.20)$$

In this form it is easy to see that when $C = 2$ the multi-class Perceptron cost reduces to the two-class version detailed in Section 6.4.

7.3.4   Regularizing the multi-class Perceptron

In deriving the fusion rule and subsequently the multi-class Perceptron cost function we have assumed that the normal vector for each two-class classifier has unit length, so that we can fairly compare the (signed) distance of each input $\mathbf{x}_p$ to each of our One-versus-All two-class decision boundaries (as first detailed in Section 7.2.4). This means that in minimizing the multi-class Perceptron cost in Equation (7.16) we should (at least formally) subject it to the constraints that all these normal vectors have unit length, giving the *constrained* minimization problem

$$\underset{b_0, \omega_0, ..., b_{C-1}, \omega_{C-1}}{\text{minimize}} \quad \frac{1}{P} \sum_{p=1}^{P} \left[\left(\max_{j=0,...,C-1} b_j + \mathbf{x}_p^T \omega_c \right) - \left(b_{y_p} + \mathbf{x}_p^T \omega_{y_p} \right) \right]$$

$$\text{subject to} \quad \left\| \omega_j \right\|_2^2 = 1, \quad j = 0, ..., C - 1 \tag{7.21}$$

where we have used the bias/feature-touching weight notation, allowing us to decompose each weight vector $\mathbf{w}_j$ as shown in Equation (7.6).

While this problem can be solved in its constrained form, it is more common-place (in the machine learning community) to *relax* such a problem (as we have seen previously with the two-class Perceptron in Section 6.4.6 and the support vector machine in Section 6.5), and solve a *regularized* version.

While in theory we could provide a distinct penalty (or regularization) parameter for each of the C constraints in Equation (7.21), for simplicity one can choose a single regularization parameter $\lambda \geq 0$ to penalize the magnitude of all normal vectors simultaneously. This way we need only provide one regularization value instead of C distinct regularization parameters, giving the regularized version of the multi-class Perceptron problem as

$$\underset{b_0, \omega_0, ..., b_{C-1}, \omega_{C-1}}{\text{minimize}} \quad \frac{1}{P} \sum_{p=1}^{P} \left[\left(\max_{j=0,...,C-1} b_j + \mathbf{x}_p^T \omega_j \right) - \left(b_{y_p} + \mathbf{x}_p^T \omega_{y_p} \right) \right] + \lambda \sum_{j=0}^{C-1} \left\| \omega_j \right\|_2^2.$$

$$\tag{7.22}$$

This regularized form does not quite match the original constrained formulation as regularizing all normal vectors together will not necessarily guarantee that $\left\| \omega_j \right\|_2^2 = 1$, for all j. However, it will generally force the magnitude of all normal vectors to "behave well" by, for instance, disallowing (the magnitude of) one normal vector to grow arbitrarily large while one shrinks to almost zero. As we see many times in machine learning, it is commonplace to make such compromises to get something that is "close enough" to the original as long as it does work well in practice. This is indeed the case here, with λ typically set to a small value (e.g., 10^{-3} or smaller).

7.3.5 The multi-class Softmax cost function

As with the two-class Perceptron (see Section 6.4.3), we are often willing to sacrifice a small amount of modeling precision, by forming a closely matching smoother cost function to the one we already have, in order to make optimization easier or expand the optimization tools we can bring to bear. As was the case with the two-class Perceptron, here too we can *smooth* the multi-class Perceptron cost employing the *Softmax* function.

Replacing the max function in each summand of the multi-class Perceptron in Equation (7.16) with its Softmax approximation in Equation (6.34) gives the following cost function

$$g\left(\mathbf{w}_0, \ldots, \mathbf{w}_{C-1}\right) = \frac{1}{P}\sum_{p=1}^{P}\left[\log\left(\sum_{j=0}^{C-1} e^{\mathring{\mathbf{x}}_p^T \mathbf{w}_j}\right) - \mathring{\mathbf{x}}_p^T \mathbf{w}_{y_p}\right]. \tag{7.23}$$

This is referred to as the *multi-class Softmax cost function*, both because it is built by smoothing the multi-class Perceptron using the Softmax function, and because it can be shown to be the direct multi-class generalization of the two-class Softmax function (see, e.g., Equation (6.37)). The multi-class Softmax cost function in Equation (7.23) also goes by *many* other names including the *multi-class Cross Entropy cost*, the *Softplus cost*, and the *multi-class logistic* cost.

7.3.6   Minimizing the multi-class Softmax

Not only is the multi-class Softmax cost function convex (see chapter's exercises) but (unlike the multi-class Perceptron) it also has infinitely many smooth derivatives, enabling us to use second-order methods (in addition to zero- and first-order methods) in order to properly minimize it. Notice also that it no longer has a trivial solution at zero, akin to its two-class Softmax analog that removes this deficiency from the two-class Perceptron.

Example 7.4   **Newton's method applied to minimizing the multi-class Softmax cost**

In Figure 7.10 we show the result of applying Newton's method to minimize the multi-class Softmax cost to a toy dataset with $C = 4$ classes, first shown in Figure 7.7.

Figure 7.10 Figure associated with Example 7.4.

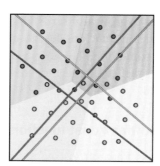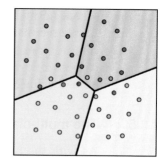

7.3.7   Alternative formulations of the multi-class Softmax

Smoothing the formulation of the multi-class Perceptron given in Equation (7.16) by replacing the max with the Softmax function gives an equivalent but different formulation of the multi-class Softmax as

$$g\left(\mathbf{w}_0, ..., \mathbf{w}_{C-1}\right) = \frac{1}{P}\sum_{p=1}^{P}\log\left(1 + \sum_{\substack{j=0\\j\neq y_p}}^{C-1} e^{\mathring{\mathbf{x}}_p^T(\mathbf{w}_j - \mathbf{w}_{y_p})}\right). \tag{7.24}$$

Visually, this formulation appears more similar to the two-class Softmax cost, and indeed does reduce to it when $C = 2$ and $y_p \in \{-1, +1\}$.

This cost function is also referred to as the *multi-class Cross Entropy cost* because it is, likewise, a natural generalization of the two-class version seen in Section 6.2. To see that this is indeed the case, first note that we can rewrite the pth summand of the multi-class Softmax cost in Equation (7.23), using the fact that $\log(e^s) = s$, as

$$\log\left(\sum_{j=0}^{C-1} e^{\mathring{\mathbf{x}}_p^T\mathbf{w}_j}\right) - \mathring{\mathbf{x}}_p^T\mathbf{w}_{y_p} = \log\left(\sum_{j=0}^{C-1} e^{\mathring{\mathbf{x}}_p^T\mathbf{w}_c}\right) - \log\left(e^{\mathring{\mathbf{x}}_p^T\mathbf{w}_{y_p}}\right). \tag{7.25}$$

Next, we can use the log property that $\log(s) - \log(t) = \log\left(\frac{s}{t}\right)$ to rewrite Equation (7.25) as

$$\log\left(\sum_{j=0}^{C-1} e^{\mathring{\mathbf{x}}_p^T\mathbf{w}_j}\right) - \log e^{\mathring{\mathbf{x}}_p^T\mathbf{w}_{y_p}} = \log\left(\frac{\sum_{j=0}^{C-1} e^{\mathring{\mathbf{x}}_p^T\mathbf{w}_j}}{e^{\mathring{\mathbf{x}}_p^T\mathbf{w}_{y_p}}}\right). \tag{7.26}$$

Finally, since $\log(s) = -\log\left(\frac{1}{s}\right)$ this can be rewritten equivalently as

$$\log\left(\frac{\sum_{j=0}^{C-1} e^{\mathring{\mathbf{x}}_p^T\mathbf{w}_j}}{e^{\mathring{\mathbf{x}}_p^T\mathbf{w}_{y_p}}}\right) = -\log\left(\frac{e^{\mathring{\mathbf{x}}_p^T\mathbf{w}_{y_p}}}{\sum_{j=0}^{C-1} e^{\mathring{\mathbf{x}}_p^T\mathbf{w}_j}}\right). \tag{7.27}$$

Altogether we can then express the multi-class Softmax in Equation (7.23) equivalently as

$$g\left(\mathbf{w}_0, ..., \mathbf{w}_{C-1}\right) = -\frac{1}{P}\sum_{p=1}^{P}\log\left(\frac{e^{\mathring{\mathbf{x}}_p^T\mathbf{w}_{y_p}}}{\sum_{j=0}^{C-1} e^{\mathring{\mathbf{x}}_p^T\mathbf{w}_j}}\right). \tag{7.28}$$

Visually, this formulation appears more similar to the two-class Cross Entropy cost in Equation (6.12), and indeed does reduce to it in quite a straightforward manner when $C = 2$ and $y_p \in \{0, 1\}$.

7.3.8 Regularization and the multi-class Softmax

As with the multi-class Percpetron cost (in Section 7.3.4), it is common to *regularize* the multi-class Softmax as

$$\frac{1}{P}\sum_{p=1}^{P}\left[\log\left(\sum_{j=0}^{C-1}e^{b_j+\mathbf{x}_p^T\boldsymbol{\omega}_j}\right)-\left(b_{y_p}+\mathbf{x}_p^T\boldsymbol{\omega}_{y_p}\right)\right]+\lambda\sum_{j=0}^{C-1}\left\|\boldsymbol{\omega}_j\right\|_2^2 \tag{7.29}$$

where, once again, we have used the bias/feature-touching weight notation allowing us to decompose each weight vector $\mathbf{w}_c$ as shown in Equation (7.6). Regularization can also help prevent local optimization methods like Newton's method (which take large steps) from diverging when dealing with perfectly separable data (see Section 6.4).

7.3.9   Python implementation

To implement either of the multi-class cost functions detailed in this section it is first helpful to rewrite our model using the matrix notation first introduced in Section 5.6. In other words we first stack the weights from our C classifiers together into a single $(N+1)\times C$ array of the form

$$\mathbf{W}=\begin{bmatrix} w_{0,0} & w_{0,1} & w_{0,2} & \cdots & w_{0,C-1} \\ w_{1,0} & w_{1,1} & w_{1,2} & \cdots & w_{1,C-1} \\ w_{2,0} & w_{2,1} & w_{2,2} & \cdots & w_{2,C-1} \\ \vdots & \vdots & \vdots & \cdots & \vdots \\ w_{N,0} & w_{N,1} & w_{N,2} & \cdots & w_{N,C-1} \end{bmatrix}. \tag{7.30}$$

Here the bias and normal vector of the cth classifier have been stacked on top of one another and make up the cth column of the array. We likewise extend our `model` notation to also denote the evaluation of our C individual linear models together as

$$\text{model}\,(\mathbf{x},\mathbf{W})=\mathring{\mathbf{x}}^T\mathbf{W}=\begin{bmatrix} \mathring{\mathbf{x}}^T\mathbf{w}_0 & \mathring{\mathbf{x}}^T\mathbf{w}_1 & \cdots & \mathring{\mathbf{x}}^T\mathbf{w}_{C-1} \end{bmatrix}. \tag{7.31}$$

This is precisely the *same* condensed linear model we used to implement multi-output regression in Section 5.6.3, which we repeat below.

```
1  # compute C linear combinations of input point, one per classifier
2  def model(x,w):
3      a = w[0] + np.dot(x.T,w[1:])
4      return a.T
```

With this `model` notation we can more conveniently implement essentially any formula derived from the fusion rule like, for example, the multi-class Perceptron. For example, we can write the fusion rule in Equation (7.12) itself equivalently as

$$y=\text{argmax}\,[\,\text{model}\,(\mathbf{x},\mathbf{W})]. \tag{7.32}$$

Likewise we can write the pth summand of the multi-class Perceptron compactly as

$$\left(\max_{c=0,\ldots,C-1} \mathring{\mathbf{x}}_p^T \mathbf{w}_c \right) - \mathring{\mathbf{x}}_p^T \mathbf{w}_{y_p} = \max \left[\text{model}\left(\mathbf{x}_p, \mathbf{W}\right) \right] - \text{model}\left(\mathbf{x}_p, \mathbf{W}\right)_{y_p} \quad (7.33)$$

where here the term $\text{model}\left(\mathbf{x}_p, \mathbf{W}\right)_{y_p}$ refers to the y_pth entry of $\text{model}\left(\mathbf{x}_p, \mathbf{W}\right)$.

Python code often runs much faster when for loops – or equivalently list comprehensions – are written equivalently using matrix-vector NumPy operations (this has been a constant theme in our implementations since linear regression in Section 5.2.4).

Below we show an example implementation of the multi-class Perceptron in Equation (7.16) that takes in the model function provided above. Note that np.linalg.fro denotes the *Frobenius* matrix norm (see Section C.5.3).

One can implement the multi-class Softmax in Equation (7.23) in an entirely similar manner.

```python
# multiclass perceptron
lam = 10**-5  # our regularization parameter
def multiclass_perceptron(w):
    # pre-compute predictions on all points
    all_evals = model(x,w)

    # compute maximum across data points
    a = np.max(all_evals,axis = 0)

    # compute cost in compact form using numpy broadcasting
    b = all_evals[y.astype(int).flatten(),np.arange(np.size(y))]
    cost = np.sum(a - b)

    # add regularizer
    cost = cost + lam*np.linalg.norm(w[1:,:],'fro')**2

    # return average
    return cost/float(np.size(y))
```

Finally, note that since any cost for multi-class classification described in this section takes in a *matrix* of parameters, when using autograd as part of your optimization process it can be very convenient to first *flatten* your chosen cost – as explained in Section B.10.3 – prior to minimization. Doing this avoids the need to explicitly loop over weights in your local optimization routine, allowing you to directly employ the basic Python implementations of, e.g., gradient descent (see Section 3.5.4) and Newton's method (see Section 4.3.6) without modification.

7.4   Which Approach Produces the Best Results?

In the previous two sections we have seen two fundamental approaches to linear multi-class classification: the One-versus-All (OvA) and the multi-class Perceptron/Softmax. Both approaches are commonly used in practice and often (depending on the dataset) produce similar results (see, e.g., [25, 26]). However, the latter approach is (at least in principle) capable of achieving higher accuracy on a broader range of datasets. This is due to the fact that with OvA we solve a sequence of C two-class subproblems (one per class), tuning the weights of our classifiers *independently* of each other. Only afterward do we combine all classifiers together to form the fusion rule. Thus the weights we learn satisfy the fusion rule *indirectly*. On the other hand, with the multi-class Perceptron or Softmax cost function minimization we are tuning all parameters of all C classifiers *simultaneously* to *directly* satisfy the fusion rule over our training dataset. This joint minimization permits potentially valuable interactions to take place in-between the two-class classifiers in the tuning of their weights that cannot take place in the OvA approach.

Example 7.5   Comparison of OvA and multi-class Perceptron
We illustrate this principal superiority of the multi-class Perceptron approach over OvA using a toy dataset with $C = 5$ classes, shown in the left panel of Figure 7.11, where points colored red, blue, green, yellow, and violet have label values $y_p = 0, 1, 2, 3$, and 4, respectively.

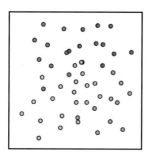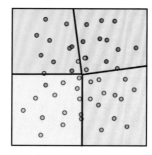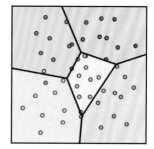

Figure 7.11 Figure associated with Example 7.5. See text for details.

Think for a moment how the OvA approach will perform in terms of the yellow colored class concentrated in the middle of the dataset, particularly how the subproblem in which we distinguish between members of this class and all others will be solved. Because this class of data is surrounded by members of the other classes, and there are fewer members of the yellow class than all other classes combined, the optimal linear classification rule for this subproblem is to classify all points as nonyellow (or in other words, to misclassify the entire yellow class). This implies that the linear decision boundary will lie outside the range of the points shown, with all points in the training data lying on its *negative side*. Since the weights of decision boundary associated with the yellow colored class are tuned solely based on this subproblem, this will lead to the entire yellow class being misclassified in the final OvA solution provided by the fusion rule, as shown in the middle panel of Figure 7.11.

On the other hand, if we employ the multi-class Perceptron or Softmax approach we will not miss this class since all $C = 5$ two-class classifiers are learned *simultaneously*, resulting in a final fused decision boundary that is far superior to the one provided by OvA, as shown in the right panel of Figure 7.11. We misclassify far fewer points and, in particular, do not misclassify the entire yellow class of data.

7.5 The Categorical Cross Entropy Cost Function

In the previous sections we employed by default the numerical label values $y_p \in \{0, 1, ..., C - 1\}$. However, as in the case of two-class classification (see Section 6.7), the choice of label values with multi-class classification is also *arbitrary*. Regardless of how we define label values we still end up with precisely the same multi-class scheme we saw in Section 7.3 and cost functions like the multi-class Softmax.

In this section we see how to use *categorical* labels, that is labels that have no intrinsic numerical order, to perform multi-class classification. This perspective introduces the notion of a *discrete probability distribution* as well as the notion of a *Categorical Cross Entropy cost*, which (as we will see) is completely equivalent to the multi-class Softmax/Cross Entropy cost function we saw in Section 7.3.

7.5.1 Discrete probability distributions

Suppose you took a poll of ten friends or work colleagues inquiring if they owned a pet cat or dog. From this group you learned that three people owned neither a cat nor a dog, one person owned a cat, and six people owned dogs. Building a corresponding data vector $\mathbf{s} = \begin{bmatrix} 3, 1, 6 \end{bmatrix}$ from this survey response, $\mathbf{s}$ can be represented visually as a *histogram* where the value of each entry is represented by a vertical bar whose height is made proportional to its respective value. A histogram of this particular vector is shown in the left panel of Figure 7.12.

It is quite common to *normalize* data vectors like this one so that they can be interpreted as a *discrete probability distribution*. To do so, the normalization must be done in a way to ensure that (i) the numerical ordering of its values (from smallest to largest) is retained, (ii) its values are nonnegative, and (iii) its values sum exactly to 1. For a vector of all nonnegative entries this can be done by

Figure 7.12 (left panel) Visual representation of a sample data vector $\mathbf{s} = \begin{bmatrix} 3, 1, 6 \end{bmatrix}$, resulting from a pet ownership survey and shown as a histogram. (right panel) Normalizing this vector by dividing it off by the sum of its entries allows us to treat it as a discrete probability distribution.

simply dividing it by the sum of its values. With the toy example we use here since its (nonnegative) values sum to $3 + 1 + 6 = 10$ this requires dividing all entries of $\mathbf{s}$ by 10 as

$$\mathbf{s} = \begin{bmatrix} 0.3, 0.1, 0.6 \end{bmatrix}. \tag{7.34}$$

This *normalized histogram* (sometimes referred to as a probability mass function) is shown in the right panel of Figure 7.12.

7.5.2   Exponential normalization

This kind of normalization can be done in general for any length C vector $\mathbf{s} = \begin{bmatrix} s_0, s_1, ..., s_{C-1} \end{bmatrix}$ with potentially negative elements. Exponentiating each element in $\mathbf{s}$ gives the vector

$$\begin{bmatrix} e^{s_0}, e^{s_1}, ..., e^{s_{C-1}} \end{bmatrix} \tag{7.35}$$

where all entries are now guaranteed to be nonnegative. Notice also that exponentiation maintains the ordering of values in $\mathbf{s}$ from small to large.[1] If we now divide off the sum of this exponentiated version of $\mathbf{s}$ from each of its entries, as

$$\sigma(\mathbf{s}) = \left[\frac{e^{s_0}}{\sum_{c=0}^{C-1} e^{s_c}}, \frac{e^{s_1}}{\sum_{c=0}^{C-1} e^{s_c}}, ..., \frac{e^{s_{C-1}}}{\sum_{c=0}^{C-1} e^{s_c}} \right] \tag{7.36}$$

we not only maintain the two aforementioned properties (i.e., nonnegativity and numerical order structure) but, in addition, satisfy the third property of a valid discrete probability distribution: all entries now sum to 1.

The function $\sigma(\cdot)$ defined in Equation (7.36) is called a *normalized exponential*

---

[1] This is because the exponential function $e^{(\cdot)}$ is always monotonically increasing.

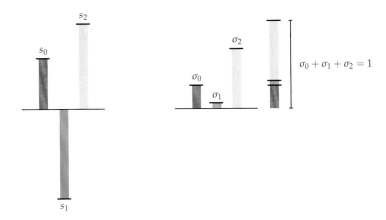

Figure 7.13 (left panel) A vector of length $C = 3$ shown as a histogram. (right panel) Taking the normalized exponential of this vector, as defined in Equation (7.36), produces a new vector of all nonnegative entries whose numerical order is preserved, and whose total value sums to 1.

function.[2] It is often used so that we may *interpret* an arbitrary vector **s** (possibly containing negative as well as positive entries) as a discrete probability distribution (see Figure 7.13), and can be thought of as a generalization of the *logistic sigmoid* introduced in Section 6.2.3.

7.5.3 Exponentially normalized signed distances

In the previous two sections we relied on a point **x**'s *signed distance* to each of the C individual decision boundaries (or something very close to it if we do not normalize feature-touching weights) to properly determine class membership. This is codified directly in the fusion rule (given in, for example, Equation (7.12)) itself.

For a given setting of our weights for all C two-class classifiers the evaluation of **x** through all decision boundaries produces C signed distance measurements

$$\mathbf{s} = \begin{bmatrix} \mathring{\mathbf{x}}^T \mathbf{w}_0 & \mathring{\mathbf{x}}^T \mathbf{w}_1 & \cdots & \mathring{\mathbf{x}}^T \mathbf{w}_{C-1} \end{bmatrix} \tag{7.37}$$

which we can think of as a *histogram*.

Because – as we have seen above – the normalized exponential function preserves numerical order we can likewise consider the exponentially normalized

---

[2] This function is sometimes referred to as *Softmax activation* in the context of neural networks. This naming convention is unfortunate, as the normalized exponential is *not* a soft version of the max function as the rightly named *Softmax function* detailed in Section 7.2.5 is, and should not be confused with it. While it is a transformation that does preserve the index of the largest entry of its input, it is *not* a soft version of the argmax function either as it is sometimes erroneously claimed to be.

signed distance in determining proper class ownership. Denoting by $\sigma(\cdot)$ the normalized exponential our generic histogram of signed distances becomes

$$\sigma(\mathbf{s}) = \left[\frac{e^{\mathring{x}^T \mathbf{w}_0}}{\sum_{c=0}^{C-1} e^{\mathring{x}^T \mathbf{w}_c}} \quad \frac{e^{\mathring{x}^T \mathbf{w}_1}}{\sum_{c=0}^{C-1} e^{\mathring{x}^T \mathbf{w}_c}} \quad \cdots \quad \frac{e^{\mathring{x}^T \mathbf{w}_{C-1}}}{\sum_{c=0}^{C-1} e^{\mathring{x}^T \mathbf{w}_c}} \right]. \tag{7.38}$$

Transforming the histogram of signed distance measurements also gives us a way of considering class ownership *probabilistically*. For example, if for a particular setting of the entire set of weights gave for a particular point $\mathbf{x}_p$

$$\sigma(\mathbf{s}) = \left[0.1,\ 0.7,\ 0.2 \right] \tag{7.39}$$

then while we would still *assign a label based on the fusion rule* (see Equation (7.12)) – here assigning the label $y_p = 1$ since the second entry 0.7 of this vector is largest – we could also add a note of confidence that "$y_p = 1$ with 70 percent probability."

Example 7.6   Signed distances as a probability distribution
In Figure 7.14 we use the prototype $C = 3$ class dataset (shown in Figure 7.1) and visualize both signed distance vectors as histograms for several points (left panel) as well as their normalized exponential versions (right panel). Note how the largest positive entry in each original histogram shown on the left panel remains the largest positive entry in the normalized version shown on the right panel.

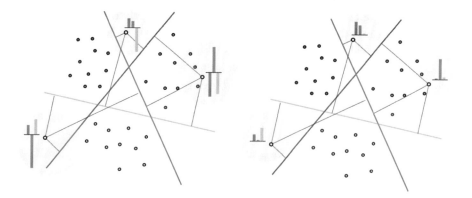

Figure 7.14 (left panel) Histogram visualizations of signed distance measurements of three exemplar points in a $C = 3$ class dataset. (right panel) Exponentially normalized signed distance measurements visualized as histograms for the same three points.

7.5.4 Categorical classification and the Categorical Cross Entropy cost

Suppose we begin with a multi-class classification dataset $\left\{\left(\mathbf{x}_p, y_p\right)\right\}_{p=1}^P$ with N-dimensional input and transform our numerical label values $y_p \in \{0, 1, ..., C - 1\}$ with *one-hot encoded vectors* of the form

$$y_p = 0 \longleftarrow \mathbf{y}_p = \left[1, 0, \cdots 0, 0\right]$$

$$y_p = 1 \longleftarrow \mathbf{y}_p = \left[0, 1, \cdots 0, 0\right] \qquad (7.40)$$

$$\vdots$$

$$y_p = C - 1 \longleftarrow \mathbf{y}_p = \left[0, 0, \cdots 0, 1\right].$$

Here each one-hot encoded *categorical label* is a length C vector and contains all zeros except a 1 in the index equal to the value of y_p (note that the first entry in this vector has index 0 and the last $C - 1$).

Each vector representation uniquely identifies its corresponding label value, but now our label values are no longer ordered numerical values, and our dataset now takes the form $\left\{\left(\mathbf{x}_p, \mathbf{y}_p\right)\right\}_{p=1}^P$, where $\mathbf{y}_p$ are one-hot encoded labels defined as above. Our goal, however, remains the same: to properly tune the weights of our C One-versus-All two-class classifiers to learn the best correspondence between the N-dimensional input and C-dimensional output of our training dataset.

With vector-valued output, instead of scalar numerical values, we can phrase multi-class classification as an instance of *multi-output regression* (see Section 5.6). In other words, denoting by $\mathbf{W}$ the $(N + 1 \times C)$ matrix of weights for all C classifiers (see Section 7.3.9) we can aim at tuning $\mathbf{W}$ so that the approximate linear relationship holds for all our points

$$\mathring{\mathbf{x}}_p^T \mathbf{W} \approx \mathbf{y}_p. \qquad (7.41)$$

However, since now our output $\mathbf{y}_p$ does not consist of continuous values but one-hot encoded vectors, a linear relationship would not represent such vectors very well at all. Entries of the left-hand side for a given p can be nonnegative, less than zero or greater than one, etc. However, taking the *normalized exponential* transform of our linear model normalizes it in such a way (by forcing all its entries to be nonnegative and sum exactly to one) so that we could reasonably propose to tune $\mathbf{W}$ so that

$$\sigma\left(\mathring{\mathbf{x}}_p^T \mathbf{W}\right) \approx \mathbf{y}_p \qquad (7.42)$$

holds as tightly as possible over our training dataset (where $\sigma\left(\cdot\right)$ is the normalized exponential). Interpreting $\sigma\left(\mathring{\mathbf{x}}_p^T \mathbf{W}\right)$ as a discrete probability distribution, this is saying that we want to tune the weights of our model so that this distribution concentrates completely at index y_p, i.e., the only nonzero entry of the one-hot encoded output $\mathbf{y}_p$.

To learn our weights properly we could employ a standard point-wise regression cost such as the Least Squares (see Section 5.2)

$$g_p(\mathbf{W}) = \left\| \sigma\left(\mathring{\mathbf{x}}_p^T \mathbf{W}\right) - \mathbf{y}_p \right\|_2^2. \tag{7.43}$$

However, as we discussed in Section 6.2, a more appropriate point-wise cost when dealing with binary output is the log error since it more heavily penalizes errors less than one in such instances. Here the log error of $\sigma\left(\mathring{\mathbf{x}}_p^T \mathbf{W}\right)$ and $\mathbf{y}_p$ can be written as

$$g_p(\mathbf{W}) = -\sum_{c=0}^{C} y_{p,c} \log \sigma\left(\mathring{\mathbf{x}}_p^T \mathbf{W}\right)_c \tag{7.44}$$

where $y_{p,c}$ is the cth entry of the one-hot encoded label $\mathbf{y}_p$. Note that this formula simplifies considerably since $\mathbf{y}_p$ is a one-hot encoded vector, hence all but one summand on the right-hand side above equals zero. This is precisely the original label integer label c of the point $\mathbf{x}_p$, thus the cth index of $\mathbf{y}_p$ equals one i.e., $y_{p,c} = 1$. This means that the above simplifies too

$$g_p(\mathbf{W}) = -\log \sigma\left(\mathring{\mathbf{x}}_p^T \mathbf{W}\right)_{y_p} \tag{7.45}$$

and from the definition of the normalized exponential this is precisely

$$g_p(\mathbf{W}) = -\log\left(\frac{e^{\mathring{\mathbf{x}}_p^T \mathbf{w}_{y_p}}}{\sum_{c=0}^{C-1} e^{\mathring{\mathbf{x}}_p^T \mathbf{w}_c}}\right). \tag{7.46}$$

If we then form a cost function by taking the average of the above over all P training data points we have

$$g(\mathbf{W}) = -\frac{1}{P}\sum_{p=1}^{P} \log\left(\frac{e^{\mathring{\mathbf{x}}_p^T \mathbf{w}_{y_p}}}{\sum_{c=0}^{C-1} e^{\mathring{\mathbf{x}}_p^T \mathbf{w}_c}}\right). \tag{7.47}$$

This is precisely a form of the standard multi-class Cross Entropy/Softmax cost function we saw in Section 7.3.7 in Equation (7.28) where we used numerical label values $y_p \in \{0, 1, ..., C-1\}$.

7.6   Classification Quality Metrics

In this section we describe simple metrics for judging the quality of a trained multi-class classification model, as well as how to make predictions using one.

7.6.1   Making predictions using a trained model

If we denote the optimal set of weights for the cth One-versus-All two-class classifier as $\mathbf{w}_c^{\star}$ – found by minimizing a multi-class classification cost function in Section 7.3 or via performing OvA as detailed in Section 7.2 – then to predict the label y' of an input $\mathbf{x}'$ we employ the *fusion rule* as

$$y' = \underset{c=0,...,C-1}{\mathrm{argmax}}\ \mathbf{x}'^{T}\mathbf{w}_c^{\star} \tag{7.48}$$

where any point lying *exactly* on the decision boundary should be assigned a label randomly based on the index of those classifiers providing maximum evaluation. Indeed this set of points forms the multi-class decision boundary – illustrated in black for a toy $C = 3$ class dataset in the left panel of Figure 7.15 – where the regions in this image have been colored based on predicted label provided by the fusion rule evaluation of each point in the input space. The right panel of this image shows the same dataset and the individual One-versus-All boundaries (where each is colored based on the class it distinguishes from the remainder of the data), and for a point $\mathbf{x}'$ depicts its fusion rule evaluation as the maximum distance to each One-versus-All boundary (which is roughly what is computed by the fusion rule evaluation – see Section 7.2.4).

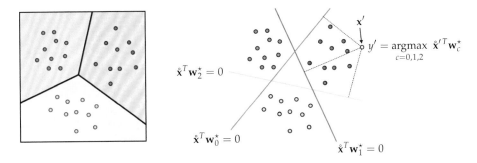

Figure 7.15 (left panel) A toy $C = 3$ class dataset, multi-class decision boundary (shown in black, along with regions colored according to the prediction provided by the fusion rule. (right panel) The same dataset shown with each One-versus-All decision boundary (with each is colored based on the class it distinguishes from the remainder of the data), which – when combined via the fusion rule – provides the multi-class decision boundary shown in the left panel. Here the label y' of a new point $\mathbf{x}'$, shown as a hollow circle in the upper-right, is predicted via the fusion rule which measures – roughly speaking – the maximum distance from this point to each One-versus-All decision boundary.

7.6.2   Confidence scoring

Once a proper decision boundary is learned, we can describe the *confidence* we have in any point based on *the point's distance to the decision boundary*, in the same way we can with two-class data (see Section 6.8.2). More specifically, we can use its exponentially normalized distance to score our confidence in the prediction, as described in Section 7.5.3.

7.6.3   Judging the quality of a trained model using accuracy

To count the number of misclassifications a trained multi-class classifier forms over our training dataset, we simply take a raw count of the number of training data points $\mathbf{x}_p$ whose true label y_p is predicted *incorrectly*. To compare the point $\mathbf{x}_p$'s predicted label $\hat{y}_p = \underset{j=0,\dots,C-1}{\operatorname{argmax}} \, \mathbf{x}_p^T \mathbf{w}_j^\star$ and true true label y_p we can use an identity function $\mathcal{I}\left(\cdot\right)$ and compute

$$\mathcal{I}\left(\hat{y}_p, y_p\right) = \begin{cases} 0 & \text{if } \hat{y}_p = y_p \\ 1 & \text{if } \hat{y}_p \neq y_p. \end{cases} \tag{7.49}$$

Taking a sum of the above over all P points gives the total number of misclassifications of our trained model

$$\text{number of misclassifications} = \sum_{p=1}^{P} \mathcal{I}\left(\hat{y}_p, y_p\right). \tag{7.50}$$

Using this we can also compute the *accuracy*, denoted $\mathcal{A}$, of a trained model. This is simply the normalized number of training data points whose labels are correctly predicted by the model, defined as

$$\mathcal{A} = 1 - \frac{1}{P} \sum_{p=1}^{P} \mathcal{I}\left(\hat{y}_p, y_p\right). \tag{7.51}$$

The accuracy ranges from 0 (no points are classified correctly) to 1 (all points are classified correctly).

Example 7.7   Comparing cost function and counting cost values

In Figure 7.16 we compare the number of misclassifications versus the value of the multi-class Softmax cost in classifying the $C = 5$ class dataset shown in Figure 7.11 over three runs of standard gradient descent using a steplength parameter $\alpha = 10^{-2}$ for all three runs.

Comparing the left and right panels of the figure we can see that the number of misclassifications and Softmax evaluations at each step of a gradient descent run do not perfectly track one another. That is, it is not the case that just because the cost function value is decreasing that so too is the number of misclassifications (very much akin to the two-class case). This occurs because our Softmax cost is

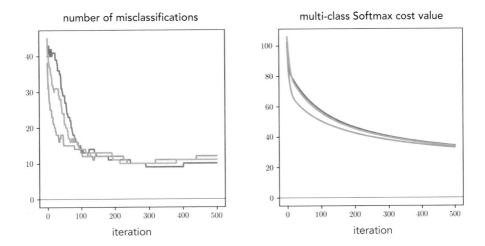

number of misclassifications   multi-class Softmax cost value

iteration   iteration

Figure 7.16 Figure associated with Example 7.7. See text for further details.

only an approximation of the true quantity we would like to minimize, i.e., the number of misclassifications.

This simple example has an extremely practical implication: while we tune the weights of our model by minimizing an appropriate multi-class cost, after minimization the best set of weights from the run are associated with the lowest *number of misclassifications* (or likewise the highest accuracy) **not** the lowest cost function value.

7.6.4 Advanced quality metrics for dealing with unbalanced classes

The advanced quality metrics we saw in the case of two-class classification to deal with severe class imbalance, including balanced accuracy (see Section 6.8.4) as well as further metrics defined by a confusion matrix (see Section 6.8.5), can be directly extended to deal with class-imbalance issues in the multi-class context. These direct extensions are explored further in this chapter's exercises.

Example 7.8   Confusion matrix for a toy dataset
In the left panel of Figure 7.17 we show the result of a fully tuned multi-class classifier trained on the dataset first shown in Figure 7.7. The *confusion matrix* corresponding to this classifier is shown in the right panel of the figure. The (i, j)th entry of this matrix counts the number of training data points that have *true* label $y = i$ and *predicted* label $\hat{y} = j$.

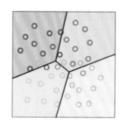

Figure 7.17 Figure associated with Example 7.9. See text for details.

$$
\begin{array}{c}
\begin{array}{cccc}
\text{red} & \text{blue} & \text{green} & \text{yellow}
\end{array} \\
\begin{array}{c}
\text{red} \\ \text{blue} \\ \text{green} \\ \text{yellow}
\end{array}
\left(
\begin{array}{cccc}
8 & 1 & 1 & 0 \\
1 & 7 & 1 & 1 \\
1 & 1 & 7 & 1 \\
0 & 1 & 1 & 8
\end{array}
\right)
\end{array}
$$

Example 7.9   The Iris dataset

In this example we explore the application of linear multi-class classification to the popular Iris dataset, taken from [27]. This dataset consists of a set of $P = 150$ statistical measurements on $C = 3$ types of flowers (the *Iris Setosa*, *Iris Versicolor*, and *Iris Virginica* species), with 50 of each type of flower represented in the data. Each input point consists of $N = 4$ features including sepal length and width as well as petal length and width (all measured in centimeters).

Minimizing the multi-class Softmax cost we can easily attain almost perfect classification of this dataset, resulting in two misclassifications and the following confusion matrix.

		Predicted		
		Setosa	*Versicolor*	*Virginica*
Actual	*Setosa*	50	0	0
	Versicolor	0	49	1
	Virginica	0	1	49

7.7   Weighted Multi-Class Classification

Weighted multi-class classification arises for precisely the same reasons described for two-class classification in Section 6.9; that is, as a way of including a notion of confidence in data points and for dealing with severe class imbalances. One can easily derive weighted versions of the multi-class Perceptron/Softmax cost functions that completely mirror the two-class analog detailed in this earlier section (see Exercise 7.10).

7.8   Stochastic and Mini-Batch Learning

In Appendix Section A.5 we describe a simple extension of the standard gradient descent algorithm (outlined in Chapter 3) called *mini-batch gradient descent*. This approach is designed to accelerate the minimization of cost functions consisting of a sum of P terms

$$g(\mathbf{w}) = \sum_{p=1}^{P} g_p(\mathbf{w}) \tag{7.52}$$

where g_1 g_2, ..., g_P are all functions of the same kind that take in the same parameters.

As we have seen, every supervised learning cost function looks like this – including those used for regression, two-class, and multi-class classification. Each g_p is what we have generally referred to as a point-wise cost that measures the error of a particular model on the pth point of a dataset. For example, with the Least Squares cost we saw in Section 5.2 that the point-wise cost took the form $g_p(\mathbf{w}) = \left(\mathring{\mathbf{x}}_p^T \mathbf{w} - y_p\right)^2$, with the two-class Softmax (in Section 6.4.3) it took the form $g_p(\mathbf{w}) = -\log\left(\sigma\left(y_p \mathring{\mathbf{x}}_p^T \mathbf{w}\right)\right)$, and with the multi-class Softmax (in Section 7.3.5) the form $g_p(\mathbf{W}) = \left[\log\left(\sum_{c=0}^{C-1} e^{\mathring{\mathbf{x}}_p^T \mathbf{w}_c}\right) - \mathring{\mathbf{x}}_p^T \mathbf{w}_{y_p}\right]$. More generally, as we will see moving forward (as with, for example, the linear Autoencoder described in Section 8.3), *every machine learning cost function takes this form* (because they always decompose over their training data) where g_p is a point-wise cost of the pth point of a dataset. Because of this we can directly apply mini-batch optimization in tuning their parameters.

7.8.1   Mini-batch optimization and online learning

As detailed in Appendix Section A.5, the heart of the mini-batch idea is to minimize such a cost *sequentially* over small mini-batches of its summands, one mini-batch of summands at a time, as opposed to the standard local optimization step where minimize in the entire set of summands at once. Now, since machine learning cost summands are inherently tied to training data points, in the context of machine learning we can think about mini-batch optimization *equivalently in terms of mini-batches of training data* as well. Thus, as opposed to a standard (also called *full batch*) local optimization that takes individual steps by sweeping through an *entire* set of training data *simultaneously*, the mini-batch approach has us take smaller steps sweeping through training data *sequentially* (with one complete sweep through the data being referred to as an *epoch*). This machine learning interpretation of mini-batch optimization is illustrated schematically in the Figure 7.18.

As with generic costs, mini-batch learning often greatly accelerates the minimization of machine learning cost functions (and thus the corresponding learn-

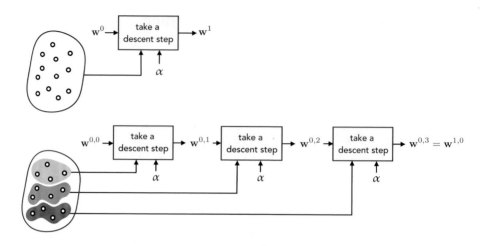

Figure 7.18 Schematic comparison of first iteration of (top panel) full batch and (bottom) stochastic gradient descent, through the lens of machine learning using (for the sake of simplicity) a small dataset of points. In the full batch sweep we take a step in all points *simultaneously*, whereas (bottom panel) in the mini-batch approach we sweep through these points *sequentially* as if we received the data in an *online fashion*.

ing taking place) – and is most popularly paired with gradient descent (Section 3.5) or one of its advanced analogs (see Appendix Section A.4). This is particularly true when dealing with *very large datasets*, i.e., when P is large (see e.g., [21, 28]). With very large datasets the mini-batch approach can also help limit the amount of active memory consumed in storing data by loading in – at each step in a mini-batch epoch – *only the data included in the current mini-batch*. The mini-batch approach can also be used (or interpreted) as a so-called *online learning* technique, wherein data actually *arises* in small mini-batches and is directly used to update the parameters of the associated model.

Example 7.10   Recognition of handwritten digits
In this example we perform handwritten digit recognition (as introduced in Example 1.10) via multi-class classification. In Figure 7.19 we illustrate the accelerated convergence of mini-batch gradient descent over the standard gradient descent method using the multi-class Softmax cost and $P = 50,000$ randomly selected training points from the MNIST dataset [29], a popular collection of handwritten images like those shown in Figure 1.13. In particular we show a comparison of the first 10 steps/epochs of both methods, using a batch of size 200 for the mini-batch version and the same steplength for both runs, where we see that the mini-batch run drastically accelerates minimization in terms of both the cost function (left panel) and number of misclassifications (right panel).

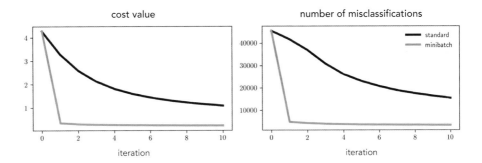

Figure 7.19 Figure associated with Example 7.10. A comparison of the progress made by standard (in black) and mini-batch (in magenta) gradient descent over the MNIST training dataset in terms of cost function value (left panel) and number of misclassifications (right panel). The mini-batch approach makes significantly faster progress than standard gradient descent in both measures.

7.9 Conclusion

In this chapter we went from two to multi-class linear classification, detailing a range of perspectives on mirroring our discussion in the previous chapter.

We began with the *One-versus-All* (or OvA for short) approach outlined in Section 7.2 which involves intelligently combining the results of several one-versus-all two-class classifiers. In Section 7.3 we then saw how to train this set of two-class classifiers *simultaneously* by minimizing *multi-class Perceptron* or *Cross Entropy/Softmax* cost functions. This simultaneous approach – at least in principle – can provide superior performance when compared to OvA, but in practice both approaches often lead to similar results (as detailed in Section 7.4). Next in Section 7.5 we discussed the arbitrary nature of choosing label values, by detailing the most arbitrary choice: categorical labels. Multi-class metrics – natural extensions of those metrics first described for two-class problems in Section 6.8 – were then detailed in Section 7.6. Finally, picking up on our discussion in Section A.5, we described mini-batch optimization in the context of machine learning in Section 7.8.

7.10 Exercises

† The data required to complete the following exercises can be downloaded from the text's github repository at `github.com/jermwatt/machine_learning_refined`

7.1 One-versus-All classification pseudo-code
Write down a psuedo-code that summarizes the One-versus-All (OvA) algorithm described in Section 7.2.

7.2   One-versus-All classification

Repeat the experiment described in Example 7.1. You need not reproduce the illustrations shown in Figure 7.7; however, you should make sure your trained model achieves a similar result (in terms of the number of misclassifications) as the one shown there (having less than ten misclassifications).

7.3   Multi-class Perceptron

Repeat the experiment described in Example 7.3. You need not reproduce the illustrations shown in Figure 7.9; however, you should make sure your trained model achieves zero misclassifications. You can use the `Python` implementation outlined in Section 7.3.9 of the multi-class Perceptron cost.

7.4   The multi-class and two-class Perceptrons

Finish the argument started in Section 7.3.3 to show that the multi-class Perceptron cost in Equation (7.16) reduces to the two-class Perceptron cost in Equation (6.33).

7.5   Multi-class Softmax

Repeat the experiment described in Example 7.4 using any local optimization method. You need not reproduce the illustrations shown in Figure 7.10; however, you should make sure your trained model achieves a small number of misclassifications (ten or fewer). You can use the `Python` implementation outlined in Section 7.3.9 as a basis for you implementation of the multi-class Softmax cost.

7.6   Show the multi-class Softmax reduces to two-class Softmax when $C = 2$

Finish the argument started in Section 7.3.7 to show that the multi-class Softmax cost in Equation (7.23) reduces to the two-class Softmax cost in Equation (6.34).

7.7   Hand-calculations with the multi-class Softmax cost

Show that the Hessian of the multi-class Softmax cost function can be computed block-wise, for $s \neq c$, as

$$\nabla^2_{\mathbf{w}_c \mathbf{w}_s} g = -\sum_{p=1}^{P} \frac{e^{\mathbf{x}_p^T \mathbf{w}_c + \mathbf{x}_p^T \mathbf{w}_s}}{\left(\sum_{d=1}^{C} e^{\mathbf{x}_p^T \mathbf{w}_d} \right)^2} \mathbf{x}_p \mathbf{x}_p^T \tag{7.53}$$

and for $s = c$, as

$$\nabla^2_{\mathbf{w}_c\mathbf{w}_c} g = \sum_{p=1}^{P} \frac{e^{\mathbf{x}_p^T \mathbf{w}_c}}{\sum_{d=1}^{C} e^{\mathbf{x}_p^T \mathbf{w}_d}} \left(1 - \frac{e^{\mathbf{x}_p^T \mathbf{w}_c}}{\sum_{d=1}^{C} e^{\mathbf{x}_p^T \mathbf{w}_d}} \right) \mathbf{x}_p \mathbf{x}_p^T. \tag{7.54}$$

7.8   The multi-class Perceptron and Softmax costs are convex

Show that the multi-class Perceptron and Softmax costs are always convex (regardless of the dataset used). To do this you can use, e.g., the zero-order definition of convexity (see Exercise 5.8).

7.9   Balanced accuracy in the multi-class setting

Extend the notion of balanced accuracy, detailed in the context of two-class classification (see Section 6.8.4), to the multi-class setting. In particular give an equation for the balanced accuracy in the context of multi-class classification that is analogous to Equation (6.79) for the two-class case.

7.10   Weighted multi-class Softmax

A general weighted version of the two-class Softmax cost is given in Equation (6.83). What will the analogous weighted multi-class Softmax cost function look like? If we set the weights the of cost function to deal with class imbalance – as detailed for the two-class case in Section 6.9.3 – how should we set the weight values?

7.11   Recognizing handwritten digits

Repeat the experiment outlined in Example 7.10 by implementing mini-batch gradient descent. You may not obtain precisely the same results shown in Figure 7.19 based on your implementation, initialization of the algorithm, etc.; however, you should be able to recreate the general result.

8   Linear Unsupervised Learning

8.1   Introduction

In this chapter we discuss several useful techniques for *unsupervised learning*, broadly introduced in Section 1.3, which are designed to reduce the dimension of a given dataset by intelligently reducing either its number of input features or data points. These techniques can be employed as a preprocessing step to supervised learning, allowing them to scale to larger datasets, or for the sake of human in-the-loop data analysis. We begin the chapter by reviewing the notion of a *spanning set* from vector algebra, and then detail the linear *Autoencoder*, our fundamental feature dimension reduction tool, and address the highly related topic of *Principal Component Analysis*. We then discuss the *K-means clustering* algorithm, *Recommender Systems*, and end with a discussion of general *matrix factorization problems*.

8.2   Fixed Spanning Sets, Orthonormality, and Projections

In this section we review the rudimentary concepts from linear algebra that are crucial to understanding unsupervised learning techniques. The interested reader needing a refresher in basic vector and matrix operations (which are critical to understanding the concepts presented here) may find a review of such topics in Sections C.2 and C.3.

8.2.1   Notation

As we have seen in previous chapters, data associated with *supervised* tasks of regression and classification always comes in as input/output pairs. Such dichotomy does not exist with *unsupervised* learning tasks wherein a typical dataset is written simply as a set of P (input) points

$$\{\mathbf{x}_1, \, \mathbf{x}_2, \ldots, \mathbf{x}_P\} \tag{8.1}$$

or $\left\{\mathbf{x}_p\right\}_{p=1}^{P}$ for short, all living in the same N-dimensional space. Throughout the remainder of this section we will assume that our dataset has been *mean-centered*:

a simple and completely reversible operation that involves subtracting off the mean of the dataset along each input dimension so that it straddles the origin.

As illustrated in Figure 8.1, when thinking about points in a multi-dimensional vector space we can picture them either as *dots* (as shown in the left panel), or as *arrows* stemming from the origin (as shown in the middle panel). The former is how we have chosen to depict our regression and classification data thus far in the book since it is just more aesthetically pleasing to picture linear regression as fitting of a line to a *scatter of dots* as opposed to a *collection of arrows*. The equivalent latter perspective, however (i.e., viewing multi-dimensional points as arrows), is the conventional way vectors are depicted, e.g., in any standard linear algebra text. In discussing unsupervised learning techniques it is often helpful to visualize points living in an N-dimensional space using both of these conventions, some as *dots* and some as *arrows* (as shown in the right panel of Figure 8.1). Those vectors drawn as arrows are particular points, often called a *basis* or *spanning set of vectors*, over which we aim to represent every other point in the space.

Notationally, we denote a spanning set by

$$\{\mathbf{c}_1, \mathbf{c}_2, \ldots, \mathbf{c}_K\} \tag{8.2}$$

or, $\{\mathbf{c}_k\}_{k=1}^{K}$ for short.

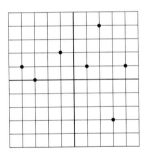 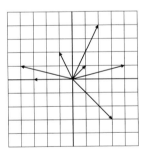 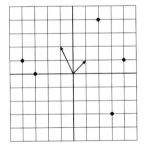

Figure 8.1 Two-dimensional points illustrated as dots (left panel), arrows (middle panel), and a mixture of both (right panel). Those shown as arrows on the right are a *basis* or *spanning set* over which we aim to represent every point in the entire space.

8.2.2   Perfect representation of data using fixed spanning sets

A spanning set is said to be capable of perfectly representing all P of our points if we can express each data point $\mathbf{x}_p$ as some linear combination of our spanning set's members, as

$$\sum_{k=1}^{K} \mathbf{c}_k w_{p,k} = \mathbf{x}_p, \qquad p = 1, \ldots, P. \tag{8.3}$$

Generally speaking two simple conditions, if met by a spanning set of vectors $\{\mathbf{c}_k\}_{k=1}^{K}$, guarantee all P equalities in Equation (8.3) to hold regardless of the dataset $\{\mathbf{x}_p\}_{p=1}^{P}$ used: (i) $K = N$, or in other words, the number of spanning vectors matches the data dimension,[1] and (ii) all spanning vectors are linearly independent.[2] For such a spanning set Equation (8.3) can be written more compactly as

$$\mathbf{C}\mathbf{w}_p = \mathbf{x}_p, \qquad p = 1, ..., P, \qquad (8.4)$$

where the spanning matrix $\mathbf{C}$ is formed by stacking the spanning vectors column-wise

$$\mathbf{C} = \begin{bmatrix} | & | & \cdots & | \\ \mathbf{c}_1 & \mathbf{c}_2 & & \mathbf{c}_N \\ | & | & \cdots & | \end{bmatrix} \qquad (8.5)$$

and where the linear combination weights are the stacked into column vectors $\mathbf{w}_p$

$$\mathbf{w}_p = \begin{bmatrix} w_{p,1} \\ w_{p,2} \\ \vdots \\ w_{p,N} \end{bmatrix} \qquad (8.6)$$

for all $p = 1, ..., P$.

To tune the weights in each $\mathbf{w}_p$ we can form an associated Least Squares cost function (as we have done multiple times previously, for example, with linear regression in Section 5.2) that, when minimized, forces the equalities in Equation (8.4) to hold

$$g(\mathbf{w}_1, ..., \mathbf{w}_P) = \frac{1}{P} \sum_{p=1}^{P} \left\| \mathbf{C}\mathbf{w}_p - \mathbf{x}_p \right\|_2^2. \qquad (8.7)$$

This Least Squares cost function can be minimized via any local optimization method. In particular we can use the first-order condition (Section 3.2) in each weight vector $\mathbf{w}_p$ *independently*, with the corresponding first-order system taking the form

$$\mathbf{C}^T \mathbf{C} \mathbf{w}_p = \mathbf{C}^T \mathbf{x}_p. \qquad (8.8)$$

This is an $N \times N$ symmetric linear system that can be easily solved numerically

---

[1] Otherwise if $K < N$, some portions of the space will definitely be out of the spanning vectors' reach.

[2] See Section C.2.4 if unfamiliar with the notion of linear independence.

(see, e.g., Section 3.2.2 and Example 3.6). Once solved, the optimally tuned weight vector $\mathbf{w}_p^\star$ for the point $\mathbf{x}_p$ is often referred to as the *encoding* of the point over the spanning matrix $\mathbf{C}$. The actual linear combination of the spanning vectors for each $\mathbf{x}_p$ (i.e., $\mathbf{C}\mathbf{w}_p^\star$) is likewise referred to as the *decoding* of the point.

Example 8.1   Data encoding
In the left panel of Figure 8.2 we show an $N = 2$ dimensional toy dataset centered at the origin, along with the spanning set $\mathbf{C} = \begin{bmatrix} 2 & 1 \\ 1 & 2 \end{bmatrix}$ shown as two red arrows. Minimizing the Least Squares cost in Equation (8.7), in the right panel we show the *encoded* version of this data, plotted in a new space whose coordinate axes are now in line with the two spanning vectors.

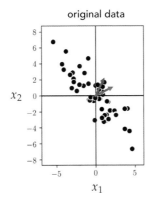
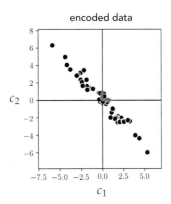

Figure 8.2 Figure associated with Example 8.1. A toy dataset (left panel) with spanning vectors shown as red arrows, and its encoding (right panel). See text for further details.

8.2.3   Perfect representation of data using fixed orthonormal spanning sets

An *orthonormal* basis or spanning set is a very special kind of spanning set whose elements (i) have unit length, and (ii) are perpendicular or orthogonal to each other. Algebraically this means that vectors belonging to an orthonormal spanning set satisfy the following condition

$$\mathbf{c}_i^T \mathbf{c}_j = \begin{cases} 1 & \text{if } i = j \\ 0 & \text{if } i \neq j \end{cases} \tag{8.9}$$

which can be expressed equivalently but more compactly, in terms of the spanning matrix $\mathbf{C}$, as

$$\mathbf{C}\,\mathbf{C}^T = \mathbf{I}_{N \times N}. \tag{8.10}$$

Because of this very special property of orthonormal spanning sets we can solve for the ideal weights $\mathbf{w}_p$ (or *encoding*) of the point $\mathbf{x}_p$ immediately, since the first-order solution in Equation (8.8) simplifies to

$$\mathbf{w}_p = \mathbf{C}^T \mathbf{x}_p. \tag{8.11}$$

In other words, encoding is enormously cheaper when our spanning set is orthonormal since there is no system of equations left to solve for, and we can get the encoding of each data point directly via a simple matrix-vector multiplication.

Substituting this form of the encoding into the set of equalities in Equation (8.4) we have

$$\mathbf{C}\,\mathbf{C}^T \mathbf{x}_p = \mathbf{x}_p, \qquad p = 1, ..., P. \tag{8.12}$$

We call this the *Autoencoder* formula since it expresses how a point $\mathbf{x}_p$ is first *encoded* (via $\mathbf{w}_p = \mathbf{C}^T \mathbf{x}_p$) and then *decoded* back to itself ($\mathbf{C}\mathbf{w}_p = \mathbf{C}\mathbf{C}^T \mathbf{x}_p$). This is because with orthonormal spanning sets we also have that $\mathbf{C}\,\mathbf{C}^T = \mathbf{I}_{N \times N}$, and the two transformations we apply to the data, the *encoding* transformation $\mathbf{C}^T$ and the *decoding* transformation $\mathbf{C}$, are inverse operations.

8.2.4   Imperfect representation of data using fixed spanning sets

In the previous two subsections, in order to be able to represent data perfectly, we assumed that the number of linearly independent spanning vectors K and the ambient input dimension N were identical. When $K < N$ we can no longer perfectly represent every possible data point in an input space. Instead we can only hope to *approximate*, as well as possible, our dataset as

$$\mathbf{C}\mathbf{w}_p \approx \mathbf{x}_p, \qquad p = 1, ..., P. \tag{8.13}$$

This is analogous to Equation (8.4) except now $\mathbf{C}$ and $\mathbf{w}_p$ are an $N \times K$ matrix and a $K \times 1$ column vector, respectively, where $K < N$.

To learn the proper encodings for our data we still aim to minimize the Least Squares cost in Equation (8.7) and can still use the first-order system to independently solve for each $\mathbf{w}_p$ as in Equation (8.8). Geometrically speaking, in solving the Least Squares cost we aim at finding the best K-dimensional subspace on which to project our data points, as illustrated in Figure 8.3. When the *encoding* $\mathbf{w}_p$ is optimally computed for the point $\mathbf{x}_p$, its *decoding* $\mathbf{C}\mathbf{w}_p$ is precisely the projection of $\mathbf{x}_p$ onto the subspace spanned by $\mathbf{C}$. This is called a *projection* because, as illustrated in the figure, the representation is given by projecting or dropping $\mathbf{x}_p$ perpendicularly onto the subspace.

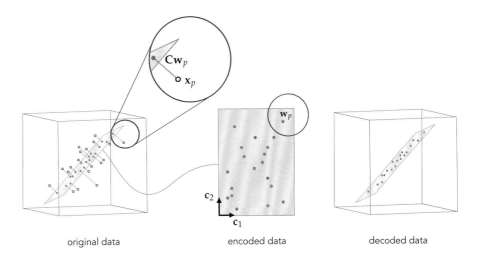

original data   encoded data   decoded data

Figure 8.3 (left panel) A dataset of points $\mathbf{x}_p$ in $N = 3$ dimensions along with a linear subspace spanned by $K = 2$ vectors $\mathbf{c}_1$ and $\mathbf{c}_2$, colored in red. (middle panel) The encoding space spanned by $\mathbf{c}_1$ and $\mathbf{c}_2$, where our encoded vectors $\mathbf{w}_p$ live. (right panel) The projected or decoded versions of each data point $\mathbf{x}_p$ shown in the subspace spanned by $\mathbf{C} = [\mathbf{c}_1 \ \mathbf{c}_2]$. The decoded version of the original point $\mathbf{x}_p$ is expressed as $\mathbf{C}\mathbf{w}_p$.

Akin to what we saw in the previous subsection, if our spanning set of K elements is orthonormal the corresponding formula for each encoding vector $\mathbf{w}_p$ simplifies to Equation (8.11), and the *autoencoder* formula shown previously in Equation (8.12) become

$$\mathbf{C}\,\mathbf{C}^T\mathbf{x}_p \approx \mathbf{x}_p \qquad\qquad p = 1, ..., P. \qquad\qquad (8.14)$$

In other words, since $K < N$ the encoding $\mathbf{C}^T$ and decoding $\mathbf{C}$ transformations are no longer quite inverse operations of one another.

8.3   The Linear Autoencoder and Principal Component Analysis

The most fundamental unsupervised learning method, known as Principal Component Analysis or PCA for short, follows directly from our discussion in the previous section regarding fixed spanning set representations with one crucial caveat: instead of just learning the proper weights to best represent input data over a given *fixed* spanning set we *learn* a proper spanning set as well.

8.3.1   Learning proper spanning sets

Imagine we returned to the previous section, but instead of assuming we were given $K \le N$ *fixed* spanning vectors over which to represent our mean-centered input as

$$\mathbf{Cw}_p \approx \mathbf{x}_p \qquad p = 1, ..., P \tag{8.15}$$

we aimed to *learn* the best spanning vectors to make this approximation as tight as possible. To do this we could simply add $\mathbf{C}$ to the set of parameters of our Least Squares function in Equation (8.7) giving

$$g\left(\mathbf{w}_1, ..., \mathbf{w}_P, \mathbf{C}\right) = \frac{1}{P} \sum_{p=1}^{P} \left\|\mathbf{Cw}_p - \mathbf{x}_p\right\|_2^2 \tag{8.16}$$

which we could then minimize to learn the best possible set of weights $\mathbf{w}_1$ through $\mathbf{w}_P$ as well as the spanning matrix $\mathbf{C}$. This Least Squares cost function, which is generally nonconvex,[3] can be properly minimized using any number of local optimization techniques including gradient descent (see Section 3.5) and coordinate descent (see Section 3.2.2).

Example 8.2   Learning a proper spanning set via gradient descent
In this example we use gradient descent to minimize the Least Squares cost in Equation (8.16) in order to learn the best $K = 2$ dimensional subspace for the mean-centered $N = 3$ dimensional dataset of $P = 100$ points shown in the left panel of Figure 8.4.

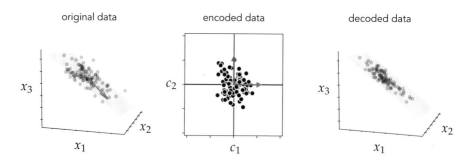

original data   encoded data   decoded data

Figure 8.4 Figure associated with Example 8.2. See text for details.

In addition to the original data, the left panel of the figure shows the learned spanning vectors as red arrows, and corresponding subspace colored in light

---

[3] However, it is convex in each $\mathbf{w}_p$ keeping all other weights and $\mathbf{C}$ fixed, and convex in $\mathbf{C}$ keeping all weight vectors $\mathbf{w}_p$ fixed.

red. This is the very best two-dimensional subspace representation for the input data. In the middle panel we show the corresponding learned encodings $\mathbf{w}_p$ of the original input $\mathbf{x}_p$ in the space spanned by the two recovered spanning vectors. In the right panel of the figure we show the original data space again as well as the decoded data, i.e., the projection of each original data point onto our learned subspace.

8.3.2 The linear Autoencoder

As detailed in Section 8.2.4, if our K spanning vectors concatenated column-wise to form the spanning matrix $\mathbf{C}$ are orthonormal, then the encoding of each $\mathbf{x}_p$ may be written simply as $\mathbf{w}_p = \mathbf{C}^T\mathbf{x}_p$. If we plug in this simple solution for $\mathbf{w}_p$ into the pth summand of the Least Squares cost in Equation (8.16), we get a cost that is a function of $\mathbf{C}$ *alone*

$$g\left(\mathbf{C}\right) = \frac{1}{P}\sum_{p=1}^{P}\left\|\mathbf{C}\,\mathbf{C}^T\mathbf{x}_p - \mathbf{x}_p\right\|_2^2. \qquad (8.17)$$

We can think of this Least Squares as enforcing the *Autoencoder* formula shown in Equation (8.14) to hold when properly minimized, and thus it is often referred to as the *linear Autoencoder*. Instead of being given an encoding/decoding scheme for each data point, by minimizing this cost function we *learn* one.

Even though we were led to the linear Autoencoder by assuming our spanning matrix $\mathbf{C}$ is orthonormal, we need not constrain our minimization of Equation (8.17) to enforce this condition because, as is shown in Section 8.9, the minima of the linear Autoencoder are *always* orthonormal (see Section 8.9.1).

Example 8.3   Learning a linear Autoencoder using gradient descent
In the left panel of Figure 8.5 we show a mean-centered two-dimensional dataset, along with a single spanning vector (i.e., $K = 1$) *learned* to the data by minimizing the linear Autoencoder cost function in Equation (8.17) using gradient descent. The optimal vector is shown as a red arrow in the left panel, the corresponding encoded data is shown in the middle panel, and the decoded data in the right panel along with the optimal subspace for the data (a line) shown in red.

8.3.3 Principal Component Analysis

The linear Autoencoder cost in Equation (8.17) may have many minimizers, of which the set of *principal components* is a particularly important one. The spanning set of principal components always provide a consistent *skeleton* for a dataset, with its members pointing in the dataset's *largest directions of orthogonal*

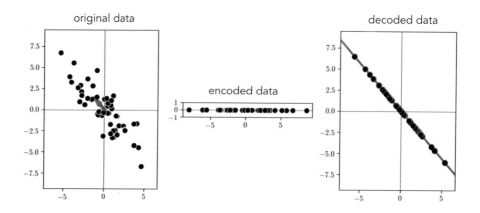

Figure 8.5 Figure associated with Example 8.3. See text for details.

variance. Employing this particular solution to the linear Autoencoder is often referred to as *Principal Component Analysis,* or PCA for short, in practice.

This idea is illustrated for a prototypical $N = 2$ dimensional dataset in Figure 8.6, where the general elliptical distribution of the data is shown in light grey. A scaled version of the first principal component of this dataset (shown as the longer red arrow) points in the direction in which the dataset is most spread out, also called its largest direction of variance. A scaled version of the second principal component (shown as the shorter of the two red arrows) points in the next most important direction in which the dataset is spread out that is *orthogonal* to the first.

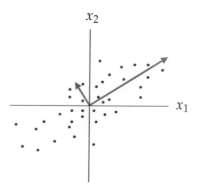

Figure 8.6 A prototypical dataset with scaled versions of its first and second principal components shown as the longer and shorter red arrows, respectively. See text for further details.

As we show in Section 8.9.2, this special orthonormal minimizer of the linear Autoencoder is given by the eigenvectors of the so-called *covariance matrix* of the data. Denoting by $\mathbf{X}$ the $N \times P$ data matrix consisting of our P *mean-centered* input points stacked column-wise

$$\mathbf{X} = \begin{bmatrix} | & | & & | \\ \mathbf{x}_1 & \mathbf{x}_2 & \cdots & \mathbf{x}_P \\ | & | & & | \end{bmatrix} \tag{8.18}$$

the covariance matrix is defined as the $N \times N$ matrix $\frac{1}{P}\mathbf{X}\mathbf{X}^T$. Denoting the eigendecomposition (see Section C.4) of the covariance matrix as

$$\frac{1}{P}\mathbf{X}\mathbf{X}^T = \mathbf{V}\mathbf{D}\mathbf{V}^T \tag{8.19}$$

the *principal components* are given by the orthonormal eigenvectors in $\mathbf{V}$, and the variance in each (principal component) direction is given precisely by the corresponding nonnegative eigenvalues in the diagonal matrix $\mathbf{D}$.

Example 8.4   Principal components
In the left panel of Figure 8.7 we show the mean-centered data first displayed in Figure 8.5, along with its two principal components (pointing in the two orthogonal directions of greatest variance in the dataset) shown as red arrows. In the right panel we show the *encoded* version of the data in a space where the principal components are in line with the coordinate axes.

Figure 8.7 Figure associated with Example 8.4.

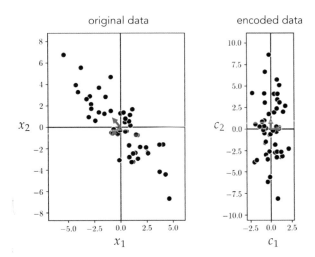

Example 8.5   A warning example!
While PCA can technically be used to reduce the dimension of data in a predictive modeling scenario (in hopes of improving accuracy, computation time, etc.) it can cause severe problems in the case of classification. In Figure 8.8

we illustrate feature space dimension reduction via PCA on a simulated two-class dataset where the two classes are linearly separable. Because the ideal one-dimensional subspace for the data in this instance runs (almost) parallel to the ideal linear classifier, projecting the complete dataset onto this subspace completely destroys the inter-class separation. For this very reason, while it is commonplace to *sphere* classification data using PCA, as detailed in Section 9.5, one needs to be extremely careful using PCA as a dimension reduction tool with classification or when the data does not natively live in or near a linear subspace.

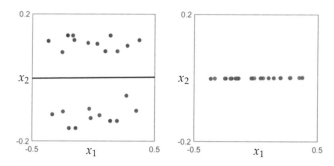

Figure 8.8 Figure associated with Example 8.5. (left panel) A toy classification dataset consisting of two linearly separable classes. The ideal one-dimensional subspace produced via PCA is shown in black. (right panel) Reducing the feature space dimension by projecting the data onto this subspace completely destroys the original separability of the data.

8.3.4   Python implementation

Below we provide a Python implementation involved in computing the principal components of a dataset – including data centering, principal component computation, and the PCA encoding. This implementation extensively leverages NumPy's linear algebra submodule.

First we center the data using the short implementation below.

```
1  # center an input dataset X
2  def center(X):
3      X_means = np.mean(X,axis=1)[:,np.newaxis]
4      X_centered = X - X_means
5      return X_centered
```

Next, we compute the principal components of the mean-centered data.

```
1  # function for computing principal components of input dataset X
2  def compute_pcs(X,lam):
3      # create the data covariance matrix
4      P = float(X.shape[1])
5      Cov = 1/P*np.dot(X,X.T) + lam*np.eye(X.shape[0])
6
7      # use numpy function to compute eigenvectors / eigenvalues
8      D,V = np.linalg.eigh(Cov)
9      return D,V
```

Note that in practice it is often helpful to slightly *regularize* a matrix prior to computing its eigenvalues/vectors to avoid natural numerical instability issues associated with their computation. Here this means adding a small weighted identity $\lambda \mathbf{I}_{N \times N}$, where $\lambda \geq 0$ is some small value (e.g., 10^{-5}), to the data covariance matrix prior to computing its eigenvalues/vectors. In short, in order to avoid computational trouble we typically compute principal components of the regularized covariance matrix $\frac{1}{P}\mathbf{X}\mathbf{X}^T + \lambda \mathbf{I}_{N \times N}$ instead of the raw covariance matrix itself. Thus the addition of the term `lam*np.eye(X.shape[0])` in line 5 of the implementation above.

8.4 Recommender Systems

In this section we discuss the fundamental linear *Recommender System*, a popular unsupervised learning framework commonly employed by businesses to help automatically recommend products and services to their customers. From the vantage of machine learning, however, the basic Recommender System detailed here is simply a slight twist on our core unsupervised learning technique: Principal Component Analysis.

8.4.1 Motivation

Recommender Systems are heavily used in e-commerce today, providing customers with personalized recommendations for products and services by using a consumer's previous purchasing and rating history, along with those of similar customers. For instance, a movie provider like Netflix with millions of users and tens of thousands of movies, records users' reviews and ratings (typically in the form of a number on a scale of 1–5 with 5 being the most favorable rating) in a large matrix such as the one illustrated in Figure 8.9. These matrices are very sparsely populated, since an individual customer has likely rated only a small portion of the movies available.

With this sort of product ratings data available, online movie and commerce sites often use the unsupervised learning technique we discuss in this section as their main tool for making personalized recommendations to customers regarding what they might like to consume next. With the technique for producing

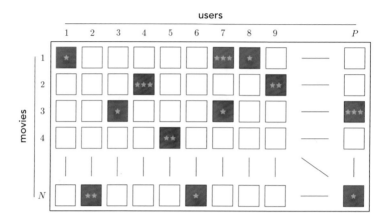

Figure 8.9 A prototypical movie rating matrix is very sparsely populated, with each user having rated only a very small number of films. In this diagram movies are listed along rows with users along columns. In order to properly recommend movies for users to watch we try to intelligently guess the missing values of this matrix, and then recommend movies that we predict users would rate highly (and therefore enjoy the most).

personalized recommendations we discuss here we aim to first intelligently guess the values of missing entries in the ratings matrix. Next, in order to recommend a new product to a given user, we examine our filled-in ratings matrix for products we have predicted the user would rate highly (and thus enjoy), and recommend those.

8.4.2   Notation and modeling

With a Recommender System we continue to use our familiar notation $\{x_1, ..., x_P\}$ to denote input data, each of which has dimension N. In this application the point x_p denotes our pth customer's rating vector of all N possible products available to be rated. The number of products N is likely quite large, so large that each customer has the chance to purchase and review only a very small sampling of them, making x_p a very sparsely populated vector (with whatever ratings user p has input into the system). We denote the index set of these nonempty entries of x_p as

$$\Omega_p = \Big\{ (j,p) \mid j\text{th entry of } x_p \text{ is filled in}\Big\}. \tag{8.20}$$

Since our goal is to fill in the missing entries of each input vector x_p we have no choice but to make assumptions about how users' tastes behave in general. The simplest assumption we can make is that every user's tastes can be expressed as a linear combination of some small set of fundamental user taste profiles. For example, in the case of movies these profiles could include the prototypical romance movie lover, prototypical comedy movie lover, action movie lover, etc. The relatively small number of such categories or user types compared to the total number of users provides a useful framework to intelligently guess missing values present in a user ratings dataset. Formally this is to say that we assume that some ideal spanning set of K fundamental taste vectors (which we can package in an $N \times K$ matrix $\mathbf{C}$) exist so that each vector $\mathbf{x}_p$ can be truly expressed as the linear combination

$$\mathbf{C}\mathbf{w}_p \approx \mathbf{x}_p, \qquad p = 1, ..., P. \tag{8.21}$$

In order to then learn both the spanning set $\mathbf{C}$ and each weight vector $\mathbf{w}_p$ we could initially propose to minimize a Least Squares cost similar to one shown in Equation (8.16). However, our input data is now *incomplete* as we only have access to the entries indexed by Ω_p for $\mathbf{x}_p$. Therefore we can only minimize that Least Squares cost over these entries, i.e.,

$$g\left(\mathbf{w}_1, ..., \mathbf{w}_P, \mathbf{C}\right) = \frac{1}{P} \sum_{p=1}^{P} \left\| \left\{\mathbf{C}\mathbf{w}_p - \mathbf{x}_p\right\}\big|_{\Omega_p} \right\|_2^2. \tag{8.22}$$

The notation $\{\mathbf{v}\}|_{\Omega_p}$ here denotes taking only those entries of $\mathbf{v}$ in the index set Ω_p. Because the Least Squares cost here is defined only over a select number of indices we cannot leverage any sort of orthonormal solutions to this cost, or construct a cost akin to the linear Autoencoder in Equation (8.17). However, we can easily use gradient (see Section 3.5) and coordinate descent (see Section 3.2.2) based schemes to properly minimize it.

8.5   K-Means Clustering

The subject of this section, the *K-means algorithm*, is an elementary example of another set of unsupervised learning methods called *clustering algorithms*. Unlike PCA, which was designed to reduce the ambient dimension (or feature dimension) of the data space, clustering algorithms are designed to (properly) reduce the number of points (or data dimension) in a dataset, and in doing so help us to better understand its structure.

8.5.1   Representing a dataset via clusters

One way to simplify a dataset is by grouping together nearby points into *clusters*. Take the following set of two-dimensional data points, shown in the left panel of Figure 8.10. When you carefully examine the data shown there you can see that it naturally falls into three groups or *clusters* because you have something along the lines of a clustering algorithms built into your brain.

In the right panel of Figure 8.10 we show a visual representation of each cluster,

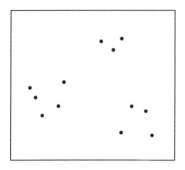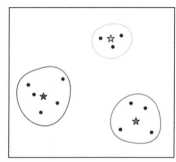

Figure 8.10 (left) A two-dimensional toy dataset with $P = 12$ data points. (right) The data shown naturally clustered into $K = 3$ clusters. Points that are geometrically close to one another belong to the same cluster, and each cluster boundary is roughly marked using a uniquely colored solid curve. Each cluster center – also called a *centroid* – is marked by a star symbol colored to match its cluster boundary.

including each cluster's boundary drawn as a uniquely colored solid curve. We also draw the center of each cluster using a star symbol that matches the unique boundary color of each cluster. These cluster centers are often referred to in the jargon of machine learning as cluster *centroids*. The centroids here allow us to think about the dataset in the big picture sense – instead of $P = 12$ points we can think of our dataset grossly in terms of these $K = 3$ cluster centroids, as each represents a chunk of the data.

How can we describe, mathematically speaking, the clustering scenario we naturally see when we view the points/clusters like those shown in Figure 8.10?

First some notation. As in the previous sections we will denote our set of P points generically as $\mathbf{x}_1, \mathbf{x}_2, ..., \mathbf{x}_P$. To keep things as generally applicable as possible we will also denote by K the number of clusters in a dataset (e.g., in the dataset of Figure 8.10, $K = 3$). Because each cluster has a centroid we need notation for this as well, and we will use $\mathbf{c}_1, \mathbf{c}_2, ..., \mathbf{c}_K$ to denote these where $\mathbf{c}_k$ is the centroid of the kth cluster. Finally we will need a notation to denote the set of points that belong to each cluster. We denote the set of indices of those points belonging to the kth cluster as

$$S_k = \left\{ p \mid \text{if } \mathbf{x}_p \text{ belongs to the } k\text{th cluster} \right\}. \tag{8.23}$$

With all of our notation in hand we can now better describe the prototype clustering scenario shown in Figure 8.10. Suppose we have identified each cluster and its centroid "by eye," as depicted in the right panel of the figure. Because the centroid denotes the center of a cluster, it seems intuitive that each one can be expressed as *the average of the points assigned to its cluster* as

$$\mathbf{c}_k = \frac{1}{|S_k|} \sum_{p \in S_k} \mathbf{x}_p. \tag{8.24}$$

This formula confirms the intuition that each centroid represents a chunk of the data – being the average of those points belonging to each cluster.

Next we can state mathematically an obvious and implicit fact about the simple clustering scenario visualized in Figure 8.10: that each point belongs to the cluster whose centroid it is closest to. To express this algebraically for a given point $\mathbf{x}_p$ is simply to say that the point must belong to the cluster whose distance to the centroid $\left\| \mathbf{x}_p - \mathbf{c}_k \right\|_2$ is *minimal*. In other words, the point $\mathbf{x}_p$ belongs to or is *assigned* to cluster $k^\star$ if

$$a_p = \operatorname*{argmin}_{k=1,\dots,K} \left\| \mathbf{x}_p - \mathbf{c}_k \right\|_2 . \tag{8.25}$$

In the jargon of machine learning these are called cluster *assignments*.

8.5.2 Learning clusters to represent data

Of course we do not want to have to rely on our visualization abilities to identify clusters in a dataset. We want an algorithm that will do this for us automatically. Thankfully we can do this rather easily using the framework detailed above for mathematically describing clusters, the resulting algorithm being called the *K-means clustering algorithm*.

To get started, suppose we want to cluster a dataset of P points into K clusters automatically. Note here that we will fix K, and address how to properly decide on its value later.

Since we do not know where the clusters nor their centroids are located we can start off by taking a random guess at the locations of our K centroids (we have to start somewhere). This "random guess" – our initialization – for the K centroids could be a random subset of K of our points, a random set of K points in the space of the data, or any number of other types of initializations. With our initial centroid locations decided on we can then determine cluster assignments by simply looping over our points and for each $\mathbf{x}_p$ finding its closest centroid using the formula we saw above

$$a_p = \operatorname*{argmin}_{k=1,\dots,K} \left\| \mathbf{x}_p - \mathbf{c}_k \right\|_2 . \tag{8.26}$$

Now we have both an initial guess at our centroids and clustering assignments. With our cluster assignments in hand we can then update our centroid locations – as the average of the points recently assigned to each cluster

$$\mathbf{c}_k = \frac{1}{|\mathcal{S}_k|} \sum_{p \in \mathcal{S}_k} \mathbf{x}_p. \tag{8.27}$$

These first three steps – initializing the centroids, assigning points to each cluster, and updating the centroid locations – are illustrated in the top row of Figure 8.11 with the dataset shown above in Figure 8.10.

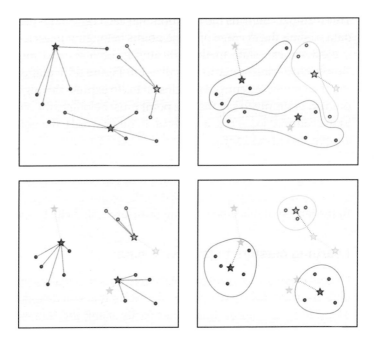

Figure 8.11 The first two iterations of the K-means algorithm illustrated using the dataset first shown in Figure 8.10. (top row) (left panel) A set of data points with random centroid initializations and assignments. (right panel) Centroid locations updated as average of points assigned to each cluster. (bottom row) (left panel) Assigning points based on updated centroid locations. (right panel) Updated location of centroids given by cluster averages.

To further refine our centroids/clusters we can now simply repeat the above two-step process of (a) reassigning points based on our new centroid locations and then (b) updating the centroid locations as the average of those points assigned to each cluster. The second iteration of the algorithm is illustrated for a particular example in the bottom row of Figure 8.11. We can halt doing so after, for example, a predefined number of maximum iterations or when the cluster centroids to not change location very much from one iteration to the next.

Example 8.6   The impact of initialization

The result of the algorithm reaching poor minima can have significant impact on the quality of the clusters learned. For example in Figure 8.12 we use a two-dimensional toy dataset with $K = 2$ clusters to find. With the initial centroid positions shown in the top panel, the K-means algorithm gets stuck in a local minimum and consequently fails to cluster the data properly. A different initialization for one of the centroids, however, leads to a successful clustering of the data, as shown in the bottom panel of the figure. To overcome this issue in prac-

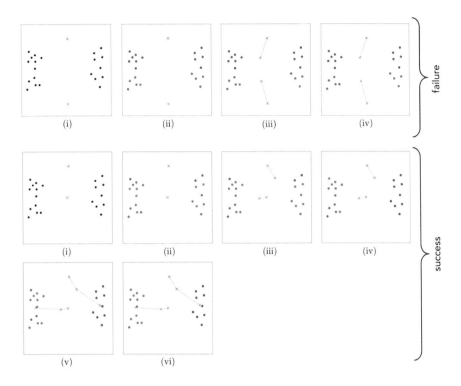

Figure 8.12 Success or failure of K-means depends on the centroids' initialization. (top) (i) Two centroids are initialized, (ii) cluster assignment is updated, (iii) centroid locations are updated, (iv) no change in the cluster assignment of the data points leads to stopping of the algorithm. (bottom) (i) Two centroids are initialized with the red one being initialized differently, (ii) cluster assignment is updated, (iii) centroid locations are updated, (iv) cluster assignment is updated, (v) centroid locations are updated, (vi) no change in the cluster assignment of the data points leads to stopping of the algorithm.

tice we often run the algorithm multiple times with different initializations, with the best solution being one that results in the smallest value of some objective value of cluster quality.

For example, one metric for determining the best clustering from a set of runs is the *average distance of each point to its cluster centroid* – called the *average intra-cluster distance*. Denoting by c_{k_p} the final cluster centroid of the pth point x_p, then the average distance from each point to its respective centroid can be written as

$$\text{average intra-cluster distance} = \frac{1}{P} \sum_{p=1}^{P} \left\| x_p - c_{k_p} \right\|_2 . \tag{8.28}$$

Computing this for each run of K-means we choose the final clustering that achieves the *smallest* such value as the best clustering arrangement.

Example 8.7 Choosing the ideal number of clusters K

To determine the optimal setting of the parameter K, i.e., the number of clusters in which to cluster the data, we typically must try a range of different values for K, run the K-means algorithm in each case, and compare the results using an appropriate metric like the average intra-cluster distance in Equation (8.28). Of course if we achieve an optimal clustering for each value of K (perhaps running the algorithm multiple times for each value of K) then the intra-cluster distance should *always go down monotonically as we increase the value of K* since we are partitioning the dataset into smaller and smaller chunks.

For example, in Figure 8.13 we show the results of running ten runs of K-means ranging the value of K from $K = 1, \ldots, 10$ and keeping the clustering that provided the lowest intra-cluster distance for each value of K for the dataset shown in the left panel of the figure. In the right panel we plot the best distance value attained for each value of K tried, a plot often referred to in the jargon of machine learning as a *scree plot*.

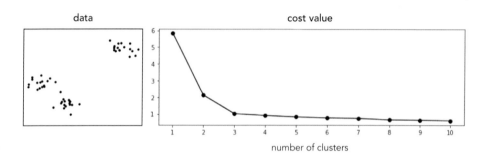

Figure 8.13 Figure associated with Example 8.7. (left panel) A dataset to cluster. (right panel) A *scree* plot. See text for further details.

As one should expect, the intra-cluster distance decreases monotonically as we increase K. Notice, however, that the scree plot above has an *elbow* at $K = 3$, meaning that increasing the number of clusters from three to four and onwards reduces the distance value by very little. Because of this we can argue that $K = 3$ is a good choice for the number of clusters for this particular dataset (which also makes sense in this instance, since we can visualize the dataset and clearly see that it has three clusters) since any fewer clusters and the intra-cluster distance is comparatively large, while adding additional clusters does not decrease the total intra-cluster distance too much.

This illustrates the typical usage of the scree plot for deciding on an ideal number of clusters K for K-means. We compute and then plot the intra-cluster distance over a range of values for K, and pick the value at the "elbow" of the plot. In practice this value is often chosen subjectively (by visual analysis of the scree plot) as was done here.

8.6   General Matrix Factorization Techniques

In this section we tie together the unsupervised learning methods described in this chapter by describing them all through the singular lens of *matrix factorization*.

8.6.1   Unsupervised learning and matrix factorization problems

If we compactly represent our P input data points by stacking them column-wise into the data matrix $\mathbf{X}$ as in Equation (8.18), we can write the Least Squares cost function in Equation (8.16) that is the basis for Principal Component Analysis (Section 8.3) compactly as

$$g\left(\mathbf{W}, \mathbf{C}\right) = \frac{1}{P} \left\| \mathbf{CW} - \mathbf{X} \right\|_F^2 . \tag{8.29}$$

Here $\|\mathbf{A}\|_F^2 = \sum_{n=1}^{N} \sum_{p=1}^{P} A_{n,p}^2$ is the Frobenius norm, and $A_{n,p}$ is the (n, p)th element of the matrix $\mathbf{A}$, which is the analog of the squared ℓ_2 norm for matrices (see Section C.5.3).

We can likewise express our set of desired approximations that motivate PCA – given in Equation (8.15) – and that the minimization of this cost function force to hold as closely as possible

$$\mathbf{CW} \approx \mathbf{X}. \tag{8.30}$$

This set of desired approximations is often referred to as a *matrix factorization* since we desire to *factorize* the matrix $\mathbf{X}$ into a product of two matrices $\mathbf{C}$ and $\mathbf{W}$. This is the matrix analog of factorizing a single digit into two "simpler" ones, e.g., as $5 \times 2 = 10$. Thus, in other words, the PCA problem can be interpreted as a basic exemplar of a *matrix factorization problem*.

PCA is not the only unsupervised learning method we can recast in this way. Recommender Systems, as we saw in Section 8.4, results in a cost function that closely mirrors PCA's – and so likewise closely mirrors its compact form given above. Here the only difference is that many entries of the data matrix are unknown, thus the factorization is restricted to only those values of $\mathbf{X}$ that are known. Denoting by Ω the set of indices of *known* values of $\mathbf{X}$, the matrix factorization involved in Recommender Systems takes the form

$$\left\{ \mathbf{CW} \approx \mathbf{X} \right\}|_\Omega \tag{8.31}$$

where the symbol $\{\mathbf{V}\}|_\Omega$ is used to denote that we care only about entries of an input matrix $\mathbf{V}$ in the index set Ω, which is a slight deviation of the PCA factorization in Equation (8.30). The corresponding Least Squares cost is similarly a slight twist on the compact PCA Least Squares cost in Equation (8.29) and is given as

$$g\left(\mathbf{W},\mathbf{C}\right)=\frac{1}{P}\,\|\{\mathbf{CW}-\mathbf{X}\}_{\Omega}\|_F^2\,. \tag{8.32}$$

Note how this is simply the matrix form of the Recommender Systems cost function given earlier in Equation (8.22).

Finally, we can also easily see that K-means (Section 8.5) falls into the same category, and can too be interpreted as a *matrix factorization problem*. We can do this by first re-interpreting our initial desire with K-means clustering, that points in the kth cluster should lie close to its centroid, which may be written mathematically as

$$\mathbf{c}_k \approx \mathbf{x}_p \quad \text{for all } p \in \mathcal{S}_k \quad k = 1, ..., K \tag{8.33}$$

where $\mathbf{c}_k$ is the centroid of the kth cluster and $\mathcal{S}_k$ the set of indices of the subset of those P data points belonging to this cluster. These desired relations can be written more conveniently in matrix notation for the centroids – denoting by $\mathbf{e}_k$ the kth standard basis vector (that is a $K \times 1$ vector with a 1 in the kth slot and zeros elsewhere) – likewise as

$$\mathbf{C}\,\mathbf{e}_k \approx \mathbf{x}_p \quad \text{for all } p \in \mathcal{S}_k \quad k = 1, ..., K. \tag{8.34}$$

Then, introducing matrix notation for the weights (here constrained to be standard basis vectors) and the data we can likewise write the above relations as

$$\mathbf{CW} \approx \mathbf{X} \quad \text{for all } p \in \mathcal{S}_k \quad k = 1, ..., K \tag{8.35}$$

where

$$\mathbf{w}_p \in \{\mathbf{e}_k\}_{k=1}^K, \quad p = 1, ..., P. \tag{8.36}$$

Figure 8.14 pictorially illustrates the compactly written desired K-means relationship above for a small prototypical dataset. Note that the location of the only nonzero entry in each column of the assignment matrix $\mathbf{W}$ determines the cluster membership of its corresponding data point in $\mathbf{X}$. So, in other words, K-means too is a matrix factorization problem (with a very particular set of constraints on the matrix $\mathbf{W}$).

Having framed the desired outcome – when parameters are set optimally – the associated cost function for K-means can then likewise be written compactly as

$$g\left(\mathbf{W},\mathbf{C}\right)=\frac{1}{P}\,\|\mathbf{CW}-\mathbf{X}\|_F^2 \tag{8.37}$$

subject to the constraint that $\mathbf{w}_p \in \{\mathbf{e}_k\}_{k=1}^K$ for $p = 1, ..., P$. In other words, we can

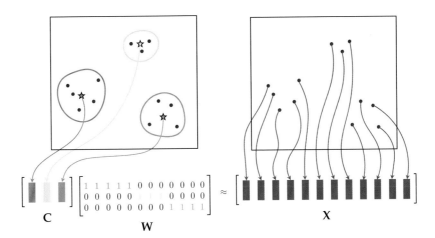

Figure 8.14 K-means clustering relations described in a compact matrix form. Cluster centroids in **C** lie close to their corresponding cluster points in **X**. The pth column of the assignment matrix **W** contains the standard basis vector corresponding to the data point's cluster centroid.

interpret the K-means algorithm described in Section 8.5 as a way of solving the constrained optimization problem

$$\underset{C,W}{\text{minimize}} \ \ \|CW - X\|_F^2$$
$$\text{subject to} \ \ \mathbf{w}_p \in \{\mathbf{e}_k\}_{k=1}^{K}, \ \ p = 1, \ldots, P. \tag{8.38}$$

One can easily show that the K-means algorithm we derived in the previous section is also the set of updates resulting from the application of the block-coordinate descent method for solving the above K-means optimization problem. This perspective on K-means is particularly helpful, since in the natural derivation of K-means shown in the previous section K-means is a somewhat heuristic algorithm (i.e., it is not tied to the minimization of a cost function, like every other method we discuss is). One practical consequence of this is that – previously – we had no framework in which to judge how a single run of the algorithm was progressing. Now we do. Now we know that we can treat the K-means algorithm precisely as we do every other optimization method we discuss – as a way of minimizing a particular cost function – and can use the cost function to understand how the algorithm is functioning.

8.6.2   Further variations

We saw in Equation (8.38) how K-means can be recast as a constrained matrix factorization problem, one where each column $\mathbf{w}_p$ of the assignment matrix **W** is constrained to be a standard basis vector. This is done to guarantee every data point $\mathbf{x}_p$ ends up getting assigned to one (and only one) cluster centroid

$\mathbf{c}_k$. There are many other popular matrix factorization problems that – from a modeling perspective – simply employ different constraints than the one given for K-means. For example, a natural generalization of K-means called *sparse coding* [30] is a clustering-like algorithm that differs from K-means only in that it allows assignment of data points to *multiple clusters*. Sparse coding is a constrained matrix factorization problem often written as

$$\underset{\mathbf{C,W}}{\text{minimize}} \ \ \|\mathbf{CW} - \mathbf{X}\|_F^2$$

$$\text{subject to} \ \ \ \left\|\mathbf{w}_p\right\|_0 \leq S, \ \ p = 1, \ldots, P \tag{8.39}$$

where the K-means constraints are replaced with constraints of the form $\left\|\mathbf{w}_p\right\|_0 \leq S$, making it possible for each $\mathbf{x}_p$ to be assigned to at most S clusters simultaneously. Recall, $\left\|\mathbf{w}_p\right\|_0$ indicates the number of nonzero entries in the vector $\mathbf{w}_p$ (see Section C.5.1).

Besides sparsity, seeking a nonnegative factorization of the input matrix $\mathbf{X}$ is another constraint sometimes put on matrices $\mathbf{C}$ and $\mathbf{W}$, giving the so-called *nonnegative matrix factorization problem*

$$\underset{\mathbf{C,W}}{\text{minimize}} \ \ \|\mathbf{CW} - \mathbf{X}\|_F^2$$

$$\text{subject to} \ \ \ \mathbf{C, W} \geq 0. \tag{8.40}$$

Nonnegative matrix factorization (see, e.g., [31])is used predominantly in situations where data is naturally nonnegative (e.g., Bag of Words representation of text data, pixel intensity representation of image data, etc.) where presence of negative entries hinders interpretability of learned solutions.

Table 8.1 shows a list of common matrix factorization problems subject to possible constraints on $\mathbf{C}$ and $\mathbf{W}$.

8.7   Conclusion

In this chapter we introduced unsupervised learning, detailing a number of popular concepts and models. Such problems differ from those we have seen in the previous three chapters in terms of their data – which contain only input (and no output).

We began in Section 8.2 by reviewing the concepts of spanning sets, as well as the notion of encodings/decodings. In Section 8.3 we then looked at how to learn optimal linear encodings via the so-called linear *Autoencoder*. In this section we also studied the special *Principal Component Analysis* (or PCA for short) solution to the linear Autoencoder, which can be conveniently computed in terms of the eigenvectors of the data covariance matrix. Next in Section 8.4 we looked at a popular twist on the notion of learning spanning sets – called Recommender Systems. *K-means clustering* was then introduced in Section 8.5. Finally in Section

Table 8.1 Common matrix factorization problems $\mathbf{CW} \approx \mathbf{X}$ subject to possible constraints on $\mathbf{C}$ and $\mathbf{W}$.

Matrix factorization problem	Constraints on C and W
Principal Component Analysis	$\mathbf{C}$ is orthonormal
Recommender systems	No constraint on $\mathbf{C}$ or $\mathbf{W}$; $\mathbf{X}$ is only partially known
K-means clustering	Each column of $\mathbf{W}$ is a standard basis vector
Sparse dictionary learning	Each column of $\mathbf{W}$ is sparse
Nonnegative matrix factorization	Both $\mathbf{C}$ and $\mathbf{W}$ are nonnegative

8.6 we introduced a *matrix factorization framework* that compactly showcases the similarities between PCA, Recommender Systems, K-means clustering, and more advanced unsupervised learning models.

8.8   Exercises

† The data required to complete the following exercises can be downloaded from the text's github repository at `github.com/jermwatt/machine_learning_refined`

8.1   The standard basis

A simple example of an orthonormal spanning set is the set of N *standard basis vectors*. The nth element of a standard basis is a vector that consist entirely of zeros, except for a 1 in its nth slot

$$(\textit{n}\text{th element of the standard basis}) \quad \mathbf{c}_n = \begin{bmatrix} 0 \\ \vdots \\ 0 \\ 1 \\ 0 \\ \vdots \\ 0 \end{bmatrix}. \tag{8.41}$$

This is also what we referred to as a *one-hot-encoded vector* in Section 7.5.

Simplify the formula for the optimal weight vector/encoding in Equation (8.11) when using a standard basis.

8.2   Encoding data

Repeat the experiment described in Example 8.1, reproducing the illustrations shown in Figure 8.2.

8.3   Orthogonal matrices and eigenvalues

Show that the $N \times K$ matrix $\mathbf{C}$ is orthogonal *if and only if* the nonzero eigenvalues of $\mathbf{C}\mathbf{C}^T$ all equal $+1$.

8.4   Nonconvexity of the linear Autoencoder

With $K = 1$ make a contour plot on the range $[-5, 5] \times [-5, 5]$ of the linear Autoencoder in Equation (8.17) over the dataset shown in the left panel of Figure 8.5. How many global minima does this contour plot appear to have? Given the concept of the linear Autoencoder and the result described in Example 8.3 describe the optimal spanning vector(s) represented by these minima and how they compare to the one shown in the left panel of Figure 8.5.

8.5   Minimizing the linear Autoencoder over a toy dataset

Repeat the experiment described in Example 8.3, reproducing the illustrations shown in Figure 8.5. Implement your code so that you can easily compute the gradient of the linear Autoencoder using the `autograd`.

8.6   Producing a PCA basis

Repeat the experiment described in Example 8.4, reproducing the illustrations shown in Figure 8.7. You may use the implementation given in Section 8.3.4 as a basis for your work.

8.7   A warning example

Repeat the experiment described in Example 8.5, reproducing the illustrations shown in Figure 8.8.

8.8   Perform K-means

Implement the K-means algorithm detailed and apply it to properly cluster the dataset shown in the left panel of Figure 8.13 using $K = 3$ cluster centroids. Visualize your results by plotting the dataset, coloring each cluster a unique color.

8.9   Making a scree plot
Repeat the experiment described in Example 8.7, reproducing the illustration shown in the right panel of Figure 8.13.

8.10   Alternating minimization
While the PCA Least Squares cost function in Equation (8.29) is not convex, it is *biconvex* in that it is convex in each matrix of parameters $\mathbf{C}$ and $\mathbf{W}$ *independently*. This realization leads to a natural extension of the coordinate descent method described in Section 3.2.2 called *alternating minimization*. Such optimization methods are widely used to solve general matrix factorization problems (like those listed in Table 8.1).

In this approach we minimize the PCA Least Squares cost sequentially by minimizing the cost (to completion) over one matrix at a time, that is $\mathbf{C}$ or $\mathbf{W}$ independently (with the other matrix fixed). This process – of fixing one of the two matrices and minimizing to completion over the other – is repeated until the state of both matrices does not change significantly, or when a maximum number of iterations is reached.

Alternatingly minimizing to completion over $\mathbf{C}$ and $\mathbf{W}$ reduces to sequentially solving two first-order systems of equations – generated checking the first-order condition of Equation (11.4) in $\mathbf{C}$ and $\mathbf{W}$, respectively. Write out these two systems of equations.

8.9   Endnotes

8.9.1   The minima of the Autoencoder are all orthogonal matrices

To show that the minima of the linear Autoencoder in Equation (8.17) are all orthogonal matrices, we first substitute the eigenvalue decomposition (see Section C.4) of $\mathbf{CC}^T$ as $\mathbf{CC}^T = \mathbf{VDV}^T$, where $\mathbf{V}$ is an $N \times N$ matrix of orthogonal eigenvectors and $\mathbf{D}$ is a $N \times N$ diagonal matrix with at most K *nonnegative* eigenvalues along the upper K entries of its diagonal (since $\mathbf{CC}^T$ is an *outer-product matrix* – see Exercise 4.2), for the matrix $\mathbf{CC}^T$ in the pth summand of the linear Autoencoder (remember we assume our data is *mean-centered*)

$$\left\| \mathbf{CC}^T\mathbf{x}_p - \mathbf{x}_p \right\|_2^2 = \left\| \mathbf{VDV}^T\mathbf{x}_p - \mathbf{x}_p \right\|_2^2 = \mathbf{x}_p^T\mathbf{VDDV}^T\mathbf{x}_p - 2\mathbf{x}_p^T\mathbf{VDV}^T\mathbf{x}_p + \mathbf{x}_p^T\mathbf{x}_p. \quad (8.42)$$

Introducing $\mathbf{I}_{N\times N} = \mathbf{VV}^T$ in between the inner product $\mathbf{x}_p^T\mathbf{x}_p = \mathbf{x}_p^T\mathbf{VV}^T\mathbf{x}_p$, denoting $\mathbf{q}_p = \mathbf{V}^T\mathbf{x}_p$, and denoting $\mathbf{A}^2 = \mathbf{AA}$ for any square matrix $\mathbf{A}$ we may rewrite the right-hand side above equivalently as

$$\mathbf{q}_p^T\mathbf{DD}\mathbf{q}_p - 2\mathbf{q}_p^T\mathbf{D}\mathbf{q}_p + \mathbf{q}_p^T\mathbf{q}_p = \mathbf{q}_p^T\left(\mathbf{D}^2 - 2\mathbf{D} + \mathbf{I}_{N\times N}\right)\mathbf{q}_p = \mathbf{q}_p^T\left(\mathbf{D} - \mathbf{I}_{N\times N}\right)^2\mathbf{q}_p \quad (8.43)$$

where the last equality follows from completing the square.

Performing this for each summand the linear Autoencoder in Equation (8.17) can be written equivalently as

$$g(\mathbf{C}) = \sum_{p=1}^{P} \mathbf{q}_p^T (\mathbf{D} - \mathbf{I}_{N \times N})^2 \, \mathbf{q}_p. \tag{8.44}$$

Since the entries of $\mathbf{D}$ are nonnegative it is easy to see that this quantity is minimized for $\mathbf{C}$ such that $g(\mathbf{C}) = (N-K) \sum_{p=1}^{P} \mathbf{x}_p^T \mathbf{x}_p$, i.e., where $\mathbf{D}_{K \times K}$ (the upper $K \times K$ portion of $\mathbf{D}$) is precisely the identity. In other words, the linear Autoencoder is minimized over matrices $\mathbf{C}$, where $\mathbf{C}\mathbf{C}^T$ has all nonzero eigenvalues equal to $+1$, and the only such matrices that have this property are *orthogonal* (see Exercise 8.3).

8.9.2   Formal derivation principal components

To begin the derivation of the classic *Principal Components Analysis* solution to the linear Autoencoder in Equation (8.17) all we must do is examine one summand of the cost, under the assumption that $\mathbf{C}$ is orthonormal. Expanding the of the pth summand we have

$$\left\| \mathbf{C}\mathbf{C}^T \mathbf{x}_p - \mathbf{x}_p \right\|_2^2 = \mathbf{x}_p^T \mathbf{C}\mathbf{C}^T \mathbf{C}\mathbf{C}^T \mathbf{x}_p - 2\mathbf{x}_p^T \mathbf{C}\mathbf{C}^T \mathbf{x}_p + \mathbf{x}_p^T \mathbf{x}_p \tag{8.45}$$

and then using our assumption that $\mathbf{C}^T \mathbf{C} = \mathbf{I}_{K \times K}$ we can see that it may be rewritten equivalently as

$$-\mathbf{x}_p^T \mathbf{C}\mathbf{C}^T \mathbf{x}_p + \mathbf{x}_p^T \mathbf{x}_p = -\left\| \mathbf{C}^T \mathbf{x}_p \right\|_2^2 + \left\| \mathbf{x}_p \right\|_2^2. \tag{8.46}$$

Since our aim is to *minimize* the summation of terms taking the form of the above, and the data point $\mathbf{x}_p$ is fixed and does not include the variable $\mathbf{C}$ we are minimizing with respect to, minimizing the original summand on the left is equivalent to minimizing *only the first term* $-\left\| \mathbf{C}^T \mathbf{x}_p \right\|_2^2$ on the right-hand side. Summing up these terms, the pth of which can be written decomposed over each individual basis element we aim to learn as

$$-\left\| \mathbf{C}^T \mathbf{x}_p \right\|_2^2 = -\sum_{n=1}^{K} \left(\mathbf{c}_n^T \mathbf{x}_p \right)^2 \tag{8.47}$$

gives us the following equivalent cost function to minimize for our ideal orthonormal basis

$$g(\mathbf{C}) = -\frac{1}{P} \sum_{p=1}^{P} \sum_{n=1}^{K} \left(\mathbf{c}_n^T \mathbf{x}_p \right)^2. \tag{8.48}$$

Studying this reduced form of our linear Autoencoder cost function we can see that it *decomposes completely over the basis vectors* c_n, i.e., there are no terms where c_i and c_j interact when $i \neq j$. This means – practically speaking – that we can optimize our orthonormal basis *one element at a time*. Reversing the order of the summands above we can isolate each individual basis element over the entire dataset, writing the above equivalently as

$$g\left(\mathbf{C}\right) = -\frac{1}{P}\sum_{n=1}^{K}\sum_{p=1}^{P}\left(\mathbf{c}_n^T\mathbf{x}_p\right)^2. \tag{8.49}$$

Now we can think about minimizing our cost function one basis element at a time. Beginning with c_1 we first isolate only those relevant terms above, which consists of: $-\frac{1}{P}\sum_{p=1}^{P}\left(\mathbf{c}_1^T\mathbf{x}_p\right)^2$. Since there is a minus sign in front of this summation, this is the same as *maximizing* its negation which we denote as

$$h\left(\mathbf{c}_1\right) = \frac{1}{P}\sum_{p=1}^{P}\left(\mathbf{c}_1^T\mathbf{x}_p\right)^2. \tag{8.50}$$

Since our basis is constrained to be orthonormal the basis element c_1 in particular is constrained to have unit-length. Statistically speaking, Equation (8.50) measures the *variance of the dataset in the direction defined by* c_1. Note: this quantity is precisely the variance because our data is assumed to have been *mean-centered*. Since we aim to maximize this quantity we can phrase our optimization in purely sample statistical terms as well: we aim to recover a form of the basis vector c_1 that points in the maximum direction of variance in the dataset.

To determine the maximum value of the function above or to determine the direction of maximum variance in the data we can rewrite the formula above by stacking the (mean-centered) data points x_p column-wise – forming the $N \times P$ *data matrix* $\mathbf{X}$ (as shown in Equation (8.18) – giving the equivalent formula in

$$h\left(\mathbf{c}_1\right) = \frac{1}{P}\mathbf{c}_1^T\mathbf{X}\mathbf{X}^T\mathbf{c}_1 = \mathbf{c}_1^T\left(\frac{1}{P}\mathbf{X}\mathbf{X}^T\right)\mathbf{c}_1. \tag{8.51}$$

Written in this form the above takes the form of a so-called *Rayleigh quotient* whose maximum is expressible algebraically in closed form based on the eigenvalue/eigenvector decomposition of the matrix $\mathbf{X}\mathbf{X}^T$ (in the middle term) or likewise the matrix $\frac{1}{P}\mathbf{X}\mathbf{X}^T$ (in the term on the right). Because the matrix $\frac{1}{P}\mathbf{X}\mathbf{X}^T$ can be interpreted statistically as the *covariance matrix* of the data it is more common to use the particular algebraic arrangement on the right.

So, denoting by v_1 and d_1 the corresponding eigenvector and largest eigenvalue of $\frac{1}{P}\mathbf{X}\mathbf{X}^T$, the maximum of the above occurs when $c_1 = v_1$, where $h\left(\mathbf{v}_1\right) = d_1$ – which is also the variance in this direction. In the jargon of machine learning v_1 is referred to as the first *principal component* of the data.

With our first basis vector in hand, we can move on to determine the second element of our ideal orthonormal spanning set. Isolating the relevant terms from above and following the same thought process we went through above results in the familiar looking function

$$h\left(\mathbf{c}_2\right) = \frac{1}{P} \sum_{p=1}^{P} \left(\mathbf{c}_2^T \mathbf{x}_p\right)^2 \tag{8.52}$$

that we aim to maximize in order to recover our second basis vector. This formula has the same sort of statistical interpretation as the analogous version of the first basis vector had above – here again it calculates the variance of the data in the direction of $\mathbf{c}_2$. Since our aim here is to maximize – given that $\mathbf{c}_1$ has already been resolved and that $\mathbf{c}_1^T \mathbf{c}_2 = 0$ due to our orthonormal assumption – the statistical interpretation here is that we are aiming to find the *second* largest orthogonal direction of variance in the data.

This formula can also be written in compact vector-matrix form as $h\left(\mathbf{c}_2\right) = \mathbf{c}_2^T \left(\frac{1}{P}\mathbf{X}\mathbf{X}^T\right)\mathbf{c}_2$, and its maximum (given our restriction to an orthonormal basis implies that we must have $\mathbf{c}_1^T \mathbf{c}_2 = 0$) is again expressible in closed form in terms of the eigenvalue/eigenvector decomposition of the covariance matrix $\frac{1}{P}\mathbf{X}\mathbf{X}^T$. Here the same analysis leading to the proper form of $\mathbf{c}_1$ shows that the maximum of the above occurs when $\mathbf{c}_2 = \mathbf{v}_2$ the eigenvector of $\frac{1}{P}\mathbf{X}\mathbf{X}^T$ associated with its second largest eigenvalue d_2, and the variance in this direction is then $h\left(\mathbf{v}_2\right) = d_2$. This ideal basis element/direction is referred to as the *second principal component of the data.*

More generally, following the same analysis for the nth member of our ideal orthonormal basis we look to maximize the familiar looking formula

$$h\left(\mathbf{c}_n\right) = \frac{1}{P} \sum_{p=1}^{P} \left(\mathbf{c}_n^T \mathbf{x}_p\right)^2. \tag{8.53}$$

As with the first two cases above, the desire to maximize this quantity can be interpreted as the quest to uncover the nth orthonormal direction of variance in the data. And following the same arguments, writing the above more compactly as $h\left(\mathbf{c}_n\right) = \mathbf{c}_n^T \left(\frac{1}{P}\mathbf{X}\mathbf{X}^T\right)\mathbf{c}_n$, etc., we can show that it takes the form $\mathbf{c}_n = \mathbf{v}_n$, where $\mathbf{v}_n$ is the nth eigenvector of $\frac{1}{P}\mathbf{X}\mathbf{X}^T$ associated with its nth largest eigenvalue d_n, and here the sample variance is expressible in terms of this eigenvalue $h\left(\mathbf{c}_n\right) = d_n$. This learned element/direction is referred to as the nth principal component of the data.

9 Feature Engineering and Selection

9.1 Introduction

In this chapter we discuss the principles of *feature engineering* and *selection*.

Feature engineering methods consist of an array of techniques that are applied to data *before* they are used by either supervised or unsupervised models. Some of these tools, e.g., the *feature scaling* techniques that we describe in Sections 9.3 through 9.5, properly *normalize* input data and provide a consistent preprocessing pipeline for learning, drastically improving the efficacy of many local optimization methods.

Another branch of feature engineering focuses on the development of data transformations that extract useful information from raw input data. For example, in the case of a two-class classification, these tools aim to extract critical elements of a dataset that ensure instances within a single class are seen as "similar" while those from different classes are "dissimilar." Designing such tools often requires significant domain knowledge and a rich set of experiences dealing with particular kinds of data. However, we will see that one simple concept, the *histogram* feature transformation, is a common feature engineering tool used for a variety of data types including categorical, text, image, and audio data. We give a high-level overview of this popular approach to feature engineering in Section 9.2.

Human beings are often an integral component of the machine learning paradigm, and it can be crucial that individuals be able to interpret and/or derive insights from a machine learning model. The *performance* of a model is a common and relatively easy metric for humans to interpret: does the model provide good[1] predictive results? Sometimes it is very useful to understand *which* input features were the most pertinent to achieving strong performance, as it helps us refine our understanding of the nature of the problem at hand. This is done through what is called *feature selection*. In Sections 9.6 and 9.7 we discuss two popular ways for performing feature selection: *boosting* and *regularization*.

Feature selection can be thought of as a supervised dimension reduction technique that reduces the total number of features involved in a regression or classification, making the resulting model more human interpretable. An abstract

---

[1] Here "good" can mean, for instance, that the learner achieves an agreed upon benchmark value for accuracy, error, etc.

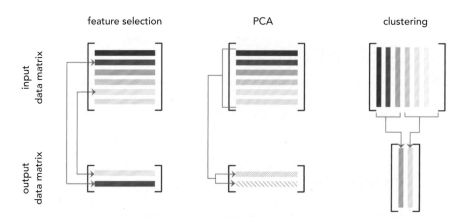

Figure 9.1 A prototypical comparison of feature selection, PCA, and clustering as dimension reduction schemes on an arbitrary data matrix, like those we have discussed in previous chapters for un/supervised learning, whose rows contain features and columns individual data points. The former two methods reduce the dimension of the feature space, or in other words the number of rows in a data matrix. However, the two methods work differently: while feature selection literally selects rows from the original matrix to keep, PCA uses the geometry of the feature space to produce a new data matrix based on a lower feature dimensional version of the data. K-means, on the other hand, reduces the dimension of the data/number of data points, or equivalently the number of columns in the input data matrix. It does so by finding a small number of new averaged representatives or "centroids" of the input data, forming a new data matrix whose fewer columns (which are not present in the original data matrix) are precisely these centroids.

illustration of this concept is shown in the left panel of Figure 9.1, and is compared visually to the result of both Principal Component Analysis (PCA) and clustering (two unsupervised learning techniques introduced in the previous chapter). In contrast to feature selection, when PCA (see Section 8.3) is applied to reduce the dimension of a dataset it does so by learning a new (smaller) set of features over which the dataset may be fairly represented. Likewise any clustering technique (like K-means detailed in Section 8.5) learns new representations to reduce the shear number of points in a dataset.

9.2   Histogram Features

A histogram is an extremely simple yet useful way of summarizing and representing the contents of an array. In this section we see how this rather simple concept is at the core of designing features for common types of input data including categorical, text, image, and audio data types. Although each of these data types differs substantially in nature, we will see how the notion of a histogram-based feature makes sense in each context. This discussion aims at giving the reader a high-level, intuitive understanding of how common histogram-based feature methods work. The interested reader is encouraged to consult specialized texts (referenced throughout this section) on each data type for further study.

9.2.1   Histogram features for categorical data

Every machine learning paradigm requires that the data we deal with consist strictly of *numerical* values. However, raw data does not always come pre-packaged in this manner. Consider, for instance, a hypothetical medical dataset consisting of several patients' vital measurements such as blood pressure, blood glucose level, and blood type. The first two features (i.e., blood pressure and blood glucose level) are naturally numerical, and hence ready for supervised or unsupervised learning. Blood type, on the other hand, is a *categorical* feature, taking on the values O, A, B, and AB. Such categorical features need to be translated into numerical values before they can be fed into any machine learning algorithm.

A first, intuitive approach to do this would be to represent each category with a distinct real number, e.g., by assigning 0 to the blood type O, 1 to A, 2 to B, and 3 to AB, as shown in the top left panel of Figure 9.2. Here the way we assign numbers to each category is important. In this particular instance, by assigning 1 to A, 2 to B, and 3 to AB, and because the number 3 is closer to number 2 than it is to number 1, we have inadvertently injected the assumption into our dataset that an individual with blood type AB is more "similar" to one with blood type B than one with blood type A. One could argue that it is more appropriate to switch the numbers used to represent categories B and AB so that AB now sits between (and hence equidistant from) A and B, as shown in the top right panel of Figure 9.2. However, with this reassignment blood type O is now interpreted as being maximally different from blood type B, a kind of assumption that *may* or *may not* be true in reality.

The crux of the matter here is that there is always a natural order to any set of numbers, and by using such values we inevitably inject assumptions about *similarity* or *dissimilarity* of the existing categories into our data. In most cases we want to avoid making such assumptions that fundamentally change the geometry of the problem, especially when we lack the intuition or knowledge necessary for determining similarity between different categories.

A better way of encoding categorical features, which avoids this problem, is to use a *histogram* whose bins are all categories present in the categorical feature of interest. Doing this in the example of blood type, an individual with blood type O is no longer represented by a single number but by a four-binned histogram that contains all zero values except the bin representing blood type O, which is set to 1. Individuals with other blood types are represented similarly, as depicted visually in the bottom panel of Figure 9.2. This way, all blood type representations (each a four-dimensional vector with a single 1 and three 0s) are

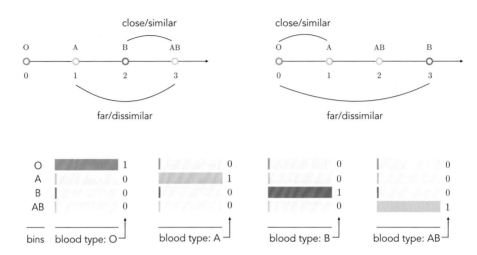

Figure 9.2 Blood type converted into numerical features (top panels) and histogram-based features (bottom panel). See text for further details.

geometrically equidistant from one another. This method of encoding categorical features is sometimes referred to as *one-hot encoding* (see Section 6.7.1).

9.2.2 Histogram features for text data

Many popular uses of machine learning, including sentiment analysis, spam detection, and document categorization or clustering are based on text data, e.g., online news articles, emails, social media posts, etc. However, with text data, the initial input (i.e., the document itself) requires a significant amount of preprocessing and transformation prior to being input into any machine learning algorithm. A very basic but widely used feature transformation of a document for machine learning tasks is called a *Bag of Words* (BoW) histogram or feature vector. Here we introduce the BoW histogram and discuss its strengths, weaknesses, and common extensions.

A BoW feature vector of a document is a simple histogram count of the different words it contains with respect to a single corpus or collection of documents, minus those nondistinctive words that do not characterize the document (in the context of the application).

To illustrate this idea let us build a BoW representation for the following corpus of two simple text documents each consisting of a single sentence.

$$
\begin{aligned}
&1.\ \texttt{dogs are the best}\\
&2.\ \texttt{cats are the worst}
\end{aligned}
\tag{9.1}
$$

To make the BoW representation of these documents we begin by parsing them,

$$\begin{array}{ll}
\text{best} & 1/\sqrt{2} \\
\text{cat} & 0 \\
\text{dog} & 1/\sqrt{2} \\
\text{worst} & 0
\end{array}
\mathbf{x}_1 \qquad
\begin{array}{ll}
0 \\
1/\sqrt{2} \\
0 \\
1/\sqrt{2}
\end{array}
\mathbf{x}_2$$

Figure 9.3 Bag of Words histogram features for the two example documents shown in Equation (9.1). See text for further details.

creating representative vectors (histograms) $\mathbf{x}_1$ and $\mathbf{x}_2$ which contain the number of times each word appears in each document. The BoW vectors for both of these documents are depicted visually in Figure 9.3.

Notice that, in creating the BoW histograms, uninformative words such as are and the, typically referred to as *stop words*, are not included in the final representation. Further, notice that we count the singular dog and cat in place of their plural, which appeared in the original documents. This preprocessing step is commonly called *stemming*, where related words with a common stem or root are reduced to (and then represented by) their common linguistic root. For instance, the words learn, learning, learned, and learner, in the final BoW feature vector are all represented by and counted as learn. Additionally, each BoW vector is *normalized* to have unit length.

Given that BoW vectors contain only nonnegative entries and all have unit length, the inner product between two BoW vectors $\mathbf{x}_1$ and $\mathbf{x}_2$ always ranges between 0 and 1, i.e., $0 \le \mathbf{x}_1^T \mathbf{x}_2 \le 1$. This inner product or *correlation* value may be used as a rough geometric measure of similarity between two BoW vectors. For example, when two documents are made up of completely different words, the correlation is exactly zero, their BoW vectors are perpendicular to one other, and the two documents can be considered maximally different. This is the case with BoW vectors shown in Figure 9.3. On the other hand, the higher the correlation between two BoW vectors the more similar their respective documents are purported to be. For example, the BoW vector of the document I love dogs would have a rather large positive correlation with that of I love cats.

Because the BoW vector is such a simple representation of a document, completely ignoring word order, punctuation, etc., it can only provide a *gross summary* of a document's contents. For example, the two documents dogs are better than cats and cats are better than dogs would be considered *exactly the same document* using BoW representation, even though they imply completely opposite meanings. Nonetheless, the gross summary provided by BoW can be distinctive enough for many applications. Additionally, while more complex representations of documents (capturing word order, parts of speech, etc.) may be employed they can often be unwieldy (see, e.g., [32]).

Example 9.1   Sentiment analysis

Determining the aggregated feelings of a large base of customers, using text-based content like product reviews, tweets, and social media comments, is commonly referred to as *sentiment analysis*. Classification models are often used to perform sentiment analysis, learning to identify consumer data of either positive or negative feelings.

For example, the top panel of Figure 9.4 shows BoW vector representations for two brief reviews of a controversial comedy movie, one with a positive opinion and the other with a negative one. The BoW vectors are rotated sideways in this figure so that the horizontal axis contains the common words between the two sentences (after stop word removal and stemming). The polar opposite sentiment of these two reviews is perfectly represented in their BoW representations, which as one can see are indeed perpendicular (i.e., they have zero correlation). In general, two documents with opposite sentiments are not and need not always be perpendicular for sentiment analysis to work effectively, even though we ideally expect them to have small correlations, as shown in the bottom panel of Figure 9.4.

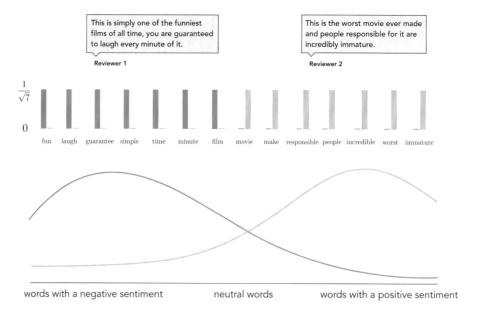

Figure 9.4 Figure associated with Example 9.1. (top panel) BoW representation of two movie review excerpts, with words (after the removal of stop words and stemming) shared between the two reviews listed along the horizontal axis. The vastly different opinion of each review is reflected very well by the BoW histograms, which have zero correlation. (bottom panel) In general, the BoW histogram of a typical document with positive sentiment is ideally expected to have small correlation with that of a typical document with negative sentiment.

Example 9.2   Spam detection

In many spam detectors (see Example 1.8) the BoW feature vectors are formed with respect to a specific list of spam words (or phrases) such as `free`, `guarantee`, `bargain`, `act now`, `all natural`, etc., that are frequently seen in spam emails. Additionally, features like the frequency of certain characters like ! and * are appended to the BoW feature, as are other spam-targeted features including the total number of capital letters in the email and the length of longest uninterrupted sequence of capital letters, as these features can further distinguish the two classes.

In Figure 9.5 we show classification results on a spam email dataset (first introduced in Example 6.10) consisting of Bag of Words (BoW), character frequencies, and other spam-focused features. Employing the two-class Softmax cost (see Section 7.3) to learn the separator, the figure shows the number of misclassifications for each step of a run of Newton's method (see Section 4.3). More specifically, these classification results are shown for the same dataset using only BoW features alone (in blue), BoW and character frequencies (in orange), and the BoW/character frequencies as well as spam-targeted features (in green). Unsurprisingly the addition of character frequencies improves the classification, with the best performance occurring when the spam-focused features are used as well.

Figure 9.5 Figure associated with Example 9.2. See text for details.

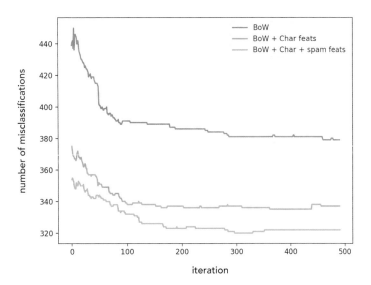

9.2.3   Histogram features for image data

To perform supervised/unsupervised learning tasks on image data, such as object recognition or image compression, the raw input data are pixel values of an image itself. The pixel values of an 8-bit grayscale image are each just a single integer in the range of 0 (black) to 255 (white), as illustrated in Figure 9.6. In other words, a grayscale image is just a *matrix* of integers ranging from 0 to 255. A color image is then just a set of three such grayscale matrices, one for each of the red, blue, and green channels.

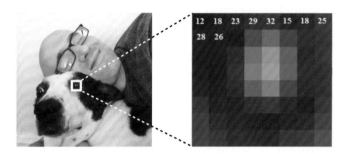

Figure 9.6 An 8-bit grayscale image consists of pixels, each taking a value between 0 (black) and 255 (white). To visualize individual pixels, a small 8 × 8 block from the original image is blown up on the right.

Although it is possible to use raw pixel values directly as features, pixel values themselves are typically not discriminative enough to be useful for machine learning tasks. We illustrate why this is the case using a simple example in Figure 9.7. Consider the three images of shapes shown in the left column of this figure. The first two are similar triangles and the third shape is a square. We would like an ideal set of features to reflect the similarity of the first two images as well as their distinctness from the last image. However, due to the difference in their relative size, position in the image, and the contrast of the image itself (the image with the smaller triangle is darker toned overall) if we were to use raw pixel values to compare the images (by taking the difference between each image pair) we would find that the square and larger triangle are more similar than the two triangles themselves.[2] This is because the pixel values of the first and third image, due to their identical contrast and location of the triangle/square, are indeed more similar than those of the two triangle images.

In the middle and right columns of Figure 9.7 we illustrate a two-step procedure that generates the sort of discriminating feature transformation we are after. In the first part we shift perspective from the pixels themselves to the edge content at each pixel. By taking edges instead of pixel values we significantly reduce the amount of information we must deal with in an image without destroying its identifying structures. In the middle column of the figure we show corresponding edge detected images, in particular highlighting eight equally (angularly) spaced edge orientations, starting from 0 degrees (horizontal edges)

[2] This is to say that if we denote by X_i the ith image then we would find that
$$\|X_1 - X_3\|_F < \|X_1 - X_2\|_F.$$

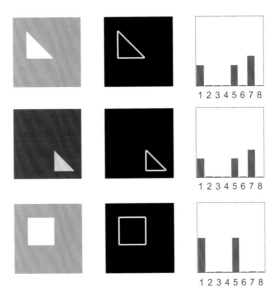

Figure 9.7 (left column) Three images of simple shapes. While the triangles in the top two images are visually similar, this similarity is not reflected by comparing their raw pixel values. (middle column) Edge detected versions of the original images, here using eight edge orientations, retain the distinguishing structural content while significantly reducing the amount of information in each image. (right column) By taking normalized histograms of the edge content we have a feature representation that captures the similarity of the two triangles quite well while distinguishing both from the square. See text for further details.

with seven additional orientations at increments of 22.5 degrees, including 45 degrees (capturing the diagonal edges of the triangles) and 90 degrees (vertical edges). Clearly the edges retain distinguishing characteristics from each original image, while significantly reducing the amount of total information in each case.

We then make normalized histogram of each image's edge content (as shown for the examples in the right column of Figure 9.7). That is, we make a vector consisting of the total amount of each edge orientation found in the image (the vertical axis of the histogram) and normalize the resulting vector to have unit length. This is completely analogous to the BoW feature representation described for text data previously, with the counting of edge orientations being the analog of counting "words" in the case of text data. Here we also have a normalized histogram that represents an image grossly while ignoring the location and ordering of its information. However, as shown in the right panel of Figure 9.7 (unlike raw pixel values) these histogram feature vectors capture characteristic information about each image, with the top two triangle images having very similar histograms and both differing significantly from that of the third image of the square.

Generalizations of this simple edge histogram concept are widely used as feature transformations for visual object recognition where the goal is to locate objects of interest (e.g., faces in a face recognition app or pedestrians in a self-driving car) in an example image or when different objects need to be distinguished from each other across multiple images (e.g., handwritten digits as in the example below, or even distinguishing cats from dogs as discussed in Section 1.2). This is due to the fact that edge content tends to preserve the structure of more complex images – like the one shown in Figure 9.8 – while drastically reducing the amount of information in the image [33, 34]. The majority of the pixels in this image do not belong to any edges, yet with just the edges we can still tell what the image contains.

Figure 9.8 (left panel) A natural image (in this instance of the two creators/writers of the television show "South Park" (this image is reproduced with permission of Jason Marck). (right panel) The edge-detected version of this image, where the bright yellow pixels indicate large edge content, still describes the scene very well (in the sense that we can still tell there are two people in the image) using only a fraction of the information contained in the original image. Note that edges have been colored yellow for visualization purposes only.

However, for such complex images, preserving local information (features of the image in smaller areas) becomes important. Thus a natural way to extend the edge histogram feature is to compute it not over the entire image, but by breaking the image into relatively small patches and computing an edge histogram of each patch, then concatenating the results. In Figure 9.9 we show a diagram of a common variation of this technique often used in practice where neighboring histograms are normalized jointly in larger blocks (see, e.g., [5, 35, 36, 37, 38] for further details).

Interestingly this edge-based histogram feature design mimics the way many animals seem to process visual information. From visual studies performed largely on frogs, cats, and primates, where a subject is shown visual stimuli while electrical impulses are recorded in a small area in the subject's brain where visual information is processed, neuroscientists have determined that individual neurons involved roughly operate by identifying edges [39, 40]. Each

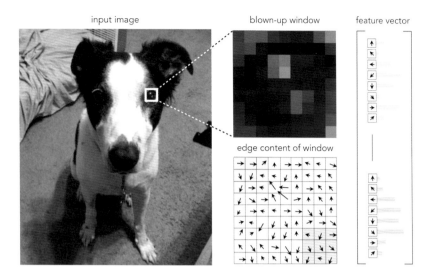

Figure 9.9 A pictorial representation of the sort of generalized edge histogram feature transformation commonly used for object detection. An input image is broken down into small (here 9×9) blocks, and an edge histogram is computed on each of the smaller nonoverlapping (here 3×3) patches that make up the block. The resulting histograms are then concatenated and normalized jointly, producing a feature vector for the entire block. Concatenating such block features by scanning the block window over the entire image gives the final feature vector.

neuron therefore acts as a small "edge detector," locating edges in an image of a specific orientation and thickness, as shown in Figure 9.10. It is thought that by combining and processing these edge detected images that humans and other mammals "see."

Example 9.3   Handwritten digit recognition

In this example we look at the problem of handwritten digit recognition (introduced in Example 1.10), and compare the training effectiveness of mini-batch gradient descent (20 steps/epochs with a learning rate of $\alpha = 10^{-2}$ and batch size of 200 applied to a multi-class Softmax cost) using $P = 50,000$ raw (pixel-based) data points from the MNIST handwritten digit recognition dataset (introduced in Example 7.10), to the effectiveness of precisely the same setup applied to edge histogram based features extracted from these same data points.

The effectiveness of this setup over the raw data in terms of both the cost function and misclassification history resulting from the optimization run is shown as the black curves in the left and right panels, respectively, of Figure 9.11. The results of the same run over the edge feature extracted version of the dataset is shown by the magenta curves in the same figure. Here we can see a massive performance gap, particularly in the misclassification history plot in

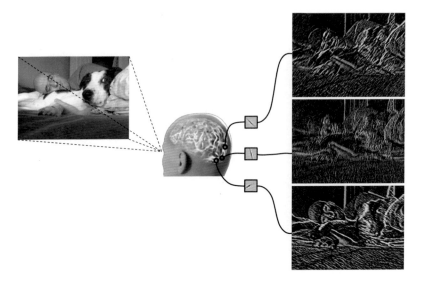

Figure 9.10 Visual information is processed in an area of the brain where each neuron detects in the observed scene edges of a specific orientation and width. It is thought that what we (and other mammals) "see" is a processed interpolation of these edge detected images.

the right panel of the figure, where the difference in performance is around 4000 misclassifications (in favor of the run over the edge-based features).

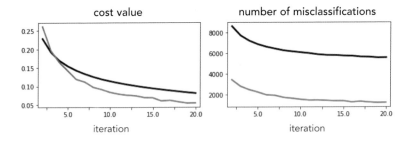

Figure 9.11 Figure associated with Example 9.3. See text for details.

9.2.4   Histogram features for audio data

Like images, audio signals in raw form are not discriminative enough to be used for audio-based classification tasks (e.g., speech recognition) and once again properly designed histogram-based features could be used. In the case of an audio signal it is the histogram of its frequencies, otherwise known as its

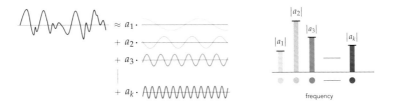

Figure 9.12 A pictorial representation of an audio signal and its representation as a frequency histogram or spectrum. (left panel) A figurative audio signal can be decomposed as a linear combination of simple sinusoids with varying frequencies (or oscillations). (right panel) The frequency histogram then contains the strength of each sinusoid in the representation of the audio signal.

spectrum, that provides a robust summary of its contents. As illustrated pictorially in Figure 9.12, the spectrum of an audio signal counts up (in histogram fashion) the strength of each level of its frequency or oscillation. This is done by decomposing the speech signal over a basis of sinusoidal waves of ever increasing frequency, with the weights on each sinusoid representing the strength of that frequency in the original signal. Each oscillation level is analogous to an edge direction in the case of an image, or an individual word in the case of a BoW text feature.

As with image data in Figure 9.9, computing frequency histograms over overlapping windows of an audio signal as illustrated pictorially in Figure 9.13, produces a feature vector that preserves important local information as well, and is a common feature transformation used for speech recognition called a *spectrogram* [41, 42]. Further processing of the windowed histograms (e.g., to emphasize the frequencies of sound best recognized by the human ear) is also commonly performed in practical implementations of this sort of feature transformation.

9.3   Feature Scaling via Standard Normalization

In this section we describe a popular method of input normalization in machine learning called *feature scaling* via *standard normalization*. This sort of feature engineering scheme provides several benefits to the learning process, including substantial improvement in learning speed when used with local optimization algorithms, and with first-order methods in particular. As such this feature engineering method can also be thought of as an *optimization trick* that substantially improves our ability to minimize virtually every machine learning model.

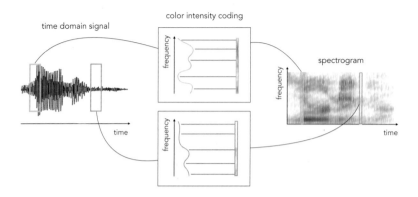

Figure 9.13 A pictorial representation of histogram-based features for audio data. The original speech signal (shown on the left) is broken up into small (overlapping) windows whose frequency histograms are computed and stacked vertically to produce a so-called *spectrogram* (shown on the right).

9.3.1   Standard normalization

Standard normalization of the input features of a dataset is a very simple two-step procedure consisting of first *mean-centering* and then *rescaling* each of its input features by the inverse of its standard deviation. Phrased algebraically, we normalize along the nth input feature of our dataset by replacing $x_{p,n}$ (the nth coordinate of the input point $\mathbf{x}_p$) with

$$\frac{x_{p,n} - \mu_n}{\sigma_n} \tag{9.2}$$

where μ_n and σ_n are the mean and standard deviation computed along the nth dimension of the data, defined respectively as

$$\mu_n = \frac{1}{P} \sum_{p=1}^{P} x_{p,n}$$
$$\sigma_n = \sqrt{\frac{1}{P} \sum_{p=1}^{P} \left(x_{p,n} - \mu_n \right)^2}. \tag{9.3}$$

This (completely invertible) procedure is done for each input dimension $n = 1, \dots, N$. Note that if $\sigma_n = 0$ for some n the standard normalization in Equation (9.2) is undefined as it involves division by zero. However, in this case the corresponding input feature is *redundant* since this implies that the nth feature is the same constant value across the entirety of data. As we discuss further in the next section, such a feature should be removed from the dataset in the beginning as nothing can be learned from its presence in any machine learning model.

Generally speaking, standard normalization alters the shape of machine learning cost functions by making their contours appear more "circular." This idea is illustrated in Figure 9.14 where in the top row we show a prototypical $N = 2$

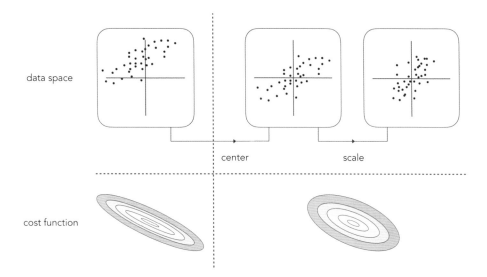

Figure 9.14 (Standard normalization illustrated. The input space of a generic dataset (top-left panel) as well as its mean-centered (top-middle panel) and scaled version (top-right panel). As illustrated in the bottom row, where a prototypical cost function corresponding to this data is shown, standard normalization results in a cost function with less elliptical and more circular contours compared to the original cost function.

dimensional dataset (top-left panel) as well as its standard normalized version (top-right panel). In the input data space, standard normalization produces a centered and more compactly confined version of the original data. Simultaneously, as shown in the bottom row of the figure, a generic cost function associated with the standard normalized version of the data has contours that are much more circular than those associated with the original, unnormalized data.

Making the contours of a machine learning cost function more circular helps speed up the convergence of local optimization schemes, particularly first-order methods like gradient descen (however, feature scaling techniques can also help to better condition a dataset for use with second-order methods, helping to potentially avoid issues with numerical instability as briefly touched on in Section 4.3.3). This is because, as detailed in Section 3.6.2, the gradient descent direction always points perpendicular to the contours of a cost function. This means that, when applied to minimize a cost function with elliptical contours like the example shown in the top panel of Figure 9.15, the gradient descent direction points *away* from the minimizer of the cost function. This characteristic naturally leads the gradient descent algorithm to take *zig-zag* steps back and forth.

In standard normalizing the input data we temper such elliptical contours, transforming them into more circular contours as shown in the bottom-left panel and (ideally) bottom-right panel of Figure 9.15. With more circular contours the gradient descent direction starts pointing more and more in the direction of the cost function's minimizer, making each gradient descent step much more effective. This often means we can use a much larger steplength parameter α when minimizing a cost function over standard normalized data.

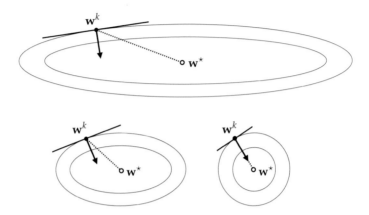

Figure 9.15 In standard normalizing input data we temper its associated cost function's often-elliptical contours, like those shown in the top panel, into more circular ones as shown in the bottom-left and bottom-right panels. This means that the gradient descent direction, which points away from the minimizer of a cost function when its contours are elliptical (leading to the common zig-zagging problem with gradient descent), points more towards the function's minimizer as its contours become more circular. This makes each gradient descent step much more effective, typically allowing the use of much larger steplength parameter values α, meaning that measurably fewer steps are required to adequately minimize the cost function.

Example 9.4   Linear regression with standard normalized data

A simple regression dataset is plotted in the top-left panel of Figure 9.16. With a quick glance at the data we can see that, if tuned properly, a linear regressor will fit this dataset exceedingly well. Since this is a low-dimensional example with only two parameters to tune (i.e., the bias and slope of a best fit line) we can visualize its associated Least Squares cost function, as illustrated in the top-middle panel of Figure 9.16. Notice how elliptical the contours of this cost function are, creating a long narrow valley along the long axis of the ellipses. In the top-middle panel of the figure we also show 100 steps of gradient descent initialized at the point $\mathbf{w}^0 = [0\ 0]^T$, using a fixed steplength parameter $\alpha = 10^{-1}$. Here the steps on the contour plot are colored from green to red as gradient descent begins (green) to when it ends (red). Examining the panel we can see that even at the end of the run we still have quite a way to travel to reach the minimizer of the cost function. We plot the line associated with the final set of weights resulting from this run in blue in the top-left panel of the figure. Because these weights lie rather far from the true minimizer of the cost function they provoke a (relatively) poor fit of the data.

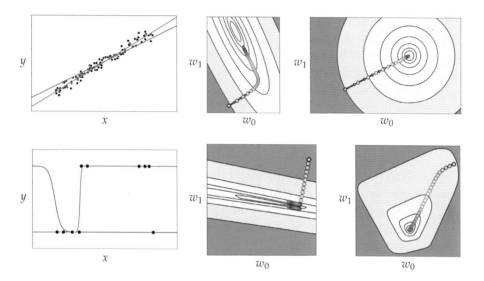

Figure 9.16 Figure associated with Examples 9.4 and 9.5, showing the result of standard normalization applied to a regression (top row) and a two-class classification (bottom row) dataset. See text for details.

We then standard normalize the data, and visualize the contour plot of the associated Least Squares cost in the top-right panel of Figure 9.16. As we can see the contours of this Least Squares cost are perfectly circular, and so gradient descent can much more rapidly minimize this cost. On top of the contour plot we show a run of 20 (instead of 100) gradient descent steps using the same initial point and steplength parameter as we used before with unnormalized data. Note that since cost function associated with the standard normalized version of the data is so much easier to optimize, we reach a point much closer to the minimizer after just a few steps, resulting in a linear model (shown in red) which fits the data in the top-left panel far better than the regressor provided by the first run (shown in blue).

Example 9.5   **Linear two-class classification with standard normalized data**
The bottom-left panel of Figure 9.16 shows a simple two-class classification dataset. Just like the previous example, since we only have two parameters to tune in learning a linear classifier for this dataset, it is possible to visualize the contour plot of the corresponding two-class Softmax cost. The contours of this cost function are plotted in the bottom-middle panel of Figure 9.16. Once

again, their extremely long and narrow shape suggests that gradient descent will struggle immensely in determining the global minimizer (located inside the smallest contour shown).

We confirm this intuition by making a run of 100 steps of gradient descent beginning at the point $\mathbf{w} = [20\ 30]^T$ and using a steplength parameter of $\alpha = 1$. As shown in the bottom-middle panel, these steps (colored from green to red as gradient descent progresses) zig-zag considerably. Moreover we can see that at the end of the run we are still a long way from the minimizer of the cost function, resulting in a very poor fit to the dataset (shown in blue in the bottom-left panel of the figure).

In the bottom-right panel of Figure 9.16 we show the result of repeating this experiment using standard normalized input. Here we use the same initialization, but only 25 steps, and (since the contours of the associated cost are so much more circular) a larger steplength value $\alpha = 10$. This rather large value would have caused the first run of gradient descent to diverge. Nonetheless even with so few steps we are able to find a good approximation to the global minimizer. We plot the corresponding tanh model (in red) in the bottom-left panel of the figure, which fits the data much better than the result of the first run (shown in blue).

9.3.2   Standard normalized model

Once a general model taking in N-dimensional standard normalized input has been properly tuned, and the optimal parameters $w_0^\star, w_1^\star, ..., w_N^\star$ have been determined, in order to evaluate any new point we must standard normalize each of its input features using the same statistics we computed on the training data.

9.4   Imputing Missing Values in a Dataset

Real-world data can contain *missing values* for various reasons including human error in collection, storage issues, faulty sensors, etc. Generally speaking, if a supervised learning data point is missing its *output* value there is little we can do to salvage it, and usually such a corrupted data point is thrown away in practice. Likewise, if a large number of input values of a data point are missing it is best discarded. However, a data point missing just a handful of its input features can be salvaged by filling in missing input features with appropriate values. This process, often called *imputation*, is particularly useful when the data is scarce.

9.4.1   Mean imputation

Suppose as usual that we have a set of P inputs, each of which is N-dimensional, and that the set Ω_n contains the indices of all data points whose nth input feature value is missing. In other words, for all $j \in \Omega_n$ the value of $x_{j,n}$ is missing in our input data. An intuitive value to fill for all missing entries along the nth input feature is the simple average (or expected value) of the dataset along this dimension. That is, for all $j \in \Omega_n$ we set $x_{j,n} = \mu_n$, where

$$\mu_n = \frac{1}{P - |\Omega_n|} \sum_{j \notin \Omega_n} x_{j,n} \tag{9.4}$$

and where $|\Omega_n|$ denotes the number of elements in Ω_n. This is often called *mean imputation*. Notice, after mean imputation, that the mean value of the entire nth feature of the input remains unchanged, since

$$\frac{1}{P} \sum_{p=1}^{P} x_{p,n} = \frac{1}{P} \sum_{j \notin \Omega_n} x_{j,n} + \frac{1}{P} \sum_{j \in \Omega_n} x_{j,n} = \frac{1}{P} \left(P - |\Omega_n| \right) \mu_n + \frac{1}{P} \sum_{j \in \Omega_n} \mu_n = \mu_n. \tag{9.5}$$

Therefore one consequence of imputing missing values of a dataset using the mean along each input dimension is that when we standard normalize this dataset (as detailed in the previous section), all values imputed with the mean become exactly zero. This is illustrated for a simple example in Figure 9.17. Thus any parameter or weight in the model that touches such a mean-imputed entry is completely nullified numerically. This is desirable given that such values were missing in the first place.

9.5 Feature Scaling via PCA-Sphering

In the Section 9.3 we saw how *feature scaling* via *standard normalization* significantly improves the topology of a machine learning cost functions, enabling much more rapid minimization via first-order methods like gradient descent (see Section 3.5). In this Section we describe how Principal Component Analysis (PCA) (detailed in Section 8.3) can be used to perform a more advanced form of input normalization, commonly called *PCA-sphering* (or sometimes *whitening*).

9.5.1 PCA-sphering: the big picture

PCA-sphering takes the idea of standard normalization described in Section 9.3 one step further by using PCA (see Section 8.3) to rotate the mean-centered dataset, so that its largest orthogonal directions of variance align with the coordinate axes, prior to scaling each input by its standard deviation. This simple adjustment typically allows us to better compactify the data, and more importantly results in a cost function whose contours are even more "circular" than that provided by standard normalization (indeed PCA-sphering regression data makes the contours of a Least Squares cost for linear regression *perfectly circular* – see Exercise 9.6 for further details). This is illustrated in Figure 9.18 where we

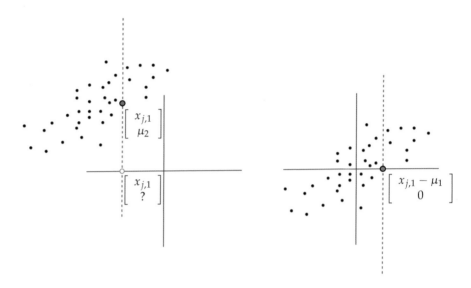

Figure 9.17 (left panel) The input of a prototypical $N = 2$ dimensional dataset where a single point $\mathbf{x}_j$, drawn as a *hollow* red dot, is missing its second entry. The mean-imputed version of this point is then shown as a *filled-in* red point. (right panel) By mean-centering such a dataset (which is the first step of standard normalization) the mean-imputed feature of $\mathbf{x}_j$ becomes exactly equal to zero.

compare pictorially the effect of standard normalization and PCA-sphering on a prototypical $N = 2$ dimensional input dataset, as well as how each scheme changes the topology of the associated cost function. As outlined in Section 9.3.1, gradient descent schemes work far better the more circular we make the contours of a cost function.

The trade-off, of course, is that while PCA-sphering makes first-order optimization considerably easier once it is enacted, we must pay an extra up-front cost of performing PCA on the data, making PCA-sphering more computationally expensive than the standard normalization procedure. Whether or not this extra up-front cost is worth it can vary in practice from problem to problem, however often times it is. This is particularly true when employing first-order optimization (see, e.g., Exercise 9.7) since – as summarized in Figure 9.15 – the more circular we make the contours of a cost function the easier gradient-based optimization methods can minimize them properly.

9.5.2   PCA-sphering: the technical details

More formally we can express the standard normalization scheme applied to a single data point $\mathbf{x}_p$ in two steps as

mean-center: for each n replace $x_{p,n} \longleftarrow \left(x_{p,n} - \mu_n \right)$

std-scale: for each n replace $x_{p,n} \longleftarrow \dfrac{x_{p,n}}{\sigma_n}$

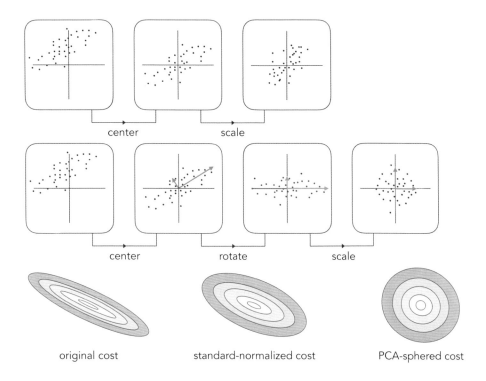

Figure 9.18 The standard normalization procedure (top row) compared to PCA-sphering (middle row) on a generic set of input data. With PCA- sphering we insert a single extra step into the standard normalization pipeline *in between* mean-centering and scaling by standard deviations, where we rotate the data using PCA. This not only shrinks the space consumed by the data more than standard normalization (compare top right and middle right panels), it also tends to make any associated cost function considerably easier to minimize by better tempering its contours, making them more circular (bottom row panels).

where "std" is short for "standard deviation," and the mean and standard deviation are defined as $\mu_n = \frac{1}{P}\sum_{p=1}^{P} x_{p,n}$ and $\sigma_n = \sqrt{\frac{1}{P}\sum_{p=1}^{P} x_{p,n}^2}$ for each n. Note here too that the notation $a \longleftarrow b$ denotes the replacement of quantity a with quantity b.

Denoting by $\mathbf{X}$ the $N \times P$ matrix of input whose pth column contains the input data point $\mathbf{x}_p$, and by $\mathbf{V}$ the set of eigenvectors of the data covariance matrix $\frac{1}{P}\mathbf{X}\mathbf{X}^T = \mathbf{V}\mathbf{D}\mathbf{V}^T$ (as detailed in Section 8.3.3), we can then write the PCA-sphering scheme applied to the same data point $\mathbf{x}_p$ in three highly related steps noting importantly that the nth eigenvalue d_n (the nth diagonal entry of the matrix $\mathbf{D}$) is precisely equal[3] to the variance σ_n^2, and equivalently $\sqrt{d_n} = \sigma_n$

---

[3] The Rayleigh quotient definition (see, e.g., Exercise 3.3) of the nth eigenvalue d_n of the data covariance matrix states that numerically speaking $d_n = \frac{1}{P}\mathbf{v}_n\mathbf{X}\mathbf{X}^T\mathbf{v}_n$, where $\mathbf{v}_n$ is the nth and

mean-center: for each n replace $x_{p,n} \longleftarrow \left(x_{p,n} - \mu_n \right)$

PCA-rotate: transform $\mathbf{x}_p \longleftarrow \mathbf{V}^T \mathbf{x}_p$

std-scale: for each n replace $x_{p,n} \longleftarrow \frac{x_{p,n}}{\sqrt{d_n}}$.

Denoting $\mathbf{D}^{-1/2}$ as the diagonal matrix whose nth diagonal element is $\frac{1}{\sqrt{d_n}}$, we can then (after mean-centering the data) express steps 2 and 3 of the PCA-sphering algorithm recipe above quite compactly as

$$\mathbf{X} \longleftarrow \mathbf{D}^{-1/2} \mathbf{V}^T \mathbf{X}. \qquad (9.6)$$

9.5.3   PCA-sphered model

Once a general model taking in N-dimensional PCA-sphered input has been properly tuned, and the optimal parameters have been determined, in order to evaluate any new point we must PCA-sphere new input features using the same statistics we computed on the training data.

9.6   Feature Selection via Boosting

In Chapters 5 through 7 we saw how the fully tuned linear model for *supervised learning* generally takes the form

$$\text{model}\,(\mathbf{x}, \mathbf{w}^\star) = \mathring{\mathbf{x}}^T \mathbf{w}^\star = w_0^\star + x_1 w_1^\star + x_2 w_2^\star + \cdots + x_N w_N^\star \qquad (9.7)$$

where the weights $w_0^\star$, $w_1^\star$ ..., $w_N^\star$ are optimally tuned via the minimization of an appropriate cost function. Understanding the intricate connections the input features of a dataset have with its corresponding output naturally boils down to human analysis of these $N + 1$ tuned weights. However, it is not always straightforward to derive meaning from such a sequence of $N + 1$ numbers, exacerbated by the fact that the idea of *human interpretability* quickly becomes untenable as the input dimension N grows. To ameliorate this issue we can use what is called a *feature selection* technique.

In this section we discuss one popular way of performing feature selection, called *boosting* or *forward stage-wise selection*. Boosting is a bottom-up approach to feature selection wherein we gradually build up our model one feature at a time by training a supervised learner *sequentially*, one weight at a time. Doing this gives human interpreters an easier way to gage the importance of individual

---

corresponding eigenvector. Now in terms of our PCA-transformed data this is equivalently written as $d_n = \frac{1}{P} \left\| \mathbf{v}_n^T \mathbf{X} \right\|_2^2 = \frac{1}{P} \sum_{p=1}^{P} \left(w_{p,n} \right)^2 = \sigma_n^2$ or in other words, it is the *variance* along the nth axis of the PCA-transformed data.

features, and likewise lets them more easily derive insight about a particular phenomenon.

9.6.1   Boosting based feature selection

In tuning a model's weights one at a time we do not want to tune them in *any* order (e.g., at random) as this will not aid human interpretation. Instead we want to tune them starting with the most important (feature-touching) weight, then tune the second most important (feature-touching) weight, then the third, and so forth. Here by "importance" we mean how each input feature contributes to the final supervised learning model as determined by its associated weight or, in other words, how each contributes to minimizing the corresponding cost (or associated metric) as much as possible.

The boosting process is started with a model, which we will denote as model$_0$, that consists of the bias w_0 alone

$$\text{model}_0\left(\mathbf{x}, \mathbf{w}\right) = w_0. \tag{9.8}$$

We then tune the bias parameter w_0 by minimizing an appropriate cost (depending on whether we are solving a regression or classification problem) over this variable *alone*. For example, if we are performing regression employing the Least Squares cost we would minimize

$$\frac{1}{P}\sum_{p=1}^{P}\left(\text{model}_0\left(\mathbf{x}, \mathbf{w}\right) - y_p\right)^2 = \frac{1}{P}\sum_{p=1}^{P}\left(w_0 - y_p\right)^2 \tag{9.9}$$

which gives the optimal value for our bias $w_0 \longleftarrow w_0^\star$. Plugging this learned weight into our starting model in Equation (9.8) gives

$$\text{model}_0\left(\mathbf{x}, \mathbf{w}\right) = w_0^\star. \tag{9.10}$$

Next, at the *first round* of boosting, in order to determine the most important feature-touching weight (among $w_1, w_2, \ldots, w_N$) we *try out each one* by minimizing an appropriate cost over each individually, having already set the bias optimally. For example, in the case of Least Squares regression the nth of these N subproblems takes the form

$$\frac{1}{P}\sum_{p=1}^{P}\left(\text{model}_0\left(\mathbf{x}_p, \mathbf{w}\right) + w_n x_{n,p} - y_p\right)^2 = \frac{1}{P}\sum_{p=1}^{P}\left(w_0^\star + x_{n,p}w_n - y_p\right)^2. \tag{9.11}$$

Notice, since the bias weight has already been set we only tune the weight w_n in the nth subproblem.

The feature-touching weight that produces the *smallest* cost (or *best* metric value in general) from these N subproblems corresponds to the individual feature that helps best explain the relationship between the input and output of our dataset. It can therefore be interpreted as the most important feature-touching weight we learn. Denoting this weight as w_{s_1}, we then *fix* it at its optimally determined value $w_{s_1}^\star$ (discarding all other weights tuned in each of these subproblems) and update our `model` accordingly. Our updated `model` at the end of the first round of boosting, which we call model$_1$, is a sum of our optimal bias and this newly determined optimal feature-touching weight

$$\text{model}_1\,(\mathbf{x},\mathbf{w}) = \text{model}_0\,(\mathbf{x},\mathbf{w}) + x_{s_1} w_{s_1}^\star = w_0^\star + x_{s_1} w_{s_1}^\star. \qquad (9.12)$$

This boosting process is then repeated sequentially. In general at the mth round of boosting the mth most important feature-touching weight is determined following the same pattern. At the beginning of the mth round (where $m > 1$) we have already determined the optimal setting of our bias as well as the top $m-1$ most important feature-touching weights, and our `model` takes the form

$$\text{model}_{m-1}\,(\mathbf{x},\mathbf{w}) = w_1^\star + x_{s_1} w_{s_1}^\star + \cdots + x_{s_{m-1}} w_{s_{m-1}}^\star. \qquad (9.13)$$

We then set up and solve $N-m+1$ subproblems, one for each feature-touching weight we have not yet chosen. For example, in the case of Least Squares regression the nth of these takes the form

$$\frac{1}{P}\sum_{p=1}^{P}\left(\text{model}_{m-1}\left(\mathbf{x}_p,\mathbf{w}\right) + w_n x_{n,p} - y_p\right)^2 \qquad (9.14)$$

where again in each case we only tune the individual weight w_n. The feature-touching weight that produces the smallest cost value corresponds to the mth most important feature. Denoting this weight w_{s_m} we then fix it at its optimal value $w_{s_m}^\star$, and add its contribution to the running model as

$$\text{model}_m\,(\mathbf{x},\mathbf{w}) = \text{model}_{m-1}\,(\mathbf{x},\mathbf{w}) + x_{s_m} w_{s_m}^\star. \qquad (9.15)$$

Given that we have N input features we can continue until $m \le N$ or when some maximum number of iterations is reached. Note too that, after M rounds of boosting, we have constructed a sequence of models $\{\text{models}_m\}_{m=0}^{M}$. This method of recursive model building via adding features one at a time and tuning only the added feature's weight (keeping all others fixed at their previously tuned values) is referred to as *boosting*.

9.6.2   **Selecting the right number of features via boosting**

Recall that feature selection is done primarily for the purposes of *human interpretation*. Therefore a benchmark value for the number of features M to select can be chosen based on the desire to explore a dataset, and the procedure halted once this number of rounds have completed. One can also halt exploration when adding additional features to the model results in very little decrease in the cost, as most of the correlation between inputs and outputs has already been explained. Finally, M can be chosen entirely based on the sample statistics of the dataset via a procedure known as *cross-validation*, which we discuss in Chapter 11.

Regardless of how we select the value of M, it is important to standard normalize the input data (as detailed in Section 9.3) before we begin the boosting procedure for feature selection. In addition to the optimization speed-up advantage, standard normalization also allows us to fairly compare each input feature's contribution by examining their tuned weight values.

Example 9.6 **Exploring predictors of housing prices via boosting**
The result of running $M = 5$ rounds of boosting on the Boston Housing dataset (first introduced in Example 5.5), using the Least Squares cost and Newton's method optimizer, is visualized in the top panel of Figure 9.19. This special kind of cost function history shows each weight/feature index added to the model at each round of boosting (starting with the bias which has index 0). As can be seen on the horizontal axis, the first two most contributing features found via boosting are LSTAT (feature 13) and the average number of rooms per dwelling (feature 6). Examining the histogram of model weights in the bottom panel of Figure 9.19, we can see (unsurprisingly) that the LSTAT weight, having a negative value, is negatively correlated with the output (i.e., home price) while the weight associated with the average number of rooms feature is positively correlated with the output.

Example 9.7 **Exploring predictors of credit risk via boosting**
In Figure 9.20 we show the results of running the boosting procedure (using a Softmax cost and Newton's method optimizer) on the German credit score dataset first introduced in Example 6.11, which consists of $P = 1000$ samples, each a set of statistics extracted from loan applications to a German bank. The $N = 20$ dimensional input features in this dataset include: the individual's current account balance with the bank (feature 1), the duration (in months) of previous credit with the bank (feature 2), the payment status of any prior credit taken out with the bank (feature 3), and the current value of their savings/stocks (feature 6). These are precisely the top four features found via boosting, most of which are positively correlated with an individual being a good credit risk (as shown in the bottom panel of Figure 9.20).

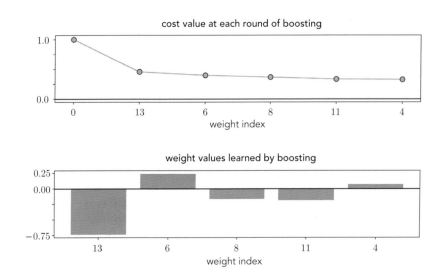

Figure 9.19 Figure associated with Example 9.6. See text for details.

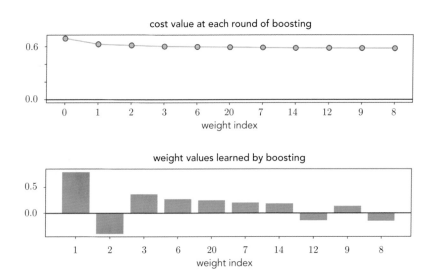

Figure 9.20 Figure associated with Example 9.7. See text for details.

9.6.3   On the efficiency of boosting as a greedy algorithm

Boosting is essentially a *greedy* algorithm, where at each stage we choose the next most important (feature-touching) weight and tune it properly by solving a set of respective subproblems. While each round of boosting demands we solve a number of subproblems, each one is a minimization with respect to *only a single weight* and is therefore cheap to solve virtually regardless of the local optimization scheme used. This makes boosting a computationally effective approach to feature selection, and allows it to scale to datasets with a large number of input dimensions N. A weakness inherent in doing this, however, is that in determining feature-touching weights one at a time, interactions between features/weights can be potentially missed.

To capture these potentially missed interactions one might naturally extend the boosting idea and try to add a group of R feature-touching weights at each round instead of just one. However, a quick calculation shows that this idea would quickly fail to scale. In order to determine the first best group of R feature-touching weights at the first stage of this approach we would need to try out every combination of R weights by solving a subproblem for each. The issue here is that there are combinatorially many subgroups of size R, more precisely $\binom{N}{R} = \frac{N!}{R!(N-R)!}$ of them. This is far too many problems to solve in practice, even for small-to-moderate values of N and R (e.g., $\binom{100}{5} = 75,287,520$).

9.6.4 The residual perspective on boosting regression

Recall the nth subproblem in the mth stage of boosting in the case of Least Squares regression aims at minimizing

$$\frac{1}{P} \sum_{p=1}^{P} \left(\text{model}_{m-1}\left(\mathbf{x}_p, \mathbf{w}\right) + w_n x_{n,p} - y_p \right)^2. \tag{9.16}$$

If we rearrange the terms in each summand and denote

$$r_p^m = \left(y_p - \text{model}_{m-1}\left(\mathbf{x}_p, \mathbf{w}\right) \right) \tag{9.17}$$

then we can write the Least Squares cost function in Equation (9.16) equivalently as

$$\frac{1}{P} \sum_{p=1}^{P} \left(w_n x_{n,p} - r_p^m \right)^2 \tag{9.18}$$

where the term r_p^m on the right-hand side of each summand is *fixed* since w_n is the only weight being tuned here. This r_p^m term is the *residual* of the original output y_p after the contribution of model model_m has been subtracted off. We can then think about each round of boosting as determining the next feature that best correlates with the residual from the previous round.

9.7   Feature Selection via Regularization

With the boosting approach to feature selection discussed in the previous section we took a greedy "bottom-up" approach to feature selection: we began by tuning the bias and then added new features to our model one at a time. In this section we introduce a complementary approach to feature selection, called *regularization*. Instead of building up a model starting at the bottom, with regularization we take a "top-down" view and start off with a complete model that includes every one of our input features, and then we gradually remove input features of inferior importance. We do this by adding a second function to our cost (called a *regularizer*) that penalizes all weights, forcing our model to shrink the weight values associated with less-important input features.

9.7.1   Regularization using weight vector norms

The simple linear combination of the cost function g and an auxiliary function h

$$f(\mathbf{w}) = g(\mathbf{w}) + \lambda\, h(\mathbf{w}) \tag{9.19}$$

is often referred to as *regularization* in the parlance of machine learning, with the function h called a *regularizer* and the parameter $\lambda \geq 0$ called a *regularization* or *penalty* parameter.

When $\lambda = 0$ the linear combination in Equation (9.19) reduces to the original cost function g. As we increase λ the two functions g and h start to *compete for dominance*, with the linear combination taking on properties of both functions. As we set λ to a larger and larger value the function h dominates the combination, eventually completely drowning out g, and we end up essentially with a scaled version of the regularizer h.

In machine learning applications it is very common to use a vector or matrix norm of model parameters (see Section C.5) as a regularizer h, with different norms used to produce different effects in the learning of machine learning models. In what follows we outline how several common vector norms affect the minimization of a generic cost function g.

Example 9.8   Regularization using the ℓ_0 norm

The ℓ_0 vector norm of $\mathbf{w}$, written as $\|\mathbf{w}\|_0$, measures its length or magnitude as

$$\|\mathbf{w}\|_0 = \text{number of nonzero entries of } \mathbf{w}. \tag{9.20}$$

By regularizing a cost g using this regularizer (i.e., $f(\mathbf{w}) = g(\mathbf{w}) + \lambda\,\|\mathbf{w}\|_0$), we penalize the regularized cost f for every nonzero entry of $\mathbf{w}$ since every such entry adds one unit to $\|\mathbf{w}\|_0$. Conversely, then, in minimizing f, the two functions g and $\|\mathbf{w}\|_0$ compete for dominance with g wanting $\mathbf{w}$ to be resolved as a point near its minimizer, while the regularizer $\|\mathbf{w}\|_0$ aims to determine a $\mathbf{w}$ that has as few nonzero elements as possible, or in other words, a weight vector $\mathbf{w}$ that is very *sparse*.

Example 9.9   Regularization using the ℓ_1 norm
The ℓ_1 vector norm, written as $\|\mathbf{w}\|_1$, measures the magnitude of $\mathbf{w}$ as

$$\|\mathbf{w}\|_1 = \sum_{n=0}^{N} |w_n|. \tag{9.21}$$

By regularizing a cost g using this regularizer (i.e., $f(\mathbf{w}) = g(\mathbf{w}) + \lambda \|\mathbf{w}\|_1$), we penalize the regularized cost based on the sum of the absolute value of the entries of $\mathbf{w}$.

   Conversely, then, in minimizing this sum, the two functions g and $\|\mathbf{w}\|_1$ compete for dominance with g wanting $\mathbf{w}$ to be resolved as a point near its minimizer, while the regularizer $\|\mathbf{w}\|_1$ aims to determine a $\mathbf{w}$ that is small in terms of the absolute value of each of its components, but also because the ℓ_1 norm is closely related to the ℓ_0 norm (see Section C.5), one that has few nonzero entries and is therefore *sparse*.

Example 9.10   Regularization using the ℓ_2 norm
The ℓ_2 vector norm of $\mathbf{w}$, written as $\|\mathbf{w}\|_2$, measures its magnitude as

$$\|\mathbf{w}\|_2 = \sqrt{\sum_{n=0}^{N} w_n^2}. \tag{9.22}$$

By regularizing a cost g using this regularizer (i.e., $f(\mathbf{w}) = g(\mathbf{w}) + \lambda \|\mathbf{w}\|_2$), we penalize the regularized cost based on the sum of squares of the entries of $\mathbf{w}$.

   Conversely in minimizing this sum, the two functions g and $\|\mathbf{w}\|_2$ compete for dominance with g wanting $\mathbf{w}$ to be resolved as a point near its minimizer, while the regularizer $\|\mathbf{w}\|_2$ aims to determine a $\mathbf{w}$ that is small in the sense that all of its entries have a small *squared* value.

   We have so far seen a number of instances of ℓ_2 regularization, e.g., in the context of Softmax classification in Section 6.4.6 and support vector machines in Section 6.5.4.

9.7.2   Feature selection via ℓ_1 regularization

In the context of machine learning by inducing the discovery of sparse weight vectors, the ℓ_0 and ℓ_1 reguarlizers help uncover the identity of a dataset's *most important features*. This is because when we employ such norms as regularizer we force the recovered model weights to be rather *sparse* (provided we have set λ appropriately), with only those weights associated with a model's most important features remaining. This makes either norm (at least in principle) quite appropriate for the task of feature selection.

Of the two sparsity-inducing norms described in Examples 9.8 and 9.9, the ℓ_0 norm (while promoting sparsity directly and to the greatest degree) is the most challenging to employ due to its *discontinuous* nature, making the minimization of an ℓ_0 regularized cost function quite difficult. While the ℓ_1 norm induces sparsity to less of a degree, it is both *convex* and *differentiable* (almost everywhere), making the use of first-order methods possible for its minimization. Because of this practical advantage the ℓ_1 norm is by far the more commonly used regularizer for feature selection in practice.

Finally, remember as detailed in the previous section that when performing feature selection we are only interested in determining the importance of *feature-touching* weights w_1, w_2, ..., w_N, thus we only need regularize them (and not the bias weight w_0). This means that our regularization will more specifically take the form

$$f(\mathbf{w}) = g(\mathbf{w}) + \lambda \sum_{n=1}^{N} |w_n|. \tag{9.23}$$

Using our individual notation for the bias and feature-touching weights (used for instance in Section 6.4.6)

$$(\text{bias}): b = w_0 \quad (\text{feature-touching weights}): \quad \omega = \begin{bmatrix} w_1 \\ w_2 \\ \vdots \\ w_N \end{bmatrix} \tag{9.24}$$

we can write this general ℓ_1 regularized cost function equivalently as

$$f(b, \omega) = g(b, \omega) + \lambda \|\omega\|_1. \tag{9.25}$$

9.7.3 Selecting the right regularization parameter

Because feature selection is done for the purposes of *human interpretation*, the value of λ can be set based on several factors. A benchmark value for λ can be chosen based on the desire to explore a dataset, finding a value that provides sufficient sparsity while retaining a low cost value. The value of λ can also be chosen entirely based on the sample statistics of the dataset via a procedure known as *cross-validation*, which we discuss in Chapter 11.

Just like boosting and regardless of how we select λ, it is important to standard normalize the input data so that we can fairly compare each input feature's contribution by examining their tuned weight values recovered from minimization of the regularized cost function.

Example 9.11   **Exploring predictors of housing prices via regularization**

In this example we form an ℓ_1 regularized Least Squares cost using the Boston Housing dataset (first introduced in Example 5.5 and used in the context of boosting-based feature selection in Example 9.6) and examine 50 evenly spaced values for λ in the range $[0, 130]$. For each value of λ in this range, starting from 0 (as illustrated in the top panel of Figure 9.21), we run gradient descent with a fixed number of steps and steplength parameter to minimize the regularized cost. By the time λ is set to the largest value in the range, three major weights remain recovered by the parameter tuning (as illustrated in the bottom panel of Figure 9.21), corresponding to feature 6, feature 13, and feature 11. The first two features (i.e., feature 6 and feature 13) were also determined to be important via boosting in Example 9.6.

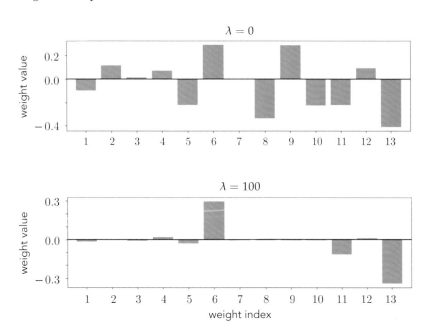

Figure 9.21   Figure associated with Example 9.11. See text for details.

Example 9.12   **Exploring predictors of credit risk via regularization**

In this example we minimize the ℓ_1 regularized two-class Softmax cost over the German credit dataset (first introduced in Example 9.20), using 50 evenly spaced values for λ in the range $[0, 130]$. For each value of λ in this range, starting from

0 (as illustrated in the top panel of Figure 9.22) we run gradient descent with a fixed number of steps and steplength parameter to minimize the regularized cost.

By the time $\lambda \approx 40$, five major weights remain, corresponding to features 1, 2, 3, 6, and 7 (as illustrated in the bottom panel of Figure 9.22). The first four of these features were also determined to be important via boosting in Example 9.7.

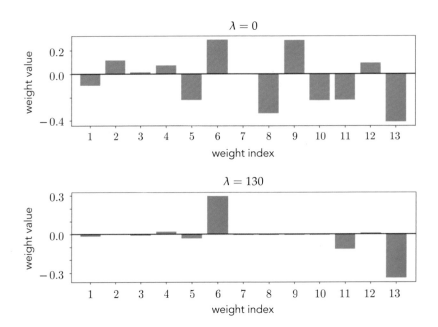

Figure 9.22 Figure associated with Example 9.12. See text for details.

9.7.4　Comparing regularization and boosting

While boosting is an efficient greedy scheme, the regularization idea detailed in this section can be computationally intensive to perform since for each value of λ tried, a full run of local optimization must be completed. On the other hand, while boosting is a "bottom-up" approach that identifies individual features one at a time, regularization takes a more "top-down" approach and identifies important features all at once. In principle this allows regularization to uncover groups of important features correlated in such an interconnected way with the output that may be missed by boosting.

9.8   Conclusion

In this chapter we reviewed fundamental techniques for *feature engineering* and *feature selection*.

We began by discussing feature engineering techniques, which are used as data preprocessing steps for virtually all machine learning problems, and which we will make use of extensively in the remainder of the text. In Section 9.2 we detailed histogram features, which neatly summarize the content in data and can be designed for virtually any data modality. In Sections 9.3–9.5 various input scaling techniques were described, including *standard normalization* and *PCA-sphering*. These methods standardize input data and improve the topology of machine learning cost functions, making them significantly easier to minimize (particularly using the first-order methods described in Chapter 3). Finally in Sections 9.6 and 9.7 we described two complementary approaches to feature selection – *boosting* and *regularization* – which enable straightforward human analysis of the stregnth of individual features included in a trained machine learning model.

9.9   Exercises

† The data required to complete the following exercises can be downloaded from the text's github repository at `github.com/jermwatt/machine_learning_refined`

9.1   Spam email
Repeat the experiment described in Example 1.8 using any local optimization scheme you desire and the two-class Softmax cost. Make sure to produce a plot like the one shown in Figure 9.5 to compare the results of using each combination of features.

9.2   MNIST classification: pixels versus edge-based features
Repeat the experiment outlined in Example 9.3 and create a pair of cost function/misclassification history plots like the ones shown in Figure 9.11. Your results may vary slightly from those reported in the example depending on the details of your implementation.

9.3   Student debt
Produce two contour plots of the Least Squares cost function over the student debt dataset [2] shown in Figure 1.8, as well as its standard normalized version. Compare the overall topology of each contour plot and describe why the plot associated with the standard normalized data will be far easier to optimize via

gradient descent. Indeed, fitting to the original dataset using gradient descent is almost impossible here. Minimize the Least Squares cost over the standard normalized version of the data using gradient descent and reproduce the plot shown in Figure 1.8.

9.4   Least Squares and perfectly circular contours: part 1

In Example 9.4 we saw how the contour plot of the Least Squares cost over an $N = 1$ regression dataset changed from highly elliptical to *perfectly* circular after the data was standard normalized (see Figure 9.16). Show that the contour plot of a Least Squares cost over standard normalized data will always be perfectly circular when $N = 1$. Then describe why this does *not* necessarily happen when $N > 1$.

9.5   Breast cancer dataset

Perform linear two-class classification on a breast cancer dataset [43] consisting of $P = 569$ data points using an appropriate cost function and local optimizer. Fill in any missing values in the data using mean-imputation and report the best misclassification rate you were able to achieve. Compare how quickly you can minimize the your cost function over the original and standard normalized data using the same parameters (i.e., the same number of steps, steplength parameter, etc.).

9.6   PCA-sphering and the Least Squares cost for linear regression

The Least Squares cost for linear regression presents the ideal example of how PCA-sphering a dataset positively effects the topology of a machine learning cost function (making it considerably easier to minimize properly). This is because – as we show formally below – PCA-sphering the input of a regression dataset leads to a Least Squares cost for linear regression that has *perfectly circular contours* and is thus very easy to minimize properly. While PCA-sphering does not improve all cost functions to such a positive degree, this example is still indicative of the effect PCA-sphering has on improving the topology of machine learning cost functions in general.

To see how PCA-sphering perfectly tempers the contours of the Least Squares cost for linear regression first note – as detailed in Section 5.9.1 – that the Least Squares cost is always (regardless of the dataset used) a convex quadratic function of the general form $g(\mathbf{w}) = a + \mathbf{b}^T\mathbf{w} + \mathbf{w}^T\mathbf{C}\mathbf{w}$, where – in particular – $\mathbf{C} = \frac{1}{P}\sum_{p=1}^{P}\mathring{\mathbf{x}}_p\mathring{\mathbf{x}}_p^T$. If the input of a regression dataset is PCA-sphered, then the lower $N \times N$ submatrix of $\mathbf{C}$ is given as $\frac{1}{P}\mathbf{S}\mathbf{S}^T$, where $\mathbf{S}$ is defined in Equation (9.6). However, because of the very way $\mathbf{S}$ is defined we have that $\frac{1}{P}\mathbf{S}\mathbf{S}^T = \frac{1}{P}\mathbf{I}_{N\times N}$, where $\mathbf{I}_{N\times N}$ is the $N \times N$ identity matrix, and thus, in general, that $\mathbf{C} = \frac{1}{P}\mathbf{I}_{(N+1)\times(N+1)}$. In other words, the Least Squares cost over PCA-sphered input is a convex quadratic with all eigenvalues equal to 1, implying that it is a quadratic with perfectly circular contours (see Section 4.2.2).

9.7   Comparing standard normalization to PCA-sphering on MNIST

Compare a run of ten gradient descent steps using the multi-class Softmax cost over $50,000$ random digits from the MNIST dataset (introduced in Example 7.10), a standard normalized version, and a PCA-sphered version of the data. For each run use the largest fixed steplength of the form $\alpha = 10^{\gamma}$ for γ an integer you find to produce descent. Create a plot comparing the progress of each run in terms of the cost function and number of misclassifications. Additionally, make sure your initialization of each run is rather small, particularly the first run where you apply no normalization at all to the input as each raw input point of this dataset is large in magnitude. In the case of the raw data initializing at a point too far away from the origin can easily cause numerical overflow, producing nan or inf values, ruining the rest of the corresponding local optimization run.

9.8   Least Squares and perfectly circular contours: part 2

In Exercise 9.4 we saw how PCA-sphering input data reshapes the topology of the Least Squares cost so that its contours become *perfectly circular*. Explain how this makes the minimization of such a PCA-sphered Least Squares cost extremely easy. In particular explain how – regardless of the dataset – such a cost can be perfectly minimized using one "simplified" Newton step – as described in Section A.8.1 – where we ignore the off-diagonal elements of the Hessian matrix when taking a Newton step.

9.9   Exploring predictors of housing prices

Implement the boosting procedure detailed in Section 9.6.1 for linear regression employing the Least Squares cost, and repeat the experiment described in Example 9.6. You need not reproduce the visualizations in Figure 9.19, but make sure you are able to reach the similar conclusions to those outlined in the example.

9.10   Predicting Miles-per-Gallon in automobiles

Run $M = 6$ rounds of boosting on the automobile MPG dataset introduced in Example 5.6, employing the Least Squares cost function, to perform feature selection. Provide an interpretation of the three most important features you find, and how they correlate with the output.

9.11   Studying important predictors of credit risk

Implement the boosting procedure detailed in Section 9.6.1 for linear classification employing the two-class Softmax cost, and repeat the experiment described in Example 9.7. You need not reproduce the visualizations in Figure 9.20, but make sure you are able to reach the same conclusions outlined in the example.

9.12   **Exploring predictors of housing prices**
Implement the regularization procedure detailed in Section 9.7 for linear regression employing the Least Squares cost, and repeat the experiment described in Example 9.11. You need not reproduce the visualizations in Figure 9.21, but make sure you are able to reach the same conclusions outlined in the example.

9.13   **Studying important predictors of credit risk**
Implement the regularization procedure detailed in Section 9.7 for linear classification employing two-class Softmax cost, and repeat the experiment described in Example 9.12. You need not reproduce the visualizations in Figure 9.22, but make sure you are able to reach the same conclusions outlined in the example.

Part III

Nonlinear Learning

10 Principles of Nonlinear Feature Engineering

10.1 Introduction

Thus far we have dealt with major supervised and unsupervised learning paradigms assuming – for the sake of simplicity – underlying *linear* models. In this chapter we drop this simplifying assumption and begin our foray into nonlinear machine learning by exploring nonlinear *feature engineering* in the context of supervised and unsupervised learning. While nonlinear feature engineering is really only feasible when a dataset (or the phenomenon generating it) is well understood, understanding the basis of nonlinear feature engineering is extremely valuable as it allows us to introduce an array of crucial concepts in a relatively simple environment that will be omnipresent in our discussion of nonlinear learning going forward. As we will see, these important concepts include a variety of formal mathematical and programmatic principles, modeling tools, and jargon terms.

10.2 Nonlinear Regression

In this section we introduce the general framework of nonlinear regression via *engineering* of nonlinear feature transformations based on visual intuition, along with a range of examples. The form of many classic natural laws from the sciences – as first discussed in Example 1.2 – can be derived via the general methods described here (as we explore in Example 10.2 and Exercises 10.2 and 10.4).

10.2.1 Modeling principles

In Chapter 5 we detailed the basic linear model for regression

$$\text{model}\,(\mathbf{x}, \mathbf{w}) = w_0 + x_1 w_1 + x_2 w_2 + \cdots + x_N w_N \tag{10.1}$$

or, more compactly

$$\text{model}\,(\mathbf{x}, \mathbf{w}) = \mathring{\mathbf{x}}^T \mathbf{w} \tag{10.2}$$

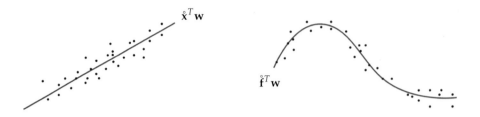

Figure 10.1 (left panel) Linear regression illustrated. Here the fit to the data is defined by the linear model $\mathring{\mathbf{x}}^T\mathbf{w}$. (right panel) Nonlinear regression is achieved by injecting nonlinear feature transformations into our model. Here the fit to the data is a nonlinear curve defined by $\mathring{\mathbf{f}}^T\mathbf{w}$. See text for further details.

where

$$\mathring{\mathbf{x}} = \begin{bmatrix} 1 \\ x_1 \\ x_2 \\ \vdots \\ x_N \end{bmatrix} \quad \text{and} \quad \mathbf{w} = \begin{bmatrix} w_0 \\ w_1 \\ w_2 \\ \vdots \\ w_N \end{bmatrix}. \tag{10.3}$$

To tune the parameters of our linear model over a generic dataset of P points $\left\{\left(\mathbf{x}_p, y_p\right)\right\}_{p=1}^{P}$ so that it represents the data (an example of which is shown in the left panel of Figure 10.1) as well as possible, or phrased algebraically so that we have[1]

$$\mathring{\mathbf{x}}_p^T\mathbf{w} \approx y_p \qquad p = 1, 2, ..., P \tag{10.4}$$

we minimize a proper regression cost function, e.g., the Least Squares cost

$$g\left(\mathbf{w}\right) = \frac{1}{P}\sum_{p=1}^{P}\left(\mathring{\mathbf{x}}_p^T\mathbf{w} - y_p\right)^2. \tag{10.5}$$

We can move from *linear* to general *nonlinear* regression, in both its principles and implementation, simply by swapping out the linear model used in the construction of our linear regression with a nonlinear one. For example, instead of using a linear model we can use a nonlinear one involving a single nonlinear function f (e.g., a quadratic, a sine wave, a logistic function, etc.) that can be parameterized or unparameterized. In the jargon of machine learning such a nonlinear function f is often called a nonlinear *feature transformation* (or just a *feature*) since it transforms our original input features $\mathbf{x}$. Our corresponding nonlinear model would then take the form

---

[1] Following the compact notation in Equation (10.3), $\mathring{\mathbf{x}}_p^T = \begin{bmatrix} 1 & x_{1,p} & x_{2,p} & \cdots & x_{N,p} \end{bmatrix}$ where $x_{n,p}$ is the nth entry in $\mathbf{x}_p$.

$$\text{model}\,(\mathbf{x}, \Theta) = w_0 + f\,(\mathbf{x})\,w_1 \tag{10.6}$$

where the set Θ contains all model parameters including the linear combination weights (here w_0 and w_1) as well as potential internal parameters of the function f itself.

We can simply extend this idea to create nonlinear models that use more than just a single nonlinear feature transformation. In general, we can form a nonlinear model as the weighted sum of B nonlinear functions of our input, as

$$\text{model}\,(\mathbf{x}, \Theta) = w_0 + f_1\,(\mathbf{x})\,w_1 + f_2\,(\mathbf{x})\,w_2 + \cdots + f_B\,(\mathbf{x})\,w_B \tag{10.7}$$

where $f_1, f_2, \ldots, f_B$ are nonlinear (parameterized or unparameterized) functions, and w_0 through w_B along with any additional weights internal to the nonlinear functions are represented in the weight set Θ.

Regardless of what nonlinear functions we choose, the steps we take to formally resolve such a model for the purposes of regression are entirely similar to what we have seen for the simple case of linear regression. In analogy to the linear case, here too it is helpful to write the generic nonlinear model in Equation (10.7) more compactly as

$$\text{model}\,(\mathbf{x}, \Theta) = \mathring{\mathbf{f}}^T \mathbf{w} \tag{10.8}$$

denoting

$$\mathring{\mathbf{f}} = \begin{bmatrix} 1 \\ f_1\,(\mathbf{x}) \\ f_2\,(\mathbf{x}) \\ \vdots \\ f_B\,(\mathbf{x}) \end{bmatrix} \quad \text{and} \quad \mathbf{w} = \begin{bmatrix} w_0 \\ w_1 \\ w_2 \\ \vdots \\ w_B \end{bmatrix}. \tag{10.9}$$

Once again to tune the parameters of our generic nonlinear model over a dataset of P points so that it represents the data (an example of which is shown in the right panel of Figure 10.1) as well as possible, or phrased algebraically so that we have[2]

$$\mathring{\mathbf{f}}_p^T \mathbf{w} \approx y_p \qquad p = 1, 2, \ldots, P \tag{10.10}$$

we minimize a proper regression cost function over Θ, e.g., the Least Squares cost

$$g\,(\Theta) = \frac{1}{P} \sum_{p=1}^{P} \left(\mathring{\mathbf{f}}_p^T \mathbf{w} - y_p \right)^2. \tag{10.11}$$

---

[2] Here $\mathring{\mathbf{f}}_p^T = \begin{bmatrix} 1 & f_1\,(\mathbf{x}_p) & f_2\,(\mathbf{x}_p) & \cdots & f_B\,(\mathbf{x}_p) \end{bmatrix}$, following the compact notation in Equation (10.9).

Despite all these structural similarities between the linear and nonlinear frameworks, one question still remains: how do we determine the appropriate nonlinear feature transformations for our model, and their number B for a generic dataset? This is indeed one of the most important challenges we face in machine learning, and is one which we will discuss extensively in the current chapter as well as several of those to come.

10.2.2   Feature engineering

Here we begin our investigation of nonlinear regression by discussing some simple instances where we can determine the sort and number of nonlinear features we need by *visualizing* the data, and by relying on our own pattern recognition abilities to determine the appropriate nonlinearities. This is an instance of what is more broadly referred to as *feature engineering* wherein the functional form of nonlinearities to use in machine learning models is determined (or engineered) by humans through their expertise, domain knowledge, intuition about the problem at hand, etc.

Example 10.1   Modeling a wave

In the left panel of Figure 10.2 we show a nonlinear regression dataset. Because of the wavy appearance of this data we can defensibly propose a nonlinear model consisting of a sine function

$$f(x) = \sin(v_0 + xv_1) \tag{10.12}$$

parameterized by tunable weights v_0 and v_1, with our regression model given as

$$\text{model}(x, \Theta) = w_0 + f(x)w_1 \tag{10.13}$$

where $\Theta = \{w_0, w_1, v_0, v_1\}$. Intuitively it seems like this model could fit the data well if its parameters were all properly tuned. In the left panel of Figure 10.2 we show the resulting model fit to the data (in green) by minimizing the Least Squares cost via gradient descent.

With our weights fully tuned, notice that our model is defined *linearly* in terms of its feature transformation. This means that if we plot the transformed version of our dataset, i.e., $\left\{ \left(f\left(x_p \right), y_p \right) \right\}_{p=1}^{P}$ wherein the internal feature weights v_0 and v_1 have been optimally tuned, our model fits this transformed data *linearly*, as shown in the right panel of Figure 10.2. In other words, in this *transformed feature space* whose input axis is given by $f(x)$ and whose output is y, our tuned nonlinear model becomes a linear one.

Finally note that, as detailed in Section 9.3 for the case of linear regression, with nonlinear regression it is still highly advantageous to employ standard

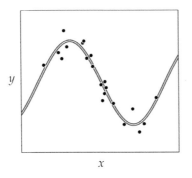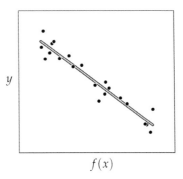

x $f(x)$

Figure 10.2 Figure associated with Example 10.1. (left panel) A nonlinear regression dataset and corresponding tuned model defined in Equations (10.12) and (10.13). (right panel) The same data and tuned model viewed in the *transformed feature space*. See text for further details.

normalization to scale our input when employing gradient descent schemes to minimize a corresponding Least Squares cost function g. In Figure 10.3 we show the cost function history plots resulting from a run of gradient descent employing the original unnormalized (in black) and standard normalized (in magenta) versions of the input. Comparing the two histories we can see that a significantly lower cost function value is found when using the standard normalized input.

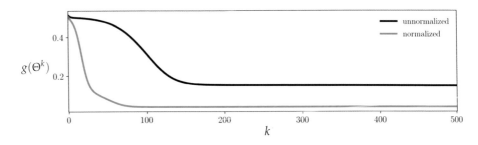

Figure 10.3 Figure associated with Example 10.1. See text for details.

Example 10.2   Galileo and gravity

In 1638 Galileo Galilei, infamous for his expulsion from the Catholic church for daring to claim that the Earth orbited the Sun and not the converse (as was the prevailing belief at the time) published his final book: *"Discourses and Mathematical Demonstrations Relating to Two New Sciences"* [44]. In this book, written as a discourse among three men in the tradition of Aristotle, he described his experimental and philosophical evidence for the notion of uniformly accelerated physical motion. Specifically, Galileo (and others) had intuition that the acceleration of an object due to (the force we now know as) gravity is uniform in time or, in other words, that the distance an object falls is directly proportional to the amount of time it has been traveling, squared. This relationship was empirically solidified using the following ingeniously simple experiment performed by Galileo.

Repeatedly rolling a metal ball down a grooved 5.5 meter long piece of wood set at an incline as shown in Figure 10.4, Galileo timed how long the ball took to get $\frac{1}{4}, \frac{1}{2}, \frac{2}{3}, \frac{3}{4}$, and all the way down the wood ramp.[3]

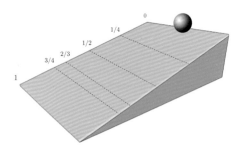

Figure 10.4 Figure associated with Example 10.2. Figurative illustration of Galileo's ramp experiment setup used for exploring the relationship between time and the distance an object falls due to gravity. To perform this experiment he repeatedly rolled a ball down a ramp and timed how long it took to get $\frac{1}{4}, \frac{1}{2}, \frac{2}{3}, \frac{3}{4}$, and all the way down the ramp. See text for further details.

Data from a modern reenactment[45] of this experiment (averaged over 30 trials) is shown in the left panel of Figure 10.5 where the input axis is time (in seconds) while the output is the portion of the ramp traveled by the ball during the experiment. The data here displays a nonlinear *quadratic relationship* between its input and output. This translates to using the quadratic model

$$\text{model}(x, \Theta) = w_0 + f_1(x)\, w_1 + f_2(x)\, w_2 \tag{10.14}$$

with two *unparameterized* feature transformations: the identity transformation $f_1(x) = x$ and the quadratic transformation $f_2(x) = x^2$. Replacing $f_1(x)$ and $f_2(x)$ in Equation (10.14) with x and x^2 gives the familiar quadratic form $w_0 + x w_1 + x^2 w_2$.

After standard normalizing the input of this dataset (see Section 9.3) we minimize the corresponding Least Squares cost via gradient descent, and plot the corresponding best nonlinear fit on the original data in the left panel of Figure 10.5. Since this model is a linear combination of its two feature transformations (plus a bias weight) we can also visualize its corresponding *linear fit* in the transformed feature space, as shown in the right panel of Figure 10.5. In this space

---

[3] Why did Galileo not simply drop the ball from some height and time how long it took to reach certain distances to the ground? Because no reliable way to measure time had yet existed. As a result he had to use a *water clock* for his ramp experiments! Interestingly, Galileo was the one who set humanity on the route towards its first reliable time-piece in his studies of the pendulum.

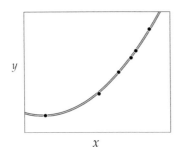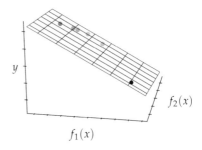

Figure 10.5 Figure associated with Example 10.2. (left panel) Data from a modern reenactment of a famous experiment performed by Galileo along with a well tuned quadratic model. (right panel) The same dataset and model viewed in the transformed feature space. See text for further details.

the input axes are given by $f_1(x)$ and $f_2(x)$ respectively, and our transformed points by the triplet $\left(f_1\left(x_p\right), f_2\left(x_p\right), y_p\right)$.

10.2.3   Python implementation

Below we show a universal way to implement the generic nonlinear model shown in Equation (10.8), generalizing our original linear implementation from Section 5.2.4.

```
1   # an implementation of our model employing a
2   # general nonlinear feature transformation
3   def model(x, theta):
4
5       # feature transformation
6       f = feature_transforms(x, theta[0])
7
8       # compute linear combination and return
9       a = theta[1][0] + np.dot(f.T, theta[1][1:])
10
11      return a.T
```

Here our generic set of engineered feature transformations are implemented in the Python function feature_transforms, and differ depending on how the features themselves are defined. We have implemented this function as generically as possible, to encompass the case where our desired feature transformations have internal parameters. That is, we package the model weights in the set Θ as theta, which is a list containing the *internal weights* of feature_transforms in its first entry theta[0], and the weights in the final linear combination of the model stored in the second entry theta[1].

For example, the feature_transforms function employed in Example 10.1 can be implemented as follows.

```
1  def feature_transforms(x, w):
2
3      # compute feature transformation
4      f = np.sin(w[0] + np.dot(x.T, w[1:])).T
5
6      return f
```

If our desired feature transformations do not have internal parameters we can either leave the parameter input to this function empty, or implement the model above slightly differently by computing our set of feature transformations as

```
1  f = feature_transforms(x)
```

and computing the linear combination of these transformed features as

```
1      a = theta[0] + np.dot(f.T, theta[1:])
```

In either case, in order to successfully perform nonlinear regression we can focus our attention solely on implementing the function feature_transforms, employing the autograd-wrapped NumPy library if we wish to employ automatic differentiation (see Section B.10). Nothing about how we implement our *regression cost functions* changes from the original context of linear regression detailed in Chapter 5. In other words, once we have implemented a given set of feature transformations correctly, employing the model above we can then tune the parameters of our nonlinear regression precisely as we have done in Chapter 5, using any regression cost function and local optimization scheme. The only caveat one must keep in mind is that when employing *parameterized* models (like the one in Example 10.1) the corresponding cost functions are generally *nonconvex*. Thus either zero- or first-order methods of optimization should be applied, or second-order methods adjusted in the manner detailed in Appendix Section A.7.

10.3   Nonlinear Multi-Output Regression

In this section we present a description of nonlinear feature engineering for multi-output regression first introduced in Section 5.6. This mirrors what we have seen in the previous section completely with one small but important difference: in the multi-output case we can choose to model each regression *separately*, employing a (potentially) different nonlinear model for each output, or *jointly*, producing a single nonlinear model for all outputs simultaneously.

10.3.1 Modeling principles

With linear multi-output regression we construct C linear models of the form $\mathring{\mathbf{x}}^T \mathbf{w}_c$, or equivalently, one joint linear model including all C regressions by stacking the weight vectors $\mathbf{w}_c$ column-wise into an $(N+1) \times C$ matrix $\mathbf{W}$ (see Section 5.6.1) and forming the multi-output linear model

$$\text{model}(\mathbf{x}, \mathbf{W}) = \mathring{\mathbf{x}}^T \mathbf{W}. \tag{10.15}$$

Given a dataset of P points $\left\{\left(\mathbf{x}_p, \mathbf{y}_p\right)\right\}_{p=1}^P$ where each paired input $\mathbf{x}_p$ and output $\mathbf{y}_p$ is $N \times 1$ and $1 \times C$ dimensional, respectively, we aim to tune the parameters in $\mathbf{W}$ to learn a linear relationship between the input and output as

$$\mathring{\mathbf{x}}_p^T \mathbf{W} \approx \mathbf{y}_p \qquad p = 1, 2, ..., P \tag{10.16}$$

by minimizing an appropriate cost function, e.g., the Least Squares cost

$$g(\mathbf{W}) = \frac{1}{P} \sum_{p=1}^P \left\| \mathring{\mathbf{x}}_p^T \mathbf{W} - \mathbf{y}_p \right\|_2^2. \tag{10.17}$$

With multi-output regression the move from linear to nonlinear modeling closely mirrors what we saw in the previous section. That is, for the cth regression problem we construct a model using (in general) B_c nonlinear feature transformations as

$$\text{model}_c(\mathbf{x}, \Theta_c) = w_{c,0} + f_{c,1}(\mathbf{x}) w_{c,1} + f_{c,2}(\mathbf{x}) w_{c,2} + \cdots + f_{c,B_c}(\mathbf{x}) w_{c,B_c} \tag{10.18}$$

where $f_{c,1}, f_{c,2}, ..., f_{c,B_c}$ are nonlinear (potentially parameterized) functions, and $w_{c,0}$ through w_{c,B_c} (along with any additional weights internal to the nonlinear functions) are represented in the weight set Θ_c.

To simplify the chore of choosing nonlinear features for each regressor we can instead choose a single set of nonlinear feature transformations and *share* them among all C regression models. If we choose the same set of B nonlinear features for all C models, the cth model takes the form

$$\text{model}_c(\mathbf{x}, \Theta_c) = w_{c,0} + f_1(\mathbf{x}) w_{c,1} + f_2(\mathbf{x}) w_{c,2} + \cdots + f_B(\mathbf{x}) w_{c,B} \tag{10.19}$$

where Θ_c now contains both the linear combination weights $w_{c,0}, w_{c,0}, ..., w_{c,B}$ as well as any weights internal to the shared feature transformations. Note that in this case the only parameters unique to the cth model are the linear combination

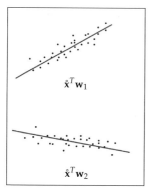
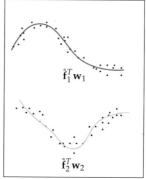
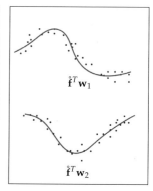

Figure 10.6 Figurative illustrations of multi-output regression with $C = 2$ outputs. (left panel) Linear multi-output regression. (middle panel) Nonlinear multi-output regression where each regressor uses its own distinct nonlinear feature transformations. (right panel) Nonlinear multi-output regression where both regressors share the same nonlinear feature transformations. See text for further details.

weights since every model shares any weights internal to the feature transformations. Employing the same compact notation for our feature transformations as in Equation (10.9) we can express each of these models more compactly as

$$\text{model}_c(\mathbf{x}, \Theta_c) = \mathring{\mathbf{f}}^T \mathbf{w}_c. \tag{10.20}$$

Figure 10.6 shows a prototypical multi-output regression using this notation. We can then express all C models together by stacking all C weight vectors $\mathbf{w}_c$ column-wise into a $(B + 1) \times C$ weight matrix $\mathbf{W}$, giving the joint model as

$$\text{model}(\mathbf{x}, \Theta) = \mathring{\mathbf{f}}^T \mathbf{W}. \tag{10.21}$$

This is a direct generalization of the original linear model shown in Equation (10.15), and the set Θ contains the linear combination weights in $\mathbf{W}$ as well as any parameters internal to our feature transformations themselves. To tune the weights of our joint model so that it represents our dataset as well as possible, or phrased algebraically so that we have

$$\mathring{\mathbf{f}}_p^T \mathbf{W} \approx \mathbf{y}_p \qquad p = 1, 2, ..., P \tag{10.22}$$

we minimize an appropriate regression cost of this model over the parameters in Θ, e.g., the Least Squares cost[4]

---

[4] Note that if these feature transformations contain no internal parameters (e.g., polynomial functions) then each individual regression model can be tuned separately. However, when employing *parameterized* features (e.g., neural networks) then the cost function does not decompose over each regressor and we must tune all of our model parameters jointly, that is, we must learn all C regressions *simultaneously*. This differs from the linear case where tuning the

$$g\left(\Theta\right) = \frac{1}{P}\sum_{p=1}^{P}\left\|\mathring{\mathbf{f}}_{p}^{T}\mathbf{W} - \mathbf{y}_{p}\right\|_{2}^{2}. \tag{10.23}$$

10.3.2 Feature engineering

With multi-output regression, determining appropriate features by visual inspection is more challenging than the basic instance of regression detailed in the previous section. Here we provide one relatively simple example of this sort of feature engineering to give a flavor of this challenge and the nonlinear modeling involved.

Example 10.3   Modeling multiple waves

In Figure 10.7 we show an example of nonlinear multi-output regression using a toy dataset with input dimension $N = 2$ and output dimension $C = 2$, where the input paired with the first and second outputs are shown in the left and right panel, respectively. Both instances appear to be *sinusoidal* in nature, with each having its own unique shape.

Figure 10.7 Figure associated with Example 10.3. See text for details and compare to the linear case shown in Figure 5.11.

From visual examination of the data we can reasonably choose to model both regressions simultaneously using $B = 2$ parameterized sinusoidal feature transformations

$$\begin{aligned} f_1\left(\mathbf{x}\right) &= \sin\left(v_{1,0} + v_{1,1}x_1 + v_{1,2}x_2\right) \\ f_2\left(\mathbf{x}\right) &= \sin\left(v_{2,0} + v_{2,1}x_1 + v_{2,2}x_2\right). \end{aligned} \tag{10.24}$$

Fitting this set of nonlinear features jointly by minimizing the Least Squares cost in Equation (10.23) using gradient descent results in the nonlinear surfaces plotted in green in Figure 10.7.

---

parameters of the linear model, either one regressor at a time or simultaneously, returns the same result (see Section 5.6.2).

10.3.3   Python implementation

As with the linear case detailed in Section 5.6.3, here likewise we can piggy-back on our general `Pythonic` implementation of nonlinear regression introduced in Section 10.2.3, and employ precisely the same model and cost function implementation as used in the single-output case. The only difference is in how we define our feature transformations and the dimensions of our matrix of linear combination weights.

10.4   Nonlinear Two-Class Classification

In this section we introduce the general framework of nonlinear classification, along with a number of elementary examples. As in the prior sections, these examples are all low-dimensional, allowing us to visually examine patterns in the data and propose appropriate nonlinearities, which we can inject into our linear supervised paradigm to produce nonlinear classifications. In doing this we are essentially performing nonlinear feature engineering for the two-class classification problem.

10.4.1   Modeling principles

While we employed a linear model in deriving linear two-class classification in Chapter 6, this linearity was simply an *assumption* about the sort of boundary that (largely) separates the two classes of data. Employing by default label values $y_p \in \{-1, +1\}$ and expressing our linear model algebraically as

$$\text{model}(\mathbf{x}, \mathbf{w}) = \mathring{\mathbf{x}}^T \mathbf{w} \tag{10.25}$$

our linear decision boundary then consists of all input points $\mathbf{x}$ where $\mathring{\mathbf{x}}^T \mathbf{w} = 0$. Likewise, label predictions are made (see Section 7.6) as

$$y = \text{sign}\left(\mathring{\mathbf{x}}^T \mathbf{w}\right). \tag{10.26}$$

To tune $\mathbf{w}$ we then minimize a proper two-class classification cost function, e.g., the two-class Softmax (or Cross Entropy) cost

$$g(\mathbf{w}) = \frac{1}{P} \sum_{p=1}^{P} \log\left(1 + e^{-y_p \mathring{\mathbf{x}}_p^T \mathbf{w}}\right). \tag{10.27}$$

We can adjust this framework to jump from linear to nonlinear classification in an entirely similar fashion as we did with regression in Section 10.2. That is, we can swap out our linear model with a nonlinear one of the generic form

$$\text{model}(\mathbf{x}, \Theta) = w_0 + f_1(\mathbf{x}) w_1 + f_2(\mathbf{x}) w_2 + \cdots + f_B(\mathbf{x}) w_B \tag{10.28}$$

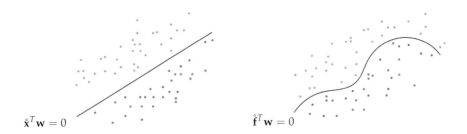

Figure 10.8 Figurative illustrations of linear and nonlinear two-class classification. (left panel) In the linear case the separating boundary is defined as $\mathring{\mathbf{x}}^T\mathbf{w} = 0$. (right panel) In the nonlinear case the separating boundary is defined as $\mathring{\mathbf{f}}^T\mathbf{w} = 0$. See text for further details.

where $f_1, f_2, \ldots, f_B$ are nonlinear parameterized or unparameterized functions, and w_0 through w_B (along with any additional weights internal to the nonlinear functions) are represented in the weight set Θ. Just as with regression, here too we can express this more compactly (see Section 10.2.1) as

$$\text{model}(\mathbf{x}, \Theta) = \mathring{\mathbf{f}}^T\mathbf{w}. \tag{10.29}$$

In complete analogy to the linear case, our decision boundary here consists of all inputs $\mathbf{x}$ where $\mathring{\mathbf{f}}^T\mathbf{w} = 0$, and likewise predictions are made as

$$y = \text{sign}\left(\mathring{\mathbf{f}}^T\mathbf{w}\right). \tag{10.30}$$

Figure 10.8 shows a prototypical two-class classification using this notation. Finally, in order to tune the parameters in Θ we must minimize a proper cost function with respect to it, e.g., the two-class Softmax cost (again in complete analogy to Equation (10.27))

$$g(\Theta) = \frac{1}{P} \sum_{p=1}^{P} \log\left(1 + e^{-y_p \mathring{\mathbf{f}}_p^T\mathbf{w}}\right). \tag{10.31}$$

10.4.2 Feature engineering

With /low-dimensional datasets we can, in certain instances, fairly easily engineer a proper set of nonlinear features for two-class classification by examining the data visually. Below we explore two such examples.

Example 10.4   When the decision boundary is just two single points
In discussing classification through the lens of logistic regression in Section 6.3 we saw how linear classification can be thought of as a specific instance

of nonlinear regression. In particular, we saw how from this perspective we aim at fitting a curve (or surface in higher dimensions) that consists of a linear combination of our input passed through the tanh function. For datasets with input dimension $N = 1$, like the one shown in the left column of Figure 10.9, this results in learning a decision boundary that is defined by a *single* point.

However, a linear decision boundary is quite inflexible in general, and fails to provide good separation even in the case of the simple example shown in the right column of Figure 10.9. For such a dataset we clearly need a model that is capable of crossing the input space (the x axis) *twice* at points separated by some distance, something a linear model can never do.

What sort of simple function crosses the horizontal axis twice? A quadratic function can. If adjusted to the right height a quadratic certainly can be made to cross the horizontal axis twice and, when passed through a tanh function, could indeed give us the sort of predictions we desire (as illustrated in the right column of Figure 10.9).

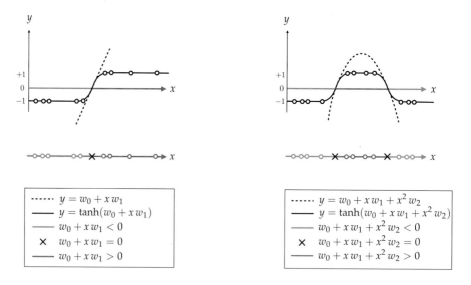

Figure 10.9 Figure associated with Example 10.4. (left column) A prototypical linear two-class classification dataset with fully tuned linear model shown from the regression perspective (top), and from the perceptron perspective (bottom) where label values are encoded as colors (red for +1 and blue for −1). (right column) A simple nonlinear two-class classification dataset that requires a decision boundary consisting of two points, something a linear model cannot provide. As can be seen here, a quadratic model can achieve this goal (provided its parameters are tuned appropriately).

A quadratic model takes the form

$$\text{model}(x, \Theta) = w_0 + x w_1 + x^2 w_2 \tag{10.32}$$

which uses two feature transformations: the identity $f_1(x) = x$ and the quadratic transformation $f_2(x) = x^2$, with the weight set Θ only containing the weights w_0, w_1, and w_2.

In the left panel of Figure 10.10 we illustrate a toy dataset like the one shown in the right column of Figure 10.9. We also show, in green, the result of fully tuning the quadratic model in Equation (10.32) by minimizing (via gradient descent) the corresponding two-class Softmax in Equation (10.31). In the right panel we show the same dataset only in the *transformed feature space* defined by our two features (as first detailed in Examples 10.1 and 10.2) wherein the nonlinear decision boundary becomes linear. This finding is true in general: a well-separating *nonlinear* decision boundary in the original space of a dataset translates to a well-separating *linear* decision boundary in the transformed feature space. This is analogous to the case of regression, as detailed in Section 10.2.2, where a good nonlinear fit in an original space corresponds to a good linear fit in the transformed feature space.

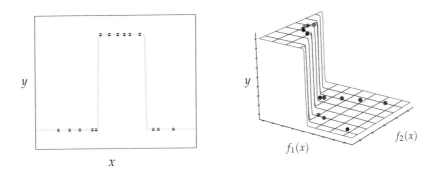

Figure 10.10 Figure associated with Example 10.4. See text for details.

Example 10.5   An elliptical decision boundary
In the left column of Figure 10.11 we show a toy two-class classification dataset with input dimension $N = 2$, shown from the perceptron perspective in the top panel and from the regression perspective in the bottom panel.

Visually examining the dataset it appears that some sort of elliptical decision boundary centered at the origin, defined by

$$\text{model}(\mathbf{x}, \Theta) = w_0 + x_1^2 w_1 + x_2^2 w_2 \qquad (10.33)$$

might do a fine job of classification. Parsing this formula we can see that we have used two feature transformations, i.e., $f_1(\mathbf{x}) = x_1^2$ and $f_2(\mathbf{x}) = x_2^2$, with the parameter set $\Theta = \{w_0, w_1, w_2\}$.

Minimizing the Softmax cost in Equation (10.31) using this model (via gradient descent) we show the resulting nonlinear decision boundary in black, from the perceptron perspective in the top panel and from the regression perspective in the bottom panel. Finally, in the right column we show the data in the *transformed*

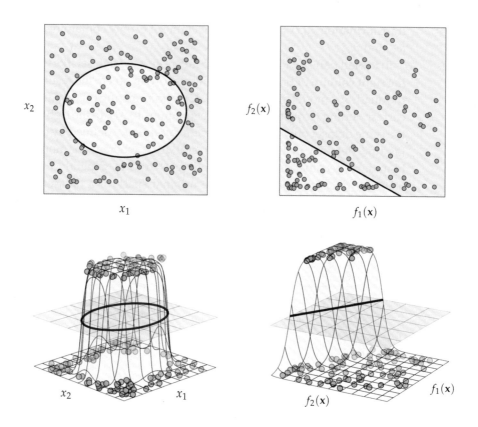

Figure 10.11 Figure associated with Example 10.5. See text for details.

feature space along with the corresponding linear decision boundary.

10.4.3   Python implementation

The general nonlinear model in Equation (10.29) can be implemented precisely as described in Section 10.2.3, since it is the same general nonlinear model we use with nonlinear regression. Therefore, just as with regression, we need not alter the implementation of any two-class classification cost function introduced in Chapter 6 to perform nonlinear classification: all we need to do is properly define our nonlinear transformation(s) in Python.

10.5   Nonlinear Multi-Class Classification

In this section we present the general nonlinear extension of linear multi-class classification first introduced in Chapter 7. This mirrors what we have seen in the previous sections very closely, and is in particular almost entirely similar to the discussion of nonlinear multi-output regression in Section 10.3.

10.5.1 Modeling principles

As we saw in Chapter 7 with linear multi-class classification we construct C linear models of the form $\mathring{\mathbf{x}}^T \mathbf{w}_c$, which we may represent jointly by stacking the weight vectors $\mathbf{w}_c$ column-wise into an $(N+1) \times C$ matrix $\mathbf{W}$ (see Section 7.3.9) and forming a single multi-output linear model of the form

$$\text{model}(\mathbf{x}, \mathbf{W}) = \mathring{\mathbf{x}}^T \mathbf{W}. \tag{10.34}$$

Given a dataset of P points $\left\{ \left(\mathbf{x}_p, y_p \right) \right\}_{p=1}^{P}$ where each input $\mathbf{x}_p$ is N-dimensional and each y_p is a label value in the set $y_p \in \{0, 1, ..., C-1\}$, we aim to tune the parameters in $\mathbf{W}$ to satisfy the fusion rule

$$y_p = \underset{c=0,1,...,C-1}{\text{argmax}} \left[\text{model} \left(\mathbf{x}_p, \mathbf{W} \right) \right] \qquad p = 1, 2, ..., P \tag{10.35}$$

by either tuning each column of $\mathbf{W}$ one at a time in a One-versus-All fashion (see Section 7.2), or by minimizing an appropriate cost, e.g., multi-class Softmax cost

$$g(\mathbf{W}) = \frac{1}{P} \sum_{p=1}^{P} \left[\log \left(\sum_{c=0}^{C-1} e^{\mathring{\mathbf{x}}_p^T \mathbf{w}_c} \right) - \mathring{\mathbf{x}}_p^T \mathbf{w}_{y_p} \right] \tag{10.36}$$

over the entire matrix $\mathbf{W}$ simultaneously (see Section 7.3).

With multi-class classification the move from linear to nonlinear modeling very closely mirrors what we saw in the case of multi-output regression in Section 10.3. That is, for the cth classifier we can construct a model using (in general) B_c nonlinear feature transformations as

$$\text{model}_c(\mathbf{x}, \Theta_c) = w_{c,0} + f_{c,1}(\mathbf{x}) w_{c,1} + f_{c,2}(\mathbf{x}) w_{c,2} + \cdots + f_{c,B_c}(\mathbf{x}) w_{c,B_c} \tag{10.37}$$

where $f_{c,1}, f_{c,2}, \ldots, f_{c,B_c}$ are nonlinear parameterized or unparameterized functions, and $w_{c,0}$ through w_{c,B_c} (along with any additional weights internal to the nonlinear functions) are represented in the weight set Θ_c.

To simplify the chore of choosing nonlinear features for each classifier we can instead choose a *single* set of nonlinear feature transformations and *share* them among all C two-class models. If we choose the same set of B nonlinear features for all C models, the cth model takes the form

$$\text{model}_c(\mathbf{x}, \Theta_c) = w_{c,0} + f_1(\mathbf{x}) w_{c,1} + f_2(\mathbf{x}) w_{c,2} + \cdots + f_B(\mathbf{x}) w_{c,B} \tag{10.38}$$

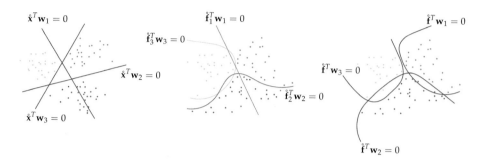

Figure 10.12 Figurative illustrations of multi-class classifiers on a generic dataset with $C = 3$ classes. (left panel) Linear multi-class classification. (middle panel) Nonlinear multi-class classification where each classifier uses its own distinct nonlinear feature transformations. (right panel) Nonlinear multi-class classification where all classifiers share the same nonlinear feature transformations. See text for further details.

where Θ_c now contains the linear combination weights $w_{c,0}$, $w_{c,1}$, ..., $w_{c,B}$ as well as any weights internal to the shared feature transformations. Employing the same compact notation for our feature transformations as in Equation (10.9) we can express each of these models more compactly as

$$\text{model}_c\,(\mathbf{x}, \Theta_c) = \mathring{\mathbf{f}}^T \mathbf{w}_c. \tag{10.39}$$

Figure 10.12 shows a prototypical multi-class classification using this notation. We can then tune the parameters of each of these models individually, taking a One-versus-All approach, or simultaneously by minimizing a single joint cost function over all of them together. To perform the latter approach it is helpful to first re-express all C models together by stacking all C weight vectors $\mathbf{w}_c$ column-wise into a $(B + 1) \times C$ weight matrix $\mathbf{W}$, giving the joint model as

$$\text{model}\,(\mathbf{x}, \Theta) = \mathring{\mathbf{f}}^T \mathbf{W} \tag{10.40}$$

where the set Θ contains the linear combination weights in $\mathbf{W}$ as well as any parameters internal to our feature transformations themselves. To tune the weights of our joint model so that the fusion rule

$$y_p = \underset{c=0,1,\ldots,C-1}{\text{argmax}}\left[\text{model}\left(\mathbf{f}_p, \mathbf{W}\right)\right] \qquad p = 1, 2, \ldots, P \tag{10.41}$$

holds as well as possible, we minimize an appropriate multi-class cost of this model over the parameters in Θ, e.g., the multi-class Softmax cost

$$g\,(\Theta) = \frac{1}{P}\sum_{p=1}^{P}\left[\log\left(\sum_{c=0}^{C-1} e^{\mathring{\mathbf{f}}_p^T \mathbf{w}_c}\right) - \mathring{\mathbf{f}}_p^T \mathbf{w}_{y_p}\right]. \tag{10.42}$$

10.5.2   Feature engineering

Determining nonlinear features via visual analysis is even more challenging in the multi-class setting than in the instance of two-class classification detailed in the previous section. Here we provide a simple example of this sort of feature engineering, noting that in general we will want to *learn* such feature transformations automatically (as we begin to detail in the next chapter).

Example 10.6   Multi-class data with elliptical boundaries
In this example we engineer nonlinear features to perform multi-class classification on the dataset shown in the top-left panel of Figure 10.13, which consists of $C = 3$ classes that appear to be (roughly) separable by elliptical boundaries. Here the points colored blue, red, and green have label values 0, 1, and 2, respectively.

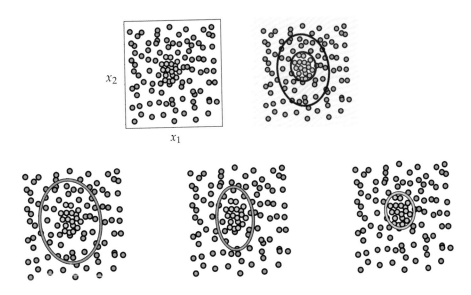

Figure 10.13 Figure associated with Example 10.6. See text for details.

Because the data is not centered at the origin we must use a full degree-two polynomial expansion of the input consisting of features of the form $x_1^i x_2^j$ where $i + j \leq 2$. This gives the degree-two polynomial model

$$\text{model}\,(\mathbf{x}, \Theta) = w_0 + x_1 w_1 + x_2 w_2 + x_1 x_2 w_3 + x_1^2 w_4 + x_2^2 w_5. \qquad (10.43)$$

Using this nonlinear model we minimize the multi-class Softmax cost in Equation (10.42) via gradient descent, and plot the corresponding fused multi-class boundary in the top-right panel of Figure 10.13 where each region is colored according to the prediction made by the final classifier.

We also plot the resulting two-class boundary produced by each individual classifier in the bottom row of Figure 10.13, coloring each boundary according to the One-versus-All classification performed in each instance. Here, as in the linear case outlined in Section 7.3, we can see that while each two-class subproblem cannot be solved correctly, when fused via Equation (10.41) the resulting multi-class classification can still be very good.

10.5.3   Python implementation

The general nonlinear model in Equation (10.40) can be implemented as described in Section 10.3.3, since it is the same general nonlinear model we use with nonlinear multi-output regression. Therefore, just as with multi-output regression, we need not alter the implementation of joint nonlinear multi-class classification cost functions introduced in Chapter 7: all we need to do is properly define our nonlinear transformation(s) in `Python`.

10.6   Nonlinear Unsupervised Learning

In this section we discuss the general nonlinear extension of our fundamental unsupervised learning technique introduced in Section 8.3: the Autoencoder.

10.6.1   Modeling principles

In Section 8.3 we described the linear Autoencoder, an elegant way to determine the best linear subspace to represent a set of mean-centered N-dimensional input data points $\left\{\mathbf{x}_p\right\}_{p=1}^{P}$. To determine the projection of our data onto the K-dimensional subspace spanned by the columns of an $N \times K$ matrix $\mathbf{C}$, we first *encode* our data on the subspace using the encoder model

$$\text{model}_e\left(\mathbf{x}, \mathbf{C}\right) = \mathbf{C}^T\mathbf{x} \tag{10.44}$$

which takes in an N-dimensional input $\mathbf{x}$ and returns a K-dimensional output $\mathbf{C}^T\mathbf{x}$. We then *decode* to produce the projection onto the subspace as

$$\text{model}_d\left(\mathbf{v}, \mathbf{C}\right) = \mathbf{C}\mathbf{v}. \tag{10.45}$$

The output of the decoder, being a projection onto a subspace lying in the original N-dimensional space, is itself N-dimensional.

The composition of these two steps gives the linear Autoencoder model

$$\text{model}\left(\mathbf{x}, \mathbf{C}\right) = \text{model}_d\left(\text{model}_e\left(\mathbf{x}, \mathbf{C}\right), \mathbf{C}\right) = \mathbf{C}\,\mathbf{C}^T\mathbf{x} \tag{10.46}$$

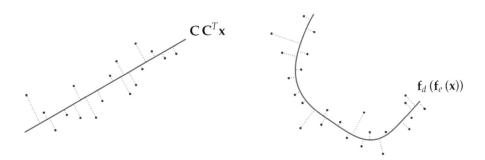

Figure 10.14 Figurative illustrations of a linear (left panel) and nonlinear (right panel) Autoencoder. See text for further details.

that, when $\mathbf{C}$ is tuned correctly, produces a linear subspace that represents the data extremely well

$$\mathbf{C}\,\mathbf{C}^T\mathbf{x}_p \approx \mathbf{x}_p \qquad p = 1, 2, ..., P \tag{10.47}$$

or, equivalently, has an effect on input as close to the identity transformation as possible, i.e.,

$$\mathbf{C}\mathbf{C}^T \approx \mathbf{I}_{N \times N}. \tag{10.48}$$

In order to recover this ideal setting for $\mathbf{C}$ we can minimize, for example, the Least Squares error measurement of the desired effect in Equation (10.47)

$$g\,(\mathbf{C}) = \frac{1}{P}\sum_{p=1}^{P}\left\|\mathbf{C}\,\mathbf{C}^T\mathbf{x}_p - \mathbf{x}_p\right\|_2^2. \tag{10.49}$$

To introduce nonlinearity here, i.e., to determine a nonlinear surface (also called a *manifold*) to project the data onto (as illustrated in the right panel of Figure 10.14) we can simply replace the linear encoder/decoder models in Equations (10.44) and (10.45) with nonlinear versions of the generic form

$$\begin{aligned} \text{model}_e\,(\mathbf{x}, \Theta_e) &= \mathbf{f}_e\,(\mathbf{x}) \\ \text{model}_d\,(\mathbf{v}, \Theta_d) &= \mathbf{f}_d\,(\mathbf{v}). \end{aligned} \tag{10.50}$$

Here, $\mathbf{f}_e$ and $\mathbf{f}_d$ are (in general) nonlinear vector-valued functions, with Θ_e and Θ_d denoting their parameter sets. With this notation we can write the general nonlinear Autoencoder model simply as

$$\text{model}\,(\mathbf{x}, \Theta) = \mathbf{f}_d\,(\mathbf{f}_e\,(\mathbf{x})) \tag{10.51}$$

where the parameter set Θ now contains all parameters of both Θ_e and Θ_d.

As with the linear version, here our aim is to properly design the encoder/decoder pair and tune the parameters of Θ properly in order to determine the appropriate nonlinear manifold for the data, as

$$\mathbf{f}_d\left(\mathbf{f}_e\left(\mathbf{x}_p\right)\right) \approx \mathbf{x}_p \qquad p = 1, 2, ..., P. \qquad (10.52)$$

To tune the parameters in Θ we can minimize, for example, the Least Squares error measurement of the desired effect in Equation (10.52)

$$g\left(\Theta\right) = \frac{1}{P} \sum_{p=1}^{P} \left\| \mathbf{f}_d\left(\mathbf{f}_e\left(\mathbf{x}_p\right)\right) - \mathbf{x}_p \right\|_2^2. \qquad (10.53)$$

10.6.2 Feature engineering

Given that both encoder and decoder models contain nonlinear features that must be determined, and the compositional manner in which the model is formed in Equation (10.51), engineering features by visual analysis can be difficult even with an extremely simple example, like the one we discuss now.

Example 10.7 A circular manifold
In this example we use a simulated dataset of $P = 20$ two-dimensional data points to learn a circular manifold via the nonlinear Autoencoder scheme detailed in the current section. This dataset is displayed in the left panel of Figure 10.15 where we can see it has an almost perfect circular shape.

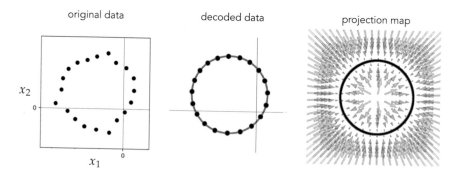

Figure 10.15 Figure associated with Example 10.7. (left panel) Original data is distributed roughly on a circle. (middle panel) The final decoding of the original data onto the determined circular manifold. (right panel) A projection mapping showing how points in the nearby space are projected onto the final learned manifold.

To engineer a nonlinear Autoencoder model for this dataset, recall that a circle in two dimensions, as illustrated in the left panel of Figure 10.16, can be fully characterized using its center point $\mathbf{w} = [w_1 \ w_2]^T$ and radius r. Subtracting off $\mathbf{w}$ from any point $\mathbf{x}_p$ on the circle then centers the data at the origin, as shown in the right panel of Figure 10.16.

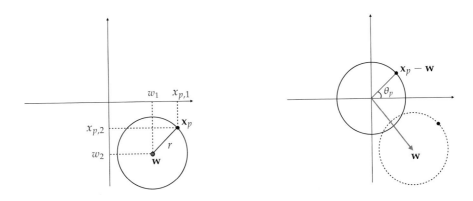

Figure 10.16 Figure associated with Example 10.7. (left panel) A circle in two dimensions is characterized by its center point $\mathbf{w}$ and its radius r. (right panel) when centered at the origin, any point on the circle can be represented using the angle created between its connector to the origin and the horizontal axis.

Once centered, any two-dimensional point $\mathbf{x}_p - \mathbf{w}$ on the circle can be encoded as the (scalar) angle θ_p between the line segment connecting it to the origin and the horizontal axis. Mathematically speaking, we have

$$\mathbf{f}_e\left(\mathbf{x}_p\right) = \theta_p = \arctan\left(\frac{x_{p,2} - w_2}{x_{p,1} - w_1}\right). \tag{10.54}$$

To design the decoder, beginning with θ_p, we can reconstruct $\mathbf{x}_p$ as

$$\mathbf{f}_d\left(\theta_p\right) = \left[\begin{array}{c} r\cos(\theta_p) + w_1 \\ r\sin(\theta_p) + w_2 \end{array} \right]. \tag{10.55}$$

Taken together this encoder/decoder pair defines an appropriate nonlinear Autoencder model of the general form given in Equation (10.51), with a set of parameters $\Theta = \{w_1, w_2, r\}$.

In the middle panel of Figure 10.15 we show the final learned manifold along with the decoded data, found by minimizing the cost function in Equation (10.53) via gradient descent. In the right panel of the figure we show the manifold recovered as a black circle (with red outline for ease of visualization), and illustrate how points in the space are attracted (or *projected*) to the recovered manifold as a vector field.

10.7   Conclusion

In this chapter we described *nonlinear feature engineering* for supervised and unsupervised learning problems. Nonlinear feature engineering involves the design of nonlinear models via philosophical reflection or visual analysis of data. While nonlinear feature engineering is itself a very useful skill-set, the greater value of this chapter is in introducing the general nonlinear modeling framework, including general nonlinear models, formalities, and concepts, that are the foundation for the remaining chapters of the text.

10.8   Exercises

† The data required to complete the following exercises can be downloaded from the text's github repository at `github.com/jermwatt/machine_learning_refined`

10.1   Modeling a wave
Repeat the experiment described in Example 10.1, including making the panels shown in Figure 10.2.

10.2   Modeling population growth
Figure 10.17 shows the population of yeast cells growing in a constrained chamber (data taken from [46]). This is a common shape found with population growth data, where the organism under study starts off with only a few members and is limited in growth by how fast it can reproduce and the resources available in its environment. In the beginning such a population grows exponentially, but the growth halts rapidly when the population reaches the maximum carrying capacity of its environment.

  Propose a *single* nonlinear feature transformation for this dataset and fit a corresponding model to it, using the Least Squares cost function and gradient descent. Make sure to *standard normalize* the input of the data (see Section 9.3). Plot the data along with the final fit provided by your model together in a single panel.

10.3   Galileo's experiment
Repeat the experiment described in Example 10.2, including making the panels shown in Figure 10.5.

Figure 10.17 Figure associated with Exercise 10.2. See text for details.

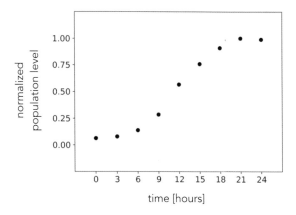

10.4   Moore's law

Gordon Moore, co-founder of Intel corporation, predicted in a 1965 paper [47] that the number of transistors on an integrated circuit would double approximately every two years. This conjecture, referred to nowadays as Moore's law, has proven to be sufficiently accurate over the past five decades. Since the processing power of computers is directly related to the number of transistors in their CPUs, Moore's law provides a trend model to predict the computing power of future microprocessors. Figure 10.18 plots the transistor counts of several microprocessors versus the year they were released, starting from Intel 4004 in 1971 with only 2300 transistors, to Intel's Xeon E7 introduced in 2014 with more than 4.3 billion transistors.

Figure 10.18 Figure associated with Exercise 10.4. See text for details.

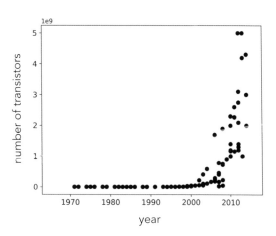

(a) Propose a single feature transformation for the Moore's law dataset shown in Figure 10.18 so that the transformed input/output data is related linearly. *Hint: to produce a linear relationship you will end up having to transform the output, not the input.*

(b) Formulate and minimize a Least Squares cost function for appropriate weights, and fit your model to the data in the original data space as shown in Figure 10.18.

10.5   Ohm's law

Ohm's law, proposed by the German physicist Georg Simon Ohm following a series of experiments made by him in the 1820s, connects the magnitude of the current in a galvanic circuit to the sum of all the exciting forces in the circuit, as well as the length of the circuit. Although he did not publish any account of his experimental results, it is easy to verify his law using a simple experimental setup shown in the left panel of Figure 10.19, that is very similar to what he then utilized. The spirit lamp heats up the circuit, generating an electromotive force which creates a current in the coil deflecting the needle of the compass. The tangent of the deflection angle is directly proportional to the magnitude of the current passing through the circuit. The magnitude of this current, denoted by I, varies depending on the length of the wire used to close the circuit (dashed curve). In the right panel of Figure 10.19 we plot the readings of the current I (in terms of the tangent of the deflection angle) when the circuit is closed with a wire of length x (in cm), for five different values of x. The data plotted here is taken from [48].

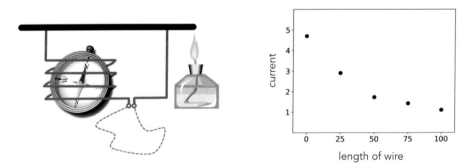

Figure 10.19 Figure associated with Exercise 10.5. (left panel) Experimental setup for verification of Ohm's law. Black and brown wires are made up of constantan and copper, respectively. (right panel) Current measurements for five different lengths of closing wire.

(a) Suggest a single nonlinear transformation of the original data so that the transformed input/output data is related linearly.

(b) Formulate a proper Least Squares cost function using your transformed data, minimize it to recover ideal parameters, fit your proposed model to the data, and display it in the original data space.

10.6 **Modeling multiple waves**

Repeat the experiment outlined in Example 10.3. You need not reproduce the illustrations shown in Figure 10.7. Instead use a cost function history plot to ensure that you are able to learn an accurate fit to the data.

10.7 **An elliptical decision boundary**

Repeat the experiment outlined in Example 10.5. You need not reproduce the illustrations shown in Figure 10.11. Instead use a cost function history plot to ensure that you are able to learn an accurate fit to the data.

10.8 **Engineering features for a two-class classification dataset**

Propose a nonlinear model for the dataset shown in Figure 10.20 and perform nonlinear two-class classification. Your model should be able to achieve perfect classification on this dataset.

Figure 10.20 Figure associated with Exercise 10.8. See text for details.

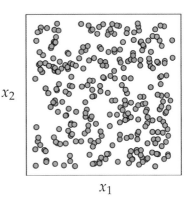

10.9 **A circular manifold**

Repeat the experiment outlined in Example 10.7 and report the optimal values you find for w_1, w_2, and r.

10.10 **Another nonlinear extension of PCA**

As we saw in Section 8.3 minimizing the PCA Least Squares cost

$$g\left(\mathbf{w}_1, \mathbf{w}_2, ..., \mathbf{w}_P, \mathbf{C}\right) = \frac{1}{P}\sum_{p=1}^{P}\left\|\mathbf{C}\mathbf{w}_p - \mathbf{x}_p\right\|_2^2 \tag{10.56}$$

over the $N{\times}K$ spanning set $\mathbf{C}$ and corresponding $K{\times}1$ weight vectors $\mathbf{w}_1$, $\mathbf{w}_2$, ..., $\mathbf{w}_P$ determines a proper K-dimensional linear subspace for the set of input points $\mathbf{x}_1$, $\mathbf{x}_2$, ..., $\mathbf{x}_P$.

As an alternative to extending Autoencoders to allow PCA to capture nonlinear subspaces (as described at length in Section 10.5), we can extend this PCA Least Squares cost. This is a more restricted version of nonlinear PCA, often termed *kernel* PCA, that is the basis for similar nonlinear extensions of other unsupervised problems including K-means. In this exercise you will investigate fundamental principles involved in this nonlinear extension.

To do this we begin by choosing a set of B nonlinear feature transformations $f_1, f_2, ..., f_B$ that have no internal parameters, and denote the transformation of $\mathbf{x}_p$ using this entire set of feature transformations as

$$\mathbf{f}_p = \begin{bmatrix} f_1(\mathbf{x}_p) \\ f_2(\mathbf{x}_p) \\ \vdots \\ f_B(\mathbf{x}_p) \end{bmatrix}. \tag{10.57}$$

Then, instead of learning a linear subspace for our input data, we learn one for these transformed inputs by minimizing

$$g(\mathbf{w}_1, \mathbf{w}_2, ..., \mathbf{w}_P, \mathbf{C}) = \frac{1}{P} \sum_{p=1}^{P} \left\| \mathbf{C}\mathbf{w}_p - \mathbf{f}_p \right\|_2^2. \tag{10.58}$$

Note here that since each $\mathbf{f}_p$ has size $B \times 1$, our spanning set $\mathbf{C}$ must necessarily be of size $B \times K$.

(a) Assuming that the set of B feature transformations chosen have no internal parameters, describe the classical PCA solution to the problem of minimizing the cost function in Equation (10.58). *Hint: see Section 8.4.*

(b) Suppose that we have a dataset of input points distributed roughly on the unit circle in a two-dimensional space (i.e., $N = 2$), and that we use the two feature transformations $f_1(\mathbf{x}) = x_1^2$ and $f_2(\mathbf{x}) = x_2^2$. What kind of subspace will we find in both the original and transformed feature space if we use just the first principle component from the classical PCA solution to represent our data? Draw a picture of what this looks like in both spaces.

10.11   A nonlinear extension of K-means

The same idea introduced in the previous exercise can also be used to extend K-means clustering (see Section 8.4) to a nonlinear setting as well – and is the basis for the so-called *kernel* K-means. This can be done by first noting that both PCA and K-means have the same Least Squares cost function. With K-means, however, the minimization of this cost function is *constrained* so that each weight vector $\mathbf{w}_p$ is a standard basis vector (see Section 8.7).

(a) Extend the K-means problem precisely as shown with PCA in the previous exercise. Compare this to the same sort of clustering performed on the original input, and describe in words what is being clustered in each instance.

(b) Suppose that we have a dataset of two-dimensional input points distributed roughly in two clusters: one cluster consists of points distributed roughly on the unit circle, and another consists of points distributed roughly on the circle of radius two, centered at the origin. Using the two feature transformations $f_1(\mathbf{x}) = x_1^2$ and $f_2(\mathbf{x}) = x_2^2$, what kind of clusters will we find upon proper execution of K-means with $K = 2$, both in the original and transformed feature space? Draw a picture of what this looks like in both spaces.

11 Principles of Feature Learning

11.1 Introduction

In Chapter 10 we saw how linear supervised and unsupervised learners alike can be extended to perform nonlinear learning via the use of nonlinear functions (or feature transformations) that we engineered ourselves by visually examining data. For example, we expressed a general nonlinear model for regression as a weighted sum of B nonlinear functions of our input as

$$\text{model}\,(\mathbf{x}, \Theta) = w_0 + f_1\,(\mathbf{x})\,w_1 + f_2\,(\mathbf{x})\,w_2 + \cdots + f_B\,(\mathbf{x})\,w_B \tag{11.1}$$

where f_1 through f_B are nonlinear parameterized or unparameterized functions (or features) of the data, and w_0 through w_B (along with any additional weights internal to the nonlinear functions) are represented in the weight set Θ.

In this chapter we detail the fundamental tools and principles of *feature learning* (or automatic feature engineering) that allow us to automate this task and *learn* proper features from the data itself, instead of *engineering* them ourselves. In particular we discuss how to choose the form of the nonlinear transformations f_1 through f_B, the number B of them employed, as well as how the parameters in Θ are tuned, *automatically* and for *any dataset*.

11.1.1 The limits of nonlinear feature engineering

As we have described in previous chapters, *features* are those defining characteristics of a given dataset that allow for optimal learning. In Chapter 10 we saw how the quality of the mathematical features we can design ourselves is fundamentally dependent on our level of knowledge regarding the phenomenon we were studying. The more we understand (both intellectually and intuitively) about the process generating the data we have at our fingertips, the better we can design features ourselves. At one extreme where we have near perfect understanding of the process generating our data, this knowledge having come from considerable intuitive, experimental, and mathematical reflection, the features we design allow near perfect performance. However, more often than not we know only a few facts, or perhaps none at all, about the data we are analyzing. The universe is an enormous and complicated place, and we have a solid understanding only of how a sliver of it all works.

Most (particularly modern) machine learning datasets have far more than two inputs, rendering visualization useless as a tool for feature engineering. But even in rare cases where data visualization is possible, we cannot simply rely on our own pattern recognition skills. Take the two toy datasets illustrated in Figure 11.1, for example. The dataset on the left is a regression dataset with one-dimensional input and the one on the right is a two-class classification dataset with two-dimensional input. The true underlying nonlinear model used to generate the data in each case is shown by the dashed black lines. We humans are typically taught only how to recognize the simplest of nonlinear patterns *by eye*, including those created by elementary functions (e.g., polynomials of low degree, exponential functions, sine waves) and simple shapes (e.g., squares, circles, ellipses). Neither of the patterns shown in the figure match such simple nonlinear functionalities. Thus, whether or not a dataset can be visualized, human engineering of proper nonlinear features can be difficult if not outright impossible.

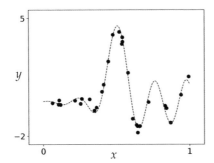 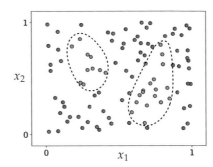

Figure 11.1 Toy (left panel) regression and (right panel) two-class classification datasets that clearly exhibit nonlinear behavior. The true underlying nonlinear function used to generate the data in each case is shown in dashed black. See text for further details.

It is precisely this challenge which motivates the fundamental *feature learning* tools described in this chapter. In short, these technologies *automate* the process of identifying appropriate nonlinear features for arbitrary datasets. With these tools in hand we no longer need to *engineer* proper nonlinearities, at least in terms of how we engineered nonlinear features in the previous chapter. Instead, we aim at *learning* their appropriate forms. Compared to our own limited nonlinear pattern recognition abilities, feature learning tools can identify virtually any nonlinear pattern present in a dataset regardless of its input dimension.

11.1.2   Chapter outline

The aim to automate nonlinear learning is an ambitious one and perhaps at first glance an intimidating one as well, for there are an infinite variety of nonlinearities and nonlinear functions to choose from. How do we, in general, parse this infinitude automatically to determine the appropriate nonlinearity for a given dataset?

The first step, as we will see in Section 11.2, is to organize the pursuit of automation by first placing the fundamental building blocks of this infinitude into *manageable collections* of (relatively simple) nonlinear functions. These collections are often called *universal approximators*, of which three strains are popularly used and which we introduce here: fixed-shape approximators, artificial neural networks, and trees. After introducing universal approximators we then discuss the fundamental concepts underlying how they are employed, including the necessity for *validation error* as a measurement tool in Section 11.3, a description of *cross-validation* and the *bias-variance trade-off* in Section 11.4, the automatic tuning of nonlinear complexity via *boosting* and *regularization* in Sections 11.5 and 11.6, respectively, as well as the notion of *testing error* in Section 11.7 and *bagging* in Section 11.9.

11.1.3   The complexity dial metaphor of feature learning

The ultimate aim of feature learning is a paradigm for the appropriate and automatic learning of features for any *any dataset* regardless of problem type. This translates – formally speaking – into the the automatic determination of both the proper *form* of the general nonlinear model in Equation (11.1) and the proper *parameter tuning* of this model regardless of training data and problem type. We can think about this challenge metaphorically as (i) the *construction* of, and (ii) the *automatic setting* of, a "complexity dial," like the one illustrated in Figure 11.2 for a simple nonlinear regression dataset (first used in Example 10.1). This complexity dial conceptualization of feature learning visually depicts the challenge of feature learning at a high level as a dial that must be built and automatically tuned to determine the appropriate amount of model complexity needed to represent the phenomenon generating a given dataset.

Setting this complexity dial all the way to the left corresponds, generally speaking, to choosing a model with lowest nonlinear complexity (i.e., a linear model, as depicted visually in the figure). As the dial is turned from left to right various models of increasing complexity are tried against the training data. If turned too far to the right the resulting model will be too complex (or too "wiggly") with respect to the training data (as depicted visually in the two small panels on the right side of the dial). When set "just right" (as depicted visually in the small image atop the complexity dial that is second to the left) the resulting model represents the data – as well as the underlying phenomenon generating it – very well.

While the complexity dial is a simplified depiction of feature learning we will see that it is nonetheless a helpful metaphor, as it will help us organize our understanding of the diverse set of ideas involved in performing it properly.

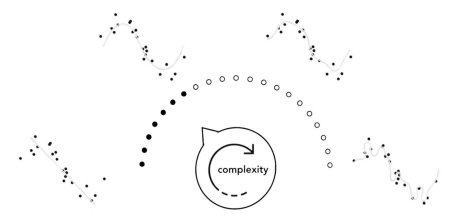

Figure 11.2 A visual depiction of feature learning as the *construction* and *automatic setting* of a "complexity dial" that – broadly speaking – controls the form the nonlinear model in Equation (11.1) as well as its parameter tuning, and thus the complexity of the model with respect to the training data.

11.2   Universal Approximators

In the previous chapter we described how to engineer appropriate nonlinear features ourselves to match the patterns we gleamed in simple datasets. However, very rarely in practice can we design perfect or even strongly-performing nonlinear features by completely relying on our own understanding of a dataset, whether this is gained by visualizing the data, philosophical reflection, or domain expertise.

In this section we jettison the unrealistic assumption that proper nonlinear features can be engineered in the manner described in the previous chapter, and replace it with an equally unrealistic assumption that has far more practical repercussions (as we will see in the forthcoming sections): that we have *complete* and *noiseless* access to the phenomenon generating our data. Here we will see, in the case where we have such unfettered access to data, that absolutely perfect features can be *learned* automatically by combining elements from a set of basic feature transformations, known as *universal approximators*. In this section we will also see elementary exemplars from the three most popular universal approximators, namely, *fixed-shape* approximators, *neural networks*, and *trees*.

For the sake of simplicity we will restrict our discussion to nonlinear regression and two-class classification, which as we saw in Chapter 10, share the same generic nonlinear model, formed as a linear combination of B nonlinear feature transformations of the input

$$\text{model}\,(\mathbf{x},\Theta) = w_0 + f_1\,(\mathbf{x})\,w_1 + f_2\,(\mathbf{x})\,w_2 + \cdots + f_B\,(\mathbf{x})\,w_B. \qquad (11.2)$$

Recall that with nonlinear two-class classification, we simply pass the nonlinear regression model in Equation (11.2) through the mathematical sign function to make binary predictions. While our focus in this section will be on these two supervised problems, because the general nonlinear model in Equation (11.2) is used in virtually all other forms of nonlinear learning including multi-class classification (see Section 10.4) and unsupervised learning (see Section 10.6), the thrust of the story unveiled here holds more generally for all machine learning problems.

11.2.1 Perfect data

We now start by imagining the impossible: a *perfect* dataset for regression. Such a dataset has two important characteristics: it is *completely noiseless* and *infinitely large*. Being completely noiseless, the first characteristic means that we could completely trust the quality of every one of its input/output pairs. Being infinitely large, means that we have unfettered access to every input/output pair $\left(\mathbf{x}_p, y_p\right)$ of the dataset that could possibly exist. Combined, such a dataset *perfectly* describes the phenomenon that generates it. In the top panels of Figure 11.3 we illustrate what such a perfect dataset would look like in the simplest instance where the input/output data is related linearly.

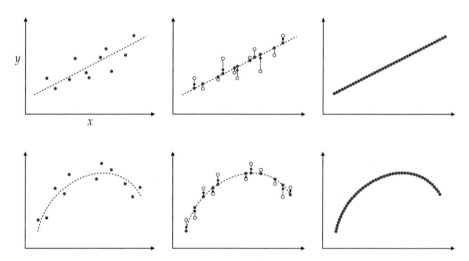

Figure 11.3 (top-left panel) A prototypical realistic linear regression dataset is a noisy and (relatively) small set of points that can be roughly modeled by a line. (top-middle panel) The same dataset with all noise removed from each output. (top-right panel) The perfect linear regression dataset where we have infinitely many points lying precisely on a line. (bottom-left panel) A prototypical realistic nonlinear regression dataset is a noisy and (relatively) small set of points that can be roughly modeled by a nonlinear curve. (bottom-middle panel) All noise removed from the output, creating a noise-free dataset. (bottom-right panel) The perfect nonlinear regression dataset where we have infinitely many points lying precisely on a curve.

Starting in the left panel we show a *realistic* dataset (the kind we deal with in practice) that is both *noisy* and *small*. In the middle panel we show the same dataset, but with the noise removed from each output. In the right panel we depict a *perfect* dataset by adding all missing points from the line to the noiseless data in the middle panel, making the data appear as a continuous line (or hyperplane, in higher dimensions). In the bottom panels of Figure 11.3 we show a similar transition for a prototypical nonlinear regression dataset wherein the perfect data (shown in the rightmost panel) carves out a *continuous* nonlinear curve (or surface, in higher dimensions).

With two-class classification a perfect dataset (using label values $y_p \in \{-1, +1\}$ by default) would share the same characteristics: it is *completely noiseless* and *infinitely large*. However, in this case, the perfect data would appear not as continuous curve or surface itself, but a step function with a *continuous* nonlinear boundary between its top and bottom steps. This is illustrated in Figure 11.4, which mirrors very closely what we saw with regression in Figure 11.3.

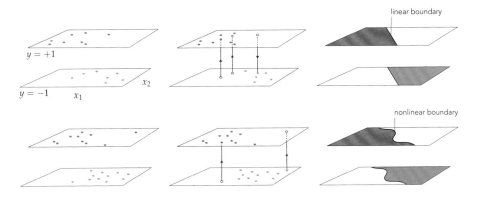

Figure 11.4 (top-left panel) A prototypical realistic linear two-class classification dataset is a noisy and (relatively) small set of points that can be roughly modeled by a step function with linear boundary. (top-middle panel) We progress to remove all noise from the data by returning the true label values to our noisy points. (top-right panel) The perfect linear two-class classification dataset where we have infinitely many points lying precisely on a step function with linear boundary. (bottom-left panel) A prototypical realistic nonlinear two-class classification dataset is a noisy and (relatively) small set of points that can be roughly modeled by a step function with nonlinear boundary. (bottom-middle panel) We progress to remove all noise from the data, creating a noise-free dataset. (bottom-right panel) The perfect nonlinear two-class classification dataset where we have infinitely many points lying precisely on a step function with nonlinear boundary.

In short, a perfect regression dataset is a continuous function with unknown equation. Because of this we will refer to our perfect data using the function notation $y(\mathbf{x})$, meaning that the data pair defined at input $\mathbf{x}$ can be written as either $(\mathbf{x}, y(\mathbf{x}))$ or likewise $(\mathbf{x}, y)$. In the same vein a perfect two-class classification dataset can be represented as a step function $\text{sign}(y(\mathbf{x}))$ with a continuous boundary – determined by $y(\mathbf{x})$. It is important to bear in mind that the function notation $y(\mathbf{x})$ does not imply that we have knowledge of a closed-form *formula* relating the input/output pairs of a perfect dataset; we do not! Indeed our aim next is to understand how such a formula can be devised to adequately represent a perfect dataset.

11.2.2   The spanning set analogy for universal approximation

Here we will leverage our knowledge and intuition about basic linear algebra concepts such as vectors, spanning sets, and the like (see Section 8.2) to better understand how we can combine nonlinear functions to model perfect regression and classification data. In particular, we will see how vectors and nonlinear functions are very much akin when it comes to the notions of linear combination and spanning sets.

Linear combinations of vectors and functions

To begin, assume we have a set of B vectors $\{\mathbf{f}_1, \mathbf{f}_2, \ldots, \mathbf{f}_B\}$, each having length N. We call this a *spanning* set of vectors. Then, given a particular set of weights w_1 through w_B, the linear combination

$$\mathbf{f}_1 w_1 + \mathbf{f}_2 w_2 + \cdots + \mathbf{f}_B w_B = \mathbf{y} \qquad (11.3)$$

defines a new N-dimensional vector $\mathbf{y}$. This is illustrated in the top row of Figure 11.5 for a particular set of vectors and weights where $B = 3$ and $N = 3$.

The arithmetic of nonlinear functions works in an entirely similar manner: given a spanning set of B nonlinear functions $\{f_1(\mathbf{x}), f_2(\mathbf{x}), \ldots, f_B(\mathbf{x})\}$ (where the input $\mathbf{x}$ is N-dimensional and output is scalar), and a corresponding set of weights, the linear combination

$$w_0 + f_1(\mathbf{x}) w_1 + f_2(\mathbf{x}) w_2 + \cdots + f_B(\mathbf{x}) w_B = y(\mathbf{x}) \qquad (11.4)$$

defines a new function $y(\mathbf{x})$. This is illustrated in the bottom row of Figure 11.5 for a particular set of functions and weights where $B = 3$ and $N = 1$.

Notice the similarity between the vector and function arithmetic in Equations (11.3) and (11.4): taking a particular linear combination of a set of vectors creates a new vector with qualities inherited from each vector $\mathbf{f}_b$ in the set, just as taking a linear combination of a set of functions creates a new function taking on qualities of each function $f_b(\mathbf{x})$ in that set. One difference between the two linear combination formulae is the presence of a *bias parameter* w_0 in Equation (11.4). This bias parameter could be rolled into one of the nonlinear functions and not made explicit (by adding a constant function to the mix), but we choose to leave it out-front of the linear combination of functions (as we did with linear models in previous chapters). The sole purpose of this bias parameter is to move our linear combination of functions vertically along the output axis.

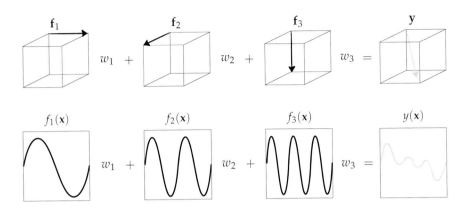

Figure 11.5 (top row) A particular linear combination of vectors $\mathbf{f}_1$, $\mathbf{f}_2$, and $\mathbf{f}_3$ (shown in black) creates a new vector $\mathbf{y}$ (shown in blue). (bottom row) In an entirely similar fashion a particular linear combination of three functions $f_1(x)$, $f_2(x)$, and $f_3(x)$ (shown in black) creates a new function $y(x)$ (shown in blue).

Capacity of spanning sets

Computing the vector $\mathbf{y}$ in Equation (11.3) for a *given* set of weights w_1 through w_B is a trivial affair. The inverse problem on other hand, i.e., finding the weights given $\mathbf{y}$, is slightly more challenging. Stated algebraically, we want to find the weights w_1 through w_B such that

$$\mathbf{f}_1 w_1 + \mathbf{f}_2 w_2 + \cdots + \mathbf{f}_B w_B \approx \mathbf{y} \tag{11.5}$$

holds as well as possible. This is illustrated for a simple example in the top row of Figure 11.6.

How well the *vector approximation* in Equation (11.5) holds depends on three crucial and interrelated factors: (i) the diversity (i.e., linear independence) of the spanning vectors, (ii) the number B of them used (in general the larger we make B the better), and (iii) how well we tune the weights w_1 through w_B via minimization of an appropriate cost.[1]

Factors (i) and (ii) determine a spanning set's *rank* or *capacity*, that is a measure for the range of vectors $\mathbf{y}$ we can possibly represent with such a spanning set. A spanning set with a *low* capacity, that is one consisting of a nondiverse and/or a small number of spanning vectors can approximate only a tiny fraction of those present in the entire vector space. On the other hand, a spanning set with a *high* capacity can represent a broader swath of the space. The notion of capacity for

---

[1] For instance, here we can use the Least Squares cost

$$g(w_1, w_2, ..., w_B) = \left\| \mathbf{f}_1 w_1 + \mathbf{f}_2 w_2 + \cdots + \mathbf{f}_B w_B - \mathbf{y} \right\|_2^2. \tag{11.6}$$

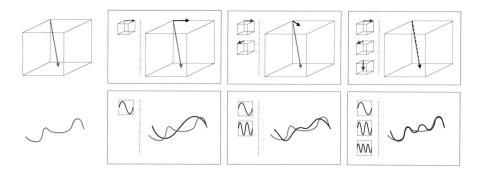

Figure 11.6 (top panels) A three-dimensional vector **y** (shown in red in the first panel from the left) is approximated using one (second panel from the left), two (third panel from the left), and three (fourth panel from the left) spanning vectors (here, standard basis vectors). As the number of spanning vectors increases we can approximate **y** with greater precision. (bottom panels) The same concept holds with functions as well. A continuous function with scalar input $y(x)$ (shown in red) is approximated using one (second panel from the left), two (third panel from the left), and three (fourth panel from the left) spanning functions (here, sine waves of varying frequency). As the number of functions increases we can approximate $y(x)$ with greater precision.

a spanning set of vectors is illustrated for a particular spanning set in the top row of Figure 11.7.

Turning our attention from vectors to functions, notice that computing the function $y(\mathbf{x})$ in Equation (11.4) for a *given* set of weights w_1 through w_B is straightforward. As with the vector case, we can reverse this problem and try to find the weights w_1 through w_B, for a given $y(\mathbf{x})$, such that

$$w_0 + f_1(\mathbf{x}) w_1 + f_2(\mathbf{x}) w_2 + \cdots + f_B(\mathbf{x}) w_B \approx y(\mathbf{x}) \qquad (11.7)$$

holds as well as possible. This is illustrated for a simple example in the bottom row of Figure 11.6.

Once again, how well this *function approximation* holds depends on three crucial and interrelated factors: (i) the diversity of the spanning functions, (ii) the number B of them used, and (iii) how well we tune the weights w_0 through w_B (as well as any parameters internal to our nonlinear functions) via minimization of an appropriate cost.[2]

In analogy to the vector case, factors (i) and (ii) determine the *capacity* of a spanning set of functions. A *low* capacity spanning set that uses a nondiverse and/or small array of nonlinear functions is only capable of representing a small

---

[2] For instance, here we can use the Least Squares cost

$$g(w_0, w_1, ..., w_B) = \int_{\mathbf{x} \in \mathcal{D}} (w_0 + f_1(\mathbf{x}) w_1 + \cdots + f_B(\mathbf{x}) w_B - y(\mathbf{x}))^2 \, d\mathbf{x} \qquad (11.8)$$

where $\mathcal{D}$ is any desired portion of the input domain.

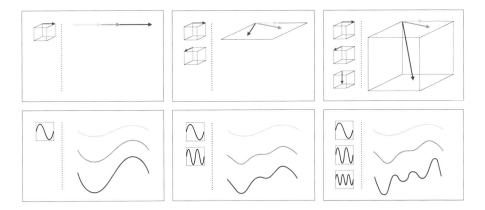

Figure 11.7 (top panels) As we increase the number of (diverse) vectors in a spanning set, from one in the left panel to two and three in the middle and right panels, respectively, we increase the capacity of the spanning set. This is reflected in the increasing diversity of sample vectors created using each spanning set (shown in different shades of blue). (bottom panels) The same concept holds with functions as well. As we increase the number of (diverse) functions in the spanning set, from a single function in the left panel to two and three functions in the middle and right panels, respectively, we increase the spanning set's capacity. This is reflected in the increasing diversity of sample functions created using each spanning set (shown in different shades of blue).

range of nonlinear functions. On the other hand, a spanning set with a *high* capacity can represent a wider swath of functions. The notion of capacity for a spanning set of functions is illustrated for a particular spanning set in the bottom row of Figure 11.7.

Sometimes the spanning functions f_1 through f_B are parameterized, meaning that they have internal parameters themselves. An unparameterized spanning function is very much akin to a spanning vector, as they are both parameter-free. A parameterized spanning function on the other hand can take on a variety of shapes alone, and thus can itself have high capacity. The same cannot be said about spanning vectors and unparameterized spanning functions. This concept is illustrated in Figure 11.8 where in the left column we show an ordinary spanning vector $\mathbf{x} = [1\ 1]^T$ (top-left panel) along with an unparameterized spanning function, i.e., $\sin(x)$ (bottom-left panel). In the bottom-right panel of the figure we show the parameterized function $\sin(wx)$, which can represent a wider range of different functions as its internal parameter w is adjusted. Thinking analogously, we can also parameterize the spanning vector $\mathbf{x}$, e.g., via multiplying it by the rotation matrix

$$R_w = \begin{bmatrix} \cos(w) & -\sin(w) \\ \sin(w) & \cos(w) \end{bmatrix} \tag{11.9}$$

that allows it to rotate in the plane and represent a range of different vectors depending on how the rotation angle w is set.

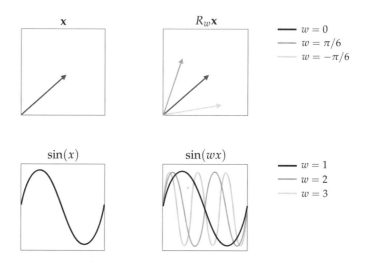

Figure 11.8 (top-left panel) An ordinary spanning vector. (bottom-left panel) An unparameterized spanning function. (bottom-right panel) A parameterized spanning function with a single internal parameter. By changing the value of this internal parameter it can be made to take on a range of shapes. (top-right panel) A parameterized spanning vector (premultiplied by the rotation matrix in Equation (11.9)) changes direction depending on how the parameter w is set.

Universal approximation

In the case of vector-based approximation in Equation (11.5) if we choose $B \geq N$ vectors for our spanning set, and at least N of them are linearly independent, then our spanning set has maximal capacity and we can therefore approximate *every* N-dimensional vector **y** to *any* given precision, provided we tune the parameters of the linear combination properly. Such a set of spanning vectors, of which there are infinitely many for an N-dimensional vector space, can approximate (or in this case perfectly represent) every vector *universally*, and is thus sometimes referred to as a *universal approximator*. For example, the simple standard basis (see Exercise 8.1) for a vector space is a common example of a spanning set that is a universal approximator. This notion of universal approximation of vectors is illustrated in the top panel of Figure 11.9.

The same concept holds with function approximation in Equation (11.7) as well. If we choose the right kind of spanning functions, then our spanning set has maximal capacity and we can therefore approximate *every* function $y(\mathbf{x})$ to *any* given precision, provided we tune the parameters of the linear combination properly. Such a set of spanning functions, of which there are infinitely many varieties, can approximate every function *universally*, and is thus often referred

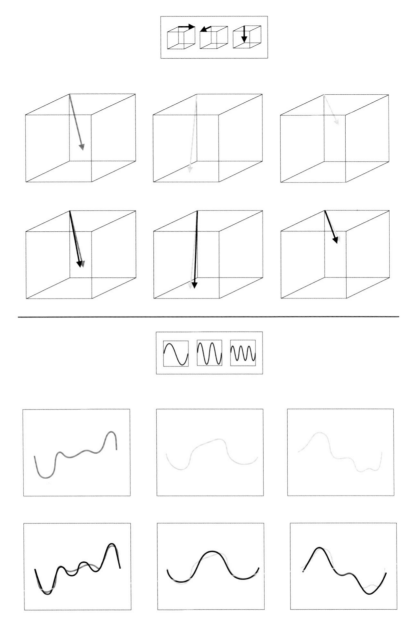

Figure 11.9 (top panel) Universal approximation illustrated in the vector case. (top row) A universal approximator spanning set consisting of three vectors, shown in black. (middle row) Three example vectors to approximate colored red, yellow, and blue, from left to right. (bottom row) The approximation of each vector in the middle row using the spanning set in the top row, shown in black in each instance. This approximation can be made perfect, but for visualization purposes is shown slightly offset here. (bottom panel) The analogous universal approximation scenario illustrated for functions. (top row) A universal approximator spanning set consisting of three functions (in practice many more spanning functions may be needed than shown here). (middle row) Three example functions to approximate colored red, yellow, and blue, from left to right. (bottom row) The approximation of each function in the middle row using the spanning set in the top row, shown in black in each instance.

to as a *universal approximator*. This notion of universal approximation of functions is illustrated in the bottom panel of Figure 11.9.

One difference between the vector and the function regime of universal approximation is that with the latter we may need infinitely many spanning functions to be able to approximate a given function to an arbitrary precision (whereas with the former it is always sufficient to set B greater than or equal to N).

11.2.3   Popular universal approximators

When it comes to approximating functions there is an enormous variety of spanning sets that are *universal approximators*. Indeed, just as in the vector case, with functions there are infinitely many universal approximators. However, for the purposes of organization, convention, as well as a variety of technical matters, universal approximators used in machine learning are often lumped into three main categories referred to as *fixed-shape* approximators, *neural networks*, and *trees*. Here we introduce only the most basic exemplar from each of these three categories, which we will reference throughout the remainder of the chapter. Each of these popular families has its own unique practical strengths and weaknesses as a universal approximator, a wide range of technical details to explore, and conventions of usage.

Example 11.1   The fixed-shape family of universal approximators
The family of *fixed-shape* functions consists of groups of nonlinear functions with no internal parameters, a popular example being *polynomials*.[3] When dealing with just one input this subfamily of fixed-shape functions consists of

$$f_1(x) = x, \quad f_2(x) = x^2, \quad f_3(x) = x^3, \quad \text{etc.,} \tag{11.10}$$

with the Dth element taking the form $f_D(x) = x^D$. A combination of the first D units from this subfamily is often referred to as a *degree-D polynomial*. There are an infinite number of these functions (one for each positive whole number D) and they are *naturally ordered* by their degree. The fact that these functions have no tunable internal parameters gives each a *fixed shape* as shown in the top row of Figure 11.10.

With two inputs x_1 and x_2, a general degree-D polynomial unit takes the analogous form

$$f_b(x_1, x_2) = x_1^p x_2^q \tag{11.11}$$

where p and q are nonnegative integers and $p + q \leq D$. Classically, a degree-D

---

[3] Polynomials were the first provable universal approximators, this having been shown in 1885 via the so-called (Stone–) Weierstrass approximation theorem (see, e.g., [49]).

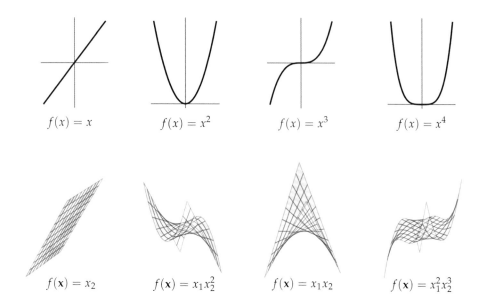

$f(x) = x$ $f(x) = x^2$ $f(x) = x^3$ $f(x) = x^4$

$f(\mathbf{x}) = x_2$ $f(\mathbf{x}) = x_1 x_2^2$ $f(\mathbf{x}) = x_1 x_2$ $f(\mathbf{x}) = x_1^2 x_2^3$

Figure 11.10 Four units from the polynomial family of fixed-shape universal approximators with $N = 1$ (top row) and $N = 2$ (bottom row) dimensional input.

polynomial is a linear combination of all such units. Furthermore, the definition in Equation (11.11) directly generalizes to higher-dimensional input as well. Fixed-shape approximators are discussed in much greater detail in Chapter 12.

Example 11.2   The neural network family of universal approximators
Another popular family of universal approximators are *neural networks*.[4] Broadly speaking neural networks consist of *parameterized* functions,[5] allowing them to take on a variety of different shapes (unlike the fixed-shape functions described previously, each of which takes on a single fixed form).

The simplest subfamily of neural networks consists of parameterized elementary functions (e.g., tanh) of the form

$$f_b(x) = \tanh\left(w_{b,0} + w_{b,1}x\right) \tag{11.12}$$

where the internal parameters $w_{b,0}$ and $w_{b,1}$ of the bth unit allow it to take on a variety of shapes. In the top row of Figure 11.11 we illustrate this fact by randomly setting the values of its two internal parameters, and plotting the result.

To construct neural network features taking in higher-dimensional input we take a linear combination of the input and pass the result through the nonlinear

---

[4] Neural networks were shown to be universal approximators in the late 1980s and early 1990s [50, 51, 52].

[5] An evolutionary step between fixed-shape and neural network units, that is a network unit whose internal parameters are randomized and fixed, are also universal approximators [53, 54].

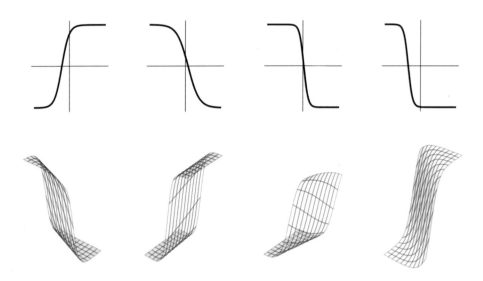

Figure 11.11 Unlike fixed-shape approximators, neural network units are flexible and can take on a variety of shapes based on how we set their internal parameters. Four such units, taking in $N = 1$ (top row) and $N = 2$ (bottom row) dimensional input, are shown whose internal parameters are set randomly in each instance.

function (here, tanh). For example, an element f_b for general N-dimensional input takes the form

$$f_b(\mathbf{x}) = \tanh\left(w_{b,0} + w_{b,1}x_1 + \cdots + w_{b,N}x_N\right). \tag{11.13}$$

As with the lower-dimensional example in Equation (11.12), each function in Equation (11.13) can take on a variety of different shapes, as illustrated in the bottom row of Figure 11.11, based on how we tune its internal parameters. Neural network approximators are described in much greater detail in Chapter 13.

Example 11.3   The trees family of universal approximators
Like neural networks, a single element from the family of tree-based universal approximators[6] can take on a wide array of shapes. The simplest sort of tree unit consists of discrete step functions or, as they are more commonly referred to, *stumps* whose break lies along a single dimension of the input space. A stump with one-dimensional input x can be written as

$$f_b(x) = \begin{cases} v_1 & x \leq s \\ v_2 & x > s \end{cases} \tag{11.14}$$

where s is called a *split point* at which the stump changes values, and v_1 and

---

[6] Trees have been long known to be universal approximators. See, e.g., [49, 55].

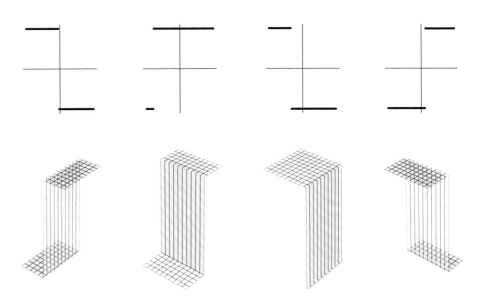

Figure 11.12 Tree-based units can take on a variety of shapes depending on how their split points and leaf values are assigned. Four instances of an $N = 1$ (top row) and $N = 2$ (bottom row) dimensional stump.

v_2 are values taken by the two sides of the stump, respectively, which we refer to as *leaves* of the stump. A tree-based universal approximator is a set of such stumps with each unit having its own unique split point and leaf values.

In the top row of Figure 11.12 we plot four instances of such a stump unit. Higher-dimensional stumps follow this one dimensional pattern. A split point s is first chosen along a *single* input dimension. Each side of the split is then assigned a single leaf value, as illustrated in the bottom row of Figure 11.12 for two-dimensional input. Tree-based approximators are described in much further detail in Chapter 14.

When forming a basic universal approximator based nonlinear model

$$\text{model} (\mathbf{x}, \Theta) = w_0 + f_1 (\mathbf{x}) w_1 + f_2 (\mathbf{x}) w_2 + \cdots + f_B (\mathbf{x}) w_B \qquad (11.15)$$

we always use units from a *single* type of universal approximator (e.g., all fixed-shape, neural network, or tree-based units). In other words, we do not "mix and match," taking a few units from each of the main families. As we will see in the present chapter as well as those following this one, by restricting a model's feature transformations to a single family we can (in each of the three cases) better optimize the learning process and better deal with each family's unique technical eccentricities relating to fundamental scaling issues associated with fixed-shape units, the nonconvexity of cost functions associated with neural network units, and the discrete nature of tree-based units.

11.2.4   The capacity and optimization dials

With any of the major universal approximators introduced previously (whether they be fixed-shape, neural networks, or trees) we can attain universal approximation to any given precision, provided that the generic nonlinear model in Equation (11.15) has sufficiently large *capacity* (which can be ensured by making B large enough), and that its parameters are tuned sufficiently well through *optimization* of an associated cost function. The notions of *capacity* and *optimization* of such a nonlinear model are depicted conceptually in Figure 11.13 as a set of two *dials*.

The *capacity dial* visually summarizes the amount of capacity we allow into a given model, with each notch on the dial denoting a distinct model constructed from units of a universal approximator. When set all the way to the left we admit as little capacity as possible, i.e., we employ a *linear* model. As we move the capacity dial from left to right (clockwise) we adjust the model, adding more and more capacity, until the dial is set all the way to the right. When set all the way to the right we can imagine admitting an infinite amount of capacity in our model (e.g., by using an infinite number of units from a particular family of universal approximators).

The *optimization dial* visually summarizes how well we minimize the cost function of a given model whose capacity is already set. The setting all the way to the left denotes the initial point of whatever local optimization technique we use. As we turn the optimization dial from left to right (clockwise) we can imagine moving further and further along the particular optimization run we use to properly tune the parameters of the model, with the final step being represented visually as the dial set all the way to the right where we imagine we have successfully minimized the associated cost function.

Note that in this conceptualization each pair of settings (of capacity and optimization dials) produces a unique tuned model: the model's overall architecture/design is decided by the capacity dial, and the set of specific values for the model parameters is determined by the optimization dial. For example, one particular setting may correspond to a model composed of $B = 10$ neural networks units, whose parameters are set by taking 1000 steps of gradient descent, while another setting corresponds to a model composed of $B = 200$ neural networks units, whose parameters are set by taking only five steps of gradient descent.

With these two dial conceptualizations in mind, we can think about the concept of universal approximation of a continuous function as turning *both* dials all the way to the right, as shown in the bottom row of Figure 11.13. That is, to approximate a given continuous function using a universal approximator, we set our model capacity as *large* as possible (possibly infinitely so) turning the capacity dial all the way to the right, and optimize its corresponding cost

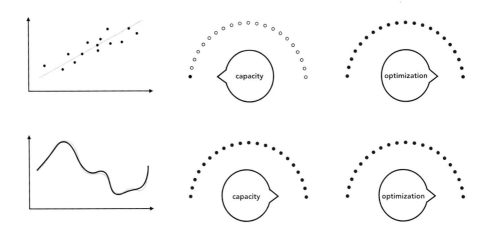

Figure 11.13 Model capacity and optimization precision visualized as two dials. When the capacity dial is set to the left we have a low-capacity linear model, when set to the right we admit maximum (perhaps infinite) capacity into the model. The optimization dial set to the left denotes the initial point of optimization, and all the way to the right denotes the final step of successful optimization. (top row) With linear regression (as we saw in previous chapters) we set the capacity dial all the way to the *left* and the optimization dial all the way to the *right* in order to find the best possible set of parameters for a low-capacity linear model (drawn in blue) that fits the given regression data. (bottom row) With universal approximation of a continuous function (drawn in black) we set both dials to the *right*, admitting infinite capacity into the model and tuning its parameters by optimizing to completion. See text for further discussion.

function as *well* as possible, turning the optimization dial all the way to the right as well.

In contrast, with the sort of linear learning we have looked at in previous chapters (as depicted in the top row of the Figure 11.13) we set our capacity dial all the way to the left (employing a linear model) but still set our optimization dial all the way to the right. By optimizing to completion we determine the proper bias and slope(s) of our linear model when performing, e.g., linear regression, as depicted in the figure.

We now examine a number of simple examples of universal approximation using various *near-perfect* regression and two-class classification datasets, where we set both the capacity and optimization dials far to the right. Here near-perfect means a very finely sampled, large dataset (as opposed to a perfect, infinitely large one). The case where a dataset is truly infinitely large ($P = \infty$) would, in theory, require infinite computing power to minimize a corresponding cost function.

Example 11.4   Universal approximation of near-perfect regression data

In Figure 11.14 we illustrate universal approximation of a near-perfect regression dataset consisting of $P = 10,000$ evenly sampled points from an underlying sinusoidal function defined over the unit interval. In the left, middle, and right columns we show the result of fitting an increasing number of polynomial, neural network, and tree units, respectively, to this data. As we increase the number of units in each case (from top to bottom) the capacity of each corresponding model increases, allowing for a better universal approximation.

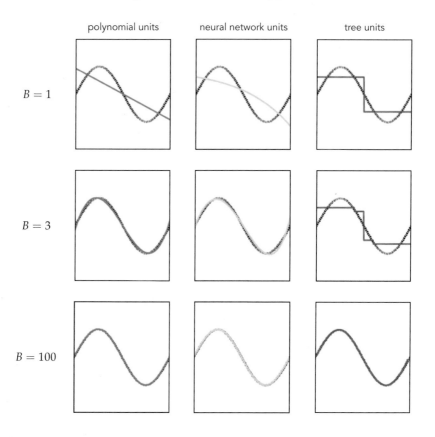

Figure 11.14 Figure associated with Example 11.4. Models built from polynomial (left column), neural network (middle column), and stump units (right column) fit to a near-perfect regression dataset. In visualizing the tree-based models in the right column we have connected each discrete step via a vertical line for visualization purposes only. As more and more units are added to the models each is able to fit the dataset with increasing precision. See text for further details.

Note here that it takes far fewer units of both the polynomial and neural network approximators to represent the data well as compared to the discrete stump units. This is because members of the former more closely resemble the smooth sinusoidal function that generated the data in the first place. This sort of phenomenon is true in general: while any type of universal approximator can be used to approximate a perfect (or near-perfect) dataset as closely as desired, some universal approximators require fewer units to do so than the others depending on the shape of the underlying function that generated the dataset.

Example 11.5   Universal approximation of near-perfect classification data
In the top row of Figure 11.15 we show four instances of near-perfect two-class classification data from the perceptron perspective (i.e., from the top) each consisting of $P = 10,000$ points. In each instance those points colored red have label value $+1$, and those colored blue have label value -1. Plotted in the second row of this figure are the corresponding datasets shown from the regression perspective (i.e., from the side).

Each of these near-perfect datasets can be approximated effectively using any of the three catalogs of universal approximators discussed in Section 11.2.3, provided that the capacity of each model is increased sufficiently and that the corresponding parameters are tuned properly. In the third and fourth rows of the figure we show the resulting fit from employing $B = 30$ polynomial approximators using a Least Squares and Softmax cost, respectively.

11.3   Universal Approximation of Real Data

In the previous section we saw how a nonlinear model built from units of a single universal approximator can be made to tightly approximate any *perfect dataset* if we increase its capacity sufficiently and tune the model's parameters properly by minimizing an appropriate cost function. In this section we will investigate how universal approximation carries over to the case of *real data*, i.e., data that is finite in size and potentially noisy. We will then learn about a new measurement tool, called *validation error*, that will allow us to effectively employ universal approximators with real data.

11.3.1   Prototypical examples

Here we explore the use of universal approximators in representing real data using two simple examples: a regression and two-class classification dataset. The problems we encounter with these two simple examples mirror those we face in general when employing universal approximator based models with real data, regardless of problem type.

Example 11.6   Universal approximation of real regression data
In this example we illustrate the use of universal approximators on a real regression dataset that is based on the near-perfect sinusoidal data presented in Example 11.4. To simulate a real version of this dataset we randomly selected

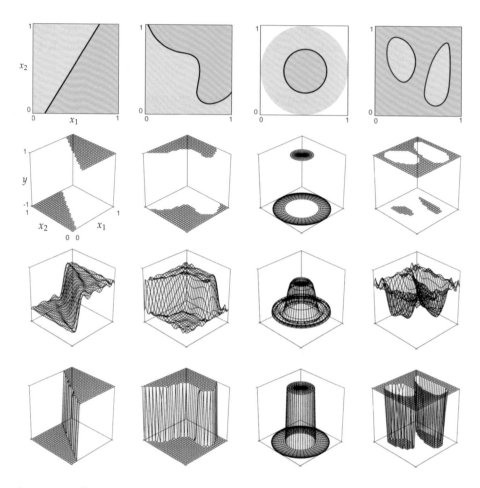

Figure 11.15 Figure associated with Example 11.5. (top row) Four instances of near-perfect two-class classification data. (second row) The corresponding data shown from a different (i.e., regression) perspective. Respective polynomial approximations to each dataset with $B = 30$ units in each instance using a Least Squares cost (third row) and a Softmax cost (fourth row). The approximations shown in the final row are passed through the tanh function before visualization. See text for further details.

$P = 21$ of its points and added a small amount of random noise to the output (i.e., y component) of each point, as illustrated in Figure 11.16.

In Figure 11.17 we illustrate the fully tuned nonlinear fit of a model employing polynomial (top row), neural network (middle row), and tree units (bottom row) to this data. Notice how, with each of the universal approximators, all three models *underfit* the data when using only $B = 1$ unit in each case (leftmost column). This underfitting of the data is a direct consequence of using low-capacity models, which produce fits that are not *complex* enough for the underlying data they are aiming to approximate. Also notice how each model improves as we increase B, but only up to a certain point after which each *tuned* model becomes far

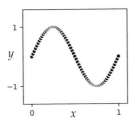 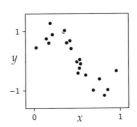

Figure 11.16 Figure associated with Example 11.6. (left panel) The original *near-perfect* sinusoidal dataset from Example 11.4. (right panel) A *real* regression dataset formed by adding random noise to the output of a small subset of the near-perfect dataset's points.

too complex and starts to look rather wild, and very much unlike the sinusoidal phenomenon that originally generated the data. This is especially visible in the polynomial and neural network cases, where by the time we reach $B = 20$ units (rightmost column) both models are extremely oscillatory and far too complex. Such *overfitting* models while representing the current data well, will clearly make for poor predictors of future data generated by the same process.

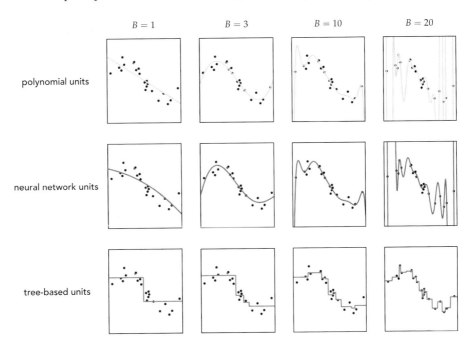

Figure 11.17 Figure associated with Example 11.6. See text for details.

In Figure 11.18 we plot several of the polynomial based models shown in Figure 11.17, along with the corresponding Least Squares cost value each attains. In adding more polynomial units we turn up the capacity of our model and, optimizing each model to completion, the resulting tuned models achieve lower and lower cost value. However, the resulting fit provided by each fully tuned model (after a certain point) becomes far too complex and starts to get *worse* in terms of how it represents the general regression phenomenon. As a measurement tool the cost value only tells us how well a tuned model fits the *training data*, but fails to tell us when our tuned model becomes too complex.

Example 11.7   Universal approximation of real classification data

In this example we illustrate the application of universal approximator-based models on a real two-class classification dataset that is based on the near-perfect dataset presented in Example 11.5. Here we simulated a realistic version of this data by randomly selecting $P = 99$ of its points, and adding a small amount of classification noise by flipping the labels of five of those points, as shown in Figure 11.19.

In Figure 11.20 we show the nonlinear decision boundaries provided by fully tuned models employing polynomial (top row), neural network (middle row), and tree units (bottom row). In the beginning where $B = 2$ (leftmost column) all three tuned models are not complex enough and thus *underfit* the data, providing a classification that in all instances simply classifies the entire space as belonging to the blue class. After that and up to a certain point the decision boundary provided by each model improves as more units are added, with $B = 5$ polynomial units, $B = 3$ neural network units, and $B = 5$ tree units providing reasonable approximations to the desired circular decision boundary. However, soon after we reach these numbers of units each tuned model becomes too complex and *overfits* the training data, with the decision boundary of each drifting away from the true circular boundary centered at the origin. As with regression in Example 11.6, both underfitting and overfitting problems occur in the classification case as well, regardless of the sort of universal approximator used.

In Figure 11.21 we plot several of the neural network based models shown in the middle row of Figure 11.20, along with the corresponding two-class Softmax cost value each attains. As expected, increasing model capacity by adding more neural network units always (upon tuning the parameters of each model by complete minimization) *decreases* the cost function value (just as with perfect or near-perfect data). However, the resulting classification, after a certain point, actually gets *worse* in terms of how it (the learned decision boundary) represents the general classification phenomenon.

In summary, Examples 11.6 and 11.7 show that, unlike the case with perfect data, when employing universal approximator based models with real data we must be careful with how we set the capacity of our model, as well as how well we tune its parameters via optimization of an associated cost. These two

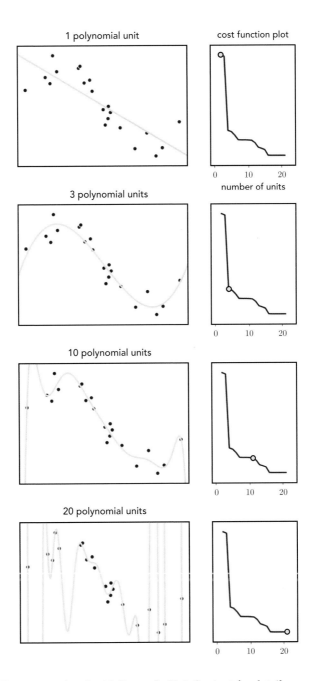

Figure 11.18 Figure associated with Example 11.6. See text for details.

simple examples also show how the cost value associated with training data (also called *training error*) fails as a reliable tool to measure how well a tuned

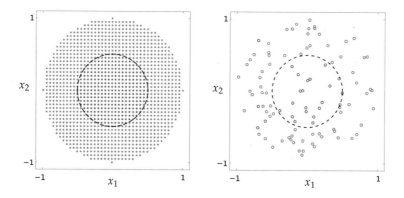

Figure 11.19 Figure associated with Example 11.7. (left panel) The original near-perfect classification dataset from Example 11.5, with the true circular boundary used to generate the data shown in dashed black (this is the boundary we hope to recover using classification). (right panel) A real dataset formed from a noisy subset of these points. See text for further details.

model represents the phenomenon underlying a real dataset. Both of these issues arise in general, and are discussed in greater detail next.

11.3.2   The capacity and optimization dials, revisited

The prototypical examples described in Section 11.3.1 illustrate how with real data we cannot (as we can in the case of perfect data) simply set our capacity and optimization dials (introduced in Section 11.2.4) all the way to the right, as this leads to overly complex models that fail to represent the underlying data-generating phenomenon well. Notice, we only control the *complexity* of a tuned model (or, roughly speaking, how "wiggly" a tuned model fit is) *indirectly* by how we set both our capacity and optimization dials, and it is not obvious *a priori* how we should set them simultaneously in order to achieve the right amount of model complexity for a given dataset. However, we can make this dial-tuning problem somewhat easier by fixing one of the two dials and adjusting only the other. Setting one dial all the way to the right imbues the other dial with the sole control over the complexity of a tuned model (and turns it into – roughly speaking – the *complexity dial* described in Section 11.1.3). That is, fixing one of the two dials all the way to the right, as we turn the unfixed dial from left to right we increase the complexity of our final tuned model. This is a general principle when applying universal approximator based models to real data that does not present itself in the case of perfect data.

To gain a stronger intuition for this principle, suppose first that we set our optimization dial all the way to the right (meaning that regardless of the dataset and model we use, we always tune its parameters by minimizing the corresponding cost function to completion). Then with perfect data, as illustrated in

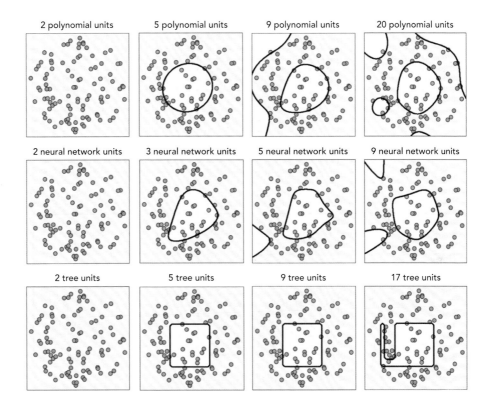

Figure 11.20 Figure associated with Example 11.7. See text for details.

the top row of Figure 11.22, as we turn our capacity dial from left to right (e.g., by adding more units) the resulting tuned model provides a better and better representation of the data.

However, with real data, as illustrated in the bottom row of Figure 11.22, starting with our capacity dial all the way to the left, the resulting tuned model is not complex enough for the phenomenon underlying our data. We say that such a tuned model *underfits*, as it does not fit the given data well.[7] Turning the capacity dial from left to right *increases* the complexity of each tuned model, providing a better and better representation of the data and the phenomenon underlying it. However, there comes a point, as we continue turning the dial from left to right, where the corresponding tuned model becomes *too complex*. Indeed past this point, where the complexity of each tuned model is wildly inappropriate for the phenomenon at play, we say that *overfitting* begins. This language is used because while such highly complex models fit the given data extremely well, they do so at the cost of not representing the underlying phenomenon well. As

---

[7] Notice that while the simple visual depiction here illustrates an underfitting model as a linear ("unwiggly") function – which is quite common in practice – it is possible for an underfitting model to be quite "wiggly." Regardless of the shape a tuned model takes, we say that it underfits if it poorly represents the training data, i.e., if it has high training error.

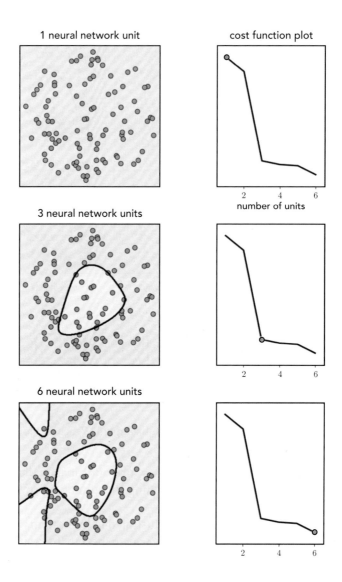

Figure 11.21 Figure associated with Example 11.7. See text for details.

we continue to turn our capacity dial to the right the resulting tuned models will become increasingly complex and increasingly less representative of the true underlying phenomenon.

Now suppose instead that we turn our capacity dial all the way to the right, using a very high-capacity model, and set its parameters by turning our optimization dial ever so slowly from left to right. In the case of perfect data, as illustrated in the top row of Figure 11.23, this approach produces tuned models that increasingly represent the data well. With real data on the other hand, as illustrated in the bottom row of Figure 11.23, starting with our optimization dial

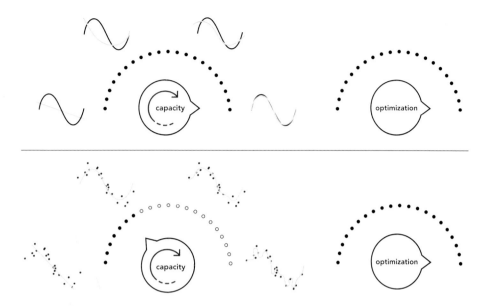

Figure 11.22 (top row) With *perfect data* if we set our optimization dial to all the way to the right, as we increase the capacity of our model by turning the capacity dial from left to right the corresponding representation gets better and better. (bottom row) With *real data* a similar effect occurs; however, here as we turn the capacity further to the right each tuned model will tend to become more and more complex, eventually overfitting the given data. See text for further details.

set all the way to the left will tend to produce low-complexity *underfitting* tuned models. As we turn our optimization dial from left to right, taking steps of a particular local optimization scheme, our corresponding model will tend to increase in complexity, improving its representation of the given data. This improvement continues only up to a point where our corresponding tuned model becomes too complex for the phenomenon underlying the data, and hence *overfitting* begins. After this point the tuned models arising from turning the optimization dial further to the right are far too complex to adequately represent the phenomenon underlying the data.

11.3.3   Motivating a new measurement tool

How we set our capacity and optimization dials in order to achieve a final tuned model that has *just the right amount of complexity* for a given dataset is the main challenge we face when employing universal approximator based models with real data. In Examples 11.6 and 11.7 we saw how training error fails to indicate when a tuned model has sufficient *complexity* for the tasks of regression and two-class classification, respectively – a fact more generally true about all nonlinear machine learning problems as well. If we cannot rely on training error to help decide on the proper amount of complexity required to address real nonlinear

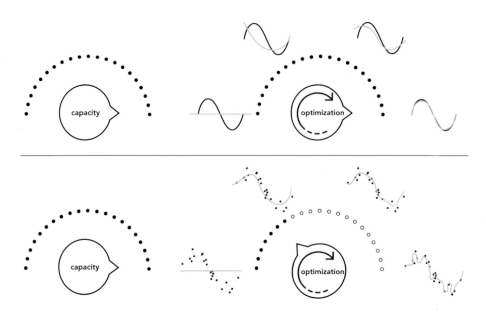

Figure 11.23 (top row) With *perfect data* if we set our capacity dial all the way to the right, as we increase the amount we optimize our model by turning the optimization dial from left to right (starting all the way on the left where for simplicity we assume all model parameters are initialized at zero) the corresponding representation gets better and better. (bottom row) With *real data* a similar effect occurs, but only up to a certain point where overfitting begins. See text for further details.

machine learning tasks, what sort of measurement tool should we use instead? Closely examining Figure 11.24 reveals the answer!

In the top row of this figure we show three instances of models presented for the toy regression dataset in Example 11.6: a fully tuned low-complexity (and underfitting) linear model in the left panel, a high-complexity (and overfitting) degree-20 polynomial model in the middle panel, and a degree-three polynomial model in the right panel that fits the data and the underlying phenomenon generating it "just right." What do both the underfitting and overfitting patterns have in common, that the "just right" model does not?

Scanning the left two panels of the figure we can see that a common problem with both the underfitting and overfitting models is that, while they differ in how well they represent data *we already have*, they will both fail to adequately represent *new data* generated via the same process by which the current data was made. In other words, we would not trust either model to predict the output of a newly arrived input point. The "just right" fully tuned model does not suffer from the same problem as it closely approximates the sort of wavy sinusoidal pattern underlying the data, and as a result would work well as a predictor for future data points.

The same story tells itself in the bottom row of Figure 11.24 with our two-

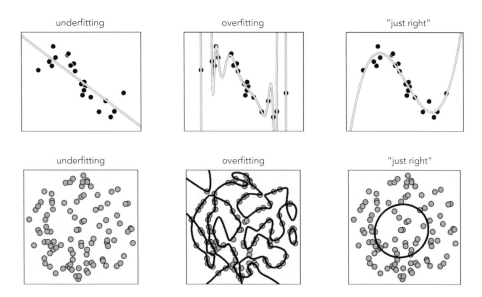

Figure 11.24 (top row) Three models for the regression dataset from Example 11.6: an underfitting model (top-left panel), an overfitting model (top-middle panel), and a "just right" one (top-right panel). (bottom row) Three models for the two-class classification dataset from Example 11.7: an underfitting model that simply classifies everything as part of the blue class (bottom-left panel), an overfitting model (bottom-middle panel), and a "just right" fit (bottom-right panel). See text for further details.

class classification dataset used previously in Example 11.7. Here we show a fully tuned low-complexity (and underfitting) linear model in the left panel, a high-complexity (and overfitting) degree-20 polynomial model in the middle panel, and a "just right" degree-two polynomial model in the right panel. As with the regression case, the underfitting and overfitting models both fail to adequately represent the underlying phenomenon generating our current data and, as a result, will fail to adequately predict the label values of *new data* generated via the same process by which the current data was made.

In summary, with both the simple regression and classification examples discussed here we can roughly qualify poorly-performing models as those that will not allow us to make accurate predictions of data we will receive in the future. But how do we quantify something we will receive in the future? We address this next.

11.3.4   The validation error

We now have an informal diagnosis for the problematic performance of under-fitting/overfitting models: such models do not accurately represent new data we might receive in the future. But how can we make use of this diagnosis? We of course do not have access to any new data we will receive in the future. To make this notion useful we need to translate it into a quantity we can always measure, regardless of the dataset/problem we are tackling or the kind of model we employ.

The universal way to do this is, in short, to *fake it*: we simply remove a random portion of our data and treat it as "new data we might receive in the future," as illustrated abstractly in Figure 11.25. In other words, we cut out a random chunk of the dataset we have, train our selection of models on only the portion of data that remains, and *validate* the performance of each model on this randomly removed chunk of "new" data. The random portion of the data we remove to validate our model(s) is commonly called the *validation data* (or validation set), and the remaining portion we use to train models is likewise referred to as the *training data* (or training set). The model providing the lowest error on the validation data, i.e., the lowest validation error, is then deemed the best choice from a selection of trained models. As we will see, validation error (unlike training error) is in fact a proper measurement tool for determining the quality of a model against the underlying data-generating phenomenon we want to capture.

Figure 11.25 Splitting the data into training and validation sets. The original data shown in the left panel as the entire round mass is split randomly in the right panel into two nonoverlapping sets. The smaller piece, typically $\frac{1}{10}$ to $\frac{1}{3}$ of the original data, is then taken as the validation set with the remaining taken as the training set.

There is no precise rule for what portion of a given dataset should be saved for validation. In practice, typically between $\frac{1}{10}$ to $\frac{1}{3}$ of the data is assigned to the validation set. Generally speaking, the larger and/or more representative (of the true phenomenon from which the data is sampled) a dataset is, the larger the portion of the original data may be assigned to the validation set (e.g., $\frac{1}{3}$). The intuition for doing this is that if the data is plentiful/representative enough, the training set still accurately represents the underlying phenomenon even after removal of a relatively large set of validation data. Conversely, in general with smaller or less representative (i.e., more noisy or poorly-distributed) datasets we usually take a smaller portion for validation (e.g., $\frac{1}{10}$) since the relatively larger training set needs to retain what little information of the underlying phenomenon was captured by the original data, and little data can be spared for validation.

11.4   Naive Cross-Validation

Validation error provides us with a concrete way of not only measuring the performance of a single tuned model, but more importantly it allows us to compare the efficacy of multiple tuned models of various levels of complexity. By carefully searching through a set of models ranging in complexity we can then easily identify the best of the bunch, the one that provides minimal error on the validation set. This comparison of models, called *cross-validation* or sometimes *model search* or *selection*, is the basis of feature learning as it provides a systematic way to *learn* (as opposed to *engineer*, as detailed in Chapter 10) the proper form a nonlinear model should take for a given dataset.

In this section we introduce what we refer to as *naive* cross-validation. This consists of a search over a set of models of varying capacity, with each model fully optimized over the training set, in search of a validation-error-minimizing choice. While it is simple in principle and in implementation, naive cross-validation is in general very expensive (computationally speaking) and often results in a rather coarse model search that can miss (or "skip over") the ideal amount of complexity desired for a given dataset.

11.4.1   The big picture

The first *organized* approach one might take to determining an ideal amount of complexity for a given dataset is to first choose a single universal approximator (e.g., one of those simple exemplars outlined in Section 11.2.3) and construct a set of M models of the general form given in Equation (11.15) by ranging the value of B from 1 to M sequentially as

$$\text{model}_1(\mathbf{x}, \Theta_1) = w_0 + f_1(\mathbf{x})w_1$$
$$\text{model}_2(\mathbf{x}, \Theta_2) = w_0 + f_1(\mathbf{x})w_1 + f_2(\mathbf{x})w_2$$
$$\vdots$$
$$\text{model}_M(\mathbf{x}, \Theta_M) = w_0 + f_1(\mathbf{x})w_1 + f_2(\mathbf{x})w_2 + \cdots + f_M(\mathbf{x})w_M.$$

(11.16)

This set of models – which we can denote compactly as $\{\text{model}_m(\mathbf{x}, \Theta_m)\}_{m=1}^{M}$ (or even more compactly as just $\{\text{model}_m\}_{m=1}^{M}$) where the set Θ_m consists of all those parameters of the mth model – naturally increases in *capacity* from $m = 1$ to $m = M$ (as first described in Section 11.2.2). If we *optimize* every one of these models to completion they will also roughly speaking – as discussed in Section 11.3.2 – increase in terms of their *complexity* with respect to training data as well. Thus, if we first split our original data randomly into training and validation portions as detailed in Section 11.3.4, and measure the error of all M fully trained models on each portion of the data, we can very easily determine which of the M models provides the ideal amount of complexity for the dataset overall by finding the one that achieves minimum validation error.

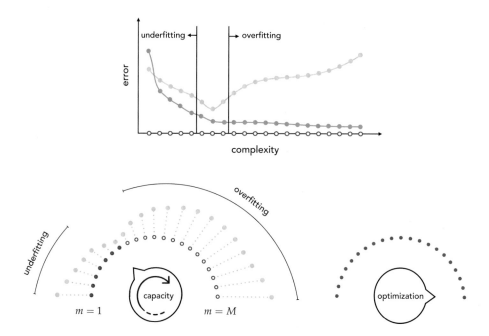

Figure 11.26 (top panel) Prototypical training (in blue) and validation (in yellow) error plots resulting from a run of naive cross-validation. Here the set of models – which generally increase in *complexity* with respect to the training set – are formed by fully optimizing a set of models of increasing *capacity*. Low-complexity models underfit the data, typically producing large training and validation errors. While the training error will monotonically decrease as model complexity increases, validation error tends to decrease only up to the point where overfitting of the training data begins. (bottom panels) Naive cross-validation using our dial conceptualization, where we turn the *capacity* dial from left to right, searching over a range of models of increasing capacity in search of a validation-error-minimizing model, while keeping the *optimization* dial set all the way to the right (indicating that we optimize each model to completion). See text for further details.

In the top panel of Figure 11.26 we show the generic sort of training (in blue) and validation (in yellow) errors we find in practice as a result of following this naive cross-validation scheme. The horizontal axis of this plot shows (roughly speaking) the complexity of each of our M fully optimized models, with the output on the vertical axis denoting error level. As can be seen in the figure, our low-complexity models *underfit* the data as reflected in their high training and validation errors. As the model complexity increases further, fully optimized models achieve lower training error since increasing model complexity allows us to constantly improve how well we can represent training data. This fact is reflected in the monotonically decreasing nature of the (blue) training error curve. On the other hand, while the validation error of our models will tend to decrease at first as we increase complexity, this trend continues only up to a point where *overfitting* of the training data begins. Once we reach a model complexity that overfits the training data our validation error starts to increase again, as our model becomes less and less a fair representation of "data we might receive in the future" generated by the same phenomenon.

Note in practice that while training error typically follows the monotonically decreasing trend shown in the top panel of Figure 11.26, validation error can oscillate up and down more than once depending on the models tested. In any event, we determine the best fully optimized model from the set by choosing the one that *minimizes* validation error. This is often referred to as solving the *bias-variance trade-off*, as it involves determining a model that (ideally) neither underfits (or has high bias) nor overfits (or has high variance).

In the bottom row of Figure 11.26 we summarize this naive approach to cross-validation using the capacity/optimization dial conceptualization first introduced in Section 11.2.2. Here we set our *optimization* dial all the way to the right – indicating that we optimize every model to completion – and in ranging over our set of M models we turn the *capacity* dial from left to right starting with $m = 1$ (on the left) and ending with $m = M$ (all the way to the right), with the value of m increasing by 1 at each notch of the dial. Since in this case the *capacity* dial roughly governs model complexity – as summarized visually in the bottom row of Figure 11.22 – our model search reduces to setting this dial correctly to the minimum validation error setting. To visually denote how this is done we wrap the prototypical validation error curve shown in the top panel of Figure 11.26 clockwise around the capacity dial. We can then imagine setting this dial correctly (and automatically) to the value of m providing minimum validation error.

Example 11.8 Naive cross-validation and regression

In this example we illustrate the use of a naive cross-validation procedure on the sinusoidal regression dataset first introduced in Example 11.6. Here we use $\frac{2}{3}$ of the original set of 21 data points for training, and the remaining $\frac{1}{3}$ for validation. The set of models we compare here are polynomials of degree $1 \leq m \leq 8$. In other words, the mth model from our set $\{model_m\}_{m=1}^{8}$ is a single-input degree-m polynomial of the form

$$model_m\,(\mathbf{x}, \Theta_m) = w_0 + xw_1 + x^2 w_2 + \cdots + x^m w_m. \tag{11.17}$$

Note how this small set of models is naturally ordered in terms of nonlinear capacity, with lower-degree models having smaller capacity and higher-degree models having larger capacity.

Figure 11.27 shows the fit of three polynomial models on the original dataset (first row), training data (second row), and validation data (third row). The errors on both the training (in blue) and validation (in yellow) data is shown in the bottom panel for all eight models. Notice, the validation error is at its lowest when the model is a degree-four polynomial. Of course as we use more polynomial units, moving from left to right in the figure, the higher-degree models fit the training data better. However, as the training error continues to decrease, the corresponding validation error starts climbing rapidly as the corresponding models provide poorer and poorer representations of the validation data (by the time $m = 7$ the validation error becomes so large that we do not plot it in the same window so that the other error values can be distinguished properly).

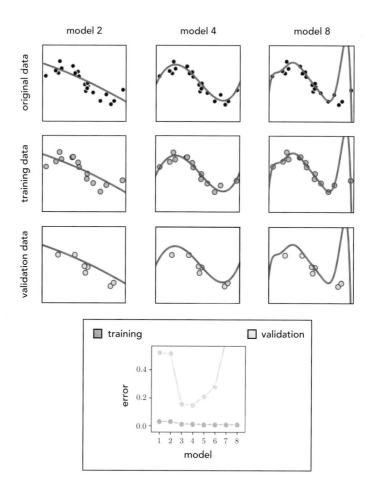

Figure 11.27 Figure associated with Example 11.8. See text for details.

Example 11.9   Naive cross-validation and classification

In this example we illustrate the use of a naive approach to cross-validation on the two-class classification dataset first shown in Example 11.7. Here we use (approximately) $\frac{4}{5}$ of the original set of 99 data points for training, and the other $\frac{1}{5}$ for validation. For the sake of simplicity we employ only a small set of polynomial models having degrees $1 \leq m \leq 7$. In other words, the mth model from our set $\{model_m\}_{m=1}^{7}$ is a degree-m polynomial (with two-dimensional input) of the form

$$model_m\,(\mathbf{x}, \Theta_m) = w_0 + \sum_{0 < i+j \leq m} x_1^i x_2^j\, w_{i,j}. \qquad (11.18)$$

These models are also naturally ordered from low to high capacity, as we increase the degree m of the polynomial.

Figure 11.28 shows the fit of three models from $\{model_m\}_{m=1}^{7}$ along with the original data (first row), the training data (second row), and the validation data (third row). The training and validation errors are likewise shown in the bottom panel for all seven models. With classification it makes more sense to use the number of misclassifications computed over the training/validation sets or some function of these misclassifications (e.g., accuracy) as our training/validation errors, as opposed to the raw evaluation of a classification cost.

In this case the degree-two polynomial model ($m = 2$) provides the smallest validation error, and hence the best nonlinear decision boundary for the entire dataset. This result does make intuitive sense as well, as we determined a circular boundary using a model of this form when engineering such features in Example 10.5 of Section 10.4.2. As the complexity goes up and training error continues to decrease, our models overfit the training data while at the same time providing a poor solution for the validation data.

11.4.2 Problems with naive cross-validation

Naive cross-validation works reasonably well for simple examples like those described above. However, since the process generally involves trying out a range of models where each model is optimized *completely* from scratch, naive cross-validation can be very expensive computationally speaking. Moreover, the *capacity* difference between even adjacent models (e.g., those consisting of m and $m + 1$ units) can be quite large, leading to huge jumps in the range of model complexities tried out on a dataset. In other words, controlling model *complexity* via adjustment of the *capacity* dial (with our *optimization* dial turned all the way to the right – as depicted in the bottom panels of Figure 11.26) often only allows for a coarse model search that can easily "skip over" an ideal amount of model complexity. As we will see in the next two sections, much more robust and fine-grained cross-validation schemes can be constructed by setting our *capacity* dial to the right and controlling model *complexity* by carefully setting our *optimization* dial.

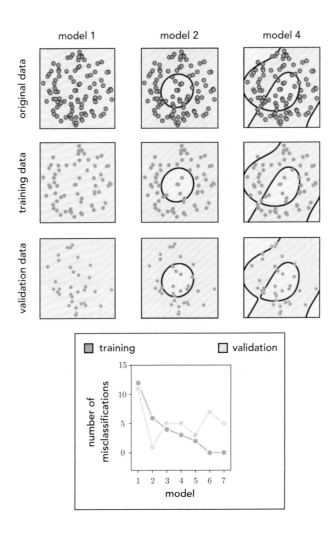

Figure 11.28 Figure associated with Example 11.9. See text for details.

11.5   Efficient Cross-Validation via Boosting

In this section we introduce *boosting*, the first of two fundamental paradigms for effective cross-validation described in this chapter. In contrast to the naive form of cross-validation described in the previous section, with boosting-based cross-validation we perform our model search by taking a single high-capacity model and optimize it *one unit at a time*, resulting in a much more efficient cross-validation procedure. While in principle any universal approximator can be used with boosting, this approach is often used as the cross-validation method of choice when employing tree-based universal approximators (as discussed further in Section 14.7).

11.5.1 The big picture

The basic principle behind boosting-based cross-validation is to progressively build a high-capacity model *one unit at a time*, using units from a single type of universal approximator (e.g., one of those simple exemplars outlined in Section 11.2.3), as

$$\text{model}\,(\mathbf{x}, \Theta) = w_0 + f_1\,(\mathbf{x})\,w_1 + f_2\,(\mathbf{x})\,w_2 + \cdots + f_M\,(\mathbf{x})\,w_M. \tag{11.19}$$

We do this sequentially in M *rounds*[8] where at each round we add one unit to the model, completely optimizing this unit's parameters alone along with its corresponding linear combination weight, and keep these parameters fixed at these optimally tuned values forever more. Alternatively, we can think of this procedure as beginning with a high-capacity model of the form in Equation (11.19) and – in M rounds – optimizing the parameters of each unit, one at a time.[9] In either case, performing boosting in this way produces a sequence of M tuned models that generally increase in *complexity* with respect to the training dataset, which we denote compactly as $[\text{model}_m]_{m=1}^{M}$ where the mth model consists of m tuned units. Since just one unit is optimized at a time, boosting tends to provide a computationally efficient fine-resolution form of model search (compared to naive cross-validation).

The general boosting procedure tends to produce training/validation error curves that generally look like those shown in the top panel of Figure 11.29. As with the naive approach detailed in the previous section, here too we tend to see training error decrease as m grows larger while validation error tends to start high where underfitting occurs, dip down to a minimum value (perhaps oscillating more than the one time illustrated here), and rise back up when overfitting begins.

Using the capacity/optimization dial conceptualization first introduced in Section 11.2.4, we can think about boosting as starting with our *capacity dial* set all the way to the *right* at some high value (e.g., some large value of M), and fidgeting with the *optimization dial* by turning it very slowly from left to right, as depicted in the bottom row of Figure 11.29. As discussed in Section 11.3.2 and summarized visually in the bottom row of Figure 11.23, with real data this general configuration allows our *optimization dial* to govern model complexity. In other words, with this configuration our optimization dial (roughly speaking) becomes the sort of fine-resolution *complexity dial* we aimed to construct at the outset of the chapter (see Section 11.1.3). With our optimization dial turned all the way to the left we begin our search with a low-complexity tuned model (called model$_1$) consisting of a single unit of a universal approximator having its parameters fully optimized. As we progress through rounds of boosting we turn the optimization dial gradually from left to right (here each notch on the

---

[8] $M + 1$ rounds if we include w_0.
[9] This is a form of coordinate-wise optimization. See, for example, Section 2.6.

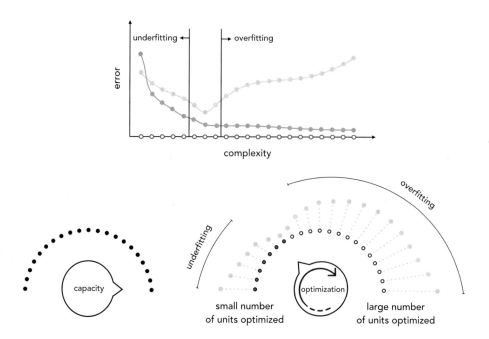

Figure 11.29 (top panel) Prototypical training and validation error curves associated with a completed run of boosting. (bottom panels) With boosting we fix our *capacity dial* all the way to the right, and begin with our *optimization dial* set all the way to the left. We then slowly turn our optimization dial from left to right, with each notch on the optimization dial denoting the complete optimization of one additional unit of the model, increasing the complexity of each subsequent model created with respect to the training set. See text for further details.

optimization dial denotes the complete optimization of one additional unit) optimizing (to completion) a single weighted unit of our original high-capacity model in Equation (11.19), so that at the mth round our tuned model (called model$_m$) consists of m individually but fully tuned units. Our ultimate aim in doing this is of course to determine a setting of the optimization (i.e., determine an appropriate number of tuned units) that minimizes validation error.

Whether we use fixed-shape, neural network, or tree-based units with boosting, we will naturally prefer units with *low capacity* so that the resolution of our model search is as fine-grained as possible. When we start adding units one at a time we turn our optimization dial clockwise from left to right. We want this dial turning to be done as smoothly as possible so that we can scan the validation error curve in a fine-grained fashion, in search of its minimum. This is depicted in the left panel of Figure 11.30. If we use *high-capacity* units at each round of boosting the resulting model search will be much coarser, as adding each additional unit results in aggressively turning the dial from left to right leaving large gaps in our model search, as depicted in the right panel of Figure 11.30. This kind of low-resolution search could easily result in us skipping over

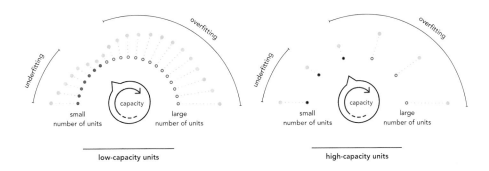

Figure 11.30 (left panel) Using low-capacity units makes the boosting procedure a high- (or fine-) resolution search for optimal model complexity. (right panel) Using high-capacity units makes boosting a low- (or coarse-) resolution search for optimal model complexity. See text for further details.

the complexity of an optimal model. The same can be said as to why we add only one unit at a time with boosting, tuning its parameters alone at each round. If we added more than one unit at a time, or if we retuned *every* parameter of *every* unit at each step of this process, not only would we have significantly more computation to perform at each step but the performance difference between subsequent models could be quite large and we might easily miss out on an ideal model.

11.5.2 Technical details

Formalizing our discussion of boosting above, we begin with a set of M nonlinear features or units from a single family of universal approximators

$$\mathcal{F} = \{f_1(\mathbf{x}), f_2(\mathbf{x}), \ldots, f_M(\mathbf{x})\}. \tag{11.20}$$

We add these units sequentially (or one at a time) building a sequence of M tuned models $[\text{model}_m]_{m=1}^M$ that increase in complexity with respect to the training data, from $m = 1$ to $m = M$, ending with a generic nonlinear model composed of M units. We will express this final boosting-made model slightly differently than in Equation (11.19), in particular reindexing the units it is built from as

$$\text{model}(\mathbf{x}, \Theta) = w_0 + f_{s_1}(\mathbf{x}) w_1 + f_{s_2}(\mathbf{x}) w_2 + \cdots + f_{s_M}(\mathbf{x}) w_M. \tag{11.21}$$

Here we have reindexed the individual units to f_{s_m} to denote the unit from the entire collection in $\mathcal{F}$ added at the mth round of the boosting process. The linear combination weights w_0 through w_M along with any additional weights internal to $f_{s_1}, f_{s_2}, \ldots, f_{s_M}$ are represented collectively in the weight set Θ.

The process of boosting is performed in a total of M rounds, one for each of the units in Equation (11.21). At each round we determine which unit, when added to the running model, best lowers its training error. We then measure the corresponding validation error provided by this update, and in the end after all rounds of boosting are complete, use the lowest validation error measurement found to decide which round provided the best overall model.

For the sake of simplicity in describing the formal details of boosting, we will center our discussion on a single problem: nonlinear regression on the training dataset $\left\{\left(\mathbf{x}_p, y_p\right)\right\}_{p=1}^{P}$ employing the Least Squares cost. However, the principles of boosting we will see remain *exactly* the same for other learning tasks (e.g., two-class and multi-class classification) and their associated costs.

Round 0 of boosting

We begin the boosting procedure by tuning

$$\text{model}_0\left(\mathbf{x}, \Theta_0\right) = w_0 \tag{11.22}$$

whose weight set $\Theta_0 = \{w_0\}$ contains a single bias weight, which we can easily tune by minimizing an appropriate cost over this variable alone. With this in mind, to find the optimal value for w_0 we minimize the Least Squares cost

$$\frac{1}{P}\sum_{p=1}^{P}\left(\text{model}_0\left(\mathbf{x}_p,\Theta_0\right)-y_p\right)^2 = \frac{1}{P}\sum_{p=1}^{P}\left(w_0-y_p\right)^2. \tag{11.23}$$

This gives the optimal value for w_0, which we denote as $w_0^{\star}$. We fix the bias weight at this value forever more throughout the process.

Round 1 of boosting

Having tuned the only parameter of model_0 we now *boost* its complexity by adding the weighted unit $f_{s_1}(\mathbf{x})w_1$ to it, resulting in a modified running model which we call model_1

$$\text{model}_1\left(\mathbf{x},\Theta_1\right) = \text{model}_0\left(\mathbf{x},\Theta_0\right) + f_{s_1}(\mathbf{x})w_1. \tag{11.24}$$

Note here the parameter set Θ_1 contains w_1 and any parameters internal to the unit f_{s_1}. To determine which unit in our set $\mathcal{F}$ best lowers the training error, we press model_1 against the data by minimizing

$$\frac{1}{P}\sum_{p=1}^{P}\left(\text{model}_0\left(\mathbf{x}_p,\Theta_0\right)+f_{s_1}\left(\mathbf{x}_p\right)w_1-y_p\right)^2 = \frac{1}{P}\sum_{p=1}^{P}\left(w_0^{\star}+f_{s_1}\left(\mathbf{x}_p\right)w_1-y_p\right)^2$$

$$\tag{11.25}$$

for every unit $f_{s_1} \in \mathcal{F}$.

Note that since the bias weight has already been set optimally in the previous round we only need tune the weight w_1 as well as the parameters internal to the nonlinear unit f_{s_1}. Also note, in particular, that with neural networks all nonlinear units take precisely the same form, and therefore we need not solve M versions of the optimization problem in Equation (11.25), one for every unit in $\mathcal{F}$, as we would do when using fixed-shape or tree-based units. Regardless of the type of universal approximator employed, round 1 of boosting ends upon finding the optimal f_{s_1} and w_1, which we denote respectively as $f_{s_1}^{\star}$ and $w_1^{\star}$, and keep fixed moving forward.

Round $m > 1$ of boosting

In general, at the mth round of boosting we begin with model_{m-1} consisting of a bias term and $m - 1$ units of the form

$$\text{model}_{m-1}(\mathbf{x}, \Theta_{m-1}) = w_0^{\star} + f_{s_1}^{\star}(\mathbf{x})w_1^{\star} + f_{s_2}^{\star}(\mathbf{x})w_2^{\star} + \cdots + f_{s_{m-1}}^{\star}(\mathbf{x})w_{m-1}^{\star}. \quad (11.26)$$

Note that the parameters of this model have been tuned sequentially, starting with the bias $w_0^{\star}$ in round 0, $w_1^{\star}$ and any internal parameters of $f_{s_1}^{\star}$ in round 1, and so forth, up to $w_{m-1}^{\star}$ and any parameters internal to $f_{s_{m-1}}^{\star}$ in round $m - 1$.

The mth round of boosting then follows the same pattern outlined in round 1, where we seek out the best weighted unit $f_{s_m}(\mathbf{x})w_m$ to add to our running model to best lower its training error on the dataset. Specifically, our mth model takes the form

$$\text{model}_m(\mathbf{x}, \Theta_m) = \text{model}_{m-1}(\mathbf{x}, \Theta_{m-1}) + f_{s_m}(\mathbf{x})w_m \quad (11.27)$$

and we determine the proper unit to add to this model by minimizing

$$\frac{1}{P}\sum_{p=1}^{P}\left(\text{model}_{m-1}(\mathbf{x}_p, \Theta_{m-1}) + f_{s_m}(\mathbf{x}_p)w_m - y_p\right)^2$$

$$= \frac{1}{P}\sum_{p=1}^{P}\left(w_0^{\star} + w_1^{\star}f_{s_1}^{\star} + \cdots + f_{s_{m-1}}^{\star}(\mathbf{x}_p)w_{m-1}^{\star} + f_{s_m}(\mathbf{x}_p)w_m - y_p\right)^2 \quad (11.28)$$

over w_m and parameters internal to f_{s_m} (if they exist), which are contained in the parameter set Θ_m.

Once again with fixed-shape or tree-based approximators, this entails solving M (or $M - m + 1$, if we decide to check only those units not used in previous rounds) such optimization problems, and choosing the one with smallest training error. With neural networks, since each unit takes the same form, we need only solve one such optimization problem.

11.5.3   Early stopping

Once all rounds of boosting are complete note how we have generated a sequence of M tuned models[10] – denoted $[\text{model}_m\,(\mathbf{x}, \Theta_m)]_{m=1}^{M}$ – which gradually increase in nonlinear complexity from $m = 1$ to $m = M$, and thus gradually decrease in training error. This gives us fine-grained control in selecting an appropriate model, as the jump in performance in terms of both the training and validation errors between subsequent models in this sequence can be quite smooth, provided we use low-capacity units (as discussed in Section 11.5.1).

Once boosting is complete we select from our set of models the one that provides the lowest validation error. Alternatively, instead of running all rounds of boosting and deciding on an optimal model after the fact, we can attempt to *halt* the procedure when the validation error first starts to increase. This concept, referred to as *early stopping*, leads to a more computationally efficient implementation of boosting, but one needs to be careful in deciding when the validation error has really reached its minimum as it can oscillate up and down multiple times (as mentioned in Section 11.4), and need not take the simple generic form illustrated in the top panel of Figure 11.29. There is no ultimate solution to this issue – thus ad hoc solutions are typically used in practice when early stopping is employed.

11.5.4   An inexpensive but effective enhancement

A slight adjustment at each round of boosting, in the form of addition of an individual bias, can significantly improve the algorithm. Formally, at the mth round of boosting instead of forming model_m as shown in Equation (11.27), we add an additional bias weight $w_{0,m}$ as

$$\text{model}_m\,(\mathbf{x}, \Theta_m) = \text{model}_{m-1}\left(\mathbf{x}, \Theta_{m-1}\right) + w_{0,m} + f_{s_m}\,(\mathbf{x})\,w_m. \tag{11.29}$$

This simple adjustment results in greater flexibility and generally better overall performance by allowing units to be adjusted "vertically" at each round (in the case of regression) at the minimal cost of adding a single variable to each optimization subproblem. Note that once tuning is done, the optimal bias weight $w_{0,m}^{\star}$ can be absorbed into the bias weights from previous rounds, creating a single bias weight $w_0^{\star} + w_{0,1}^{\star} + \cdots + w_{0,m}^{\star}$ for the entire model.

This enhancement is particularly useful when using fixed-shape or neural network units for boosting, as it is redundant when using tree-based approximators because they already have individual bias terms baked into them that always allow for this kind of vertical adjustment at each round of boosting.[11]

---

[10] We have excluded model_0 as it does not use any universal approximator units.

[11] In the jargon of machine learning boosting with tree-based learners is often referred to as *gradient boosting*. See Section 14.5 for further details.

To see this note that while Equation (11.14) shows the most common way of expressing a stump taking in one-dimensional input, repeated for convenience here, as

$$f(x) = \begin{cases} v_1 & x \leq s \\ v_2 & x > s \end{cases} \tag{11.30}$$

it is also possible to express $f(x)$ equivalently as

$$f(x) = w_0 + w_1 \, h(x) \tag{11.31}$$

where w_0 denotes an individual bias parameter for the stump and w_1 is an associated weight that scales $h(x)$, which is a simple step function with fixed levels and a split at $x = s$

$$h(x) = \begin{cases} 0 & x \leq s \\ 1 & x > s. \end{cases} \tag{11.32}$$

Expressing the stump in this equivalent manner allows us to see that every stump unit does indeed have its own individual bias parameter, making it redundant to add an individual bias at each round when boosting with stumps. The same concept holds for stumps taking in general N-dimensional input as well.

Example 11.10   Boosting regression using tree units
In this example we use the sinusoidal regression dataset first shown in Example 11.6 consisting of $P = 21$ data points, and construct a set of $B = 20$ tree (stump) units for this dataset (see Section 11.2.3). In Figure 11.31 we illustrate the result of $M = 50$ rounds of boosting (meaning many of the stumps are used multiple times). We split the dataset into $\frac{2}{3}$ training and $\frac{1}{3}$ validation, which are color-coded in light blue and yellow, respectively. Depicted in the figure are resulting regression fits and associated training/validation errors for several rounds of boosting. This example is discussed further in Section 14.5.

Example 11.11   Boosting classification using neural network units
In this example we illustrate the same kind of boosting as previously shown in Example 11.10, but now for two-class classification using a dataset of $P = 99$ data points that has a (roughly) circular decision boundary. This dataset was first used in Example 11.7. We split the data randomly into $\frac{2}{3}$ training and $\frac{1}{3}$ validation, and employ neural network units for boosting. In Figure 11.32 we illustrate the result of $M = 30$ rounds of boosting in terms of the nonlinear decision boundary and resulting classification, as well as training/validation errors.

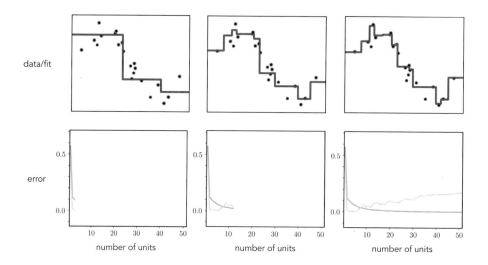

Figure 11.31 Figure associated with Example 11.10. See text for details.

11.5.5   Similarity to feature selection

The careful reader will notice how similar the boosting procedure is to the one introduced in Section 9.6 in the context of feature selection. Indeed principally the two approaches are entirely similar, except with boosting we do not select from a set of given input features but create them ourselves based on a chosen universal approximator family. Additionally, unlike feature selection where our main concern is *human interpretability*, we primarily use boosting as a tool for cross-validation. This means that unless we specifically prohibit it from occurring, we can indeed select the same feature multiple times in the boosting process as long as it contributes positively towards finding a model with minimal validation error.

   These two use-cases for boosting, i.e., feature selection and cross-validation, can occur together, albeit typically in the context of linear modeling as detailed in Section 9.6. Often in such instances cross-validation is used with a linear model as a way of automatically selecting an appropriate number of features, with human interpretation of the resulting selected features still in mind. On the other hand, rarely is feature selection done when employing a nonlinear model based on features from a universal approximator due to the great difficulty in the human interpretability of nonlinear features. The rare exception to this rule is when using tree-based units which, due to their simple structure, can in particular instances be readily interpreted by humans.

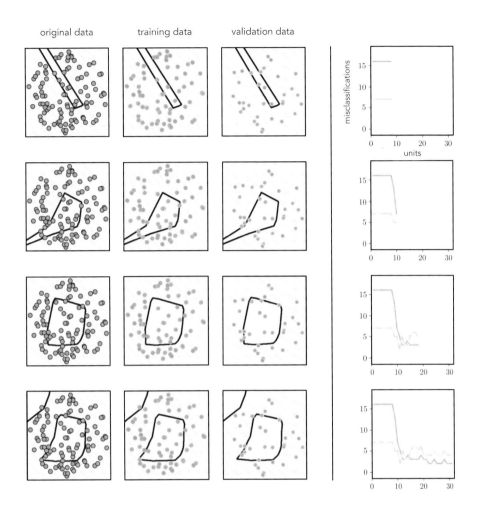

Figure 11.32 Figure associated with Example 11.11. See text for details.

11.5.6 The residual perspective with regression

Here we describe a common interpretation of boosting in the context of regression, that of sequentially fitting to the *residual* of a regression dataset. To see what this means, consider the following Least Squares cost function where we have inserted a boosted model at the mth round of its development

$$g\left(\Theta_m\right) = \frac{1}{P} \sum_{p=1}^{P} \left(\text{model}_m\left(\mathbf{x}_p, \Theta_m\right) - y_p\right)^2. \tag{11.33}$$

We can write our boosted model recursively as

$$\text{model}_m\left(\mathbf{x}_p, \Theta_m\right) = \text{model}_{m-1}\left(\mathbf{x}_p, \Theta_{m-1}\right) + f_m\left(\mathbf{x}_p\right) w_m \tag{11.34}$$

where all of the parameters of the $(m-1)$th model (i.e., model_{m-1}) are already tuned. Combining Equations (11.33) and (11.34) we can rewrite the Least Squares cost as

$$g\left(\Theta_m\right) = \frac{1}{P}\sum_{p=1}^{P}\left(f_m\left(\mathbf{x}_p\right)w_m - \left(y_p - \text{model}_{m-1}\left(\mathbf{x}_p\right)\right)\right)^2. \tag{11.35}$$

By minimizing this cost we look to tune the parameters of a single additional unit so that

$$f_m\left(\mathbf{x}_p\right)w_m \approx y_p - \text{model}_{m-1}\left(\mathbf{x}_p\right) \tag{11.36}$$

for all p or, in other words, so that this fully tuned unit approximates our original output y_p minus the contribution of the previous model. This quantity, the difference between our original output and the contribution of the $(m-1)$th model, is often called the *residual*: it is what is left to represent after subtracting off what was learned by the $(m-1)$th model.

Example 11.12   Boosting from the perspective of fitting to the residual
In Figure 11.33 we illustrate the process of boosting $M = 20$ neural network units to a toy regression dataset. In the top panels we show the dataset along with the fit provided by model_m at the mth step of boosting for select values of m. In the corresponding bottom panels we plot the *residual* at the same step, as well as the fit provided by the corresponding mth unit f_m. As boosting progresses, the fit on the original data improves while (simultaneously) the residual shrinks.

11.6   Efficient Cross-Validation via Regularization

In the previous section we saw how with boosting based cross-validation we automatically learn the proper level of model complexity for a given dataset by optimizing a general high-capacity model one unit at a time. In this section we introduce what are collectively referred to as *regularization* techniques for efficient cross-validation. With this set of approaches we once again start with a single high-capacity model, and once again adjust its complexity with respect to a training dataset via careful optimization. However, with regularization we tune all of the units *simultaneously*, controlling how well we *optimize* its associated cost so that a minimum validation instance of the model is achieved.

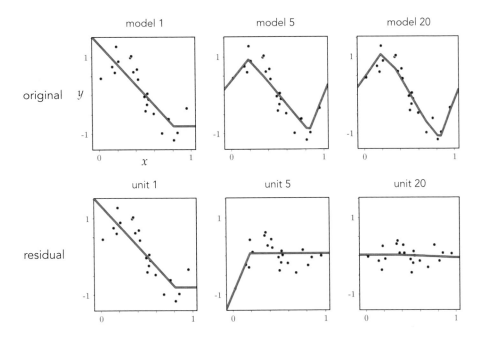

Figure 11.33 Figure associated with Example 11.12. See text for details.

11.6.1   The big picture

Imagine for a moment that we have a simple nonlinear regression dataset, like the one shown in the top-left panel of the Figure 11.34, and we use a high-capacity model (relative to the nature of the data) made up of a sum of *universal approximators* of a single kind to fit it as

$$\text{model}\,(\mathbf{x}, \Theta) = w_0 + f_1\,(\mathbf{x})\,w_1 + f_2\,(\mathbf{x})\,w_2 + \cdots + f_M\,(\mathbf{x})\,w_M. \qquad (11.37)$$

Suppose that we partition this data into training and validation portions, and then train our high-capacity model by *completely* optimizing the Least Squares cost over the training portion of the data. In other words, we determine a set of parameters for our high-capacity model that lie very close to a global minimum of its associated cost function. In the top-right panel of the figure we draw a hypothetical two-dimensional illustration of the cost function associated with our high-capacity model over the training data, denoting the global minimum by a blue dot and its evaluation on the function by a blue x.

Since our model has high capacity, the resulting fit provided by the parameters lying at the global minimum of our cost will produce a tuned model that is overly complex and *severely* overfits the training portion of our dataset. In the bottom-left panel of the Figure 11.34 we show the tuned model fit (in blue) provided by such a set of parameters, which wildly overfits the training data. In the top-

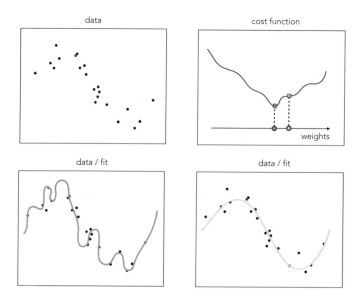

Figure 11.34 (top-left panel) A generic nonlinear regression dataset. (top-right panel) A figurative illustration of the cost function associated with a high-capacity model over the training portion of this data. The global minimum is marked here with a blue dot (along with its evaluation by a blue **x**) and a point nearby is marked in yellow (whose evaluation is shown as a yellow **x**). (bottom-left panel) The original data and fit (in blue) provided by the model using parameters from the global minimum of the cost function severely overfits the training portion of the data. (bottom-right panel) The parameters corresponding to the yellow dot shown in the top-right panel minimize the cost function over the validation portion of the data, and thus provide a much better fit (in yellow) to the data. See text for further details.

right panel we also show a set of parameters lying relatively near the global minimum as a yellow dot, and whose evaluation of the function is shown as a yellow **x**. This set of parameters lying in the general neighborhood of the global minimum is where the cost function is minimized over the *validation* portion of our data. Because of this the corresponding fit (shown in the bottom-right panel in yellow) provides a much better representation of the data.

This toy example is illustrative of a more general principle we have seen earlier in Section 11.3.2: that overfitting is due both to the *capacity* of an untuned model being too high *and* its corresponding cost function (over the training data) being *optimized* too well, leading to an overly complex tuned model. This phenomenon holds true for all machine learning problems (including regression, classification, and unsupervised learning techniques like the Autoencoder) and is the motivation for general regularization based cross-validation strategies: if proper optimization of *all parameters* of a high-capacity model leads to overfitting, it can be avoided by optimizing said model *imperfectly* when validation error (not training error) is at its lowest. In other words, *regularization* in the context of cross-validation constitutes a set of approaches to cross-validation wherein we carefully tune all parameters of a high-capacity model by setting them purposefully away from the global minimum of its associated cost function. This can be done in a variety of ways, and we detail the two most popular approaches next.

11.6.2 Early stopping based regularization

With *early stopping* based regularization[12] we properly tune a high-capacity model by making a run of local optimization (tuning all parameters of the model), and by using the set of weights from this run where the model achieves minimum validation error. This idea is illustrated in the left panel of Figure 11.35 where we employ the same prototypical cost function first shown in the top-right panel of Figure 11.34. During a run of local optimization we frequently compute training and validation errors (e.g., at each step of the optimization procedure) so that a set of weights providing minimum validation error can be determined with fine resolution.

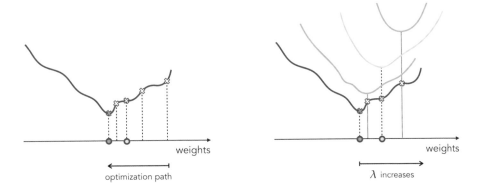

Figure 11.35 (left panel) A figurative illustration of early stopping regularization applied to a prototypical cost function of a high-capacity model. We make a run of optimization – here shown to completion at the global minimum in blue – and choose the set of weights that provide a minimum validation error (shown in yellow). (right panel) A figurative illustration of regularizer based regularization. By adding a regularizer function to the cost associated with a high-capacity model we change its shape, in particular dragging its global minimum (where overfitting behavior occurs) away from its original location. The regularized cost function can then be *completely* minimized to recover weights as close to/far away from the true global minimum of the original cost function, depending on the choice of the regularization parameter λ. Proper setting of this parameter allows for the recovery of validation-error-minimizing weights. See text for further details.

Whether one literally stops the optimization run when minimum validation

---

[12] This regularization approach is especially popular when employing deep neural network models as detailed in Section 13.7.

error has been reached (which can be challenging in practice given the somewhat unpredictable behavior of validation error as first noted in Section 11.4.2) or one runs the optimization to completion (picking the best set of weights afterwards), in either case we refer to this method as early stopping regularization. Note that the method itself is analogous to the early stopping procedure outlined for boosting based cross-validation in Section 11.5.3, in that we sequentially increase the complexity of a model until minimum validation is reached. However, here (unlike boosting) we do this by controlling how well we optimize a model's parameters *simultaneously*, as opposed to one unit at a time.

Supposing that we begin our optimization with a small initial value (which we typically do) the corresponding training and validation error curves will, in general,[13] look like those shown in the top panel of Figure 11.36. At the start of the run the complexity of our model (evaluated at the initial weights) is quite small, providing a large training and validation error. As minimization proceeds, and we continue optimizing one step at a time, error in both training and validation portions of the data decreases while the complexity of the tuned model increases. This trend continues up until a point when the model complexity becomes too great and overfitting begins, and validation error increases.

In terms of the capacity/optimization dial conceptualization detailed in Section 11.3.2, we can think of (early stopping based) regularization as beginning with our capacity dial set all the way to the *right* (since we employ a high-capacity model) and our optimization dial all the way to the *left* (at the initialization of our optimization). With this configuration – summarized visually in the bottom panel of Figure 11.36 – we allow our optimization dial to directly govern the amount of complexity our tuned models can take. In other words, with this configuration our optimization dial becomes (roughly speaking) the ideal complexity dial described at the start of the chapter in Section 11.1.3. With early stopping we turn our optimization dial from left to right, starting at our initialization making a run of local optimization one step at a time, seeking out a set of parameters that provide minimum validation error for our (high-capacity) model.

There are a number of important engineering details associated with implementing an effective early stopping regularization procedure, which we discuss below.

- **Different optimization runs may lead to different tuned models.** The cost function topology associated with high-capacity models can be quite complicated. Different initializations can thus produce different trajectories towards potentially different minima of the cost function, and produce corresponding validation-error-minimizing models that differ in shape – as illustrated pictorially in the top row of Figure 11.37. However, in practice these differences

---

[13] Note that both can oscillate in practice depending on the optimization method used.

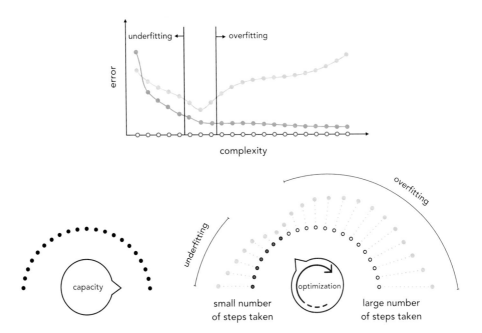

Figure 11.36 (top panel) A prototypical pair of training/validation error curves associated with a generic run of early stopping regularization. (bottom panels) With early stopping we set our capacity dial all the way to the *right* and our optimization dial all the way to the *left*. We then slowly move our optimization dial from left to right, iteratively improving the fit of our of model to the training data, adjusting all of its parameters simultaneously one step at a time. As each step of optimization progresses we slowly turn the optimization dial clockwise from left to right, gradually increasing the complexity of our tuned model, in search of a tuned model with minimum validation error. Here each notch on the optimization dial abstractly denotes a step of local optimization. See text for further details.

tend not to effect performance, and the resulting models can be easily combined or *bagged* together (see Section 11.9) to average out any major differences in their individual performance.

- **How high should capacity be set?** How do we know how high to set the capacity of our model when using early stopping (or any other form of) regularization based cross-validation? In general there is no single answer. It must simply be set at least "high" enough that the model overfits if optimized completely. This can be achieved by adjusting M (the number of units in the model) and/or the capacity of individual units (by, for example, using shallow versus deep neural network or tree based units, as we detail in Chapters 13 and 14, respectively).

- **Local optimization must be carefully performed.** One must be careful with the sort of local optimization scheme used with early stopping cross-validation.

As illustrated in the bottom-left panel of Figure 11.37, ideally we want to turn our optimization dial smoothly from left to right, searching over a set of model complexities with a fine resolution. This means, for example, that with early stopping we often avoid local optimization schemes that take very large steps (e.g., Newton's method – as detailed in Chapter 4) as this can result in a coarse and low-resolution search over model complexity that can easily skip over minimum-validation models, as depicted in the bottom-right panel of the figure. Local optimizers that take smaller, high-quality steps – like the advanced first-order methods detailed in Appendix A – are often preferred when employing early stopping. Moreover, when employing mini-batch/stochastic first-order methods (see Appendix Section A.5) validation error should be measured *several times per epoch* to avoid taking too many steps without measuring validation error.

- **When is validation error really at its lowest?** While generally speaking validation error decreases at the start of an optimization run and eventually increases (making somewhat of a "U" shape) it can certainly fluctuate up and down during optimization. Therefore it is not all together obvious when the validation error has indeed reached its lowest point unless the optimization process is performed to completion. To deal with this peculiarity, often in practice a reasonable engineering choice is made as to when to stop based on how long it has been since the validation error has *not* decreased. Moreover, as mentioned earlier, one need not truly halt a local optimization procedure to employ the thrust of early stopping, and can simply run the optimizer to completion and select the best set of weights from the run after completion.

The interested reader can see Example 13.14 for a simple illustration of early stopping based regularization.

11.6.3   Regularizer based methods

A *regularizer* is a simple function that can be added to a machine learning cost for a variety of purposes, e.g., to prevent unstable learning (as we saw in Section 6.4.6), as a natural part of relaxing the Support Vector Machine (Section 6.5.4) and multi-class learning scenarios (Section 7.3.4), and for feature selection (Section 9.7). As we will see, the latter of these applications (feature selection) is very similar to our use of the regularizer here.

Adding a simple regularizer function like one of those we have seen in previous applications (e.g., the ℓ_2 norm) to the cost of a high-capacity model, we can alter its shape and, in particular, move the location of its global minimum away from its original location. In general if our high-capacity model is given as model $(\mathbf{x}, \Theta)$, its associated cost function given by g, and a regularizer h, then the regularized cost is given as the linear combination of g and h as

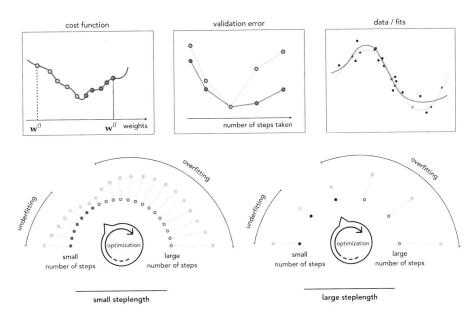

Figure 11.37 Two subtleties associated with early stopping based regularization. (top-left panel) A prototypical cost function associated with a high-capacity model, with two optimization paths (shown in red and green, respectively) resulting from two local optimization runs beginning at different starting points. (top-middle panel) The validation error histories corresponding to each optimization run. (top-right panel) While each run produces a different set of optimal weights, and a different fit to the data (here shown in green and red respectively, corresponding to each run), these fits are generally equally representative. (bottom-left panel) Taking optimization steps with a small steplength makes the early stopping procedure a fine-resolution search for optimal model complexity. With such small steps we smoothly turn the optimization dial from left to right in search of a validation-error-minimizing model. (bottom-right panel) Using steps with a large steplength makes early stopping a coarse-resolution search for optimal model complexity. With each step taken we aggressively turn the dial from left to right, performing a coarser model search that potentially skips over the optimal model.

$$g(\Theta) + \lambda h(\Theta) \tag{11.38}$$

where λ is referred to as the *regularization parameter*. The regularization parameter is always nonnegative $\lambda \geq 0$ and controls the mixture of the cost and regularizer. When it is set small and close to zero the regularized cost is essentially just g, and conversely when set very large the regularizer h dominates in the linear combination (and so upon minimization we are really just minimizing it alone). In the right panel of Figure 11.35 we show how the shape of a figurative regularized cost (and consequently the location of its global minimum) changes with the value of λ.

Supposing that we begin with a large value of λ and try progressively smaller

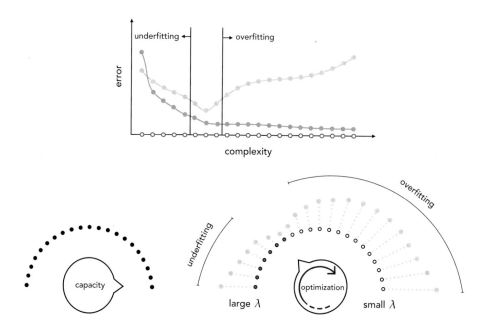

Figure 11.38 (top panel) A prototypical pair of training/validation error curves associated with a generic run of regularizer based cross-validation. (bottom panels) With regularizer-based cross-validation we set our capacity dial all the way to the *right* and our optimization dial all the way to the *left* (beginning with a large value for our regularization parameter λ). We then slowly move our optimization dial from left to right by *decreasing* the value of λ, where here each notch on the optimization dial represents the complete minimization of the corresponding regularized cost function in Equation (11.38), improving the fit our of model to the training data. By adjusting the value of λ (and completely minimizing each corresponding regularized cost) we slowly turn the optimization dial clockwise from left to right, gradually increasing the complexity of our tuned model, in search of a model with minimum validation error. See text for further details.

values (completely optimizing each regularized cost), the corresponding training and validation error curves will in general look something like those shown in the top panel of Figure 11.38 (remember in practice that *validation error* can oscillate, and need not take just one dip down). At the start of this procedure, using a large value of λ, the complexity of our model is quite small as the regularizer completely dominated in the regularized cost, and thus the associated minimum recovered belongs to the regularizer and not the cost function itself. Since the set of weights is virtually unrelated to the data we are training over, the corresponding model will tend to have large training and validation errors. As λ is decreased the parameters provided by complete minimization of the regularized cost will be closer to the global minimum of the original cost itself, and so error on both training and validation portions of the data decreases while (generally speaking) the complexity of the tuned model increases. This trend continues up until a point when the regularization parameter is small enough that the recovered parameters lie too close to that of the original cost, so that the corresponding model complexity becomes too great. Here overfitting begins and validation error increases.

In terms of the capacity/optimization dial scheme detailed in Section 11.3.2, we can think of regularizer based cross-validation as beginning with our capacity dial set to the *right* (since we employ a high-capacity model) and our optimization dial all the way to the *left* (employing a large value for λ in our regularized cost). With this configuration (summarized visually in the bottom panel of Figure 11.38) we allow our optimization dial to directly govern the amount of complexity our tuned models can take. As we turn our optimization dial from left to right we *decrease* the value of λ and *completely* minimize the corresponding regularized cost, seeking out a set of parameters that provide minimum validation error for our (high-capacity) model.

There are a number of important engineering details associated with implementing an effective regularizer based cross-validation procedure, which we discuss below.

- **Bias weights are often not included in the regularizer.** As with linear models as discussed in Section 9.7, often only the nonbias weights of a general model are included in the regularizer. For example, suppose that we employ fixed-shape universal approximator units and hence our parameter set Θ contains a single bias w_0 and feature-touching weights $w_1, w_2, ..., w_B$. If we then regularize our cost function $g(\Theta)$ using the squared ℓ_2 norm, our regularized cost would then take the form $g(\Theta) + \lambda \sum_{b=1}^{B} w_b^2$. When employing neural network units we follow the same pattern, but here we have far more bias terms to avoid including in the regularizer. For example, if we use units of the form $f_b(\mathbf{x}) = \tanh(w_{b,0} + x_1 w_{b,1} + \cdots + x_N w_{b,N})$ the term $w_{b,0}$ – internal to the unit – is a bias term we also do not want included in our regularizer. Thus, to regularize a cost function including these units using the squared ℓ_2 norm we have $g(\Theta) + \lambda \left(\sum_{b=1}^{B} w_b^2 + \sum_{b=1}^{B} \sum_{n=1}^{N} w_{b,n}^2 \right)$.

- **Choice of regularizer function.** Note that while the ℓ_2 norm is a very popular regularizer, one can – in principle – use any simple function as a regularizer. Other popular choices of regularizer functions include the ℓ_1 norm regularizer $h(\mathbf{w}) = \|\mathbf{w}\|_1 = \sum_{n=1}^{N} |w_n|$, which tends to produce *sparse* weights, and the total variation regularizer $h(\mathbf{w}) = \sum_{n=1}^{N-1} |w_{n+1} - w_n|$, which tends to produce *smoothly-varying* weights. We often use the simple quadratic regularizer (ℓ_2 norm squared) to incentivize weights to be *small*, as we naturally do with two-class and multi-class logistic regression. Each of these different kinds of regularizers tends to pull the global minimum of the sum towards different portions of the input space – as illustrated in Figure 11.39 for the quadratic (top-left panel), ℓ_1 norm (top-middle panel), and total variation norm (top-right panel).

- **Choosing the range of λ values.** Analogously to what we saw with early stopping and boosting procedures previously, with regularization we want to perform our search as carefully as possible, turning our optimization dial as smoothly as possible from left to right in search of our perfect model. This desire translates directly to both the range and number of values for λ that we test out. For instance, the more values we try within a given range, the smoother we turn our optimization dial (as depicted visually in the bottom-left panel of Figure 11.39). The limit on how many values we can try is often dictated by computation and time restrictions, since for *every* value of λ tried a complete minimization of a corresponding regularized cost function must be performed. This can make regularizer based cross-validation very computationally expensive. On the other hand, trying too few of values can result in a coarse search for weights providing minimum validation error, increasing the possibility that such weights are skipped over entirely (as depicted in the bottom-right panel of Figure 11.39).

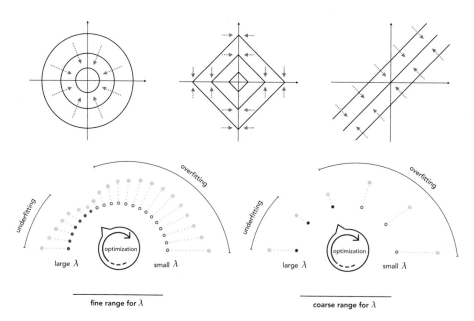

Figure 11.39 (top row) A visual depiction of where the ℓ_2 (top-left panel), ℓ_1 (top-middle panel), and total variation (top-right panel) functions pull the global minimum of a cost function – when used as a regularizer. These functions pull the global minimum towards the origin, the coordinate axes, and diagonal lines where consecutive entries are equal, respectively. (bottom-left panel) Testing out a large range and number of values for the regularization parameter λ results in a fine-resolution search for validation-error-minimizing weights. (bottom-right panel) A smaller number (or a poorly chosen range) of values can result in a coarse search that can skip over ideal weights. See text for further details.

Example 11.13   Tuning λ for a two-class classification problem

In this example we use a quadratic regularizer to find a proper nonlinear classifier for the two-class classification dataset shown in the left column of Figure
11.40 where the training set is shown with their perimeter colored in light blue,
and the validation points have their perimeter colored yellow. Here we use
$B = 20$ neural network units – a high-capacity model with respect to this data –
and try out 6 values of λ uniformly distributed between 0 and 0.5 (completely
minimizing the corresponding regularized cost in each instance). As the value of
λ changes the fit provided by the weights recovered from the global minimum of
each regularized cost function is shown in the left column, while the corresponding training and validation errors are shown in blue and yellow, respectively, in
the right column. In this simple experiment, a value somewhere around $\lambda \approx 0.25$
appears to provide the lowest validation error and corresponding best fit to the
dataset overall.

11.6.4   Similarity to regularization for feature selection

Akin to the boosting procedure detailed in the previous section, here the careful
reader will notice how similar the regularizer based framework described here
is to the concept of regularization detailed for feature selection in Section 9.7.
The two approaches are very similar in theme, except here we do not select
from a set of given input features but *create* them ourselves based on a universal
approximator. Additionally, instead of our main concern with regularization
being *human interpretability* of a machine learning model, as it was in Section 9.7,
here we use regularization as a tool for cross-validation.

11.7   Testing Data

In Section 11.3.4 we saw how, in place of training error, *validation error* is an
appropriate measurement tool that enables us to accurately identify an appropriate model/parameter tuning for generic data. However, like the training error,
choosing a model based on minimum validation error can also potentially lead
to models that *overfit* our original dataset. In other words, at least in principle, we
can overfit to validation data as well. This can make validation error a poor indicator of how well a cross-validated model will perform in general. As we will
see in this brief section, the potential dangers of this reality can be ameliorated,
provided the dataset is large enough, by splitting our original training data into
not two sets (training and validation), but *three*: training, validation, and testing
sets. By measuring a cross-validated model's performance on the *testing set* we
not only gain a better measure of its ability to capture the true nature of the

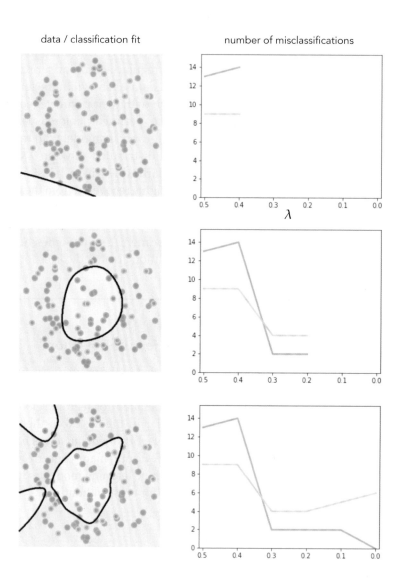

Figure 11.40 Figure associated with Example 11.13. See text for details.

phenomenon generating the data, but we also gain a reliable measurement tool for comparing the efficacy of multiple cross-validated models.

11.7.1 Overfitting validation data

In Section 11.3.4 we learned how, as a measurement tool, training error fails to help us properly identify when a tuned model has sufficient complexity to properly represent a given dataset. There we saw how an overfitting model, that is one that achieves minimum training error but is far too complex, represents the training data we currently have incredibly well, but simultaneously represents the phenomenon underlying the data (as well as any future data similarly generated by it) very poorly. While not nearly as prevalent in practice, it is possible for a properly cross-validated model to *overfit validation data*.

To see how this is possible let us analyze an extreme two-class classification dataset. As shown in the left panel of Figure 11.41 this dataset shows no meaningful relationship whatsoever between the input and output (labels). Indeed we created it by choosing the coordintes of the two-dimensional input points randomly over the input space, and then assigning label value +1 (red class) to half of the points (which are selected, once again, at random) and label value −1 (blue class) to the other half.

and points randomly (uniformly) on the unit square and assigned labels to the points at random.

Because we know that the underlying phenomenon generating this dataset is *completely random*, no model, whether it has been found via cross-validation or otherwise, should ever allow us to correctly predict the label of future points with an accuracy that is substantially greater than 50 percent. In other words, no model should truly provide better-than-chance accuracy on random data such as this. However, this reality need not be reflected in an appropriately cross-validated model (i.e., one with minimum validation error for some split of the data). Indeed in the right panel of Figure 11.41 we show the decision boundary of a naively cross-validated model for this dataset, where $\frac{1}{5}$ of the original data was used as validation, and color the regions according to the model's predictions. This particular cross-validated model provides 70 percent accuracy on the validation data, which perhaps at first glance is mysterious given our understanding of the underlying phenomenon. However, this is because, even though it was chosen as the validation-error-minimizing model, this model still *overfits* the original data. While it is not as prevalent or severe as the overfitting that occurs with training-error-minimized models, overfitting to validation data like this is still a danger that in practice should be avoided when possible.

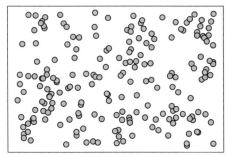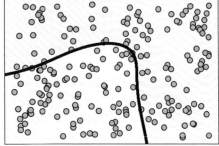

Figure 11.41 (left panel) A randomly generated two-class classification dataset. (right panel) The decision boundary of a cross-validated model providing 70 percent accuracy on the validation data, which is meaningfully greater than random chance (50 percent).

11.7.2   Testing data and testing error

Up until now we have used validation data *both* to select the best model for our data (i.e., cross-validation) *and* to measure its quality. However, much like when the notion of validation data was first introduced in Section 11.3.4, we see that using the same set of data to perform both of these tasks can lead to the selection of an overfitting model and can diminish the utility of validation error as a measure of model quality. The solution to this problem (again much like when we introduced the concept of validation data to begin with) is to split up the two tasks we now assign to validation data by introducing a *second* validation set. This "second validation set" is often called a *test set* or *testing set*, and is used solely to measure the quality of our final cross-validated model.

By splitting our data into three chunks (as illustrated in Figure 11.42) we still use training and validation portions precisely as they have been used thus far (i.e., for performing cross-validation). However, after the cross-validated model is constructed its quality is measured on the distinct *testing set* of data, on which it has been neither trained nor validated. This *testing error* gives an "unbiased" estimate of the cross-validated model's performance, and is generally closer to capturing the true error of our model on future data generated by the same phenomenon.

Figure 11.42 The original dataset (left panel) is split randomly into three nonoverlapping subsets: training, validation, and testing sets (right panel).

In the case of our random two-class data introduced in Section 11.7.1, such a testing set provides a far more accurate picture of how well our cross-validated model will work in general. In Figure 11.43 we again show this dataset (with validation data points highlighted with yellow boundaries), which is now augmented by the addition of a testing portion (those points highlighted with green boundaries) that are generated precisely the same way we created the original dataset in Figure 11.41. Note importantly that this testing portion was not used during training/cross-validation. While our cross-validated model achieved 70 percent accuracy on the validation set (as mentioned previously), it achieves only a 52 percent accuracy on the testing set, which is a more realistic indicator of our model's true classification ability, given the nature of this data.

What portion of the original dataset should we assign to our testing set? As with the portioning of training and validation (detailed in Section 11.3.4), there is no general rule here, save perhaps one: the use of testing data is a luxury we

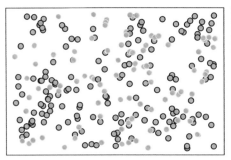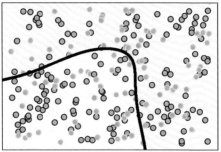

Figure 11.43 (left panel) The same dataset first shown in Figure 11.41 augmented by data points highlighted in green that were removed from training and validation procedures, and left out for testing. (right panel) The cross-validated model only achieves a 52 percent accuracy on the testing set, which is a much better estimate of a machine learning model's ability to classify random data.

can indulge in only when we have a large amount of data. When data is scarce we must leverage it all just to build a "halfway reasonable" cross-validated model. When data is plentiful, however, often the size of validation and testing sets are chosen similarly. For example, if $\frac{1}{5}$ of a dataset is used for validation, often for simplicity the same portion is used for testing as well.

11.8   Which Universal Approximator Works Best in Practice?

Beyond special circumstances such as those briefly discussed below, it is virtually never clear *a priori* which, if any, of the universal approximators will work best. Indeed cross-validation (as outlined in the previous sections) is *the* toolset one uses in practice to decide which type of universal approximator based model works best for a particular problem. Using these techniques one can create a range of different cross-validated models, each built from a distinct type of universal approximator, and compare their efficacy on a testing set (described in Section 11.7) afterwards to see which universal approximator works best. Alternatively, one can cross-validate a range of universal approximator based models and average them together afterwards, as discussed next in Section 11.9, leading to an averaged model that consists of representatives from multiple universal approximators.

In some instances broad understanding of a dataset can direct the choice of universal approximator. For example, because oftentimes business, census, and (more generally) *structured* datasets consist of broad mixtures of continuous and discontinuous categorical input features (see Section 6.7.1), tree-based universal approximators, with their discontinuous step-like shapes, often provide stronger results on average than other universal approximator types. On the other hand, data that is naturally continuous (e.g., data generated by natural processes or sensor data) is often better matched with a continuous universal approximator: fixed-shape or neural network. Understanding whether future predictions need be made *inside* or *outside* the input domain of the original dataset can also help guide the choice of approximator. In such cases fixed-shape or neural network approximators can be preferred over trees – the latter by their very nature always creating perfectly flat predictions outside of the original data's input domain (see Exercise 14.9 for further details).

When *human interpretability* is of primary importance, this desire (in certain circumstances) can drive the choice of universal approximator. For example, due to their discrete branching structure (see Section 14.2), tree-based universal approximators can often be much easier to interpret than other approximators (particularly neural networks). For analogous reasons fixed-shape approximators (e.g., polynomials) are often employed in the natural sciences, like the gravitational phenomenon underlying the Galileo's ramp dataset discussed in Example 11.17.

11.9   Bagging Cross-Validated Models

As we discussed in detail in Section 11.3.4, validation data is the portion of our original dataset we exclude at random from the training process in order to determine a proper tuned model that will faithfully represent the phenomenon generating our data. The validation error generated by our tuned model on this "unseen" portion of data is the fundamental measurement tool we use to determine an appropriate cross-validated model for our entire dataset (besides, perhaps, a testing set – see Section 11.7). However, the random nature of splitting data into training and validation poses an obvious flaw to our cross-validation process: what if the random splitting creates training and validation portions which are not desirable representatives of the underlying phenomenon that generated them? In other words, in practice what do we do about potentially bad training–validation splits, which can result in poorly representative cross-validated models?

Because we *need* cross-validation in order to choose appropriate models in general, and because we can do nothing about the (random) *nature* by which we split our data for cross-validation (what better method is there to simulate the "future" of our phenomenon?), the practical solution to this fundamental problem is to simply create several different training–validation splits, determine an appropriate cross-validated model on each split, and then *average* the resulting cross-validated models. By averaging a set of cross-validated models, also referred to as *bagging* in the jargon of machine learning, we can very often "average out" the potentially undesirable characteristics of each model while synergizing their positive attributes. Moreover, with *bagging* we can also effectively combine cross-validated models built from *different* universal approximators. Indeed this is the most reasonable way of creating a single model built from different types of universal approximators in practice.

Here we will walk through the concept of bagging or model averaging for regression, as well as two-class and multi-class classification by exploring an array of simple examples. With these simple examples we will illustrate the superior performance of bagged models visually, but in general we confirm this using the notion of testing error (see Section 11.7) or an estimate of testing error (often employed when bagging trees – see Section 14.6). Regardless, the principles detailed here can be employed more widely as well to any machine learning problem. As we will see, the best way to average/bag a set of cross-validated regression models is by taking their *median* and cross-validated classification models by computing the *mode* of their predicted labels.

11.9.1   Bagging regression models

Here we explore several ways of bagging a set of cross-validated models for the nonlinear regression dataset first described in Example 11.6. As we will see, more often than not the best way to bag (or average) cross-validated regression models is by taking their *median* (as opposed to their *mean*).

Example 11.14   Bagging cross-validated regression models

In the set of small panels in the left side of Figure 11.44 we show ten different training–validation splits of a prototypical nonlinear regression dataset, where $\frac{4}{5}$ of the data in each instance has been used for training (colored light blue) and $\frac{1}{5}$ is used for validation (colored yellow). Plotted with each split of the original data is the corresponding cross-validated model found via naive cross-validation (see Section 11.4.2) of the full range of polynomial models of degree 1 to 20. As we can see, while *many* of these cross-validated models perform quite well, several of them (due to the particular training–validation split on which they are based) severely *underfit* or *overfit* the original dataset. In each instance the poor performance is completely due to the particular underlying (random) training–validation split, which leads cross-validation to a validation-error-minimizing tuned model that still does not represent the true underlying phenomenon very well. By taking an *average* (here the *median*) of the ten cross-validated models shown in these small panels we can average out the poor performance of this handful of bad models, leading to a final bagged model that fits the data quite well – as shown in the large right panel of Figure 11.44.

Why average our cross-validated models using the *median* as opposed to the *mean*? Simply because the mean is far more sensitive to *outliers* than is the median. In the top row of Figure 11.45 we show the regression dataset shown previously along with the individual cross-validated fits (left panel), the median bagged model (middle panel), and the mean bagged model (right panel). Here the mean model is highly affected by the few overfitting models in the group,

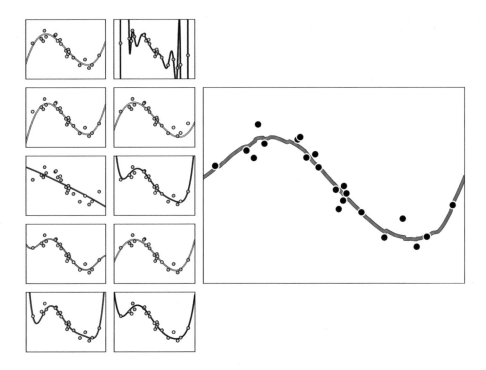

Figure 11.44 Figure associated with Example 11.14. (left columns) The results of applying cross-validation to ten random training–validation splits of a nonlinear regression dataset, with each resulting model shown in one of the ten small panels. Here the training and validation portions in each instance are colored light blue and yellow, respectively. (right column) The fit, shown in red, resulting from the bagging of the ten models whose fits are shown on the left. See text for further details.

and ends up being far too oscillatory to fairly represent the phenomenon underlying the data. The median is not affected in this way, and is therefore a much better representative.

When we bag we are simply averaging various cross-validated models with the desire to both avoid bad aspects of poorly-performing models, and jointly leverage strong elements of the well-performing ones. Nothing in this notion prevents us from bagging together cross-validated models built using different universal approximators, and indeed this is the most organized way of combining different types of universal approximators in practice.

In the bottom row of Figure 11.45 we show the result of a cross-validated polynomial model (left panel), a cross-validated neural network model (second to the left panel), and a cross-validated tree-based model (second to the right panel) built via boosting (see Section 11.5). Each cross-validated model uses a different training–validation split of the original dataset, and the bagged median of these models is shown in the right panel.

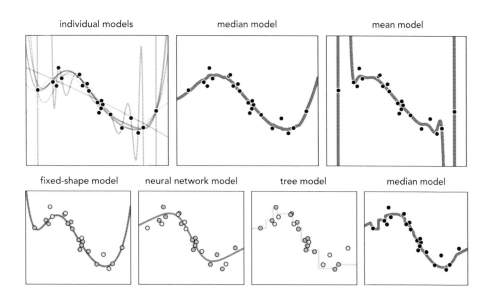

Figure 11.45 Figure associated with Example 11.14. The ten individual cross-validated models first shown in Figure 11.45 are plotted together in the top-left panel. The *median* and *mean* of these models are shown in the top-middle and top-right panel, respectively. With regression, bagging via the median tends to produce better results as it is less sensitive to outliers. (bottom row) Cross-validated fixed-shape polynomial (left panel), neural network (second panel from the left), and tree-based (second panel from the right) models. The median of these three models is shown in the right panel. See text for further details.

11.9.2   Bagging classification models

The principle behind bagging cross-validated models holds analogously for classification tasks, just as it does with regression. Because we cannot be certain whether or not a particular (randomly chosen) validation set accurately represents the "future data" from a given phenomenon well, the averaging (or bagging) of a number of cross-validated classification models provides a way of averaging out poorly representative portions of some models while combining the various models' positive characteristics.

Because the predicted output of a (cross-validated) classification model is a *discrete* label, the average used to bag cross-validated classification models is the *mode* (i.e., the most popularly predicted label).

Example 11.15   Bagging cross-validated two-class classification models
In the set of small panels in the left column of Figure 11.46 we show five different training–validation splits of the prototypical two-class classification dataset first described in Example 11.7, where $\frac{2}{3}$ of the data in each instance is used for training and $\frac{1}{3}$ is used for validation (the boundaries of these points

are colored yellow). Plotted with each split of the original data is the nonlinear decision boundary corresponding to each cross-validated model found via naive cross-validation of the full range of polynomial models of degree 1 to 8. Many of these cross-validated models perform quite well, but some of them (due to the particular training–validation split on which they are based) severely *overfit* the original dataset. By bagging these models using the most popular prediction to assign labels (i.e., the *mode* of these cross-validated model predictions) we produce an appropriate decision boundary for the data shown in the right panel of the figure.

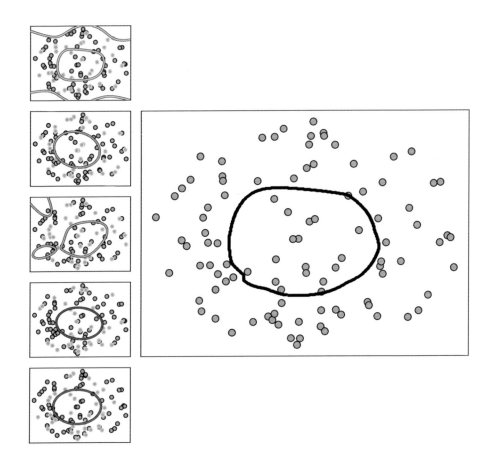

Figure 11.46 Figure associated with Example 11.15. (left column) Five models cross-validated on random training–validation splits of the data, with the validation data in each instance highlighted with a yellow outline. The corresponding nonlinear decision boundary provided by each model is shown in each panel. Some models, due to the split of the data on which they were built, severely overfit. (right column) The original dataset with the decision boundary provided by the bag (i.e., mode) of the five cross-validated models. See text for further details.

In the top-middle panel of Figure 11.47 we illustrate the decision boundaries of five cross-validated models, each built using $B = 20$ neural netwok units trained on different training–validation splits of the dataset shown in the top-left panel of the figure. In each instance $\frac{1}{3}$ of the dataset is randomly chosen as validation (highlighted in yellow). While some of the learned decision boundaries (shown in the top-middle panel) separate the two classes quite well, others do a poorer job. In the top-right panel we show the decision boundary of the bag, created by taking the mode of the predictions from these cross-validated models, which performs quite well.

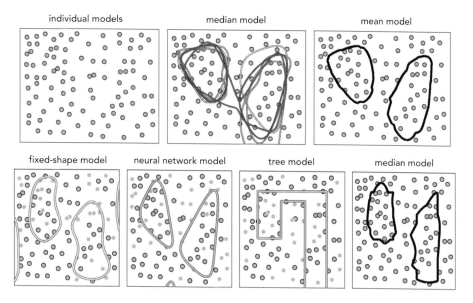

Figure 11.47 Figure associated with Example 11.15. (top-left panel) A toy two-class classification dataset first described in Example 11.7. (top-middle panel) The decision boundaries, each shown in a different color, resulting from five models cross-validated on different training-validation splits of the data. (top-right panel) The decision boundary resulting from the mode (the *modal model*) of the five individual cross-validated models. (bottom row) The decision boundaries provided by a cross-validated fixed-shape polynomial model (left panel), neural network model (second from the left panel), and tree-based model (third from the left panel). In each instance the validation portion of the data is highlighted in yellow. (right panel) The decision boundary provided by the mode of these three models. See text for further details.

As with regression, with classification we can also combine cross-validated models built from different universal approximators. We illustrate this in the bottom row of Figure 11.47 using the same dataset. In particular, we show the result of a cross-validated polynomial model (left panel), a cross-validated neural network model (in the second to the left panel), and a cross-validated tree-based model (second to the right panel). Each cross-validated model uses a different training–validation split of the original data, and the bag (mode) of these models shown in the right panel performs quite well.

Example 11.16   Bagging cross-validated multi-class classification models
In this example we illustrate the bagging of various cross-validated multi-class models on the two different datasets shown in the left column of Figure 11.48. In each case we naively cross-validate polynomial models of degree 1 through 5, with five cross-validated models learned in total. In the middle column of the figure we show the decision boundaries provided by each cross-validated model in distinct colors, while the decision boundary of the final *modal model* is shown in the right column for each dataset.

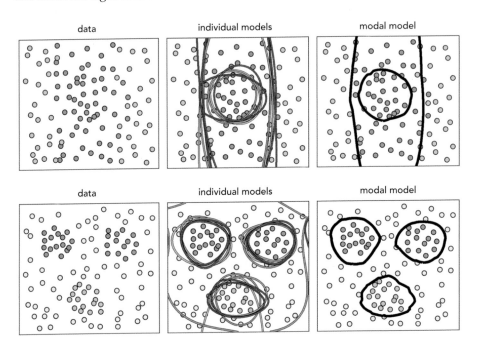

Figure 11.48 Figure associated with Example 11.16. (left column) Two multi-class classification datasets. (middle column) The decision boundaries resulting from five cross-validated models, each shown as a distinct color. (right column) The decision boundary provided by bagging. See text for further details.

11.9.3   How many models should we bag in practice?

Note that in the examples of this section the exact number of cross-validated models bagged were set somewhat arbitrarily. Like other important parameters involved with cross-validation (e.g., the portion of a dataset to reserve for validation) there is no magic number (of cross-validated models) used generally in practice for bagging. Ideally, if we knew that any random validation portion of a dataset generally represented it well, which is often true with very large datasets, there would be less of a need to ensemble multiple cross-validated models where each was trained on a different training–validation split of the original data. Indeed in such instances we could instead bag a range of models trained on a single training–validation split in order to achieve similar improvements over a single model. On the other hand, the less we could trust in the faithfulness of a random validation portion to represents a phenomenon at large, the less we could trust an individual cross-validated model, and hence we might wish to bag more of them to help average our poorly-performing models resulting from bad splits of the data. Often in practice considerations such as computation power and dataset size determine if bagging is used, and if so, how many models are employed in the average.

11.9.4   Ensembling: Bagging versus Boosting

The bagging technique described here wherein we combine a number of different models, each cross-validated independently of the others, is a primary example of what is referred to as *ensembling* in the jargon of machine learning. An *ensembling method* (as the name "ensemble" implies) generally refers to any method of combining different models in a machine learning context. Bagging certainly falls into this general category, as does the general *boosting* approach to cross-validation described in Section 11.5. However, these two ensembling methods are very different from one another.

With boosting we build up a *single* cross-validated model by gradually *adding* together simple models consisting of a single universal approximator unit (see Section 11.5.4). Each of the constituent models involved in boosting are trained in a way that makes each individual model *dependent* on its predecessors (which are trained first). On the other hand, with bagging (as we have seen) we *average* together *multiple cross-validated* models that have been trained *independently* of each other. Indeed any one of those cross-validated models in a bagged ensemble can itself be a boosted model.

11.10   K-Fold Cross-Validation

In this section we detail a twist on the notion of ensembling, called K-fold cross-validation, that is often applied when human interpretability of a final model is of significant importance. While ensembling often provides a better-fitting averaged predictor that avoids the potential pitfalls of any individual cross-validated model, *human interpretability* is typically lost as the final model is an average of many potentially very different nonlinearities.[14] Instead of *averaging* a set of cross-validated models over many splits of the data, each of which provides minimum validation error over a respective split, with K-fold cross-validation we choose a single model that has minimum *average validation*

---

[14] Stumps/tree-based approximators are sometimes an exception to this general rule, as detailed in Section 14.2.

error over all splits of the data. This produces a potentially less accurate final model, but one that is significantly simpler (than an ensembled model) and can be more easily understood by humans. As we will see, in special applications K-fold cross-validation is used with *linear* models as well.

11.10.1   The K-folds cross-validation procedure

K-fold cross-validation is a method for determining robust cross-validated models via an ensembling-like procedure that constrains the complexity of the final model so that it is more human interpretable. Instead of averaging a group of cross-validated models, each of which achieves a minimum validation error over a random training–validation split of the data, with K-fold cross-validation we choose a single final model that achieves the *lowest* average validation error over all of the splits together. By selecting a *single* model to represent the entire dataset, as opposed to an *average* of different models (as is done with ensembling), we make it easier to interpret the selected model.

Of course the desire for any nonlinear model to be interpretable means that its fundamental building blocks (universal approximators of a certain type) need to be interpretable as well. Neural networks, for example, are almost never human interpretable while fixed-shape (most commonly polynomials) and tree-based approximators (commonly stumps) can be interpreted depending on the problem at hand. Thus the latter two types of universal approximators are more commonly employed with the K-fold technique.

To further simplify the final outcome of this procedure, instead of using completely random training–validation splits (as done with ensembling) we split the data randomly into a set of K nonoverlapping pieces. This is depicted visually in Figure 11.49 where the original data is split into $K = 3$ nonoverlapping sets.

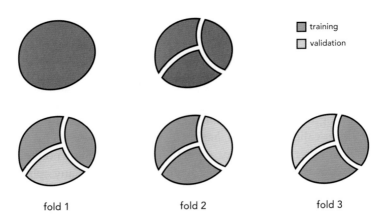

Figure 11.49 Schematic illustration of K-fold cross-validation for $K = 3$.

We then cycle through K training–validation splits of the data that consist of $K - 1$ of these pieces as training, with the final portion as validation, which allows for each point in the dataset to belong to a validation set precisely one time. Each such split is referred to as a *fold*, of which there are K in total, hence the name "K-fold" cross-validation. On each fold we cross-validate the same set of models and record the validation score of each. Afterwards, we choose the single best model that produced the lowest *average validation error*. Once this is done the chosen model is retrained over the entire dataset to provide a final tuned predictor of the data.

Since no models are combined/averaged together with this procedure, it can very easily produce less accurate models (in terms of *testing error* described in Section 11.7) for general learning problems when compared to ensembling. However, when human interpretability of a model overshadows the needs for exceptional performance, K-fold cross-validation produces a stronger-performing model than a single cross-validated model that can still be understood by human beings. This is somewhat analogous to the story of feature selection detailed in Sections 9.6 and 9.7, where human interpretability is the guiding motivator (and not simply prediction power).

Example 11.17   Galileo's gravity experiment

In this example we use K-fold cross-validation on the Galileo dataset detailed in Example 10.2 to recover the quadratic rule that was both engineered there, and that Galileo himself divined from a similar dataset. Since there are only $P = 6$ points in this dataset, intuition suggests that we use a large value for K as described in Section 11.3.4. In this instance we can set K as high as possible, i.e., $K = P$, meaning that each fold will contain only a single data point for validation purposes. This setting of K-fold cross-validation – sometimes referred to as *leave-one-out* cross-validation – is usually employed when the size of data is extremely small.

Here we search over polynomial models of degree 1 through 6, since they are not only easily interpretable, but are appropriate for data gleaned from physical experiments (which often trace out smooth rules). As shown in Figure 11.50, while not all of the models over the six folds fit the data well, the model chosen by K-fold is indeed the quadratic polynomial fit originally proposed by Galileo.

11.10.2   K-fold cross-validation and high-dimensional linear modeling

Suppose for a moment we have a high-capacity model (e.g., a polynomial of degree D where D is very large) which enables several kinds of overfitting behavior for a nonlinear regression dataset, with each overfitting instance of the model provided by different settings of the linear combination weights of the model. We illustrate such a scenario in the left panel of Figure 11.51, where two

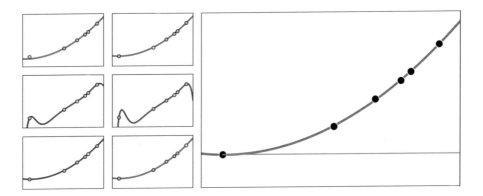

Figure 11.50 Figure associated with Example 11.17. (small panels) Six cross-validated models, each trained on all but one point from the dataset. Here the validation portion of each fold (i.e., a single data point) is highlighted in yellow. (large panel) The model with lowest average validation error is a quadratic. See text for further details.

settings of such a model provide two distinct overfitting predictors for a generic nonlinear regression dataset. As we learned in Section 10.2, any[15] *nonlinear* model in the original space of a regression dataset corresponds to a *linear* model in the transformed feature space (i.e., the space where each individual input axis is given by one of the chosen nonlinear features). Since our model easily overfits the original data, in the transformed feature space our data lies along a *linear subspace* that can be perfectly fit using many different hyperplanes. Indeed the two nonlinear overfitting models shown in the left panel of the figure correspond one-to-one with the two linear fits in the transformed feature space – illustrated symbolically[16] in the right panel of the figure.

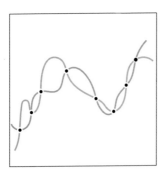
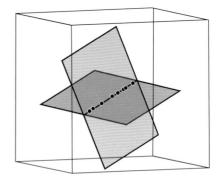

Figure 11.51 (left panel) Two instances of overfitting by a high-capacity model to a nonlinear regression dataset. (right panel) These two models – as viewed in the *feature transformed space* – are linear. See text for further details.

[15] Suppose any parameters internal to the features (if they exist) are fixed.
[16] In reality we could not visualize this space, as it would be too high-dimensional.

The general scenario depicted in the right panel of Figure 11.51 is precisely where we begin when faced with small datasets that have very high input dimension: in such scenarios even a linear model has extremely high capacity and can easily overfit, virtually ruling out the use of more complicated nonlinear models. Thus in such scenarios, in order to properly tune the parameters of a (high-capacity) linear model we often turn to *regularization* to block capacity in high-capacity models (as described in Section 11.6.3). Given the small amount of data at play to determine the best setting of the regularization parameter, K-fold cross-validation is commonly employed to determine the proper regluarization parameter value and ultimately the parameters of the linear model.

This scenario provides an interesting point of intersection with the notion of *feature selection via regularization* detailed in Section 9.7. Employing the ℓ_1 regularizer we can block the capacity of our high-capacity linear model while *simultaneously* selecting important input features, facilitating human interpretability of the learned model.

Example 11.18   Genome-wide association studies

Genome-wide association studies (GWAS) aim at understanding the connections between tens of thousands of genetic markers (input features), taken from across the human genome of several subjects, with medical conditions such as high blood pressure, high cholesterol, heart disease, diabetes, various forms of cancer, and many others (see Figure 11.52). These studies typically involve a relatively small number of patients with a given affliction (as compared to the very large dimension of the input). As a result, regularization based cross-validation is a useful tool for learning meaningful (linear) models for such data. Moreover, using a (sparsity-inducing) regularizer like the ℓ_1 norm can help researchers identify the handful of genes critical to the affliction under study, which can both improve our understanding of it and perhaps provoke development of gene-targeted therapies. See Exercise 11.10 for further details.

Example 11.19   fMRI studies

Neuroscientists believe that only a small number of active brain regions are involved in performing any given cognitive task. Therefore limiting the number of input features allowed in the classification model, via ℓ_1 *regularized feature selection*, is commonly done in order to produce high-performing and human-interpretable results. Figure 1.12 illustrates the result of applying a classification model with sparse feature selection to the problem of diagnosing patients with ADHD. The sparsely distributed regions of color represent activation areas uncovered by the learning algorithm that significantly distinguish between individuals with and without ADHD.

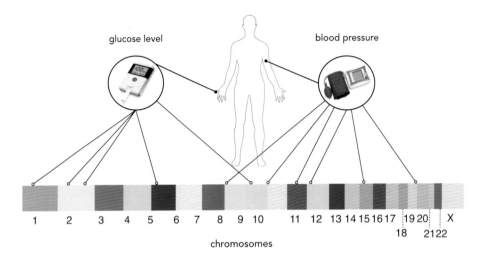

Figure 11.52 Conceptual illustration of a genome-wide association study wherein a quantitative biological trait (e.g., blood pressure or glucose level) is to be associated with specific genomic locations.

11.11 When Feature Learning Fails

Supposing we implement the tools outlined in this chapter correctly, when does cross-validation, ensembling, and (more broadly) feature learning fail? The simple answer is, feature learning fails when our data fails to sufficiently reflect the underlying phenomenon that generated it. Nonlinear feature engineering (outlined in Chapter 10) also fails in such circumstances. This can happen when one or more of the following occur.

- **When a dataset has no inherent structure:** if there is little or no relationship present in the data (due to improper measurement, experimentation, or selection of inputs) the nonlinear model learned via feature learning will be useless. For example, in the left panel of Figure 11.53 we show a small two-class dataset formed by randomly choosing points on the unit square and randomly assigning each point one of two class labels. No classification boundary learned from this dataset can ever yield value, as the data itself contains no meaningful pattern.

- **When a dataset is too small:** when a dataset is too small to represent the true underlying phenomenon feature learning can inadvertently determine an incorrect nonlinearity. For example, in the middle panel of Figure 11.53 we show a simple example of this occurrence. The phenomenon underlying this two-class dataset has a nonlinear boundary (shown in dashed black). However, because we have sampled too few points, the data we do have is linearly separable and cross-validation will recover a linear boundary (shown in solid black) that does not reflect the true nature of the underlying phenomenon. Because of the small data size this problem is unavoidable.

- **When a dataset is poorly distributed:** even if a dataset is large it can still fail to reflect the true nature of the underlying phenomenon that generated it. When this happens feature learning will fail. For example, in the right panel of Figure 11.53 we show a simple two-class dataset whose two classes are separated by a perfectly circular boundary shown in dashed black. While the dataset is relatively large, the data samples have all been taken from the top portion of the input space. Viewed on its own this reasonably large dataset does not represent the true underlying phenomenon very well. While cross-validation produces a model that perfectly separates the two classes, the corresponding parabolic decision boundary (shown in solid black) does not match the true circular boundary. Such a circumstance is bound to happen when data is poorly distributed and fails to reflect the phenomenon that generated it.

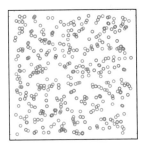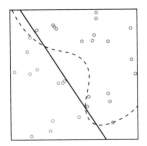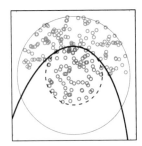

Figure 11.53 Feature learning fails when our data fails to sufficiently reflect the underlying phenomenon that generated it. This can occur when a dataset is poorly structured, too small, or poorly distributed like the data shown in the left, middle, and right panels, respectively. See text for further details.

11.12 Conclusion

This chapter outlined a range of fundamental and extremely important concepts for *feature learning*, or automatic feature engineering, that will echo repeatedly throughout the remainder of the text.

Foremost among these principles is the notion of *universal approximators* introduced in Section 11.2, which are the analog of spanning sets (from vector algebra) for the case of *perfect data*. Here we learned about basic instances of three fundamental families of universal approximators – *fixed-shape*, *neural networks*, and *trees* – each of which has unique attributes and technical eccentricities that are explored in great detail in the chapters following this one. Unlike the case of perfect data, when dealing with *real data* great care must be taken in properly setting the *capacity* of a model built with units of a universal approximator, and in *optimizing* its parameters appropriately (see Section 11.3). These two "dials" (as they are described in Section 11.3.2) constitute the two main controls we have when properly applying any universal approximator based model to real data. Indeed feature learning is in essence the appropriate setting of the capacity and optimization dials (automatically) via the methods of *cross-validation* detailed in Sections 11.4–11.6. Here we saw that it is easier to fix *capacity* at a high level and carefully *optimize* than vice versa, leading to the *boosting* and *regularization* procedures outlined in Sections 11.5 and 11.6, respectively. Finally, *bagging* – the careful combination of a collection of trained/cross-validated models – was described in Section 11.9 (along with the analogous K-fold cross-validation scheme for smaller, higher-dimensional datasets in Section 11.10), which generally leads to better-preforming (bagged) models.

11.13   Exercises

† The data required to complete the following exercises can be downloaded from the text's github repository at `github.com/jermwatt/machine_learning_refined`

11.1   Naive cross-validation I
Repeat the experiment described in Example 11.8, splitting the original dataset at random into training and validation portions. You need not reproduce the panels in Figure 11.27, but make a plot showing the training and validation errors for your range of models tested, and visualize the model you find (along with the data) that provides the lowest validation error. Given your particular training–validation split your results may be different than those presented in the example.

11.2   Naive cross-validation II
Repeat the experiment described in Example 11.9, splitting the original dataset at random into training and validation portions. You need not re-produce the panels in Figure 11.28, but make a plot showing the training and validation errors for your range of models tested. Given your particular training–validation split your results may be different than those presented in the example.

11.3   Boosting based cross-validation I
Repeat the experiment described in Example 11.11. You need not reproduce the panels in Figure 11.32, but make a plot showing the training and validation errors for your range of models tested.

11.4   **Boosting based cross-validation II**

Perform 20 rounds of boosting based cross-validation using neural network units (defined in Equation (11.12)), employing the breast cancer dataset discussed in Exercise 9.5, and randomly splitting the original dataset into 80 percent training and 20 percent validation.

11.5   **Regularization based cross-validation**

Repeat the experiment described in Example 11.13. You need not reproduce the panels in Figure 11.40, but make a plot showing the training and validation errors for your range of models tested.

11.6   **Bagging regression models**

Repeat the first experiment outlined in Example 11.14, producing ten naively cross-validated polynomial models to fit different training–validation splits of the regression dataset shown in Figure 11.44. Produce a set of plots like the ones shown in Figure 11.44 that show how each individual model fits to the data, as well as how the bagged median model fits.

11.7   **Bagging two-class classification models**

Repeat the first experiment outlined in Example 11.15, producing five naively cross-validated polynomial models to fit different training–validation splits of the two-class classification dataset shown in Figure 11.46. Compare the efficacy – in terms of number of misclassifications over the entire dataset – of each individual model and the final bagged model.

11.8   **Bagging multi-class classification models**

Repeat the second experiment outlined in Example 11.16, whose results are shown in the bottom row of Figure 11.48. Compare the efficacy – in terms of number of misclassifications over the entire dataset – of each individual model and the final bagged model.

11.9   **K-fold cross-validation**

Repeat the experiment outlined in Example 11.17, reproducing the plots shown in Figure 11.50.

11.10   **Classification of diabetes**

Perform K-fold cross-validation using a linear model and the ℓ_1 regularizer over a popular two-class classification genomics dataset consisting of $P = 72$ datapoints, each of which has input dimension $N = 7128$. This will tend to produce a sparse predictive linear model – as detailed in Example 11.18 – which is helpful in determining a small number of genes that correlate with the output of this two-class classification dataset (which is whether each individual represented in the dataset has diabetes or not).

12 Kernel Methods

12.1 Introduction

In this chapter we continue our discussion of fixed-shape universal approximators, which began back in Section 11.2.3. This will very quickly lead to the notion of *kernelization* as a clever way of representing fixed-shape features so that they scale more gracefully when applied to vector-valued input.

12.2 Fixed-Shape Universal Approximators

Using the classic polynomial as our exemplar, in Section 11.2.3 we introduced the family of fixed-shape universal approximators as collections of various non-linear functions which have no internal (tunable) parameters. In this section we pick up the discussion of fixed-shape approximators, beginning a deeper dive into the technicalities associated with these universal approximators and challenges we have to address when employing them in practice.

12.2.1 Trigonometric universal approximators

What generally characterizes fixed-shape universal approximators are their lack of internal parameters and straightforward organization, with very often the units of a particular family of fixed-shape approximators being organized in terms of degree or some other natural index (see Section 11.2.3). These simple characteristics have made fixed-shape approximators, such as the polynomials (which we have seen previously) as well as the sinusoidal and Fourier examples (which we discuss now) extremely popular in areas adjacent to machine learning, e.g., mathematics, physics, and engineering.

Example 12.1   Sinusoidal approximators
The collection of sine waves of increasing frequency is a common example of a classic fixed-shape approximator, with units of the form

$$f_1(x) = \sin(x), \quad f_2(x) = \sin(2x), \quad f_3(x) = \sin(3x), \text{ etc.,} \tag{12.1}$$

where the mth element in general is given as $f_m(x) = \sin(mx)$. The first four members of this family of functions are plotted in Figure 12.1.

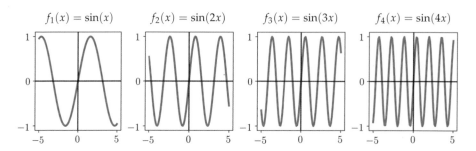

$$f_1(x) = \sin(x) \qquad f_2(x) = \sin(2x) \qquad f_3(x) = \sin(3x) \qquad f_4(x) = \sin(4x)$$

Figure 12.1 Figure associated with Example 12.1. From left to right, the first four units of a sinusoidal universal approximator. See text for further details.

As with polynomials, notice how each of these elements has no *tunable* parameter inside (and thus a *fixed* shape), and that the elements are naturally ordered in terms of their complexity from low to high. Also as with polynomials, we can easily generalize this catalog of functions to higher-dimensional input. In general, for N-dimensional input a sinusoidal unit takes the form

$$f_m(\mathbf{x}) = \sin(m_1 x_1)\sin(m_2 x_2)\cdots\sin(m_N x_N) \tag{12.2}$$

where $m_1, m_2, \ldots, m_N$ are nonnegative integers.

Example 12.2   The Fourier basis
Similarly to the sinusoidal family described in Example 12.1, the Fourier basis [56] – so named after its inventor Joseph Fourier who first used these functions in the early 1800s to study heat diffusion – consists of the set of paired sine and cosine waves (with ever-increasing frequency) of the form

$$f_{2m-1}(x) = \sin(2\pi mx) \quad \text{and} \quad f_{2m}(x) = \cos(2\pi mx) \tag{12.3}$$

for all $m \geq 1$. It is also common to write the Fourier units in the compact complex exponential form (see Exercise 12.1)

$$f_m(x) = e^{2\pi i mx}. \tag{12.4}$$

For a general N-dimensional input each multi-dimensional Fourier unit then takes the form

$$f_m(\mathbf{x}) = e^{2\pi i m_1 x_1} e^{2\pi i m_2 x_2} \cdots e^{2\pi i m_N x_N} \tag{12.5}$$

where $m_1, m_2, \ldots, m_N$ are integers.

12.2.2   The scaling challenge of fixed-shape approximators with large input dimension

As we saw in Chapter 11, when using polynomial units very often we use complete packages of monomials as a polynomial of a certain degree. For example, a polynomial model of degree D when the input is two-dimensional consists of all monomial units of the form $f_m(\mathbf{x}) = x_1^p x_2^q$ where p and q are nonnegative integers such that $0 < p + q \leq D$.

More generally with N-dimensional input a polynomial unit takes the form

$$f_m(\mathbf{x}) = x_1^{m_1} x_2^{m_2} \cdots x_N^{m_N} \tag{12.6}$$

and to construct a polynomial model of degree D we collect all such terms where $0 < m_1 + m_2 + \cdots + m_N \leq D$ and $m_1, m_2, ..., m_N$ are nonnegative integers. Unless used in tandem with boosting (see Section 11.3) we virtually always use polynomial units as a *complete package* of units of a certain degree, not *individually*. One reason for doing this is that since the polynomial units are naturally ordered (from low to high complexity), when including a unit of a particular complexity it makes sense, organizationally speaking, to include all other units in the family having lesser complexities. For instance, it usually does not make much sense to define quadratic polynomials free of linear terms. Nonetheless, packaging polynomial units in this way is not something that we *must* do when employing them, but is a sensible and common practice.

Like polynomials, it is also common when employing sinusoidal and Fourier units to use them as complete packages, since they too are ordered in terms of their individual complexities from low to high. For example, in analogy to a degree-D polynomial, we can package a degree-D Fourier model consisting of *all units* of the form given in Equation (12.5) where $0 < \max(|m_1|, |m_2|, \ldots, |m_N|) \leq D$. This choice of packaging is largely a convention.

However, a very serious practical issue presents itself when employing fixed-shape approximators like polynomials and trigonometric bases, when used in complete packages of units: even with a moderate-sized input dimension N the corresponding number of units in the package M grows rapidly with N, quickly becoming prohibitively large in terms of storage and computation. In other words, the number of units of a typical fixed-shape approximator in a model employing a complete package of such units *grows exponentially* with the dimension of input.

Example 12.3   Number of units in a degree-D polynomial approximator
The precise number M of units in a degree-D polynomial of an input with dimension N can be computed precisely as

$$M = \binom{N + D}{D} - 1 = \frac{(N + D)!}{N!D!} - 1. \tag{12.7}$$

Even if the input dimension N is of small to moderate size, e.g., $N = 100$ or $N = 1000$, then just the associated degree $D = 5$ polynomial feature map of these input dimensions has $M = 96,560,645$ and $M = 8,459,043,543,950$ monomial terms, respectively. In the latter case we cannot even hold the feature vectors in memory on a modern computer.

Example 12.4   Number of units in a degree-D Fourier approximator
The corresponding number of units M in a package of degree-D Fourier basis elements is even more gargantuan than that of a degree-D polynomial: the degree-D Fourier feature collection of arbitrary input dimension N has precisely

$$M = (2D + 1)^N - 1 \tag{12.8}$$

units. When $D = 5$ and $N = 80$ this is $11^{80} - 1$, a number larger than current estimates of the number of atoms in the visible universe!

Our cursory analyses in Examples 12.3 and 12.4 indicate that since the total number of units of a fixed-shape approximator grows *exponentially* in the input dimension, any approach to selecting fixed-shape units for a nonlinear model is problematic in general. For example, with polynomials, even if we chose a smaller set of just those units with the exact same degree, i.e., all units where $m_1 + m_2 + \cdots + m_N = D$, we would still end up with a combinatorially large number of units to employ.

This serious scaling issue motivates the so-called *kernel trick* described in the next section, that extends the use of classic fixed-shape approximators (when employing complete packages of units) to problems with high-dimensional input.

12.3   The Kernel Trick

This crucial issue, of not being able to effectively store and compute with high-dimensional fixed-shape feature transformations, motivates the search for more efficient representations. In this section we introduce the notion of *kernelization*, also commonly called *the kernel trick*, as a clever way of constructing fixed-shape features for virtually any machine learning problem. Kernelization not only allows us to avoid the combinatorial explosion problem detailed at the end of the previous section, but also provides a way of generating new fixed-shape features defined solely through such a kernelized representation.

12.3.1   A useful fact from the fundamental theorem of linear algebra

Before discussing the concept of kernelization, it will be helpful to first recall a useful proposition from the fundamental theorem of linear algebra about decomposition of any M-dimensional vector ω over the columns of a given $M \times P$ matrix $\mathbf{F}$. Denoting the pth column of $\mathbf{F}$ as $\mathbf{f}_p$, in the case where ω happens to lie *inside* the column space of $\mathbf{F}$, we can express it via a linear combination of these columns as

$$\omega = \sum_{p=1}^{P} \mathbf{f}_p \, z_p \tag{12.9}$$

where z_p is the linear combination weight or coefficient associated with $\mathbf{f}_p$. By stacking these weights into a $P \times 1$ column vector $\mathbf{z}$, we can write this relationship more compactly as

$$\omega = \mathbf{Fz}. \tag{12.10}$$

If, on the other hand, ω happens to lie *outside* the column space of $\mathbf{F}$, as illustrated pictorially in Figure 12.2, we can decompose it into two pieces – the portion of ω belonging to the subspace spanned by the columns of $\mathbf{F}$, and an orthogonal component $\mathbf{r}$ – and write it as

$$\omega = \mathbf{Fz} + \mathbf{r}. \tag{12.11}$$

Note that $\mathbf{r}$ being orthogonal to the span of columns in $\mathbf{F}$ means algebraically that $\mathbf{F}^T \mathbf{r} = \mathbf{0}_{P \times 1}$. Moreover when ω is *in* the column space of $\mathbf{F}$, we can still decompose it using the more general form given in Equation (12.11) by setting $\mathbf{r} = \mathbf{0}_{M \times 1}$ without violating the orthogonality condition $\mathbf{F}^T \mathbf{r} = \mathbf{0}_{P \times 1}$.

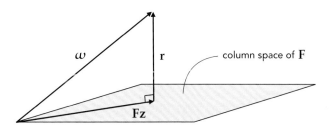

Figure 12.2 An illustration of a useful fact from the fundamental theorem of linear algebra which states that any vector ω in an M-dimensional space can be decomposed as $\omega = \mathbf{Fz} + \mathbf{r}$ where the vector $\mathbf{Fz}$ belongs in the column space of the matrix $\mathbf{F}$, and $\mathbf{r}$ is orthogonal to this subspace.

In sum, any vector ω in an M-dimensional space can be decomposed over the column space of a given matrix $\mathbf{F}$ as $\omega = \mathbf{Fz} + \mathbf{r}$. The vector $\mathbf{Fz}$ belongs in the subspace determined by the columns of $\mathbf{F}$, while $\mathbf{r}$ is orthogonal to this subspace. As we will now see this simple decomposition is the key to representing fixed-shape features more effectively.

12.3.2   Kernelizing machine learning cost functions

Here we provide several fundamental examples of how to kernelize standard supervised machine learning problems and their cost functions, including the Least Squares cost for regression and the Softmax cost for two-class classification. Virtually all machine learning cost functions can be kernelized following arguments similar to these, including the multi-class Softmax, Principal Component Analysis, and K-means clustering (see chapter's exercises).

Example 12.5   Kernelizing regression via the Least Squares cost

Suppose we want to perform a generic nonlinear regression using our M units belonging to a degree-D fixed-shape approximator, with our corresponding model evaluated at the pth input $\mathbf{x}_p$ taking the form

$$\text{model}\left(\mathbf{x}_p, \mathbf{w}\right) = w_0 + f_1\left(\mathbf{x}_p\right)w_1 + f_2\left(\mathbf{x}_p\right)w_2 + \cdots + f_M\left(\mathbf{x}_p\right)w_M. \tag{12.12}$$

For convenience we will write this more compactly, exposing the feature-touching weights and the bias separately, as

$$\text{model}\left(\mathbf{x}_p, b, \boldsymbol{\omega}\right) = b + \mathbf{f}_p^T \boldsymbol{\omega} \tag{12.13}$$

where we have used the bias/feature-touching weight notation (previously introduced in, e.g., Section 6.4.5)

$$b = w_0 \quad \text{and} \quad \boldsymbol{\omega} = \begin{bmatrix} w_1 \\ w_2 \\ \vdots \\ w_M \end{bmatrix} \tag{12.14}$$

as well as a shorthand for our entire set of M feature transformations of the training input $\mathbf{x}_p$ as

$$\mathbf{f}_p = \begin{bmatrix} f_1(\mathbf{x}_p) \\ f_2(\mathbf{x}_p) \\ \vdots \\ f_M(\mathbf{x}_p) \end{bmatrix}. \tag{12.15}$$

In this notation our Least Squares cost for regression takes the form

$$g(b, \omega) = \frac{1}{P} \sum_{p=1}^{P} \left(b + \mathbf{f}_p^T \omega - y_p \right)^2. \qquad (12.16)$$

Now, denote by $\mathbf{F}$ the $M \times P$ matrix formed by stacking the vectors $\mathbf{f}_p$ column-wise. Employing the fundamental theorem of linear algebra discussed in the previous section, we may write ω as

$$\omega = \mathbf{F}\mathbf{z} + \mathbf{r} \qquad (12.17)$$

where $\mathbf{r}$ satisfies $\mathbf{F}^T\mathbf{r} = \mathbf{0}_{P\times 1}$. Substituting this representation of ω back into the cost function in Equation (12.16) gives

$$\frac{1}{P} \sum_{p=1}^{P} \left(b + \mathbf{f}_p^T (\mathbf{F}\mathbf{z} + \mathbf{r}) - y_p \right)^2 = \frac{1}{P} \sum_{p=1}^{P} \left(b + \mathbf{f}_p^T \mathbf{F}\mathbf{z} - y_p \right)^2. \qquad (12.18)$$

Finally, denoting the symmetric $P \times P$ matrix $\mathbf{H} = \mathbf{F}^T\mathbf{F}$ (and its pth column $\mathbf{h}_p = \mathbf{F}^T\mathbf{f}_p$), referred to as a *kernel matrix* or just a *kernel* for short, our original cost function can be expressed equivalently as

$$g(b, \mathbf{z}) = \frac{1}{P} \sum_{p=1}^{P} \left(b + \mathbf{h}_p^T \mathbf{z} - y_p \right)^2 \qquad (12.19)$$

with our corresponding model evaluated at the pth input now taking the equivalent form

$$\text{model}\left(\mathbf{x}_p, b, \mathbf{z} \right) = b + \mathbf{h}_p^T \mathbf{z}. \qquad (12.20)$$

Note that in kernelizing the original regression model in Equation (12.13) and its associated cost function in Equation (12.16) we have changed their arguments (due to our substitution of ω), arriving at completely equivalent *kernelized* model in Equation (12.20) and *kernelized* cost function in Equation (12.19).

Example 12.6   Kernelizing two-class classification via the Softmax cost
Following the pattern shown in Example 12.5, here we essentially repeat the same argument employing the two-class Softmax cost.

Writing our generic two-class Softmax cost using the same notation as employed in Example 12.5 we have

$$g(b, \omega) = \frac{1}{P} \sum_{p=1}^{P} \log \left(1 + e^{-y_p \left(b + \mathbf{f}_p^T \omega \right)} \right). \qquad (12.21)$$

We then write the representation of ω over $\mathbf{F}$ as $\omega = \mathbf{F}\mathbf{z} + \mathbf{r}$ where $\mathbf{F}^T\mathbf{r} = \mathbf{0}_{P\times 1}$. Making this substitution into Equation (12.21) and simplifying gives

$$g\left(b, \mathbf{z}\right) = \frac{1}{P}\sum_{p=1}^{P} \log\left(1 + e^{-y_p\left(b + \mathbf{f}_p^T \mathbf{F}\mathbf{z}\right)}\right). \tag{12.22}$$

Denoting the $P \times P$ kernel matrix $\mathbf{H} = \mathbf{F}^T\mathbf{F}$ (where $\mathbf{h}_p = \mathbf{F}^T\mathbf{f}_p$ is the pth column of $\mathbf{H}$) we can then write the cost function in Equation (12.22) in *kernelized* form as

$$g\left(b, \mathbf{z}\right) = \frac{1}{P}\sum_{p=1}^{P} \log\left(1 + e^{-y_p\left(b + \mathbf{h}_p^T \mathbf{z}\right)}\right). \tag{12.23}$$

This kernelized form of the two-class Softmax is often referred to as *kernelized logistic regression*.

Using the same sort of argument given in Examples 12.5 and 12.6 we may kernelize virtually any machine learning problem discussed in this text including multi-class classification, Principal Component Analysis, K-means clustering, as well as any ℓ_2 regularized version of these models. For easy reference, we show both the original and kernelized forms of popular supervised learning cost functions in Table 12.1.

Table 12.1 Popular supervised learning cost functions and their kernelized versions.

Cost function	Original version	Kernelized version
Least Squares	$\frac{1}{P}\sum_{p=1}^{P}\left(b + \mathbf{f}_p^T\boldsymbol{\omega} - y_p\right)^2$	$\frac{1}{P}\sum_{p=1}^{P}\left(b + \mathbf{h}_p^T\mathbf{z} - y_p\right)^2$
Two-class Softmax	$\frac{1}{P}\sum_{p=1}^{P}\log\left(1 + e^{-y_p\left(b + \mathbf{f}_p^T\boldsymbol{\omega}\right)}\right)$	$\frac{1}{P}\sum_{p=1}^{P}\log\left(1 + e^{-y_p\left(b + \mathbf{h}_p^T\mathbf{z}\right)}\right)$
Squared-margin SVM	$\frac{1}{P}\sum_{p=1}^{P}\max^2\left(0, 1 - y_p\left(b + \mathbf{f}_p^T\boldsymbol{\omega}\right)\right)$	$\frac{1}{P}\sum_{p=1}^{P}\max^2\left(0, 1 - y_p\left(b + \mathbf{h}_p^T\mathbf{z}\right)\right)$
Multi-class Softmax	$\frac{1}{P}\sum_{p=1}^{P}\log\left(1 + \sum_{\substack{j=0 \\ j\neq y_p}}^{C-1} e^{\left(b_j - b_{y_p}\right) + \mathbf{f}_p^T\left(\boldsymbol{\omega}_j - \boldsymbol{\omega}_{y_p}\right)}\right)$	$\frac{1}{P}\sum_{p=1}^{P}\log\left(1 + \sum_{\substack{j=0 \\ j\neq y_p}}^{C-1} e^{\left(b_j - b_{y_p}\right) + \mathbf{h}_p^T\left(\mathbf{z}_j - \mathbf{z}_{y_p}\right)}\right)$
ℓ_2 regularizer [a]	$\lambda \|\boldsymbol{\omega}\|_2^2$	$\lambda\, \mathbf{z}^T\mathbf{H}\mathbf{z}$

[a] The ℓ_2 regularizer can be added to any cost function $g(b, \boldsymbol{\omega})$ in the middle column and the resulting kernelized form of the sum $g(b, \boldsymbol{\omega}) + \lambda \|\boldsymbol{\omega}\|_2^2$ will be the sum of the kernelized cost and the kernelized regularizer, i.e., $g(b, \mathbf{z}) + \lambda\, \mathbf{z}^T\mathbf{H}\mathbf{z}$.

12.3.3   Popular kernels in machine learning

The real value of kernelizing any machine learning cost is that for many fixed-shape units, including polynomial and Fourier features, the kernel matrix $\mathbf{H} = \mathbf{F}^T\mathbf{F}$ may be constructed *without* first building the matrix $\mathbf{F}$ (which often has prohibitively large row dimension). Instead, as we will see through a number of examples, this matrix may be constructed *entry-wise* via simple formulae. Moreover, thinking about constructing kernel matrices in this way leads to the construction of fixed-shape universal approximators starting with the definition of the kernel matrix itself (and not by beginning with an explicit feature transformation). In either case, by constructing the kernel matrix without first computing $\mathbf{F}$ we completely avoid the exponential scaling problem with fixed-shape universal approximators discussed in Section 12.2.2.

Example 12.7   The polynomial kernel

Consider the following degree $D = 2$ polynomial mapping from $N = 2$ to $M = 5$ dimensional space given by the feature transformation vector

$$\mathbf{f} = \begin{bmatrix} x_1 \\ x_2 \\ x_1^2 \\ x_1 x_2 \\ x_2^2 \end{bmatrix}. \tag{12.24}$$

Note that multiplying some or all of the entries in $\mathbf{f}$ by a constant value like $\sqrt{2}$, as in

$$\mathbf{f} = \begin{bmatrix} \sqrt{2}\,x_1 \\ \sqrt{2}\,x_2 \\ x_1^2 \\ \sqrt{2}\,x_1 x_2 \\ x_2^2 \end{bmatrix} \tag{12.25}$$

does not change this feature transformation for our modeling purposes, since the $\sqrt{2}$ attached to several of the terms can be absorbed by their associated weights in ω when forming model $(\mathbf{x}, b, \omega) = b + \mathbf{f}^T \omega$. Denoting briefly by $\mathbf{u} = \mathbf{x}_i$ and $\mathbf{v} = \mathbf{x}_j$ the ith and jth input data points, respectively, the (i, j)th element of the kernel matrix $\mathbf{H} = \mathbf{F}^T\mathbf{F}$ for a degree $D = 2$ polynomial is written as

$$h_{i,j} = \mathbf{f}_i^T \mathbf{f}_j$$

$$= \begin{bmatrix} \sqrt{2}\,u_1 & \sqrt{2}\,u_2 & u_1^2 & \sqrt{2}\,u_1u_2 & u_2^2 \end{bmatrix} \begin{bmatrix} \sqrt{2}\,v_1 \\ \sqrt{2}\,v_2 \\ v_1^2 \\ \sqrt{2}\,v_1v_2 \\ v_2^2 \end{bmatrix} \tag{12.26}$$

$$= \left(1 + 2\,u_1v_1 + 2\,u_2v_2 + u_1^2v_1^2 + 2\,u_1u_2v_1v_2 + u_2^2v_2^2\right) - 1$$

$$= (1 + u_1v_1 + u_2v_2)^2 - 1$$

$$= \left(1 + \mathbf{u}^T\mathbf{v}\right)^2 - 1.$$

In other words, the kernel matrix $\mathbf{H}$ may be built without first constructing the explicit features in Equation (12.25), by simply defining it entry-wise as

$$h_{i,j} = \left(1 + \mathbf{x}_i^T\mathbf{x}_j\right)^2 - 1. \tag{12.27}$$

This way of defining the polynomial kernel matrix is very useful since we only require access to the original input data, not the explicit polynomial features themselves.

Although the kernel construction rule in Equation (12.27) was derived specifically for $N = 2$ and a degree $D = 2$ polynomial, one can show that a polynomial kernel can be defined entry-wise, in a similar manner, for general N and D as

$$h_{i,j} = \left(1 + \mathbf{x}_i^T\mathbf{x}_j\right)^D - 1. \tag{12.28}$$

Example 12.8   The Fourier kernel
A degree-D Fourier feature transformation from $N = 1$ to $M = 2D$ dimensional space may be written as the $2D \times 1$ feature vector

$$\mathbf{f} = \begin{bmatrix} \sqrt{2}\cos{(2\pi x)} \\ \sqrt{2}\sin{(2\pi x)} \\ \vdots \\ \sqrt{2}\cos{(2\pi Dx)} \\ \sqrt{2}\sin{(2\pi Dx)} \end{bmatrix} \tag{12.29}$$

where, as explained in the previous example, the multiplication of its entries by $\sqrt{2}$ does not alter the original transformation defined in Equation (12.3) for our modeling purposes. In this case the corresponding (i, j)th element of the kernel matrix $\mathbf{H} = \mathbf{F}^T\mathbf{F}$ can be written as

$$h_{i,j} = \mathbf{f}_i^T \mathbf{f}_j = \sum_{m=1}^{D} 2 \left[\cos\left(2\pi m x_i\right) \cos\left(2\pi m x_j\right) + \sin\left(2\pi m x_i\right) \sin\left(2\pi m x_j\right) \right]. \quad (12.30)$$

Using the simple trigonometric identity $\cos(\alpha)\cos(\beta) + \sin(\alpha)\sin(\beta) = \cos(\alpha - \beta)$, this may be written equivalently as

$$h_{i,j} = \sum_{m=1}^{D} 2\cos\left(2\pi m(x_i - x_j)\right). \quad (12.31)$$

Employing the complex definition of cosine, i.e., $\cos(\alpha) = \frac{e^{i\alpha} + e^{-i\alpha}}{2}$, we can rewrite this as

$$h_{i,j} = \sum_{m=1}^{D} \left[e^{2\pi i m(x_i - x_j)} + e^{-2\pi i m(x_i - x_j)} \right] = \left[\sum_{m=-D}^{D} e^{2\pi i m(x_i - x_j)} \right] - 1. \quad (12.32)$$

If $x_i - x_j$ is an integer then $e^{2\pi i m(x_i - x_j)} = 1$, and the summation expression inside brackets in Equation (12.32) sums to $2D + 1$. Supposing this is not the case, examining the summation alone we may write

$$\sum_{m=-D}^{D} e^{2\pi i m(x_i - x_j)} = e^{-2\pi i D(x_i - x_j)} \sum_{m=0}^{2D} e^{2\pi i m(x_i - x_j)}. \quad (12.33)$$

Noticing that the sum on the right-hand side is a geometric series, we can further simplify the above as

$$e^{-2\pi i D(x_i - x_j)} \frac{1 - e^{2\pi i(x_i - x_j)(2D+1)}}{1 - e^{2\pi i(x_i - x_j)}} = \frac{\sin\left((2D+1)\pi\left(x_i - x_j\right)\right)}{\sin\left(\pi\left(x_i - x_j\right)\right)} \quad (12.34)$$

where the final equality follows from the complex definition of sine, i.e., $\sin(\alpha) = \frac{e^{i\alpha} - e^{-i\alpha}}{2i}$.

Because in the limit, as t approaches any integer value, we have $\frac{\sin((2D+1)\pi t)}{\sin(\pi t)} = 2D + 1$ (which one can show using L'Hospital's rule from basic calculus), we may therefore generally write, in conclusion, that

$$h_{i,j} = \frac{\sin\left((2D+1)\pi\left(x_i - x_j\right)\right)}{\sin\left(\pi\left(x_i - x_j\right)\right)} - 1. \quad (12.35)$$

Similar Fourier kernel derivation can be made for a general N-dimensional input (see Exercise 12.11).

Example 12.9   The Radial Basis Function (RBF) kernel
Another popular choice of kernel is the Radial Basis Function (RBF) kernel defined entry-wise over the input data as

$$h_{i,j} = e^{-\beta \|x_i - x_j\|_2^2} \tag{12.36}$$

where $\beta > 0$ is a *hyperparameter* that must be tuned to the data. While the RBF kernel is typically defined directly as the kernel matrix in Equation (12.36), it can be traced back to an explicit feature transformation as with the polynomial and Fourier kernels. That is, we can find the explicit form of the fixed-shape feature transformation **f** such that

$$h_{i,j} = \mathbf{f}_i^T \mathbf{f}_j \tag{12.37}$$

where $\mathbf{f}_i$ and $\mathbf{f}_j$ are the feature transformations of the input points x_i and x_j, respectively. The RBF feature transformation is different from polynomial and Fourier transformations in that its associated feature vector **f** is *infinite-dimensional*. For example, when $N = 1$ the feature vector **f** takes the form

$$\mathbf{f} = \begin{bmatrix} f_1(x) \\ f_2(x) \\ f_3(x) \\ \vdots \end{bmatrix} \tag{12.38}$$

where the mth entry (or feature) is defined as

$$f_m(x) = e^{-\beta x^2} \sqrt{\frac{(2\beta)^{m-1}}{(m-1)!}} \, x^{m-1} \quad \text{for all } m \geq 1. \tag{12.39}$$

When $N > 1$ the corresponding feature vector takes on an analogous form which is also infinite in length, making it impossible to even construct and store such a feature vector (regardless of the input dimension).

Notice that the shape (and hence fitting behavior) of RBF kernels depends on the setting of their hyperparameter β. In general, the larger β is set the more complex an associated model employing an RBF kernel becomes. To illustrate this, in Figure 12.3 we show three examples of supervised learning: regression (top row), two-class classification (middle row), and multi-class classification (bottom row), using the RBF kernel with three distinct settings of β in each instance. This creates underfitting (left column), reasonable predictive behavior (middle column), and overfitting behavior (right column). In each instance Newton's method was used to minimize each corresponding cost, and consequently tune each model's parameters. In practice β is set via *cross-validation* (see, e.g., Example 12.10).

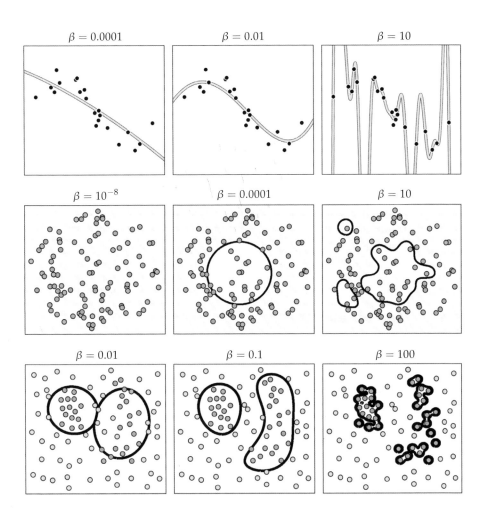

Figure 12.3 Figure associated with Example 12.9. See text for details.

While we have presented some of the most commonly used kernels in practice here, the reader can see, e.g., [57, 58] for a more exhaustive list of kernels and their properties.

12.3.4   Making predictions with kernelized models

As we saw in Examples 12.5 and 12.6, the kernelized form of a general supervised model evaluated at a point **x** takes the form

$$\text{model}\,(\mathbf{x}, b, \mathbf{z}) = b + \mathbf{h}^T \mathbf{z} \tag{12.40}$$

where the parameters b and $\mathbf{z}$ must be tuned by the minimization of an appropriate kernelized cost. In this framework the kernelization $\mathbf{h}$ of the generic input **x** involves evaluation against *every* point $\mathbf{x}_p$ in the (training) dataset. For example with a degree-D polynomial kernel, **h** is given as the P-dimensional vector

$$\mathbf{h} = \begin{bmatrix} \left(1 + \mathbf{x}_1^T \mathbf{x}\right)^D - 1 \\ \left(1 + \mathbf{x}_2^T \mathbf{x}\right)^D - 1 \\ \vdots \\ \left(1 + \mathbf{x}_P^T \mathbf{x}\right)^D - 1 \end{bmatrix}. \tag{12.41}$$

This necessity of employing every (training) data point in evaluating a trained model is virtually unique[1] to kernelized learners, as we will not see this requirement when employing other universal approximators in the chapters to come.

12.4   Kernels as Measures of Similarity

If we look back at the form of the polynomial, Fourier, and RBF kernels in Examples 12.7 through 12.9 we can see that in each instance the (i, j)th entry of the kernel matrix is a function defined on the pair $\left(\mathbf{x}_i, \mathbf{x}_j\right)$ of input data. For example, studying the RBF kernel

$$h_{i,j} = e^{-\beta \|\mathbf{x}_i - \mathbf{x}_j\|_2^2} \tag{12.42}$$

we can see that, as a function of $\mathbf{x}_i$ and $\mathbf{x}_j$, it measures the similarity between these two inputs via the ℓ_2 norm of their difference. The more similar $\mathbf{x}_i$ and $\mathbf{x}_j$ are in the input space the larger $h_{i,j}$ becomes, and vice versa. In other words, the RBF kernel can be interpreted as a *similarity measure* that describes how closely two inputs resemble each other. This interpretation of kernels as similarity measures also applies to other previously introduced kernels including the polynomial and Fourier kernels, even though these kernels clearly encode similarity in different ways.

In Figure 12.4 we visualize our three exemplar kernels (polynomial, Fourier, and RBF) as similarity measures by fixing $\mathbf{x}_i$ at $\mathbf{x}_i = [0.5 \ 0.5]^T$ and plotting $h_{i,j}$ for a fine range of $\mathbf{x}_j$ values over the unit square $[0, 1]^2$, producing a color-coded surface showing how each kernel treats points near $\mathbf{x}_i$. Analyzing this figure we can obtain a general sense of how these three kernels define *similarity* between points. Firstly, we can see that a polynomial kernel treats data points $\mathbf{x}_i$ and $\mathbf{x}_j$ similarly if their inner product is high or, in other words, they highly correlate with each other. Likewise, the points are treated as dissimilar when they are

---

[1] The evaluation of a K-nearest-neighbors classifier also involves employing the entire training set.

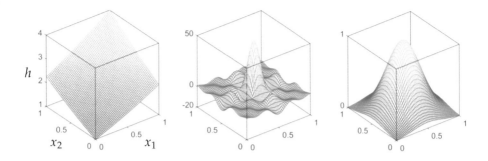

Figure 12.4 Surfaces generated by polynomial, Fourier, and RBF kernels centered at $\mathbf{x}_i = [0.5\ 0.5]^T$. Each surface point is color-coded based on its magnitude, which can be thought of as a measure of similarity between $\mathbf{x}_i$ and its corresponding input. (left panel) A degree $D = 2$ polynomial kernel, (middle panel) degree $D = 3$ Fourier kernel, and (right panel) RBF kernel with $\beta = 10$. See text for further details.

orthogonal to one another. On the other hand, the Fourier kernel treats points as similar if they lie close together, but their similarity differs like a sinc function as their distance from each other grows. Finally, an RBF kernel provides a smooth similarity between points: if they are close to each other in a *Euclidean* sense they are deemed highly similar, but once the distance between them passes a certain threshold they become rapidly dissimilar.

12.5 Optimization of Kernelized Models

As discussed previously, virtually any machine learning model (supervised or unsupervised) can be kernelized. The real value in kernelization is that, for a large range of kernel types, we can actually construct the kernel matrix $\mathbf{H}$ *without* explicitly defining the associated feature transformations. As we have seen this allows us to get around the scaling issue associated with fixed-shape approximators with large input dimension (see Section 12.2.2). Moreover, because the final kernelized model remains linear in its parameters, corresponding kernelized cost functions are quite "nice" in terms of their general geometry. For example, any convex cost function for regression and classification *remains* convex when kernelized, including popular cost functions for regression, two-class, and multi-class classification (detailed in Chapters 5–7). This allows virtually any optimization method to be used to tune a kernelized supervised learner, from zero- to first-order and even powerful second-order approaches like Newton's method (detailed in Chapters 2–4).

However, because a generic kernel matrix $\mathbf{H}$ is a square matrix of size $P \times P$ (where P is the number of data points in the training set) kernelized models inherently scale quadratically (and thus very poorly) in the size of training data. This not only makes training kernelized models extremely challenging on large datasets, but also predictions using such models (which as we saw in Section 12.3.4 require the evaluation of *every* training data point) become increasingly challenging as the size of training data increases.

Most standard ways of dealing with this crippling scaling issue in the size of the training data revolve around avoiding the creation of the entire kernel matrix $\mathbf{H}$ at once, especially during training. For example, one can use first-order methods such as stochastic gradient descent so that only a small number of training data points are dealt with at a time, meaning that only a small subset of columns of $\mathbf{H}$ are ever created concurrently when training. Sometimes the structure of certain problems can be used to avoid explicit kernel construction as well.

12.6   Cross-Validating Kernelized Learners

In general, there is a large difference between the *capacity* of subsequent degrees D and $D + 1$ in models employing polynomial and Fourier kernels. With polynomials for instance, the difference between the number of units encapsulated in a degree-D polynomial kernel and that of a degree-$(D + 1)$ polynomial kernel can be calculated, using Equation (12.7), as

$$\left[\left(\begin{array}{c} N + D + 1 \\ D + 1 \end{array} \right) - 1 \right] - \left[\left(\begin{array}{c} N + D \\ D \end{array} \right) - 1 \right] = \left(\begin{array}{c} N + D \\ D + 1 \end{array} \right). \tag{12.43}$$

When $N = 500$, for example, there are $20,958,500$ more polynomial units encapsulated in a degree $D = 3$ kernel matrix than a degree $D = 2$ kernel matrix. Because of this enormous combinatorial leap in capacity between subsequent degree kernels, cross-validation via *regularization* with the ℓ_2 norm (as detailed in Section 11.4) is common practice when employing polynomial and Fourier kernels. Since the hyperparameter β of the RBF kernel is continuous, models employing an RBF kernel can (in addition to the regularization approach) be cross-validated in principle by comparing various values of β directly.

Example 12.10   Breast cancer classification using an RBF kernel

In this example we use naive cross-validation (as first detailed in Section 11.4.2) to determine an ideal parameter β for an RBF kernel over the breast cancer dataset first described in Exercise 6.13. In this set of experiments we use the Softmax cost and set aside 20 percent of this two-class dataset (randomly) for validation purposes (with the same portion being set aside for validation for each value of β used). We try out a range of 50 evenly spaced values for β on the interval $[0, 1]$, minimize the corresponding cost using Newton's method, and plot the number of misclassifications on both the training (in blue) and validation (in yellow) sets in Figure 12.5. The minimum number of misclassifications on the validation set occured when β was set close to the value 0.2, which resulted in one and five misclassifications on the training and validation sets, respectively. A simple linear classifier trained on the same portion of data provided seven and 22 misclassifications on training and validation sets, respectively.

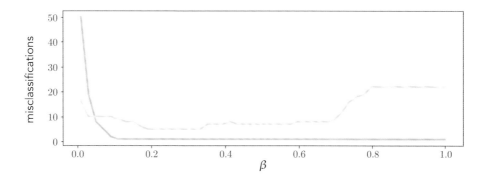

Figure 12.5 Figure associated with Example 12.10. See text for details.

12.7 Conclusion

In this chapter we continued our description of fixed-shape models, continuing from Section 11.2.3 where they were first introduced. We began with a review of several popular examples of fixed-shape universal approximators in Section 12.2. The fact that these universal approximators scale extremely poorly with the input dimension of a dataset, as described in Section 12.2.2, naturally led us to discuss their extension as *kernels* in Section 12.3. Using the "kernel trick" we can not only extend popular fixed-shape approximators to more easily deal with high-dimensional input, but can create a range of new approximators directly as kernels. However, while *kernelizing* a fixed-shape approximator helps it overcome scaling issues in the input dimension of a dataset, it introduces a scaling issue in the *dataset size*. This issue can be somewhat ameliorated via clever kernel matrix construction and optimization (as outlined in Section 12.5). Lastly, in Section 12.6 we briefly touched on the use of regularizer based cross-validation – which was previously discussed at length in Section 11.6.

12.8 Exercises

† The data required to complete the following exercises can be downloaded from the text's github repository at github.com/jermwatt/machine_learning_refined

12.1   Complex Fourier representation

Verify that using complex exponential definitions of cosine and sine functions, i.e., $\cos(\alpha) = \frac{1}{2}\left(e^{i\alpha} + e^{-i\alpha}\right)$ and $\sin(\alpha) = \frac{1}{2i}\left(e^{i\alpha} - e^{-i\alpha}\right)$, we can write the partial Fourier expansion model

$$\text{model}(x, \mathbf{w}) = w_0 + \sum_{m=1}^{M} [\cos(2\pi mx)\, w_{2m-1} + \sin(2\pi mx)\, w_{2m}] \qquad (12.44)$$

equivalently as

$$\text{model}(x, \mathbf{v}) = \sum_{m=-M}^{M} e^{2\pi\, imx}\, v_m \qquad (12.45)$$

where the complex weights $v_{-M}, ..., v_0, ..., v_M$ are given in terms of the real weights $w_0, w_1, ..., w_{2M}$ as

$$v_m = \begin{cases} \frac{1}{2}\left(w_{2m-1} - i\, w_{2m}\right) & \text{if } m > 0 \\ w_0 & \text{if } m = 0 \\ \frac{1}{2}\left(w_{1-2m} + i\, w_{-2m}\right) & \text{if } m < 0. \end{cases} \qquad (12.46)$$

12.2   Combinatorial explosion in monomials

Confirm that the number of monomial units in a degree-D polynomial grows combinatorially in input dimension as given in Equation (12.7).

12.3   Polynomial kernel regression

Reproduce polynomial kernel fits of degree $D = 1$, $D = 3$, and $D = 12$ to the nonlinear dataset shown in Figure 12.6.

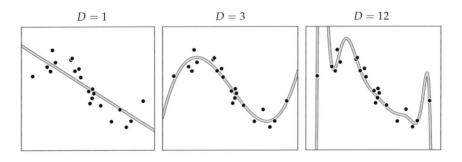

Figure 12.6 Figure associated with Exercise 12.3.

12.4   **Kernelize the ℓ_2 regularized Least Squares cost**

Use the kernelization argument made in Examples 12.5 and 12.6 to kernelize the ℓ_2 regularized Least Squares cost function.

12.5   **Kernelize the multi-class Softmax cost**

Use the kernelization argument made in Examples 12.5 and 12.6 to kernelize the multi-class Softmax cost.

12.6   **Regression with the RBF kernel**

Implement the RBF kernel in Example 12.9 and perform nonlinear regression on the dataset shown in the top row of Figure 12.3 using $\beta = 10^{-4}$, $\beta = 10^{-2}$, and $\beta = 10$ to reproduce the respective fits shown in the figure.

12.7   **Two-class classification with the RBF kernel**

Implement the RBF kernel in Example 12.9 and perform nonlinear two-class classification on the dataset shown in the middle row of Figure 12.3 using $\beta = 10^{-8}$, $\beta = 10^{-4}$, and $\beta = 10$. For each case produce a misclassification history plot to show that your results match what is shown in the figure.

12.8   **Multi-class classification with the RBF kernel**

Implement the RBF kernel in Example 12.9 and perform nonlinear multi-class classification on the dataset shown in the bottom row of Figure 12.3 using $\beta = 10^{-2}$, $\beta = 10^{-1}$, and $\beta = 100$. For each case produce a misclassification history plot to show that your results match, respectively, what is shown in the figure.

12.9   **Polynomial kernels for arbitrary degree and input dimension**

Show that a polynomial kernel can be defined entry-wise, as given in Equation (12.28), for general degree D and input dimension N.

12.10   **An infinite-dimensional feature transformation**

Verify that the infinite-dimensional feature transformation defined in Equation (12.39) indeed yields the entry-wise form of the RBF kernel in Equation (12.36).

12.11   **Fourier kernel for vector-valued input**

For a general N-dimensional input each Fourier unit takes the form

$$f_{\mathbf{m}}(\mathbf{x}) = e^{2\pi i m_1 x_1} e^{2\pi i m_2 x_2} \cdots e^{2\pi i m_N x_N} = e^{2\pi i \mathbf{m}^T \mathbf{x}} \tag{12.47}$$

where the vector $\mathbf{m}$

$$\mathbf{m} = \begin{bmatrix} m_1 \\ m_2 \\ \vdots \\ m_N \end{bmatrix} \tag{12.48}$$

contains integer-valued entries. Further, a degree-D Fourier expansion contains all such units satisfying $0 < \|\mathbf{m}\|_\infty \le D$ (see Appendix Section C.5 if not familiar with the infinity norm). Calculate the corresponding (i, j)th entry of the kernel matrix $\mathbf{H}$, i.e., $h_{i,j} = \mathbf{f}_i^T \overline{\mathbf{f}_j}$ where $\overline{\mathbf{f}_j}$ denotes the complex conjugate of $\mathbf{f}_j$.

12.12   Kernels and a cancer dataset

Repeat the experiment described in Example 12.10, and produce a plot like the one shown in Figure 12.5. You may achieve different results based on your random training–validation split of the original data.

13 Fully Connected Neural Networks

13.1 Introduction

As we first saw in Section 11.2.3, artificial neural networks, unlike polynomials and other fixed-shape approximators, have internal parameters that allow each of their units to take on a variety of shapes. In this chapter we expand on that introduction extensively, discussing general multi-layer neural networks, also referred to as *fully connected networks*, *multi-layer perceptrons*, and *deep feed-forward neural networks*.

13.2 Fully Connected Neural Networks

In this section we describe general fully connected neural networks, which are recursively built generalizations of the sort of units we first saw throughout Chapter 11. As this is an often confused subject we describe fully connected networks progressively (and with some redundancy that will hopefully benefit the reader), layer by layer, beginning with *single-hidden-layer* units first described in Section 11.2.3, providing algebraic, graphical, and computational perspectives on their construction. Afterwards, we briefly touch on the biological plausibility of fully connected networks, and end this section with an in-depth description of how to efficiently implement them in Python.

13.2.1 Single-hidden-layer units

The general algebraic representation (i.e., the formula) of a single-hidden-layer unit, also called a single-layer unit for short, is something we first saw in Equation (11.12) and is quite simple: a linear combination of input passed through a nonlinear *activation* function, which is typically an elementary mathematical function (e.g., tanh). Here we will denote such units in general as

$$f^{(1)}(\mathbf{x}) = a\left(w_0^{(1)} + \sum_{n=1}^{N} w_n^{(1)} x_n\right) \tag{13.1}$$

where $a(\cdot)$ denotes an activation function, and the superscripts on f and w_0

through w_N indicate they represent a single- (i.e., one-) layer unit and its internal weights, respectively.

Because we will want to extend the single-layer idea to create multi-layer networks, it is helpful to pull apart the sequence of two operations used to construct a single-layer unit: the linear combination of input, and the passage through a nonlinear activation function. We refer to this manner of writing out the unit as the *recursive recipe* for creating single-layer neural network units, and summarize it below.

Recursive recipe for single-layer units

---

1. Choose an activation function $a\left(\cdot\right)$
2. Compute the linear combination $v = w_0^{(1)} + \sum_{n=1}^{N} w_n^{(1)} x_n$
3. Pass the result through activation and form $a\left(v\right)$

Example 13.1   Illustrating the capacity of single-layer units

In the top panels of Figure 13.1 we plot four instances of a single-layer unit using tanh as our nonlinear activation function. These four nonlinear units take the form

$$f^{(1)}(x) = \tanh\left(w_0^{(1)} + w_1^{(1)}x\right) \tag{13.2}$$

where in each instance the internal parameters of the unit (i.e., $w_0^{(1)}$ and $w_1^{(1)}$) have been set randomly, giving each instance a distinct shape. This roughly illustrates the *capacity* of each single-layer unit. Capacity (a concept first introduced in Section 11.2.2) refers to the range of shapes a function like this can take, given all the different settings of its internal parameters.

In the bottom row of Figure 13.1 we swap out tanh for the ReLU[1] activation function, forming a single-layer unit of the form

$$f^{(1)}(x) = \max\left(0, w_0^{(1)} + w_1^{(1)}x\right). \tag{13.3}$$

Once again the internal parameters of this unit allow it to take on a variety of shapes (distinct from those created by tanh activation), four instances of which are illustrated in the bottom panels of the figure.

---

[1] The Rectified Linear Unit or ReLU function was first introduced in the context of two-class classification and the Perceptron cost in Section 6.4.2.

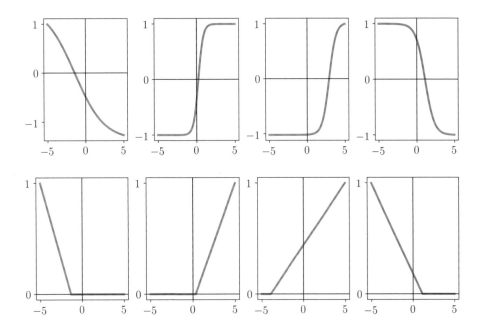

Figure 13.1 Figure associated with Example 13.1. Four instances of a single-layer neural network unit with tanh (top row) and ReLU activation (bottom row). See text for further details.

If we form a general nonlinear model using $B = U_1$ such single-layer units as

$$\text{model}\,(\mathbf{x}, \Theta) = w_0 + f_1^{(1)}\,(\mathbf{x})\,w_1 + \cdots + f_{U_1}^{(1)}\,(\mathbf{x})\,w_{U_1} \tag{13.4}$$

whose jth unit takes the form

$$f_j^{(1)}\,(\mathbf{x}) = a\left(w_{0,j}^{(1)} + \sum_{n=1}^{N} w_{n,j}^{(1)} x_n\right) \tag{13.5}$$

then the parameter set Θ contains not only the weights of the final linear combination w_0 through w_{U_1}, but all parameters internal to each $f_j^{(1)}$ as well. This is precisely the sort of model we used in the neural network examples throughout Chapter 11.

The left panel of Figure 13.2 shows a common graphical representation of the single-layer model in Equation (13.4), and visually unravels the individual algebraic operations performed by such a model. A visual representation like this is often referred to as a *neural network architecture* or just an *architecture*. Here the bias and input of each single-layer unit composing the model is shown as a sequence of dots all the way on the left of the diagram. This layer is "visible" to us since this is where we inject the input data to our network which we ourselves can "see," and is also often referred to as the *first* or *input* layer of the

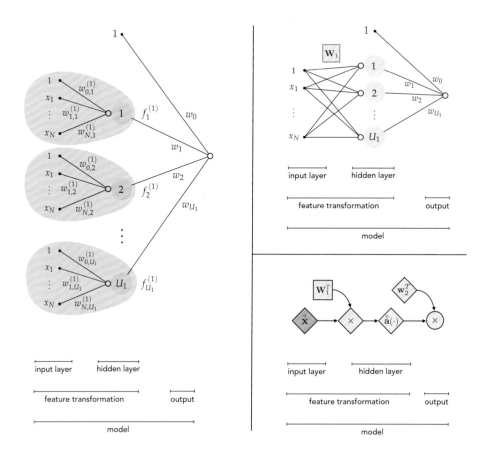

Figure 13.2 (left panel) Graphical representation of a single-layer neural network model, given in Equation (13.4), which is composed of U_1 single-layer units. (top-right panel) A condensed graphical representation of a single-layer neural network. (bottom-right panel) This network can be represented even more compactly, illustrating in a simple diagram all of the computation performed by a single-layer neural network model. See text for further details.

network. The linear combination of input leading to each unit is then shown visually by edges connecting the input to a hollow circle (summation unit), with the nonlinear activation then shown as a larger blue circle (activation unit). In the middle of this visual depiction (where the blue circles representing all U_1 activations align) is the *hidden* layer of this architecture. This layer is called "hidden" because it contains internally processed versions of our input that we do not "see." While the name hidden is not entirely accurate (as we can visualize the internal state of these units if we so desire) it is a commonly used convention, hence the name single-*hidden*-layer unit. The output of these U_1 units is then collected in a linear combination, and once again visualized by edges connecting each unit to a final summation shown as a hollow circle. This is the final output of the model, and is often called the *final* or *output* layer of the network, which is again "visible" to us (not hidden).

Compact representation of single-layer neural networks

Because we will soon wish to add more hidden layers to our rudimentary model in detailing multi-layer networks, the sort of visual depiction in the left panel of Figure 13.2 quickly becomes unwieldy. Thus, in order to keep ourselves organized and better prepared to understand *deeper* neural network units, it is quite helpful to compactify this visualization. We can do so by first using a more compact notation to represent our model algebraically, beginning by more concisely representing our input, placing a 1 at the top of our input vector $\mathbf{x}$, which we denote by placing a hollow ring symbol over $\mathbf{x}$ as[2]

$$\mathring{\mathbf{x}} = \begin{bmatrix} 1 \\ x_1 \\ \vdots \\ x_N \end{bmatrix}. \tag{13.6}$$

Next we collect all of the *internal parameters* of our U_1 single-layer units. Examining the algebraic form for the jth unit in Equation (13.5) we can see that it has $N+1$ such internal parameters. Taking these parameters we form a column vector, starting with the bias $w_{0,j}^{(1)}$ and then input-touching weights $w_{1,j}^{(1)}$ through $w_{N,j}^{(1)}$, and place them into the jth column of an $(N + 1) \times U_1$ matrix

$$\mathbf{W}_1 = \begin{bmatrix} w_{0,1}^{(1)} & w_{0,2}^{(1)} & \cdots & w_{0,U_1}^{(1)} \\ w_{1,1}^{(1)} & w_{1,2}^{(1)} & \cdots & w_{1,U_1}^{(1)} \\ \vdots & \vdots & \ddots & \vdots \\ w_{N,1}^{(1)} & w_{N,2}^{(1)} & \cdots & w_{N,U_1}^{(1)} \end{bmatrix}. \tag{13.7}$$

With this notation note how the matrix-vector product $\mathbf{W}_1^T \mathring{\mathbf{x}}$ contains every linear combination *internal* to our U_1 nonlinear units. In other words, $\mathbf{W}_1^T \mathring{\mathbf{x}}$ has dimension $U_1 \times 1$, and its jth entry is precisely the linear combination of the input data internal to the jth unit

$$\left[\mathbf{W}_1^T \mathring{\mathbf{x}}\right]_j = w_{0,j}^{(1)} + \sum_{n=1}^{N} w_{n,j}^{(1)} x_n \qquad j = 1, 2, ..., U_1. \tag{13.8}$$

Next, we extend our notation for the arbitrary activation function $a(\cdot)$ to

---

[2] This notation was introduced and employed previously in Section 5.2.

handle such a vector. More specifically, we define $\mathbf{a}(\cdot)$ as the vector function that takes in a general $d \times 1$ vector $\mathbf{v}$ and returns – as output – a vector of the same dimension containing activation of each of its input's entries, as

$$\mathbf{a}(\mathbf{v}) = \begin{bmatrix} a(v_1) \\ \vdots \\ a(v_d) \end{bmatrix}. \tag{13.9}$$

Notice how with this notation, the vector activation $\mathbf{a}\left(\mathbf{W}_1^T \overset{\circ}{\mathbf{x}}\right)$ becomes a $U_1 \times 1$ vector containing all U_1 single-layer units, the jth of which is given as

$$\left[\mathbf{a}\left(\mathbf{W}_1^T \overset{\circ}{\mathbf{x}}\right)\right]_j = a\left(w_{0,j}^{(1)} + \sum_{n=1}^{N} w_{n,j}^{(1)} x_n\right) \qquad j = 1, 2, ..., U_1. \tag{13.10}$$

Using another compact notation to denote the weights of the final linear combination as

$$\mathbf{w}_2 = \begin{bmatrix} w_0 \\ w_1 \\ \vdots \\ w_{U_1} \end{bmatrix} \tag{13.11}$$

and extending our vector $\mathbf{a}$ by tacking a 1 on top of it – denoting the resulting $(U_1 + 1) \times 1$ vector $\overset{\circ}{\mathbf{a}}$ – we can finally write out the model in Equation (13.4) quite compactly as

$$\text{model}(\mathbf{x}, \Theta) = \mathbf{w}_2^T \overset{\circ}{\mathbf{a}}\left(\mathbf{W}_1^T \overset{\circ}{\mathbf{x}}\right). \tag{13.12}$$

This more compact algebraic formulation lends itself to much more easily digestible visual depictions. In the top-right panel of Figure 13.2 we show a slightly condensed version of our original graph in the left panel, where the linear weights attached to each input are now shown more compactly as the set of crisscrossing line segments connecting the input to each unit, with the matrix $\mathbf{W}_1$ jointly representing all U_1 of the weighted combinations. In the bottom-right panel of Figure 13.2 we compactify our original visual representation even further. In this more compact representation we can more easily visualize the computation performed by the general single-layer neural network model in Equation (13.12), where we depict the scalars, vectors, and matrices in this formula symbolically as circles, diamonds, and squares, respectively.

13.2.2   Two-hidden-layer units

To create a *two-hidden-layer* neural network unit, or a two-layer unit for short, we *recurse* on the idea of the single-layer unit detailed in the previous section.

We do this by first constructing a set of U_1 single-layer units and treat them as input to another nonlinear unit. That is, we take their linear combination and pass the result through a nonlinear activation.

The algebraic form of a general two-layer unit is given as

$$f^{(2)}(\mathbf{x}) = a \left(w_0^{(2)} + \sum_{i=1}^{U_1} w_i^{(2)} f_i^{(1)}(\mathbf{x}) \right) \tag{13.13}$$

which reflects the recursive nature of constructing two-layer units using single-layer ones. This recursive nature can be also seen in the *recursive recipe* given below for building two-layer units.

Recursive recipe for two-layer units

---

1. Choose an activation function $a(\cdot)$
2. Construct U_1 single-layer units $f_i^{(1)}(\mathbf{x})$ for $i = 1, 2, ..., U_1$
3. Compute the linear combination $v = w_0^{(2)} + \sum_{i=1}^{U_1} w_i^{(2)} f_i^{(1)}(\mathbf{x})$
4. Pass the result through activation and form $a(v)$

Example 13.2 Illustrating the capacity of two-layer units
In the top row of Figure 13.3 we plot four instances of a two-layer neural network unit – using tanh activation – of the form

$$f^{(2)}(x) = \tanh \left(w_0^{(2)} + w_1^{(2)} f^{(1)}(x) \right) \tag{13.14}$$

where

$$f^{(1)}(x) = \tanh \left(w_0^{(1)} + w_1^{(1)} x \right). \tag{13.15}$$

The wider variety of shapes taken on by instances of this unit, as shown in the figure, reflects the increased capacity of two-layer units over their single-layer analogs shown in Figure 13.1.

In the bottom row of the figure we show four exemplars of the same sort of unit, only now we use ReLU activation instead of tanh in each layer.

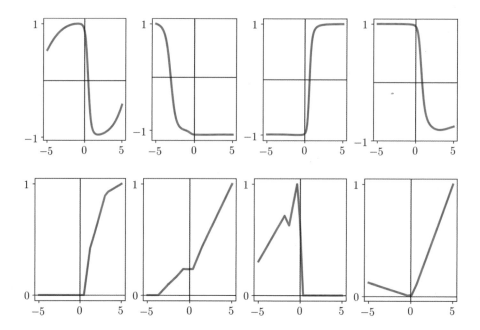

Figure 13.3 Figure associated with Example 13.2. Four instances of a two-layer neural network unit with tanh (top row) and ReLU activation (bottom row). See text for further details.

In general, if we wish to create a model using $B = U_2$ two-layer neural network units we write

$$\text{model}\,(\mathbf{x}, \Theta) = w_0 + f_1^{(2)}\,(\mathbf{x})\,w_1 + \cdots + f_{U_2}^{(2)}\,(\mathbf{x})\,w_{U_2} \tag{13.16}$$

where

$$f_j^{(2)}\,(\mathbf{x}) = a\left(w_{0,j}^{(2)} + \sum_{i=1}^{U_1} w_{i,j}^{(2)}\,f_i^{(1)}\,(\mathbf{x})\right) \qquad j = 1, 2, ..., U_2 \tag{13.17}$$

and where the parameter set Θ, as always, contains those (superscripted) weights internal to the neural network units as well as the final linear combination weights. Importantly note that while each two-layer unit $f_j^{(2)}$ in Equation (13.17) has unique internal parameters – denoted by $w_{i,j}^{(2)}$ where i ranges from 0 to U_1 – the weights internal to each single-layer unit $f_i^{(1)}$ are the same across all the two-layer units themselves.

Figure 13.4 shows a graphical representation (or architecture) of a generic two-layer neural network model whose algebraic form is given in Equation (13.16). In the left panel we illustrate each input single-layer unit precisely as shown previously in the top-right panel of Figure 13.2. The input layer, all the way to the left, is first fed into each of our U_1 single-layer units (which still constitutes

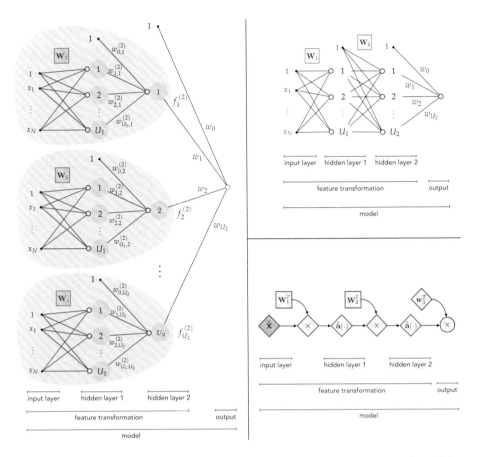

Figure 13.4 (left panel) Graphical representation of a two-layer neural network model, given in Equation (13.16), which is composed of U_2 two-layer units. (top-right panel) A condensed graphical representation of a two-layer neural network. (bottom-right panel) This network can be represented more compactly, providing a simpler depiction of the computation performed by a two-layer neural network model. See text for further details.

the first *hidden* layer of the network). A linear combination of these single-layer units is then fed into each of the U_2 two-layer units, referred to by convention as the second *hidden* layer, since its computation is also not immediately "visible" to us. Here we can also see why this sort of architecture is referred to as *fully connected*: every dimension of input is connected to every unit in the first hidden layer, and each unit of the first hidden layer is connected to every unit of the second hidden layer. At last, all the way to the right of this panel, we see a linear combination of the U_2 two-layer units which produces the final (visible) layer of the network: the output of our two-layer model.

Compact representation of two-layer neural networks

As with single-layer models, here it is also helpful to compactify both our notation and the corresponding visualization of a two-layer model in order to simplify our understanding and make the concept easier to wield. Using the same notation introduced in Section 13.2.1, we can compactly designate the output of our U_1 single-layer units as

$$\text{output of first hidden layer:} \quad \mathring{\mathbf{a}}\left(\mathbf{W}_1^T \mathring{\mathbf{x}}\right). \tag{13.18}$$

Following the same pattern as before we can then condense all internal weights of the U_2 units in the second layer column-wise into a $(U_1 + 1) \times U_2$ matrix of the form

$$\mathbf{W}_2 = \begin{bmatrix} w_{0,1}^{(2)} & w_{0,2}^{(2)} & \cdots & w_{0,U_2}^{(2)} \\ w_{1,1}^{(2)} & w_{1,2}^{(2)} & \cdots & w_{1,U_2}^{(2)} \\ \vdots & \vdots & \ddots & \vdots \\ w_{U_1,1}^{(2)} & w_{U_1,2}^{(2)} & \cdots & w_{U_1,U_2}^{(2)} \end{bmatrix} \tag{13.19}$$

which mirrors precisely how we defined the $(N + 1) \times U_1$ internal weight matrix $\mathbf{W}_1$ for our single-layer units in Equation (13.7). This allows us to likewise express the output of our U_2 two-layer units compactly as

$$\text{output of second hidden layer:} \quad \mathring{\mathbf{a}}\left(\mathbf{W}_2^T \mathring{\mathbf{a}}\left(\mathbf{W}_1^T \mathring{\mathbf{x}}\right)\right). \tag{13.20}$$

The recursive nature of two-layer units is on full display here. Remember that we use the notation $\mathring{\mathbf{a}}(\cdot)$ somewhat loosely as a vector-valued function in the sense that it simply represents taking the nonlinear activation $a(\cdot)$, *element-wise*, of whatever vector is input into it as shown in Equation (13.9), with a 1 appended to the top of the result.

Concatenating the final linear combination weights into a single vector as

$$\mathbf{w}_3 = \begin{bmatrix} w_0 \\ w_1 \\ \vdots \\ w_{U_2} \end{bmatrix} \tag{13.21}$$

allows us to write the full two-layer neural network model as

$$\text{model}\,(\mathbf{x}, \Theta) = \mathbf{w}_3^T \mathring{\mathbf{a}}\left(\mathbf{W}_2^T \mathring{\mathbf{a}}\left(\mathbf{W}_1^T \mathring{\mathbf{x}}\right)\right). \tag{13.22}$$

As with its single-layer analog, this compact algebraic formulation of a two-layer neural network lends itself to much more easily digestible visual depictions. In the top-right panel of Figure 13.4 we show a slightly condensed version of the original graph in the left panel, where the redundancy of showing every single-layer unit has been reduced to a single visual representation. In doing so we remove all the weights assigned to crisscrossing edges connecting the first and second hidden layers, and place them in the matrix $\mathbf{W}_2$ defined in Equation (13.19) to avoid cluttering the visualization. In the bottom-right panel of Figure 13.4 we condense this two-layer depiction even further where scalars, vectors, and matrices are depicted symbolically as circles, diamonds, and squares, respectively. This greatly compacted graph provides a simplified visual representation of the total computation performed by a general two-layer neural network model.

13.2.3 General multi-hidden-layer units

Following the same pattern we have seen thus far in describing single- and two-layer units we can construct general fully connected neural network units with an arbitrary number of hidden layers. With each hidden layer added we increase the capacity of a neural network unit, as we have seen previously in graduating from single-layer units to two-layer units, as well as a model built using such units.

To construct a general L-hidden-layer neural network unit, or L-layer unit for short, we simply recurse on the pattern we have established previously $L - 1$ times, with the resulting L-layer unit taking as input a number U_{L-1} of $(L - 1)$-layer units, as

$$f^{(L)}(\mathbf{x}) = a \left(w_0^{(L)} + \sum_{i=1}^{U_{L-1}} w_i^{(L)} f_i^{(L-1)}(\mathbf{x}) \right). \tag{13.23}$$

As was the case with single- and two-layer units, this formula is perhaps easier to digest if we think about it in terms of the *recursive recipe* given below.

Recursive recipe for L-layer units

---

1. Choose an activation function $a(\cdot)$
2. Construct U_{L-1} number of $(L - 1)$-layer units $f_i^{(L-1)}(\mathbf{x})$ for $i = 1, 2, ..., U_{L-1}$
3. Compute the linear combination $v = w_0^{(L)} + \sum_{i=1}^{U_{L-1}} w_i^{(L)} f_i^{(L-1)}(\mathbf{x})$
4. Pass the result through activation and form $a(v)$

Note that while in principle the same activation function need not be used for all hidden layers of an L-layer unit, for the sake of simplicity, a single kind of activation is almost always used.

Example 13.3   Illustrating the capacity of three-layer units
In the top row of Figure 13.5 we show four instances of a three-layer unit with tanh activation. The greater variety of shapes shown here, as compared to single- and two-layer analogs in Examples 13.1 and 13.2, reflects the increased capacity of these units. In the bottom row we repeat the same experiment, only using the ReLU activation function instead of tanh.

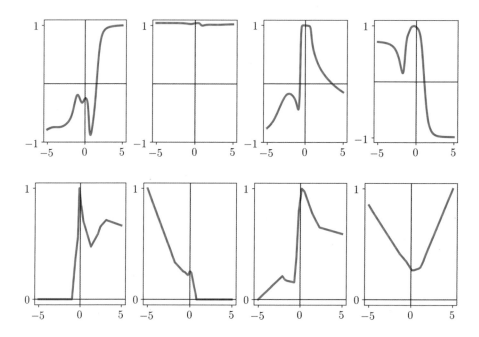

Figure 13.5 Figure associated with Example 13.3. Four instances of a three-layer network unit with tanh (top row) and ReLU activation (bottom row). See text for further details.

In general we can produce a model consisting of $B = U_L$ such L-layer units as

$$\text{model}\,(\mathbf{x}, \Theta) = w_0 + f_1^{(L)}\,(\mathbf{x})\,w_1 + \cdots + f_{U_L}^{(L)}\,(\mathbf{x})\,w_{U_L} \qquad (13.24)$$

where

$$f_j^{(L)}(\mathbf{x}) = a\left(w_{0,j}^{(L)} + \sum_{i=1}^{U_{L-1}} w_{i,j}^{(L)} f_i^{(L-1)}(\mathbf{x})\right) \qquad j = 1, 2, ..., U_L \qquad (13.25)$$

and where the parameter set Θ contains both those weights internal to the neural network units as well as the final linear combination weights.

Figure 13.6 shows an unraveled graphical representation of this model, which is a direct generalization of the kinds of visualizations we have seen previously with single- and two-layer networks. From left to right we can see the input layer to the network, its L hidden layers, and the output. Often models built using three or more hidden layers are referred to as *deep* neural networks in the jargon of machine learning.

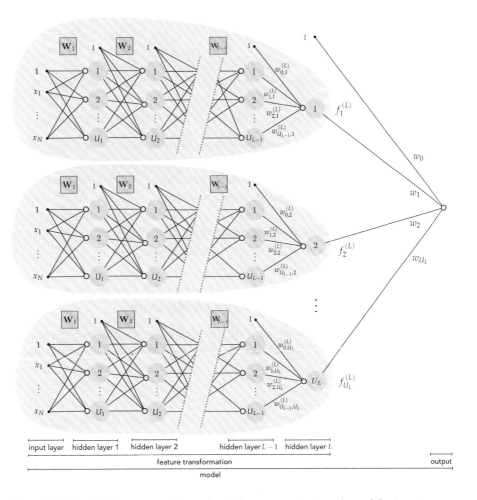

Figure 13.6 Graphical representation of an L-layer neural network model, given in Equation (13.24), which is composed of U_L L-layer units.

Compact representation of multi-layer neural networks

To simplify our understanding of a general multi-layer neural network architecture we can use precisely the same compact notation and visualizations we
have introduced in the simpler contexts of single- and two-layer neural networks. In complete analogy to the way we compactly represented two-layer
neural networks, we denote the output of the Lth hidden layer compactly as

$$\text{output of } L\text{th hidden layer:} \quad \mathring{a}\left(\mathbf{W}_L^T \mathring{a}\left(\mathbf{W}_{L-1}^T \mathring{a}\left(\cdots \mathring{a}\left(\mathbf{W}_1^T \mathring{x}\right)\right)\right)\right). \tag{13.26}$$

Denoting the weights of the final linear combination as

$$\mathbf{w}_{L+1} = \begin{bmatrix} w_0 \\ w_1 \\ \vdots \\ w_{U_L} \end{bmatrix} \tag{13.27}$$

we can express the L-layer neural network model compactly as

$$\text{model}\left(\mathbf{x}, \Theta\right) = \mathbf{w}_{L+1}^T \mathring{a}\left(\mathbf{W}_L^T \mathring{a}\left(\mathbf{W}_{L-1}^T \mathring{a}\left(\cdots \mathring{a}\left(\mathbf{W}_1^T \mathring{x}\right)\right)\right)\right). \tag{13.28}$$

Once again this compact algebraic formulation of an L-layer neural network
lends itself to much more easily digestible visual depictions. The top panel of
Figure 13.7 shows a condensed version of the original graph in Figure 13.6,
where the redundancy of showing every $(L-1)$-layer unit has been reduced to
a single visualization. The succinct visual depiction shown in the bottom panel
of Figure 13.7 represents this network architecture even more compactly, with
scalars, vectors, and matrices are shown symbolically as circles, diamonds, and
squares, respectively.

13.2.4   Selecting the right network architecture

We have now seen a general and recursive method of constructing arbitrarily
"deep" neural networks, but many curiosities and technical issues remain to
be addressed that are the subject of subsequent sections in this chapter. These
include the choice of activation function, popular cross-validation methods for
models employing neural network units, as well as a variety of optimization-
related issues such as the notions of *backpropagation* and *batch normalization*.

However, one fundamental question can (at least in general) be addressed
now, which is how we choose the "right" number of units and layers for a neural
network architecture. As with the choice of proper universal approximator in
general (as detailed in Section 11.8), typically we do not know *a priori* what sort
of architecture will work best for a given dataset (in terms of number of hidden
layers and number of units per hidden layer). In principle, to determine the
best architecture for use with a given dataset we must cross-validate an array

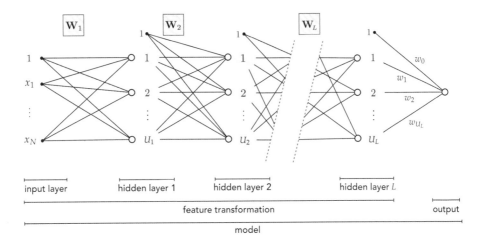

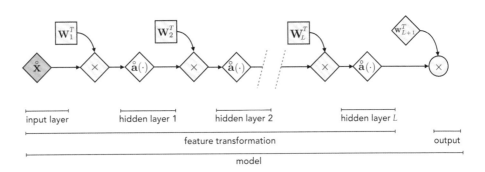

Figure 13.7 (top panel) A condensed graphical representation of an L-layer neural network model shown in Figure 13.6. (bottom panel) A more compact version, succinctly describing the computation performed by a general L-layer neural network model. See text for further details.

of choices. In doing so we note that, generally speaking, the capacity gained by adding new individual units to a neural network model is typically much smaller relative to the capacity gained by the addition of new hidden layers. This is because appending an additional *layer* to an architecture grafts an additional recursion to the computation involved in each unit, which significantly increases their capacity and that of any corresponding model, as we have seen in Examples 13.1, 13.2, and 13.3. In practice, performing model search across a variety of neural network architectures can be expensive, hence compromises must be made that aim at determining a high-quality model using minimal computation. To this end, early stopping based regularization (see Sections 11.6.2 and 13.7) is commonly employed with neural network models.

13.2.5   Neural networks: the biological perspective

The human brain contains roughly 10^{11} biological neurons which work together in concert when we perform cognitive tasks. Even when performing relatively minor tasks we employ a sizable series of interconnected neurons – called *biological neural networks* – to perform the task properly. For example, somewhere between 10^5 to 10^6 neurons are required to render a realistic image of our visual surroundings. Most of the basic jargon and modeling principles of neural network universal approximators we have seen thus far originated as a (very) rough mathematical model of such biological neural networks.

An individual biological neuron (shown in the top-left panel of Figure 13.8) consists of three main parts: *dendrites* (the neuron's receivers), *soma* (the cell body), and *axon* (the neuron's transmitter). Starting around the 1940s psychologists and neuroscientists, with a shared desire to understand the human brain better, became interested in modeling neurons mathematically. These early models, later dubbed as artificial neurons (a basic exemplar of which is shown in the top-right panel of Figure 13.8), culminated in the introduction of the *Perceptron* model in 1957 [59]. Closely mimicking a biological neuron's structure, an artificial neuron comprises a set of dendrite-like edges that connect it to other neurons, each taking an input and multiplying it by a (synaptic) weight associated with that edge. These weighted inputs are summed up after going through a *summation unit* (shown by a small hollow circle). The result is subsequently fed to an *activation unit* (shown by a large blue circle) whose output is then transmitted to the outside via an axon-like projection. From a biological perspective, neurons are believed to remain inactive until the net input to the cell body (soma) reaches a certain threshold, at which point the neuron gets *activated* and fires an electro-chemical signal, hence the name *activation function*.

Stringing together large sets of such artificial neurons in layers creates a more mathematically complex model of a biological neural network, which still remains a *very* simple approximation to what goes on in the brain. The bottom panel of Figure 13.8 shows the kind of graphical representation (here of a two-layer network) used when thinking about neural networks from this biological perspective. In this rather complex visual depiction every multiplicative operation of the architecture is shown as an edge, creating a mesh of intersecting lines connecting each layer. In describing fully connected neural networks in Sections 13.2.1 through 13.2.3 we preferred simpler, more compact visualizations that avoid this sort of complex visual mesh.

13.2.6   `Python` implementation

In this section we show how to implement a generic model consisting of L-layer neural network units in `Python` using `NumPy`. Because such models can have a large number of parameters, and because associated cost functions employing them can be highly nonconvex (as further discussed in Section 13.5), first-order

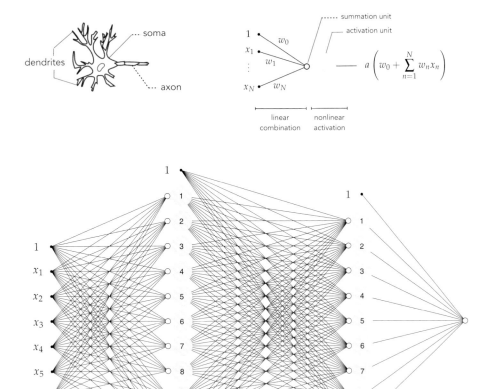

Figure 13.8 (top-left panel) A typical biological neuron. (top-right panel) An artificial neuron, that is a simplistic mathematical model of the biological neuron, consisting of: (i) weighted edges that represent the individual multiplications (of 1 by w_0, x_1 by w_1, etc.), (ii) a summation unit shown as a small hollow circle representing the sum $w_0 + w_1 x_1 + \cdots + w_N x_N$, and (iii) an activation unit shown as a larger blue circle representing the sum evaluated by the nonlinear activation function $a\,(\cdot)$. (bottom panel) An example of a fully connected two-layer neural network as commonly illustrated when detailing neural networks from a biological perspective.

local optimization methods (e.g., gradient descent and its variants detailed in Chapter 3 and Appendix A) are the most common schemes used to tune the parameters of general neural network models. Moreover, because computing derivatives of a model employing neural network units "by hand" is extremely tedious (see Section 13.4) the use of automatic differentiation is not only helpful but in most cases necessary. Thus in the Python implementation detailed here we strongly recommend using autograd – the automatic differentiator used throughout this text (see Section B.10.2) – to make computing derivatives clean and simple.

Below we provide an implementation of a neural network feature transformation called `feature_transforms` (a notation we first introduced in Chapter 10).

```
1  # neural network feature transformation
2  def feature_transforms(a, w):
3
4      # loop through each layer
5      for W in w:
6
7          # compute inner-product with current layer weights
8          a = W[0] + np.dot(a.T, W[1:])
9
10         # pass through activation
11         a = activation(a).T
12
13     return a
```

This `Python` function takes in the input **x**, written as the variable a, and the entire set of weight matrices $\mathbf{W}_1$ through $\mathbf{W}_L$ internal to our neural network architecture, written as the variable w where w = [W_1, W_2, ..., W_L]. The output of our feature transformation function is the output of the final layer of the network, expressed algebraically in Equation (13.26). We compute this output recursively, looping *forward* through the network architecture, starting with the first hidden layer using matrix $\mathbf{W}_1$ and ending with the computation of the final hidden layer using $\mathbf{W}_L$. This results in a `Python` function consisting of a simple `for` loop over the weight matrices of the hidden layers. Notice, in line 11 of the implementation above, that `activation` can refer to any elementary function built using `NumPy` operations. For example, the tanh activation can be written as `np.tanh(a)`, and the ReLU as `np.maximum(0,a)`.

With our feature transformation function complete we can now implement our `model`, which is a simple variation on the implementations we have seen in previous chapters. Here the inputs to this `Python` function are x – our input data – and a `Python` list `theta` of length two whose first entry contains our list of internal weight matrices, and whose second entry contains the weights of the final linear combination.

```
1  # neural network model
2  def model(x, theta):
3
4      # compute feature transformation
5      f = feature_transforms(x, theta[0])
6
7      # compute final linear combination
8      a = theta[1][0] + np.dot(f.T, theta[1][1:])
```

```
 9
10        return a.T
```

This implementation of a fully connected neural network model can be easily paired with Pythonic implementation of the generic machine learning cost functions detailed in previous chapters.

Finally, we provide a Python function called network_initializer that creates initial weights for a general neural network model, and also provides a simple interface for creating general architectures: it is precisely this initializer that determines the shape of our implemented network. To create a desired network we simply input a comma-separated list called layer_sizes of the following form

```
1  layer_sizes = [N, U_1, ..., U_L, C]
```

where N is the input dmension, U_1 through U_L are the number of desired units in the hidden layers 1 through L, respectively, and C is the output dimension.

The initializer will then automatically create initial weight matrices (of the proper dimensions) as well as the final weights for the linear combination, packaged together as the output theta_init.

```
 1  # create initial weights for a neural network model
 2  def network_initializer(layer_sizes, scale):
 3
 4      # container for all tunable weights
 5      weights = []
 6
 7      # create appropriately-sized initial
 8      # weight matrix for each layer of network
 9      for k in range(len(layer_sizes)-1):
10
11          # get layer sizes for current weight matrix
12          U_k = layer_sizes[k]
13          U_k_plus_1 = layer_sizes[k+1]
14
15          # make weight matrix
16          weight = scale*np.random.randn(U_k+1, U_k_plus_1)
17          weights.append(weight)
18
19      # repackage weights so that theta_init[0] contains all
20      # weight matrices internal to the network, and theta_init[1]
21      # contains final linear combination weights
22      theta_init = [weights[:-1], weights[-1]]
23
24      return theta_init
```

Next, we provide several examples using this implementation.

Example 13.4   **Nonlinear classification using multi-layer neural networks**
In this example we use a multi-layer architecture to perform nonlinear classification, first on the two-class dataset shown previously in Example 11.9.2. Here, we arbitrarily choose the network to have four hidden layers with ten units in each layer, and the tanh activation. We then tune the parameters of this model by minimizing an associated two-class Softmax cost (see Equation (10.31)) via gradient descent, visualizing the nonlinear decision boundary learned in the top row of Figure 13.9 along with the dataset itself.

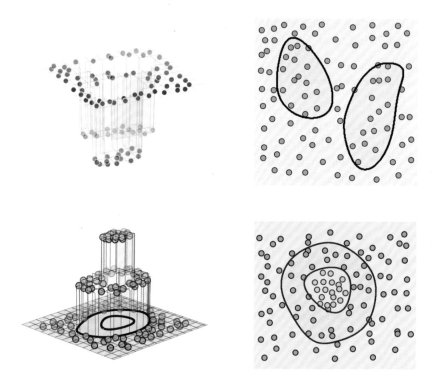

Figure 13.9 Figure associated with Example 13.4. The resulting decision boundary learned by a fully connected neural network on a two-class dataset (top row) and multi-class dataset (bottom row), from both the regression perspective (left column) and perceptron perspective (right column). See text for further details.

Next, we perform multi-class classification on the multi-class dataset first shown in Example 10.6 ($C = 3$), using a model consisting of two hidden layers, choosing the number of units in each layer arbitrarily as $U_1 = 12$ and $U_2 = 5$, respectively, the tanh activation, and using a shared scheme (that is, the network architecture is shared by each classifier, as detailed in Section 10.5 for general feature transformations). We then tune the parameters of this model by minimizing the corresponding multi-class Softmax cost (as shown in Equation

(10.42)), and show the resulting learned decision boundary in the bottom row of Figure 13.9.

Example 13.5   Random Autoencoder manifolds

In Section 10.6.1 we introduced the general nonlinear Autoencoder, which consists of two nonlinear functions: an encoder $\mathbf{f}_e$ and decoder $\mathbf{f}_d$, whose parameters we tune so that (ideally) the composition $\mathbf{f}_d(\mathbf{f}_e(\mathbf{x}))$ forms the best *nonlinear manifold* on which an input dataset rests. In other words, given a set of input points $\{\mathbf{x}_p\}_{p=1}^{P}$ we aim to tune the parameters of our encoder/decoder pair so that

$$\mathbf{f}_d\left(\mathbf{f}_e\left(\mathbf{x}_p\right)\right) \approx \mathbf{x}_p \qquad p = 1, 2, ..., P. \tag{13.29}$$

In Figure 13.10 we visualize nine instances of the function $\mathbf{f}_d(\mathbf{f}_e(\mathbf{x}))$, each of which is technically called a manifold, where both $\mathbf{f}_d$ and $\mathbf{f}_e$ are five-hidden-layer neural networks with ten units in each layer, and the sinc function as activation. All weights are set randomly in each instance to show the kind of nonlinear manifolds we could potentially discover using such encoding/decoding functions. Because of the exceedingly high capacity of this nonlinear Autoencoder model the instances shown in the figure are quite diverse in shape.

Figure 13.10 Figure associated with Example 13.5. Nine random manifolds generated by a multi-layer neural network Autoencoder. See text for further details.

Example 13.6   Nonlinear Autoencoder using multi-layer neural networks

In this example we illustrate the use of a multi-layer neural network Autoencoder for learning a nonlinear manifold over the dataset shown in the top-left panel of Figure 13.11. Here for both the encoder and decoder functions we arbitrarily use a three-hidden-layer fully connected network with ten units in each layer, and the tanh activation. We then tune the parameters of both functions together by minimizing the Least Squares cost given in Equation (10.53), and uncover the proper nonlinear manifold on which the dataset rests. In Figure 13.11 we show the learned manifold (top-right panel), the decoded version of the original dataset, i.e., the original dataset projected onto our learned manifold (bottom-left panel), and a *projection map* visualizing how all of the data in this space is projected onto the learned manifold via a vector-field plot (bottom-right panel).

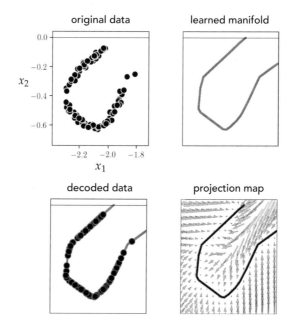

Figure 13.11 Figure associated with Example 13.6. See text for details.

13.3   Activation Functions

In principle one can use any (nonlinear) function as an activation for a fully connected neural network. For some time after their invention activations were chosen largely based on their biological plausibility, since this is the perspective in which neural networks themselves were at first largely regarded (as detailed in Section 13.2.5). Today activation functions are chosen based on practical considerations including our ability to properly optimize models which employ them as well as (of course) the general performance they provide. In this section we briefly review popular historical and modern activation functions through several examples.

Example 13.7   The step and sigmoid activations

As broadly discussed in Section 13.2.5 the concept of neural networks was first introduced from a biological perspective where each unit of an architecture mimics a biological neuron in the human brain. Such neurons were thought to act somewhat like digital switches, being either completely "on" or "off" to transmitting information to connected cells. This belief naturally led to the use of a *step function* taking on just two values: 0 (off) and 1 (on). However, this kind of step function (which we discuss in the context of logistic regression in Section 6.2) leads to piece-wise flat cost functions (see, e.g., the left panel of Figure 6.3), which are extremely difficult to optimize using any local optimization technique. In the context of logistic regression this sort of problem is what led to the *logistic sigmoid*, and for the same practical reason the sigmoid function was one of the first popularly used activation functions. As a smooth approximation to the step function, the logistic sigmoid was viewed as a reasonable compromise between the desired neuronal model and the practical need to tune parameters properly.

While the logistic sigmoid performs very well in the comparatively simpler context of linear classification, when used as an activation function it often leads to a technical issue known as the *vanishing gradient* problem. Note how the logistic sigmoid function (shown in the top-left panel of Figure 13.12) maps almost all negative input values (except those near the origin) to output values very close to zero, and its derivative (shown in the bottom-left panel of the figure) maps input values away from the origin to ouput values very close to zero. These characteristics can cause the gradient of a neural network model employing sigmoid activations to shrink undesirably, preventing proper parameter tuning – a problem that balloons as more hidden layers are added.

In practice, neural network models employing the hyperbolic tangent function (tanh) typically perform better than the same network employing logistic sigmoid activations, because the function itself centers its output about the origin. However, since the derivative of tanh likewise maps input values away from the origin to output values very close to zero, neural networks employing the tanh activation can also suffer from the vanishing gradient problem.

Example 13.8   The Rectified Linear Unit (ReLU) activation

For decades after fully connected neural networks were first introduced, researchers employed almost exclusively logistic sigmoid activation functions, based on their biologically plausible nature (outlined in Section 13.2.5). Only in the early 2000s did some researchers begin to break away from this tradition, entertaining and testing alternative activation functions [60]. The simple ReLU function (see Example 13.1) was the first such activation function to be popularized.

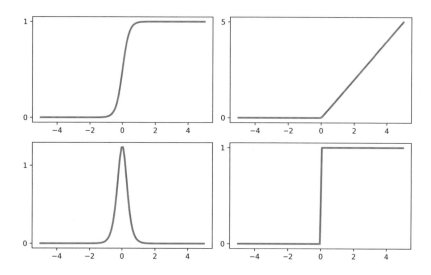

Figure 13.12 Figure associated with Examples 13.7 and 13.8. The logistic sigmoid function (top-left panel) and its derivative (bottom-left panel). The derivative here maps most input values away from the origin to output values very close to zero, which can cause the gradient of a network employing sigmoid activation to vanish during optimization, and thus prevent adequate parameter tuning. The ReLU function (top-right panel) and its derivative (bottom-right panel). While not as susceptible to vanishing gradient problems, neural networks employing the ReLU need to be initialized away from zero to prevent units from disappearing. See text for further details.

A computationally simpler function in comparison to the logistic sigmoid (which involves the use of both a log and exponential function), the ReLU has quickly become the most popular activation function in use today. Because the derivative of the ReLU function (plotted in the bottom-right panel of Figure 13.12) only maps negative input values to zero, networks employing this activation tend not to suffer (as severely) from the sort of vanishing gradient problem commonly found with the logistic sigmoid activation (detailed in the previous example). However, because of the shape of the ReLU itself (shown in the top-right panel of the figure) care still must be taken when initializing and training a network employing ReLU activations, as ReLU units themselves vanish as well at nonpositive inputs. For example, akin to the need to initialize the ReLU cost function detailed in Section 6.4 away from the origin where the optimization process will stall, a fully connected neural network employing ReLU activations should be initialized away from the origin to avoid too many of the units (and their gradients) from disappearing.

Example 13.9   The maxout activation

The *maxout* activation, defined as

$$a(x) = \max(v_0 + v_1 x, w_0 + w_1 x) \tag{13.30}$$

is a relative of the ReLU that takes the maximum of two linear combinations of the input (instead of one linear combination and zero, as is the case with ReLU). Four instances of such a maxout unit are plotted in Figure 13.13, where in each instance the parameters v_0, v_1, w_0, and w_1 are set at random. While this change seems algebraically rather minor, multi-layer neural network architectures employing the maxout activation tend to have certain advantages over those employing tanh and ReLU activations, including (i) fewer issues with problematic initializations, (ii) fewer issues with gradients vanishing, and (iii) empirically faster convergence with far fewer gradient descent steps [61]. These advantages come with a simple price: the maxout activation has twice as many internal parameters as either the ReLU or tanh, hence architectures built using them have roughly twice as many parameters to tune.

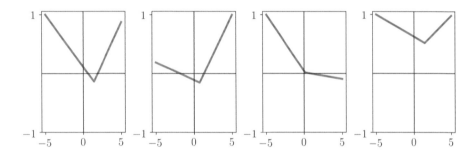

Figure 13.13 Figure associated with Example 13.9. Four instances of a maxout unit in Equation (13.30), with parameters randomly set in each case. See text for further details.

13.4   The Backpropagation Algorithm

The backpropagation algorithm, often referred to simply as backpropagation or backprop for short, is the jargon phrase used in machine learning to describe an approach to computing gradients algorithmically via a computer program that is especially effective for models employing multi-layer neural networks. Backprop is a special case of a more general scheme generally referred to as the *reverse mode of automatic differentiation*. The interested reader is encouraged to see Appendix Sections B.6 and B.7 for more on automatic differentiation.

Automatic differentiation allows us to turn over the tedious burden of computing gradients "by hand" to the tireless laborer that is the modern computer.

In other words, an automatic differentiator is an effective *calculator* that makes computing gradients of virtually any cost function a simple chore. Like many technical advances throughout history, automatic differentiation was discovered and rediscovered by different researchers in different areas of science and engineering at different times. This is precisely why this universally applicable concept (automatic differentiation) is referred to as *backpropagation* in the machine learning community, as this was the name given to it by its discoverers in this field.

13.5   Optimization of Neural Network Models

Typically cost functions associated with fully connected neural network models are highly nonconvex. In this section we address this issue, highlighting the local optimization techniques discussed in Appendix A of the text that are especially useful at properly tuning such models.

13.5.1   Nonconvexity

Models employing multi-layer neural networks are virtually always nonconvex. However, they often exhibit a variety of nonconvexity that we can fairly easily deal with using advanced optimization methods such as those detailed in Appendix A of the text. For example, in Section 6.2.4 we studied the shape of a Least Squares cost for logistic regression that employs a sigmoid based model. With one-dimensional input this model, written as

$$\text{model}(x, \Theta) = \sigma(w_0 + w_1 x) \tag{13.31}$$

can be interpreted, through the lens of neural networks, as a single unit of a single-hidden-layer neural network with scalar input and logistic sigmoid activation, where the weights of the final linear combination are fixed (with the bias set to zero and the weight touching $\sigma(\cdot)$ set to one). In the middle panel of Figure 6.3 we showed the surface of the Least Squares cost of this model over the simple dataset shown in Figure 6.2.

Examining the general shape of this cost function we can clearly see that it is nonconvex. Several portions of the cost surface, on either side of the *long narrow valley* containing the cost's global minimum, are almost completely *flat*. In general, cost functions employing neural network models have nonconvex shapes that share the kinds of basic characteristics seen in this figure: long narrow valleys, flat areas, and many saddle points and local minima.

These sorts of nonconvexities are problematic for *basic* first- and second-order optimization schemes. However, both frameworks can be extended to better deal with such eccentricities. This is especially true for gradient-based methods, as powerful modifications to the standard gradient descent method allow it to easily deal with *long narrow valleys* and *flat areas* of nonconvex functions. These modifications include *momentum-based* (see Appendix Section A.2) and *normalized* gradient methods (see Appendix Section A.3). Combining these two modifications (see Appendix Section A.4), in addition to mini-batch optimization (see Section 7.8 and Appendix Section A.5), can further enhance gradient-based methods so that they can more easily minimize cost functions like the one shown in the middle panel of Figure 6.3, and neural network models in general. In short, the sort of nonconvexity encountered with neural network models is often manageable using advanced optimization tools.

Even when neural network cost functions have many *local minima*, these tend to lie at a depth close to that of their global minima, and thus tend to provide similar performance if found via local optimization. An abstract depiction of this sort of prototypical cost function, the kind often encountered with neural network models, is shown in the top panel of Figure 13.14.

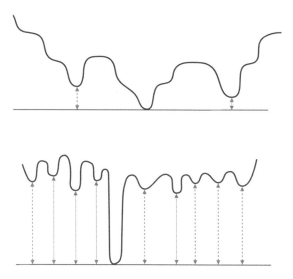

Figure 13.14 (top panel) An abstract depiction of the sort of prototypical cost function seen when employing neural network models. Such a cost may contain many saddle points, long narrow valleys, and local minima whose depth closely match its global minima. With advanced local optimization techniques we can easily traverse saddle points and long narrow valleys, determining such local minima. (bottom panel) An abstract depiction of the sort of worst-case scenario nonconvex function rarely encountered when using neural network models. Here the difference in depth between the cost's local and global minima is substantial, and therefore models employing parameters from each area will vary substantially in quality. Given the vast number of poor local minima, such a cost function is very difficult to minimize properly using local optimization.

This is starkly different than a kind of hypothetical, worst-case scenario nonconvex cost (not often encountered with neural networks) whose global minima lie considerably lower than its local minima. An abstract depiction of such a function is shown in the bottom panel of Figure 13.14. There are no enhancements one can make to any local optimization method to enable it to effectively minimize such a cost function, with the only practical salve being to make many runs starting from random initial points to see if a global minimum can be reached by some individual run.

Example 13.10   Comparing first-order optimizers on a multi-layer neural network model

In this example we use $P = 50,000$ randomly selected data points from the MNIST dataset (see Example 7.10) to perform multi-class classification ($C = 10$) using a four-hidden-layer neural network with ten units per layer, and a tanh activation (this particular architecture was chosen arbitrarily for the purposes of this example). Here we compare the efficacy of three first-order optimizers: the standard gradient descent scheme (see Section 3.5), its component-normalized version (see Section A.3.2), and RMSProp (see Section A.4).

Each optimizer is used in both batch and mini-batch (using batch size of 200) regimes (see Section 7.8) to minimize the multi-class Softmax cost over this data. For all runs we initialize at the same starting point, and in each instance use the largest fixed steplength value of the form $\alpha = 10^\gamma$ (for integer γ) that produced convergence.

In Figure 13.15 we show the results of the batch (top row) and mini-batch (bottom row) versions, showing both cost and accuracy histories measuring the efficacy of each optimizer during the first ten epochs (or full sweeps through the dataset). In both the batch and mini-batch runs we can see how the component-normalized version of gradient descent and RMSProp significantly outperform the standard algorithm in terms of speed of convergence.

13.6   Batch Normalization

Previously, in Section 9.3, we saw how normalizing each input feature of a dataset significantly aids in speeding up parameter tuning, particularly with first-order optimization methods, by improving the shape of a cost function's contours (making them more "circular").

With our generic linear model

$$\text{model}(\mathbf{x}, \mathbf{w}) = w_0 + x_1 w_1 + \cdots + x_N w_N \tag{13.32}$$

standard normalization involves normalizing the distribution of each input dimension of a dataset $\{\mathbf{x}_p\}_{p=1}^{P}$ by making the substitution

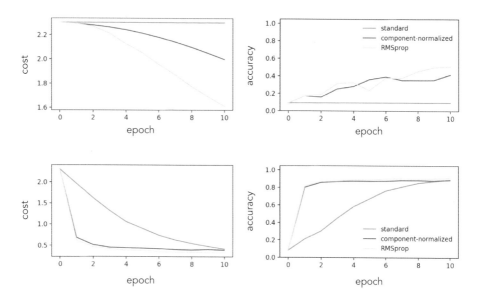

Figure 13.15 Figure associated with Example 13.10. Cost function (left column) and accuracy (right column) history plots comparing the efficacy of three first order optimizers in tuning the parameters of a four-hidden-layer neural network model over the MNIST dataset, using the batch (top row) and mini-batch (bottom row) versions of each optimizer. See text for further details.

$$x_{p,n} \longleftarrow \frac{x_{p,n} - \mu_n}{\sigma_n} \tag{13.33}$$

for the nth input dimension, where μ_n and σ_n are the mean and standard deviation along this dimension, respectively.

In this section we will learn how grafting a standard normalization step on to *each* hidden layer of an L-layer neural network model

$$\text{model}\,(\mathbf{x}, \Theta) = w_0 + f_1^{(L)}(\mathbf{x})\,w_1 + \cdots + f_{U_L}^{(L)}(\mathbf{x})\,w_{U_L} \tag{13.34}$$

where $f_1^{(L)}$ through $f_{U_L}^{(L)}$ are L-layer units as described in Section 13.2, similarly makes tuning the parameters of such a model significantly easier.

With this extended standard normalization technique, called *batch normalization* [62], we normalize not just the input to our fully connected network but the distribution of every unit in every hidden layer of the network as well. As we will see, doing this provides even greater optimization speed-up (than that provided by standard normalization of the input alone) for fully connected neural network models.

13.6.1 Batch normalization of single-hidden-layer units

Suppose first, for simplicity, that we are dealing with a single-layer neural network model. Setting $L = 1$ in Equation (13.34) gives this model as

$$\text{model}\,(\mathbf{x}, \Theta) = w_0 + f_1^{(1)}(\mathbf{x})\,w_1 + \cdots + f_{U_1}^{(1)}(\mathbf{x})\,w_{U_1}. \tag{13.35}$$

We now extend the basic standard normalization scheme, applying the same normalization concept to every "weight-touching" distribution of this model. Of course here the *input features* no longer touch the weights of the final linear combination (i.e., $w_1, w_2, ..., w_{U_1}$). They instead touch the weights internal to the single-layer units themselves. We can see this more easily by analyzing the jth single-layer unit in this network

$$f_j^{(1)}(\mathbf{x}) = a\left(w_{0,j}^{(1)} + \sum_{n=1}^{N} w_{n,j}^{(1)} x_n\right) \tag{13.36}$$

wherein the nth dimension of the input x_n only touches the internal weight $w_{n,j}^{(1)}$. Thus in standard normalizing the input we directly affect the contours of a cost function only along the weights internal to the single-layer units. To affect the contours of a cost function with respect to weights external to the first hidden layer (here the weights of the final linear combination) we must naturally normalize the *output* of the first hidden layer.

Putting these output values in a set – and denoting it by $\left\{f_j^{(1)}\left(\mathbf{x}_p\right)\right\}_{p=1}^{P}$ – we would naturally want to standard normalize *its* distribution as

$$f_j^{(1)}(\mathbf{x}) \longleftarrow \frac{f_j^{(1)}(\mathbf{x}) - \mu_{f_j^{(1)}}}{\sigma_{f_j^{(1)}}} \tag{13.37}$$

where the mean $\mu_{f_j^{(1)}}$ and standard deviation $\sigma_{f_j^{(1)}}$ are given as

$$\mu_{f_j^{(1)}} = \frac{1}{P}\sum_{p=1}^{P} f_j^{(1)}\left(\mathbf{x}_p\right) \quad \text{and} \quad \sigma_{f_j^{(1)}} = \sqrt{\frac{1}{P}\sum_{p=1}^{P}\left(f_j^{(1)}\left(\mathbf{x}_p\right) - \mu_{f_j^{(1)}}\right)^2}. \tag{13.38}$$

Note importantly that, unlike our input features, the output of the single-layer units (and hence their distributions) change every time the internal parameters of our model are changed, e.g., during each step of gradient descent. The constant alteration of these distributions is referred to as *internal covariate shift* in the jargon of machine learning, or just *covariate shift* for short, and implies that if we are to carry over the principle of standard normalization completely we will need to normalize the output of the first hidden layer at *every* step of parameter tuning. In other words, we need to graft standard normalization directly into the hidden layer of our architecture itself.

Below we show a generic recipe for doing just this, a simple extension of the recursive recipe for single-layer units given in Section 13.2.1.

Recursive recipe for batch-normalized single-layer units

---

1. Choose an activation function $a(\cdot)$

2. Compute the linear combination $v = w_0^{(1)} + \sum_{n=1}^{N} w_n^{(1)} x_n$

3. Pass the result through activation and form $f^{(1)}(\mathbf{x}) = a(v)$

4. Standard normalize $f^{(1)}$ as $f^{(1)}(\mathbf{x}) \longleftarrow \dfrac{f^{(1)}(\mathbf{x}) - \mu_{f^{(1)}}}{\sigma_{f^{(1)}}}$

Example 13.11   Visualizing internal covariate shift in a single-layer network
In this example we illustrate the internal covariate shift in a single-layer neural network model using two ReLU units $f_1^{(1)}$ and $f_2^{(1)}$, applied to performing two-class classification of the toy dataset introduced in Example 11.7. We run 5000 steps of gradient descent to minimize the two-class Softmax cost using this single-layer network, where we standard normalize the input data.

In Figure 13.16 we show the progression of this gradient descent run, plotting the tuples $\left\{ \left(f_1^{(1)}(\mathbf{x}_p), f_2^{(1)}(\mathbf{x}_p) \right) \right\}_{p=1}^{P}$ at three of the steps taken during the run. The top and bottom panels respectively show the covariate shift and the complete cost function history curve, where the current step of the optimization is marked on the curve with a red dot.

As can be seen in the top row of Figure 13.16, the distribution of these tuples change dramatically as the gradient descent algorithm progresses. We can intuit (from our previous discussions on input normalization) that this sort of shifting distribution negatively affects the speed at which gradient descent can properly minimize our cost function.

Next, we repeat this experiment using the same gradient descent settings but now with batch-normalized single-layer units, and plot the results in a similar manner in Figure 13.17. Notice how the distribution of activation outputs stays considerably more stable as gradient descent progresses.

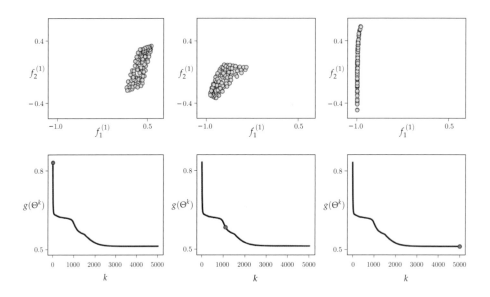

Figure 13.16 Figure associated with Example 13.11. See text for details.

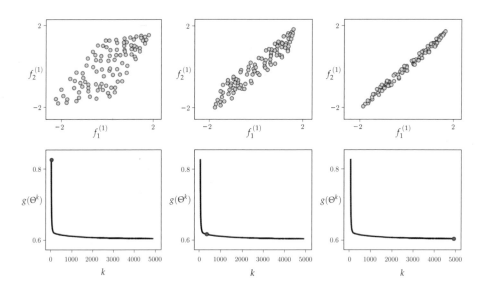

Figure 13.17 Figure associated with Example 13.11. See text for details.

13.6.2   Batch normalization of multi-hidden-layer units

Suppose for a moment that our fully connected neural network has just two hidden layers (i.e., $L = 2$ in Equation (13.34)), and that we have grafted a standard normalization step onto the first hidden layer of our network as described in the previous section, so that now our two-hidden-layer units touch the final linear combination weights (i.e., $w_1, w_2, ..., w_{U_2}$). To temper an associated cost function with respect to w_j we normalize the associated distribution of our jth unit via

$$f_j^{(2)}(\mathbf{x}) \longleftarrow \frac{f_j^{(2)}(\mathbf{x}) - \mu_{f_j^{(2)}}}{\sigma_{f_j^{(2)}}} \tag{13.39}$$

where the mean $\mu_{f_j^{(2)}}$ and standard deviation $\sigma_{f_j^{(2)}}$ are defined as

$$\mu_{f_j^{(2)}} = \frac{1}{P} \sum_{p=1}^{P} f_j^{(2)}\left(\mathbf{x}_p\right) \quad \text{and} \quad \sigma_{f_j^{(2)}} = \sqrt{\frac{1}{P} \sum_{p=1}^{P} \left(f_j^{(2)}\left(\mathbf{x}_p\right) - \mu_{f_j^{(2)}}\right)^2}. \tag{13.40}$$

As was the case in studying the single-layer case in the previous section, here too we will need to graft this step on to the second hidden layer of our network so that whenever its parameters change (e.g., during each step of a local optimizer) the distribution of this unit remains normalized.

Extending this concept to a general L-hidden-layer neural network model in Equation (13.34) we will normalize the output of every hidden layer of the network. Thus in general, for L-layer units once we have standard normalized the output of every layer preceding it we standard normalize the jth unit of the Lth hidden layer

$$f_j^{(L)}(\mathbf{x}) = a\left(w_{0,j}^{(L)} + \sum_{i=1}^{U_{L-1}} w_{i,j}^{(L)} f_i^{(L-1)}(\mathbf{x})\right) \tag{13.41}$$

via the substitution

$$f_j^{(L)}(\mathbf{x}) \longleftarrow \frac{f_j^{(L)}\left(\mathbf{x}_p\right) - \mu_{f_j^{(L)}}}{\sigma_{f_j^{(L)}}} \tag{13.42}$$

where the mean $\mu_{f_j^{(L)}}$ and standard deviation $\sigma_{f_j^{(L)}}$ are defined as

$$\mu_{f_j^{(L)}} = \frac{1}{P} \sum_{p=1}^{P} f_j^{(L)}\left(\mathbf{x}_p\right) \quad \text{and} \quad \sigma_{f_j^{(L)}} = \sqrt{\frac{1}{P} \sum_{p=1}^{P} \left(f_j^{(L)}\left(\mathbf{x}_p\right) - \mu_{f_j^{(L)}}\right)^2}. \tag{13.43}$$

As with the single-layer case, we can still construct each batch-normalized unit recursively since all we must do is insert a standard normalization step into the end of each layer as summarized in the following recipe (akin to the recipe given for a general L-layer unit in Section 13.2.3).

Recursive recipe for batch-normalized L-layer units

1. Choose an activation function $a(\cdot)$

2. Construct U_{L-1} batch-normalized $(L-1)$-layer units $f_i^{(L-1)}(\mathbf{x})$
   for $i = 1, 2, ..., U_{L-1}$

3. Compute the linear combination $v = w_0^{(L)} + \sum_{i=1}^{U_{L-1}} w_i^{(L)} f_i^{(L-1)}(\mathbf{x})$

4. Pass the result through activation and form $f^{(L)}(\mathbf{x}) = a(v)$

5. Standard normalize $f^{(L)}$ via $f^{(L)}(\mathbf{x}) \longleftarrow \dfrac{f^{(L)}(\mathbf{x}) - \mu_{f^{(L)}}}{\sigma_{f^{(L)}}}$

When employing a stochastic or mini-batch first-order method for optimization (see Section 7.8), normalization of the architecture is performed precisely as detailed here, on each individual mini-batch. Also note that, in practice, the batch normalization formula in Equation (13.42) is often parameterized as

$$f^{(L)}(\mathbf{x}) \longleftarrow \alpha \frac{f^{(L)}(\mathbf{x}) - \mu_{f^{(L)}}}{\sigma_{f^{(L)}}} + \beta \qquad (13.44)$$

where inclusion of the tunable parameters α and β (which are tuned along with the other parameters of a batch-normalized network) allows for greater flexibility. However, even without these extra parameters we can achieve significant improvement in optimization speed when tuning fully connected neural network models with the employment of (unparameterized) batch normalization.

Example 13.12   Visualizing internal covariate shift in a multi-layer network
In this example we illustrate the covariate shift in a four-hidden-layer network with two units per layer, using the ReLU activation and the same dataset employed in Example 13.11. We then compare this to the covariate shift present in the batch-normalized version of the same network. We use just two units per layer so that we can visualize the distribution of activation outputs of each layer.

Beginning with the unnormalized version of the network we can see that – as with the single-layer case in Example 13.11 – the covariate shift of this network (shown in the top row of Figure 13.18) is considerable. The distribution of each hidden layer's units is shown here, with the output tuples of the ℓth hidden layer $\left\{ \left(f_1^{(\ell)}(\mathbf{x}_p), f_2^{(\ell)}(\mathbf{x}_p) \right) \right\}_{p=1}^{P}$ colored in cyan for $\ell = 1$, in magenta for $\ell = 2$, in lime green for $\ell = 3$, and in orange for $\ell = 4$.

Performing batch normalization on each layer of this network helps considerably in taming this covariate shift. In Figure 13.19 we show the result of running the same experiment, using the same initialization, activation, and dataset, but

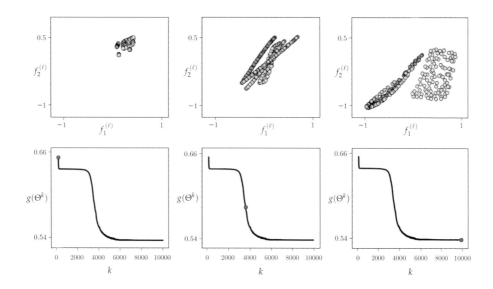

Figure 13.18 Figure associated with Example 13.12. See text for details.

this time using the batch-normalized version of the network. Studying the figure from left to right, as gradient descent progresses, we can see once again that the distribution of each layer's activation outputs remains much more stable than previously.

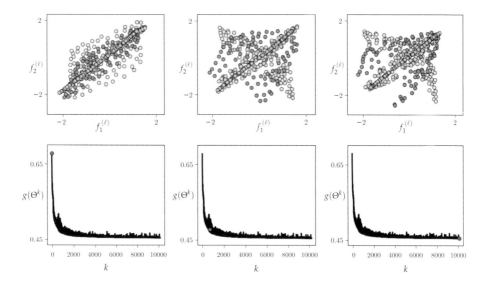

Figure 13.19 Figure associated with Example 13.12. See text for details.

Example 13.13   Standard versus batch normalization on MNIST

In this example we illustrate the benefit of batch normalization in terms of speeding up optimization via graident descent on a dataset of $P = 50,000$ randomly chosen handwritten digits from the MNIST dataset (introduced in Example 7.11). In Figure 13.20 we show cost (left panel) and classification accuracy (right panel) histories of ten epochs of gradient descent, using the largest steplength of the form 10^γ (for integer γ) we found that produced adequate convergence. We compare the standard and batch-normalized version of a four-hidden-layer neural network with ten units per layer and ReLU activation. Here we can see, both in terms of cost function value and number of misclassifications (accuracy), that the batch-normalized version allows for much more rapid minimization via gradient descent.

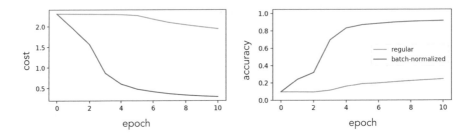

Figure 13.20 Figure associated with Example 13.13. See text for details.

13.6.3   Evaluation of new data points in batch-normalized networks

An important point to remember when employing a batch-normalized neural network is that we must treat new data points (not used in training) precisely as we treat training data. This means that the final normalization constants determined during training (i.e., the various means and standard deviations of the input as well as those for each hidden layer output) must be saved and reused in order to properly evaluate new data points. More specifically, all normalization constants in a batch-normalized network should be *fixed* to the values computed at the final step of training (e.g., at the best step of gradient descent) when evaluating new data points.

13.7   Cross-Validation via Early Stopping

Being highly parameterized, the optimization of cost functions associated with fully connected neural networks, particularly those employing many hidden layers, can require significant computation. Because of this *early stopping based regularization* (as described in Section 11.6.2), which involves learning parameters to minimize validation error during a single run of optimization, is a popular cross-validation technique when employing fully connected multi-layer networks. The notion of early stopping is also the basis for specialized ensembling techniques which aim at producing a set of neural network models for bagging (introduced in Section 11.9) at minimal computational cost (see, e.g., [63, 64]).

Example 13.14   Early stopping and regression

In this example we illustrate the early stopping procedure using a simple nonlinear regression dataset (split into $\frac{2}{3}$ training and $\frac{1}{3}$ validation), and a three-hidden-layer neural network with ten units per layer, and with tanh activation. Three different steps from a single run of gradient descent (for a total of 10,000 steps) is illustrated in Figure 13.21, one per each column, with the resulting fit at each step shown over the original (first row), training (second row), and validation data (third row). Stopping the gradient descent early after taking (around) 2000 steps provides, for this training-validation split of the original data, a fine nonlinear model for the entire dataset.

Example 13.15   Early stopping and handwritten digit classification

In this example we use early stopping based regularization to determine the optimal settings of a two-hidden-layer neural network, with 100 units per layer and ReLU activation, over the MNIST dataset of handwritten digits first described in Example 7.10. This multi-class dataset ($C = 10$) consists of $P = 50,000$ points in the training and 10,000 points in the validation set. With a batch size of 500 we run 100 epochs of the standard mini-batch gradient descent scheme, resulting in the training (blue) and validation (yellow) cost function (left panel) and accuracy (right panel) history curves shown in Figure 13.22. Employing the multi-class Softmax cost, we found the optimal epoch with this setup achieved around 99 percent accuracy on the training set, and around 96 percent accuracy on the validation set. One can introduce enhancements like those discussed in the previous sections of this chapter to improve these accuracies further. For comparison, a linear classifier – trained/validated on the same data – achieved 94 and 92 percent training and validation accuracies, respectively.

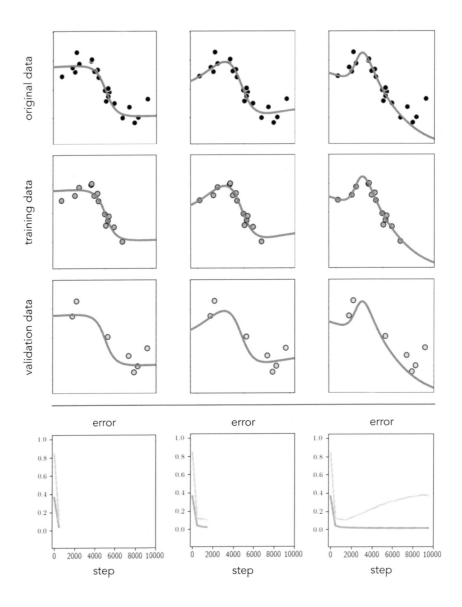

Figure 13.21 Figure associated with Example 13.14. See text for details.

13.8   Conclusion

In this chapter we described a range of technical matters associated with fully connected neural networks, which were first introduced in Section 11.2.3.

We began by carefully describing single- and multi-layer neural network architectures in Section 13.2, followed by a discussion of activation functions in Section 13.3, backpropogation in Section 13.4, and the nonconvexity of cost

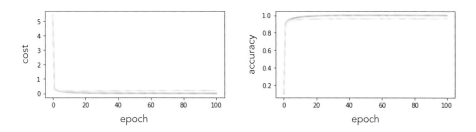

Figure 13.22 Figure associated with Example 13.15. See text for details.

functions over neural network models in Section 13.5. Batch normalization – a natural extension of the standard normalization procedure described in Section 9.4 – was then explored in Section 13.6. Finally, in Section 13.7 we discussed the use of regularization (and in particular, early stopping) based cross-validation – first described in great detail in Section 11.6 – with fully connected neural network models.

13.9 Exercises

† The data required to complete the following exercises can be downloaded from the text's github repository at `github.com/jermwatt/machine_learning_refined`

13.1 Two-class classification with neural networks
Repeat the two-class classification experiment described in Example 13.4 beginning with the implementation outlined in Section 13.2.6. You need not reproduce the result shown in the top row of Figure 13.9, but can verify your result via checking that you can achieve perfect classification of the data.

13.2 Multi-class classification with neural networks
Repeat the multi-class classification experiment described in Example 13.4 beginning with the implementation outlined in Section 13.2.6. You need not reproduce the result shown in the bottom row of Figure 13.9, but can verify your result via checking that you can achieve perfect classification of the data.

13.3 Number of weights to learn in a neural network
(a) Find the total number Q of tunable parameters in a general L-hidden-layer neural network, in terms of variables expressed in the `layer_sizes` list in Section 13.2.6.

(b) Based on your answer in part (a), explain how the input dimension N and number of data points P each contributes to Q. How is this different from what you saw with kernel methods in the previous chapter?

13.4   Nonlinear Autoencoder using neural networks
Repeat the Autoencoder experiment described in Example 13.6 beginning with the implementation outlined in Section 13.2.6. You need not reproduce the projection map shown in the bottom-right panel of Figure 13.11.

13.5   The maxout activation function
Repeat Exercise 13.4 using the maxout activation (detailed in Example 13.9).

13.6   Comparing advanced first-order optimizers I
Repeat the first set of experiments described in Example 13.10, and produce plots like those shown in the top row of Figure 13.15. Your plots may not look precisely like those shown in this figure (but they should look similar).

13.7   Comparing advanced first-order optimizers II
Repeat the second set of experiments described in Example 13.10, and produce plots like those shown in the bottom row of Figure 13.15. Your plots may not look precisely like those shown in this figure (but they should look similar).

13.8   Batch normalization
Repeat the experiment described in Example 13.13, and produce plots like those shown in Figure 13.20. Your plots may not look precisely like those shown in this figure (but they should look similar).

13.9   Early stopping cross-validation
Repeat the experiment described in Example 13.14. You need not reproduce all the panels shown in Figure 13.21. However, you should plot the fit provided by the weights associated with the minimum validation error on top of the entire dataset.

13.10   Handwritten digit recognition using neural networks
Repeat the experiment described in Example 13.15, and produce cost/accuracy history plots like the ones shown in Figure 13.22. You may not reproduce exactly what is reported based on your particular implementation. However, you should be able to achieve similar results as reported in Example 13.15.

14 Tree-Based Learners

14.1 Introduction

In this chapter we greatly expound on our discussion of *tree-based* learners, first introduced in Section 11.2.3, which are wildly popular due to their great effectiveness particularly with structured data (see, e.g., the discussion in Section 11.8). In this chapter we explore the technical eccentricities associated with tree-based learners, describe the so-called *regression* and *classification trees*, and explain their particular usage with boosting based cross-validation and bagged ensembles (first introduced in Sections 11.5 and 11.9, respectively) where they are referred to as *gradient boosting* and *random forests* in the jargon of machine learning.

14.2 From Stumps to Deep Trees

In Section 11.2.3 we introduced the simplest exemplar of a tree-based learner: the stump. In this section we discuss how, using simple stumps, we can define more general and complex tree-based universal approximators.

14.2.1 The stump

The most basic tree-based universal approximator, the stump, is a simple step function of the form

$$f(x) = \begin{cases} v_1 & x \leq s \\ v_2 & x > s \end{cases} \tag{14.1}$$

with three tunable parameters: two step levels or *leaf* parameters denoted by v_1 and v_2 (whose values are set independently of one another), and a split point parameter s defining the boundary between the two levels. This simple stump is depicted in the top-left panel of Figure 14.1. In the top-right panel of this figure we show another graphical representation of the generic stump in Equation (14.1), which helps explain the particular nomenclature (i.e., trees, leaves, etc.) used in the context of tree-based approximators. Represented this

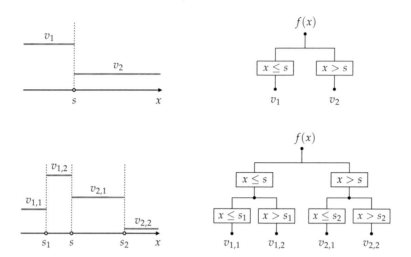

Figure 14.1 (top-left panel) A simple stump, defined in Equation (14.1). (top-right panel) A graphical illustration of a stump function as a binary tree. (bottom-left panel) A depth-two tree formed by recursing on each leaf of the stump, replacing it with a new stump. (bottom-right panel) A graphical illustration of the depth-two tree.

way, the stump can be thought of as a binary tree structure of *depth one*, with $f(x)$ as its root node, and v_1 and v_2 as its leaf nodes.

The stump defined in Equation (14.1) takes in one-dimensional (i.e., scalar) input. When the input is N-dimensional in general, the stump cuts along a single dimension (or coordinate axis). For example, when defined along some nth dimension a stump taking in N-dimensional input $\mathbf{x}$ is defined as

$$f(\mathbf{x}) = \begin{cases} v_1 & x_n \le s \\ v_2 & x_n > s \end{cases} \tag{14.2}$$

where x_n here denotes the nth dimension of $\mathbf{x}$.

14.2.2   Creating deep trees via recursion

Recursing we can construct deeper trees by applying the same concept used to build a stump to each of its leaves, i.e., by splitting each leaf in two. This recursion results in a tree of *depth two*, with three split points and four distinct leaves, as shown in the bottom row of Figure 14.1. A depth-two tree has significantly greater capacity (see Section 11.2) than a stump, since the location of the split points and the leaf values can be set in a multitude of different ways, as can be seen in the top row of Figure 14.2.

This recursive idea can then be continually applied to each leaf of a depth-two tree to create a depth-three tree, and so forth. The deeper a tree becomes the more capacity it gains, with each unit being able to take on a wider variety

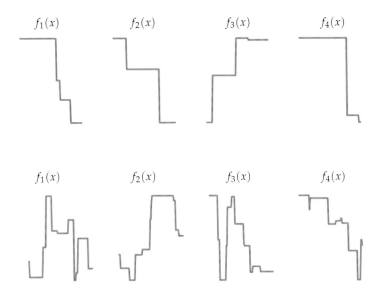

Figure 14.2 Four instances of a depth $D = 2$ tree (top row) and a depth $D = 10$ tree (bottom row), where in each instance all the parameters (i.e., split points and leaf values) are set at random. The latter is clearly capable of generating a wider swath of shapes given different settings of its parameters, and thus has higher capacity. Note here that the leaves are connected by vertical lines in order to give each tree instance a continuous appearance, which is done for visualization purposes only.

of different shapes, as can be seen in the bottom row of Figure 14.2. Indeed this is reflected in the fact that trees become exponentially more parameterized the deeper they are made: one can easily show that a tree of general depth D (with scalar input) will have $2^D - 1$ split points and 2^D leaves, thus $2^{D+1} - 1$ tunable parameters in total. This recursive procedure is often referred to as the *growing of a tree* in the jargon of machine learning.

Note importantly that *unlike* fixed-shape and neural network universal approximators, tree-based units are defined *locally*. This means that when we adjust one parameter of a polynomial or a neural network unit it can *globally* affect the shape of the function over the entire input space. However, when we split any leaf of a tree we are only affecting the shape of the tree locally at that leaf. This is why tree-based universal approximators are sometimes called *local function approximators*.

14.2.3   Creating deep trees via addition

Deeper, more flexible trees can also be constructed via *addition* of shallower trees. For instance, Figure 14.3 illustrates how a depth-two tree can be created by adding together three depth-one trees (i.e., three stumps). Again, it is easy

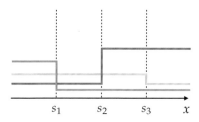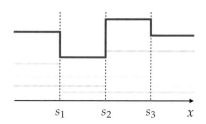

Figure 14.3 (left panel) Three stumps, each depicted in a distinct color. For ease of visualization we have connected the leaves by vertical lines in order to give each stump instance a continuous appearance. (right panel) A depth-two tree (in black) created by adding together the three stumps shown in the left panel.

to show that in general adding $2^D - 1$ stumps (with scalar input) together will create a depth-D tree (provided that the stumps do not share any split points).

14.2.4   Human interpretability

Given their particularly simple structure, shallow tree-based units (such as the depth-one and depth-two trees shown in Figure 14.1) are often easy to interpret by humans, in comparison to their fixed-shape and neural network counterparts. However, this feature of tree-based units quickly dissipates as the depth of a tree is increased (see, e.g., depth $D = 10$ trees depicted in the bottom row of Figure 14.2) as well as when trees are combined or *ensembled* together.

14.3   Regression Trees

In this section we discuss the use of general tree-based universal approximators for the problem of regression, often called *regression trees*. Unlike fixed-shape or neural network universal approximators, cost functions imbued with tree-based units create highly nonconvex, staircase-like functions that cannot be easily minimized by any local optimization method. To see why this is the case via an example, let us take the simple regression dataset shown in the top-left panel of Figure 14.4, and try to fit a nonlinear regressor to it using a model composed of a single stump, by minimizing an appropriate cost function over this model (e.g., the Least Squares cost).

14.3.1   Determining an optimal split point when leaf values are fixed

Fitting a single-stump model to our dataset entails tuning its three parameters: the location of the stump's split point, as well as its two leaf values. To make matters easy, here we fix the two leaf parameters associated with our model to two arbitrary values so that only the split point parameter s remains to be

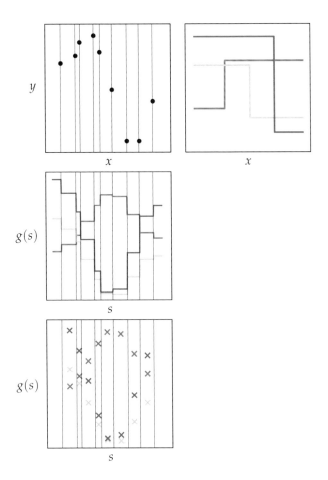

Figure 14.4 (top-left panel) A prototypical nonlinear regression dataset. (top-right panel) Three stumps with fixed leaf values (whose split point can vary). (middle panel) Each stump instance is slid horizontally across the input of the data by varying its split point value, creating three corresponding staircase-like Least Squares costs. (bottom panel) Each cost in the middle panel is constant in between consecutive inputs, implying that we need only test one split point per flat region, e.g., the mid-point, as shown in this panel. See text for further details.

optimally tuned, and hence we can now visualize the *one-dimensional* Least Squares cost function $g(s)$ involving the split point parameter alone. In the top-right panel of Figure 14.4 we show three stump instances, colored red, green, and blue, with distinct but fixed leaf values. Now we take each stump and sweep it over the dataset horizontally, trying out for each all possible split points in the input space of the dataset. The three Least Squares cost functions resulting from this exercise are shown in the middle panel of the figure and are colored to match their corresponding stumps shown in the top-right panel. Each one-dimensional cost, as we can see, looks like a staircase consisting of many perfectly flat regions. These problematic flat regions are a direct consequence of the shape of our nonlinearity (i.e., the stump). Recall, we saw similar behavior when dealing with step functions in the context of logistic regression back in Sections 6.2 and 6.3. The existence of such flat regions is massively undesirable because no local optimization algorithm can navigate them effectively.

However, notice when the leaf values of a stump are fixed the corresponding Least Squares cost remains constant for all split point values *in between* consecutive inputs. In other words, all three cost functions in the middle panel of Figure 14.4 take on a staircase shape with their flat step areas located in the same locations: the regions in between consecutive input values.

This fact has a very practical repercussion: while we cannot properly tune the split point parameter using local optimization (due to the staircase shape of the cost function over this parameter), we can, however, find one by simply testing a *single value* (e.g., the mid-point) in each of the flat areas of the cost since all split points there produce the same regression quality. This collection of mid-point evaluations for each of our three example stumps is illustrated in the bottom panel of Figure 14.4.

14.3.2 Determining optimal leaf values when split point is fixed

Contrary to the task of determining the optimal split point of a stump with fixed leaf values, determining the optimal leaf values for a stump with a fixed split point is exceedingly straightforward. Since the leaves of a stump are *constant-valued*, and we want them both to be set so that together our stump represents the data as well as possible, it makes intuitive sense simply to set the value of each leaf to the *mean output* of those points it will represent. This choice is shown in red in the right panel of Figure 14.5 for our toy regression dataset, where for a given split point (illustrated by the vertical blue dashed line in the left panel) the leaf value on the *left* is set to the mean of the output of those points lying to the *left* of the split point, and the leaf value on the *right* is set to the mean of the output of those points to the *right* of the split point.

This intuitive choice of leaf value can be completely justified via the *first-order optimality condition* introduced in Section 3.2. To see this, let us first formalize the general scenario we are investigating. In the context of regression, our fixed split point s is defined along the nth input dimension of a regression dataset – denoted by $\left\{\left(\mathbf{x}_p, y_p\right)\right\}_{p=1}^{P}$ – and splits the data into two sections. We can keep track of these two subsets of our data via index sets Ω_L and Ω_R, which denote the input/output pairs of our dataset lying on either side of the split to the "left" and "right" of it, expressed formally as

$$\Omega_L = \left\{p \mid x_{p,n} \le s\right\} \qquad \text{and} \qquad \Omega_R = \left\{p \mid x_{p,n} > s\right\}. \qquad (14.3)$$

A general stump using this split point (echoing Equation (14.2)) can then be written as

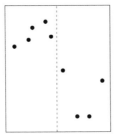 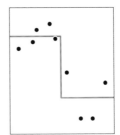

Figure 14.5 (left panel) The same regression dataset shown in Figure 14.4, along with the fixed split point shown via a vertical dashed blue line that divides the input space of data into two subspaces lying to the left and right of this line. (right panel) The stump with optimally set leaf values, determined as the mean of the output of all data points to the left and right of the split point. See text for further details.

$$f(\mathbf{x}) = \begin{cases} v_L & x_n \leq s \\ v_R & x_n > s \end{cases} \tag{14.4}$$

where x_n is the nth dimension of the input $\mathbf{x}$, and v_L and v_R are leaf values we will determine.

To determine the optimal values of v_L and v_R we can minimize two *one-dimensional* Least Squares costs, defined over the points belonging to the index set Ω_L and Ω_R, respectively, as

$$g(v_L) = \frac{1}{|\Omega_L|} \sum_{p \in \Omega_L} \left(v_L - y_p\right)^2 \quad \text{and} \quad g(v_R) = \frac{1}{|\Omega_R|} \sum_{p \in \Omega_R} \left(v_R - y_p\right)^2 \tag{14.5}$$

where $|\Omega_L|$ and $|\Omega_R|$ denote the number of points belonging to the index sets Ω_L and Ω_R, respectively.

Each of these cost functions is exceptionally simple. Setting the derivative of each to zero (with respect to its corresponding leaf value) and solving gives the optimal leaf values $v_L^{\star}$ and $v_R^{\star}$, respectively, as

$$v_L^{\star} = \frac{1}{|\Omega_L|} \sum_{p \in \Omega_L} y_p \quad \text{and} \quad v_R^{\star} = \frac{1}{|\Omega_R|} \sum_{p \in \Omega_R} y_p. \tag{14.6}$$

14.3.3 Optimization of regression stumps

Combining the two ideas discussed previously provides a reasonable work-around for tuning all three parameters of a stump for the purposes of regression (as an alternative to tuning all three together via local optimization, which we cannot do). That is, first we create a set of candidate split point values by recording every mid-point between our input data, along each of its input dimensions. For each candidate split point we then determine the stump's leaf values optimally, setting them to the mean of the training data output to the left and right of the split point, and compute its (Least Squares) cost value. After doing this for all candidate split points, we find the very best stump (with optimal split point and leaf values) as one that provides the lowest cost value.

Example 14.1   Fitting the parameters of a simple regression tree

In this example we fit a single-stump model to the toy regression dataset shown in Figures 14.4 and 14.5, illustrating the entire range of candidate stumps whose split points are formed by taking the mid-point between each consecutive pair of inputs and whose corresponding leaf values are set to the mean of the output on either side of each split. Scanning the panels of Figure 14.6 from the top-left to the bottom-right we illustrate the entire range of candidate stumps for this dataset, scanning from left to right across the input of the dataset. In the top of each panel we show the candidate stump, with the Least Squares cost values associated with that particular stump as well as those that preceded it plotted underneath. Once all candidates have been tested (as shown in the bottom-right panel), the particular stump providing the lowest possible cost value (here, the fifth stump in the bottom-left panel) is found optimal.

In general, for a dataset of P points, each of input dimension N, there are a total of $N(P-1)$ split points to choose from over the entire input space of the problem. When computation becomes a primary concern (mainly with very large P and/or N) sampling strategies derived from this basic scheme may be used, including testing split points along only a random selection of all input dimensions, testing a coarser selection of split points, and so on.

14.3.4   Deeper regression trees

To fit a depth-two tree to a regression dataset we first fit a stump as described in the previous section, and then recurse on the same idea on each of the stump's leaves. In other words, we can think of the first fitted stump as dividing our original dataset into two nonoverlapping subsets, one belonging to each leaf. Thinking recursively we can then fit a stump to each of these subsets in precisely the same way as we fit the stump to the original dataset, splitting each of the leaves of our original stump in two and creating a depth-two tree. We can go on further and repeat this process, splitting each leaf of our depth-two tree to create a depth-three tree, and so on.

Note that in the process of *growing* the tree, there are certain conditions in which we should *not* split a leaf. For example, there is no reason to split a leaf that contains just a single data point, or one where data points contained in

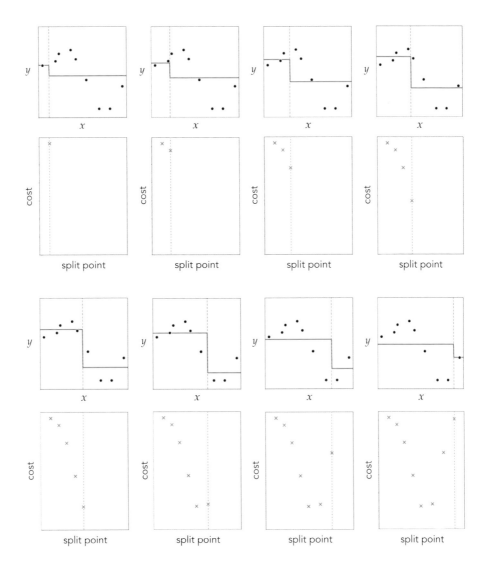

Figure 14.6 Figure associated with Example 14.1. See text for details.

the input space of the leaf have precisely the same output value (since in both instances our current leaf represents the data contained in it perfectly). Both of these practical conditions are often generalized when implementing recursive tree-based regressors in practice. For instance, progress on a leaf may be halted if it contains less than a predetermined number of points (as opposed to just a singleton point). Thus, in practice, a regression tree (with scalar input) of depth D may not end up with precisely 2^D leaves. Instead, certain branches stemming from the root of the tree may halt sooner than others, with some branches of the tree possibly growing to the defined depth. Therefore when applying binary trees to regression (and, as we will see, classification) we refer to the trees as having a *maximum depth*, i.e., the largest depth that a branch of the tree can possibly grow to.

Example 14.2   Growing a maximum-depth regression tree

The recursive procedure for growing a deep regression tree is illustrated in Figure 14.7. We begin (on the left) by fitting a stump to the original dataset. As we move from left to right the recursion proceeds, with each leaf of the preceding tree split in order to create the next, deeper tree. As can be seen in the rightmost panel, a tree with maximum depth of four is capable of representing the training data perfectly.

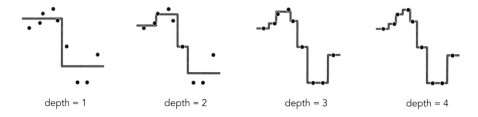

depth = 1   depth = 2   depth = 3   depth = 4

Figure 14.7 Figure associated with Example 14.2. See text for details.

14.4   Classification Trees

In this section we discuss the application of tree-based universal approximators to the problem of classification, often referred to as *classification trees*. Thankfully, virtually everything that we have previously seen regarding regression trees in the previous section caries over directly to the problem of classification. However, as we will see, the fact that classification data has *discrete* output naturally provokes different approaches to determining appropriate leaf values. As in the prior section, we begin here by discussing the proper construction of a stump, employing a toy dataset to illustrate key concepts, and recurse on this idea to create deeper trees.

14.4.1   Determining an optimal split point when leaf values are fixed

Imagine we are now dealing with a classification dataset, for example, the toy dataset shown in the left panel of Figure 14.8, and suppose that we aim to properly fit a stump to this data. If we attempt to set the *split point* of our stump via local optimization we run into precisely the same problem we came upon in Section 14.3.1 with regression. That is, not only will any corresponding classification cost function be nonconvex, but it will consist of completely flat, staircase-like sections that no local optimization algorithm can navigate effectively. Thus determining the optimal split point value in the case of classification must naturally result in the same approach we saw with regression in the previous section: we must test out an array of split point candidates to determine which works best. Once again, for the same practical purposes we saw with regression, we can simply test the mid-points between each consecutive pair of input values along each of their input dimensions (or a subset of these points if their number becomes prohibitively large).

14.4.2   Determining optimal leaf values when split point is fixed

In Section 14.3.2 we saw with regression that the leaf values for a single-stump model can be intuitively set to the *mean* output of those points belonging to each leaf of the stump. We backed up this intuitive choice by showing that these settings are precisely what we find by solving the first-order condition for a set of appropriately defined Least Squares cost functions. Here in the classification scenario, let us follow this logic in reverse and begin by optimizing an appropriate cost function (e.g., the Perceptron or Cross Entropy/Softmax costs). We then follow by presenting an intuitive choice based on a different statistic of the output: the *mode*.

Suppose we are tasked with classification of a two-class dataset $\left\{\left(\mathbf{x}_p, y_p\right)\right\}_{p=1}^{P}$ with label values $y_p \in \{-1, +1\}$, and that the split point of our stump is fixed. We define index sets Ω_L and Ω_R as in Equation (14.3) to denote the indices of all points lying to the "left" and "right" of our split point. To determine the optimal values of v_L and v_R we can minimize two *one-dimensional* classification costs (e.g., the Softmax) defined over the points belonging to the index set Ω_L and Ω_R, respectively, as

$$g(v_L) = \frac{1}{|\Omega_L|} \sum_{p \in \Omega_L} \log\left(1 + e^{-y_p v_L}\right) \quad \text{and} \quad g(v_R) = \frac{1}{|\Omega_R|} \sum_{p \in \Omega_R} \log\left(1 + e^{-y_p v_R}\right)$$

$$(14.7)$$

where again, as in Equation (14.5), $|\Omega_L|$ and $|\Omega_R|$ denote the number of points belonging to the index sets Ω_L and Ω_R, respectively. One can also weight the summands of such cost functions (as detailed in Section 6.9.3) in order to better deal with potential *class imbalance* in the leaves.

In either case, unlike the analogous pair of Least Squares costs in Equation (14.5), here we cannot solve the corresponding first-order conditions in closed form and must rely on local optimization techniques. However, because of the simplicity of each problem such optimizations are especially easy to solve iteratively. Indeed, often these sorts of costs are approximately minimized by applying just a *single* step of Newton's method. Doing this substantially minimizes the costs while keeping computation overhead low[1] and preventing potential numerical issues associated with the Softmax cost and Newton's method (introduced in the context of linear two-class classification in Section 6.6).

Note that, as with any other approach to classification, once appropriate leaf values have been determined, to make valid predictions the output of a classification stump must be passed through an appropriate discretizer, e.g., the sign function in the case of two-class classification using label values ±1 (see Section 6.8.1).

As an alternative to the *cost function* based approach detailed thus far, one can also choose optimal leaf values based on simple statistics of the output. Since the output of classification data is *discrete*, we would naturally avoid using the *mean* as our statistic of choice (as we did with regression), and instead lean towards using the *mode* (i.e., the most popular output label), also called the *majority vote*. Using the mode will keep our leaf values constrained to the discrete labels of our data, providing more appropriate stumps.

However, the standard mode statistic can lead to undesirable consequences, an example of which is illustrated for a simple two-class dataset in Figure 14.8. For this particular dataset because of the distribution of the majority class (here those points with label value $y_p = -1$) the statistical mode on *both* sides of *every* stump will always equal −1, and thus all stumps will be entirely flat and identical. The lack of stump diversity in this simple example does not invalidate the use of the standard mode. However, it does highlight its inefficiency in that deeper trees (which are more costly to create) are needed to capture the nonlinearity of such a toy dataset.

To compensate for class imbalances like the one shown here we can, in complete analogy to the concept of weighting cost functions to better handle class imbalance (see Section 6.8.1), choose leaf values based on the *balanced mode* or *balanced majority vote*. To compute the standard mode on one leaf of a stump we simply count up the number of points belonging to each class in the leaf, and determine the mode by picking the class associated with the largest count. To compute the *balanced* mode on one leaf of a stump we first count up the number of points belonging to each class on the leaf and then weight each count inversely based on the number of points in each class belonging to both leaves of the stump, determining the balanced mode by choosing the largest resulting weighted count. For a general multi-class dataset with C classes, the weighted count for the cth class on one leaf can be written as

---

[1] As detailed in Section 4.3, a single Newton step involves minimizing the best *quadratic approximation* to a cost function provided by its second-order Taylor series expansion, and for a general single-input cost $g(w)$ results in a simple update of the form

$$w^\star = w^0 - \frac{\frac{d}{dw}g(w^0)}{\frac{d^2}{dw^2}g(w^0) + \lambda} \tag{14.8}$$

where w^0 is some initial point, $\lambda \geq 0$ is a regularization parameter used to prevent possible division by zero (as discussed in Section 4.3.3), and $w^\star$ is the optimal update.

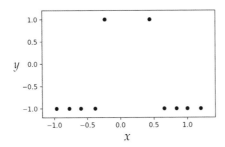 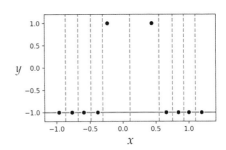

Figure 14.8 (left panel) A simple two-class classification dataset. (right panel) For each of the nine split points denoted by vertical dashed blue lines, assigning leaf values based on the *standard* mode leads to a completely flat stump, shown here in red. See text for further details.

$$\frac{\text{number of points from class } c \text{ in leaf}}{\text{number of points from class } c \text{ in both leaves of the stump}}. \quad (14.9)$$

Figure 14.9 shows the result of using this strategy (of setting the leaf values based on balanced mode instead of the mode itself) on the same dataset illustrated previously in Figure 14.8. Using the balanced mode here we produce a greater variety of stumps (when compared with using the standard mode), which allows us to capture the nonlinearity present in this dataset more effectively. To see how the balanced majority was used to define the leaf values in this instance let us examine one of the stumps (the sixth one in the middle-right panel) more closely. This stump has six data points on its left and four data points on its right side. Of the six data points lying to the left of its split point, four points have a label value of −1 and two points a label value of +1, resulting in a balanced majority vote of $\frac{4}{8}$ and $\frac{2}{2}$ for the two classes, respectively (noting that in this dataset there are a total of eight data points in the −1 class and two in the +1 class). Since $\frac{2}{2} > \frac{4}{8}$ the leaf value on the left is set to +1. Likewise, the balanced majority votes to the right of the split point for the −1 and +1 classes are calculated similarly as $\frac{4}{8}$ and $\frac{0}{2}$, respectively, and hence the right leaf value is set to −1.

14.4.3   Optimization of classification stumps

Putting everything together, to determine an optimal stump (consisting of optimal split point and leaf values) we can range over a set of reasonably chosen split points and construct corresponding leaf values for each stump using either the *cost function* based or *majority vote* based approaches described in Section 14.4.2. To determine which stump is ideal for our dataset we can then compute an appropriate classification metric over every stump instance and choose the one that provides the best performance. For example, with two-class classification we can employ an accuracy metric like those introduced in Section 6.8,

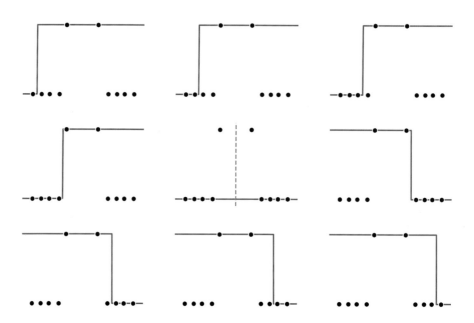

Figure 14.9 The same dataset and set of split points shown in Figure 14.8, only here the *balanced* mode calculation in Equation (14.9) is used to create the leaf values. See text for further details.

with the *balanced accuracy* discussed in Section 6.8.4 being the safest choice given the class imbalance we might encounter in practice, or more specialized metrics such as *information gain* [65].

Note that, in the context of classification trees, quality metrics are often referred to as *purity* metrics, since they measure how pure each leaf of the stump is in terms of class representation. Ideally, the stump is chosen that best represents the data while its leaves remain as "pure" as possible, each containing (largely) members of a single class if possible.

Example 14.3   Fitting the parameters of a simple classification tree

In Figure 14.10 we illustrate the resulting balanced accuracy of the stumps shown in Figure 14.9. Because of the symmetry of this particular dataset only the first five stumps are shown here, of which the fourth one provides the minimum cost, and thus is optimal for our dataset.

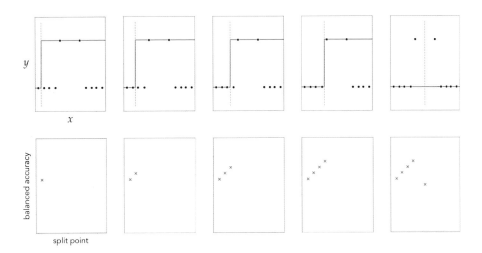

Figure 14.10 Figure associated with Example 14.3. See text for details.

14.4.4 Deeper classification trees

To build deeper classification trees we recurse on the two leaves of a stump and treat each like we did the original data, building a stump out of each. Just as with regression trees (see Section 14.3.4) this often results in binary trees of a *maximum depth*, as certain branches halt under obvious and/or user-defined conditions. With classification, one natural halting condition is when a leaf is completely *pure*, that is, it contains only members of a single class. In such a case there is no reason to continue splitting such a leaf. Other common halting conditions often used in practice include halting growth when the number of points on a leaf falls below a certain threshold and/or when splits do not sufficiently increase accuracy.

Example 14.4   Growing two maximum-depth classification trees

In Figure 14.11 we illustrate the growth of a tree to a maximum depth of seven on a two-class classification dataset. In Figure 14.12 we do the same for a multi-class classification dataset with $C = 3$ classes. In both cases as the tree grows note how many parts of the input space do not change as leaves on the deeper branches become *pure*. By the time we reach a maximum depth of seven we have considerably overfit both datasets.

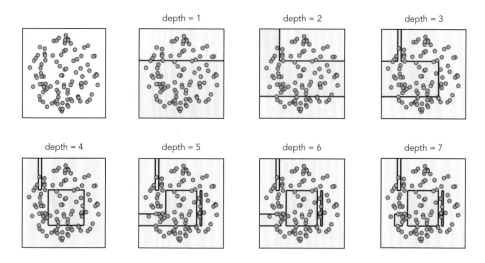

Figure 14.11 Figure associated with Example 14.4. See text for details.

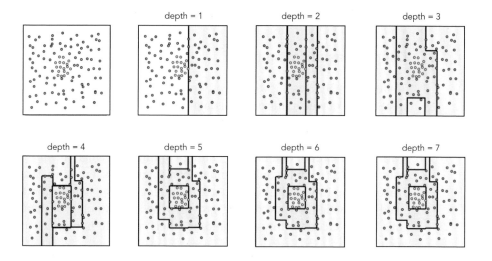

Figure 14.12 Figure associated with Example 14.4. See text for details.

14.5   Gradient Boosting

As mentioned in Section 14.2.3 deep tree-based universal approximators can be built via addition of shallow trees. The most popular way to build deeper regression and classification trees via addition is by summing together shallower ones, with each shallow tree constructed as detailed in Sections 14.3 and 14.4, and growing the tree sequentially one shallow member at a time. This scheme is an instance of the general *boosting* method introduced in Section 11.5. Moreover, trees are indeed the most popular universal approximator used when applying boosting based cross-validation, with this pairing very often referred to as *gradient boosting* [66, 67]. The principles of boosting outlined in Section 11.5 remain unchanged in the context of tree-based learners. However, with the specific knowledge of how to appropriately fit regression and classification trees to data we can now expound on important details related to gradient boosting that we could not delve into previously.

14.5.1 Shallow trees rule

As described in Section 11.5.1 *low-capacity* units (of any universal approximator) are most often used with boosting in order to provide a fine-resolution model search. In the context of tree-based units this leads to the use of *shallow trees*, with stumps and trees of depth two being especially popular. One can of course use higher-capacity tree units (of depth three and beyond) in constructing deeper cross-validated trees via boosting. However, boosting with such high-capacity units can easily lead to skipping over of optimal models (as depicted abstractly in Figure 11.30).

14.5.2 Boosting with tree-based learners

As detailed in Section 11.5, at the mth round of boosting we begin with a model consisting of a fully tuned linear combination of $m - 1$ units of a universal approximator (see Equation (11.26). In the case of tree-based learners we can dispense with the bias and weights of the linear combination (since they are naturally "baked in" to tree-based units, as detailed in Section 11.5.4) and write our model as

$$\text{model}_{m-1}(\mathbf{x}, \Theta_{m-1}) = f_{s_1}^{\star}(\mathbf{x}) + f_{s_2}^{\star}(\mathbf{x}) + \cdots + f_{s_{m-1}}^{\star}(\mathbf{x}) \tag{14.10}$$

where each function in this sum is a tree-based unit (e.g., a stump) whose split point(s) and leaf values have been chosen optimally. The mth round of boosting involves a search over a range of suitable candidates (here, various trees with differing split points) and a corresponding optimization of each candidate's leaf values. To construct the next candidate model we add a prospective unit $f_{s_m}(\mathbf{x})$ to $\text{model}_{m-1}(\mathbf{x}, \Theta_{m-1})$, forming

$$\text{model}_m(\mathbf{x}, \Theta_m) = \text{model}_{m-1}\left(\mathbf{x}, \Theta_{m-1}\right) + f_{s_m}(\mathbf{x}) \tag{14.11}$$

and optimize the leaf values of $f_{s_m}(\mathbf{x})$ using an appropriate cost function (e.g., the Least Squares cost for regression and Softmax cost for classification) with respect to a training dataset. This leaf-value optimization very closely mirrors the approaches described in Sections 14.3.2 and 14.4.2 in the case of regression and classification, respectively.

For example, suppose f_{s_m} is a stump and we are dealing with the regression case with a Least Squares cost and a dataset of P points denoted by $\left\{\left(\mathbf{x}_p, y_p\right)\right\}_{p=1}^{P}$. In complete analogy to Equation (14.5) we must then minimize the following pair of Least Squares costs

$$g(v_L) = \frac{1}{|\Omega_L|} \sum_{p \in \Omega_L} \left(\text{model}_{m-1}\left(\mathbf{x}_p, \Theta_{m-1}\right) + v_L - y_p\right)^2$$

$$g(v_R) = \frac{1}{|\Omega_R|} \sum_{p \in \Omega_R} \left(\text{model}_{m-1}\left(\mathbf{x}_p, \Theta_{m-1}\right) + v_R - y_p\right)^2 \qquad (14.12)$$

to properly determine our two leaf values v_L and v_R, where Ω_L and Ω_R are index sets as defined in Equation (14.3), and $|\Omega_L|$ and $|\Omega_R|$ denote their sizes. Like those cost functions in Equation (14.5), these simple costs can each be minimized perfectly by checking the first-order condition for optimality (or equivalently by taking a single step of Newton's method).

Similarly, if dealing with two-class classification with a Softmax cost and label values $y_p \in \{-1, +1\}$ we set leaf values of a stump by minimizing two costs – analogous to Equation (14.7) – of the form

$$g(v_L) = \frac{1}{|\Omega_L|} \sum_{p \in \Omega_L} \log\left(1 + e^{-y_p\left(\text{model}_{m-1}\left(\mathbf{x}_p, \Theta_{m-1}\right) + v_L\right)}\right)$$

$$g(v_R) = \frac{1}{|\Omega_R|} \sum_{p \in \Omega_R} \log\left(1 + e^{-y_p\left(\text{model}_{m-1}\left(\mathbf{x}_p, \Theta_{m-1}\right) + v_R\right)}\right) \qquad (14.13)$$

which (in both cases) cannot be minimized in closed form, but must be solved via local optimization. Often, as discussed in Section 14.4.2, this is done by simply taking a single step of Newton's method as it provides a positive trade-off between the minimization quality and computation effort.

Example 14.5   Regression via gradient boosting

In Figure 14.13 we illustrate the use of boosting with regression stumps, which is often (as discussed in Section 11.5.6) interpreted as successive rounds of fitting to the *residual* of a regression dataset. We can see this in the case of a simple stump by rearranging terms in Equation (14.12). For example, $g(v_L)$ in Equation (14.12) can be rewritten as

$$g(v_L) = \frac{1}{|\Omega_L|} \sum_{p \in \Omega_L} \left(v_L - r_p\right)^2 \qquad (14.14)$$

where r_p is the residual of the pth point defined as $r_p = y_p - \text{model}_{m-1}\left(\mathbf{x}_p, \Theta_{m-1}\right)$.

In the top row of this figure we show the original dataset along with the resulting fit provided by a model constructed from multiple rounds of stump-based boosting. Simultaneously in the bottom row we show each subsequent stump-based fit to the residual provided by the most recent stump added to the running model.

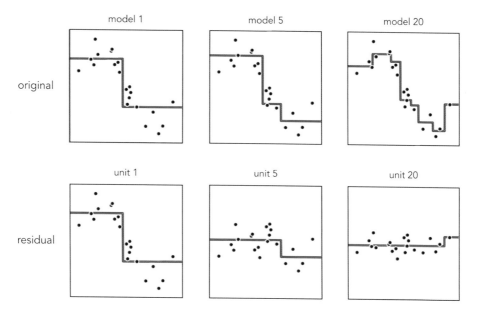

Figure 14.13 Figure associated with Example 14.5. See text for details.

Example 14.6   Spam detection via gradient boosting

In this example we use gradient boosting employing stumps, and cross-validate to determine an ideal number of rounds of boosting, using the spam dataset first described in Example 6.10. In this set of experiments we use the Softmax cost and set aside 20 percent of this two-class dataset (randomly) for validation purposes. We run 100 rounds of boosting and take a single step of Newton's method to tune each stump function. In Figure 14.14 we plot the number of misclassifications on both the training (in blue) and validation (in yellow) sets. The minimum number of misclassifications on the validation set occured at the sixty-fifth round of boosting, which resulted in 220 and 50 misclassifications on the training and validation sets, respectively. A simple linear classifier trained on the same portion of data provided 277 and 67 misclassifications on training and validation sets, respectively.

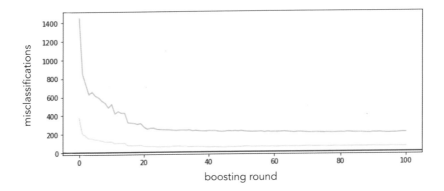

Figure 14.14 Figure associated with Example 14.6. See text for details.

14.6   Random Forests

Unless human interpretability of the final model is of primary concern, one virtually never uses a single recursively defined regression or classification tree, but a *bagged ensemble* of them. Generally speaking, bagging (as detailed in Section 11.9) involves combining multiple cross-validated models to produce a single higher-performing model. One can easily do this with recursively defined trees, employing the cross-validation techniques outlined in Section 14.7. However, in practice it is often unnecessary to grow each tree using cross-validation to temper their complexity. Instead, each tree can be trained on a random portion of training data taken from the original dataset and grown to a predetermined maximum depth, and afterwards bagged together.

This can be done with any universal approximator in principle but is especially practical with tree-based learners. This is both because trees are cheap to produce, and also because as *locally* defined approximators (see Section 14.2.2), it is natural to employ basic leaf-split halting protocols while growing the individual trees themselves (see Sections 14.3.4 and 14.4.4). While trees can certainly overfit, even when not cross-validated they are naturally prevented from exhibiting the sort of wild oscillatory overfitting behavior that is readily possible with fixed-shape or neural network models.[2] Thus bagging a set of overfitting trees can often successfully combat the sort of overfitting each tree presents, resulting in very effective models. Moreover, because each fully grown tree in such an ensemble can be learned efficiently, the computational trade-off, that is, training a large number of fully grown trees compared with a smaller number of cross-validated ones (each of which require more resources to construct), is often advantageous in practice.

Such an ensemble of recursively defined trees is often called a *random forest* [68] in the jargon of machine learning. The "random" part of the name *random forest*

---

[2] Compare, for instance, the overfitting behavior exhibited by each universal approximator in Figure 11.17.

refers both to the fact that each tree uses a random portion of the original data as training (which, by convention, is often sampled from the original dataset *with replacement*), and that often only a random subset of input feature dimensions are sampled for viable split points at each node in the trees produced. For each tree in such a forest often something like $\lfloor \sqrt{N} \rfloor$ of N features are chosen at random to determine split points.

Example 14.7   Random forest classification

In Figure 14.15 we show the result of bagging a set of five fully grown classification trees trained on different random portions of a simple two-class dataset (in each instance $\frac{2}{3}$ of the original dataset was used for training and the final $\frac{1}{3}$ was used for validation). Each of the five splits are illustrated in the small panels on the left along with the decision boundary provided by each trained model, with the validation data points in each case highlighted with a yellow boundary. Note that while most of the individual trees overfit the data, their ensemble (shown in the large panel on the right) does not. This ensembled model, as detailed in Section 11.9, is built by taking the *mode* of the five classification trees on the left.

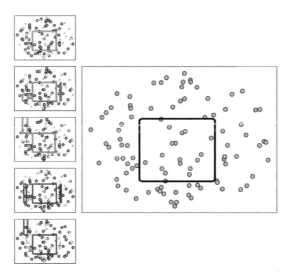

Figure 14.15 Figure associated with Example 14.7. (left column) The decision boundaries given by five fully grown classification trees, each grown on a different subset of the original data. Each individual tree tends to overfit the data but their bagged ensemble (shown in the right panel) compensates for this, and does not overfit. See text for further details.

14.7   Cross-Validation Techniques for Recursively Defined Trees

The basic principles of cross-validation, outlined in Sections 11.3 through 11.6, generally apply to the proper construction of recursively grown regression and classification trees in practice, with some technical differences arising due to the unique way such models are built. For example, we can begin with a low-capacity depth-one tree and grow it until minimum validation error is achieved (a form of *early stopping* specific to trees). Alternatively, we can begin by fitting a deep high-capacity tree to the data and gradually decrease its complexity by *pruning* leaves that do not contribute to low validation error (a form of *regularization* specific to trees).

Because recursively defined trees are typically ensembled as random forests, with each tree fully grown to a random training portion of the original dataset (as detailed in Section 14.6), the cross-validation techniques described here are often used to temper the complexity of a single regression or classification tree when *human interpretability* of a tree-based model is of crucial importance – something that is virtually always lost when ensembling multiple nonlinear models together.

14.7.1   Early stopping

We can easily use cross-validation to dictate the proper maximum depth of a tree by growing a tree of large depth, measuring validation error at each depth of the tree, and (after the fact) determine which depth produced minimal validation error. Alternatively, we can stop the growth early when we are confident[3] that (something approximating) minimum validation error has been achieved. This approach, while used in practice, translates to a relatively coarse model search since the capacity of a tree grows exponentially from one depth to the next.

As detailed in the previous two sections, practical considerations are often used to halt the leaf splitting (regardless of whether cross-validation is being performed). These include halting splitting if a leaf contains a singleton data point or a predecided (small) number of points, if all data points belong to the same class (in the case of classification) or have approximately the same output value (in the case of regression). To create a finer-resolution cross-validation search we can add validation-error-focused criteria to halt the growth of individual leaves as well. The simplest such criterion is to check whether or not splitting a leaf will result in *lowering* validation error (or lowering training error past a predetermined threshold): if yes, the leaf is split, otherwise growth of the leaf is halted. This approach to cross-validation is unique in that validation error

---

[3] As with any form of early stopping, determining when validation error is at its minimum "on the fly" is not always a straightforward affair as validation error does not always fall and rise monotonically (as discussed in the context of boosting and regularization in Sections 11.5.3 and 11.6.2, respectively).

will always monotonically decrease as the maximum depth of a tree is increased, but can result in *underfitting* models due to leaves halting growth prematurely.

Example 14.8   Early stopping by depth and leaf growth
In Figure 14.16 we show an example of cross-validating the maximum depth of a tree in the range of one through five for a simple regression dataset. The dataset, with training data colored in blue and validation data in yellow, is shown along with each subsequent fit in the top row of Figure 14.16, while the corresponding training/validation error is shown in its bottom row.

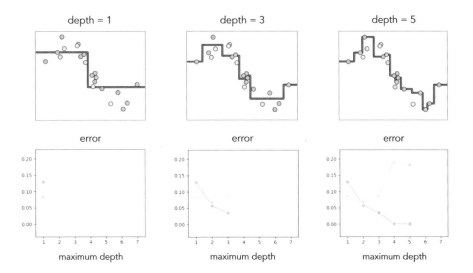

Figure 14.16 Figure associated with Example 14.8. See text for details.

In Figure 14.17 we show two examples of cross-validating the maximum depth of a tree in the range of one through five for another regression dataset, where leaf growth is now halted when validation error does not improve. In each instance the dataset, with training data colored in blue and validation data in yellow, is shown along with each subsequent fit on top, with the corresponding training/validation error is shown directly underneath it. In the first run (shown in the top two rows of Figure 14.17), because of the particular split of training and validation data, growth of each leaf is halted immediately, resulting in an underfitting, depth-one representation. With the second run on a different training–validation split of the data (shown in the bottom two rows of Figure 14.17), tree growth continues to improve validation error up to the maximum depth tested, resulting in a significantly better representation.

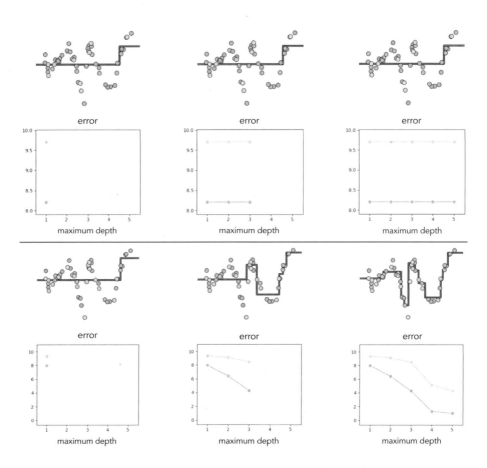

Figure 14.17 Figure associated with Example 14.8. See text for details.

14.7.2   Pruning

In contrast to beginning with a low-capacity (shallow) tree and growing it via early stopping, we can instead begin by fitting a high-capacity (deep) tree and remove leaves that do not improve validation error, until a minimum-validation tree structure remains. This technique – illustrated pictorially in Figure 14.18 – is called *pruning* because it entails examining an initially overly complicated tree and cutting off its leaves, akin to the way pruning of natural trees is done by snipping off redundant leaves and branches. Pruning is a tree-specific form of *regularization* based cross-validation, discussed previously in Section 11.6.

While early stopping is often more computationally efficient than pruning, the latter provides a finer-resolution model search in determining the tree structure with minimal validation error, since tree/leaf growth is unhindered and is only cut back after the fact.

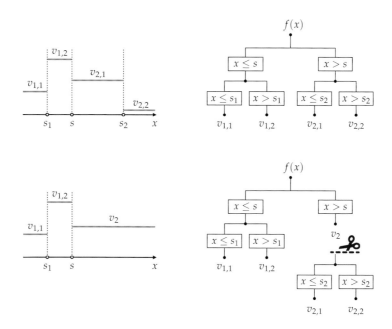

Figure 14.18 Pruning illustrated. (top panels) A fully grown tree of depth two with four leaves. (bottom) A pruned version of the original tree wherein the leaves $v_{2,1}$ and $v_{2,2}$ are pruned and replaced by a single leaf.

14.8 Conclusion

In this chapter we discussed a range of important technical matters related to tree-based universal approximators, which were first introduction in Section 11.2.3. We began in Section 14.2 by providing a more formal description of stumps as well as deeper trees, which as detailed in this section can be formed via recursion or summation. Recursively defined regression and classification trees were then gently motivated and detailed in Sections 14.3 and 14.4. Gradient boosting – the specific application of boosting based cross-validation (described in detail in Section 11.5) to tree-based learners – was touched upon in Section 14.5. Similarly, random forests – the specialized application of bagging (detailed in Section 11.9) to tree-based learners – was described in Section 14.6. Finally, the use of cross-validation with recursively defined tree-based learners – both from a naive and regularization perspective – were explored in Section 14.7.

14.9 Exercises

† The data required to complete the following exercises can be downloaded from the text's github repository at github.com/jermwatt/machine_learning_refined

14.1   **Growing deep trees by addition**

Show that in general adding $2^D - 1$ stumps (with scalar input) together will create a depth-D tree (provided that the stumps do not share any split points).

14.2   **Fitting the parameters of a simple regression tree**

Repeat the experiment described in Example 14.1, and reproduce the plots shown in Figure 14.6.

14.3   **Code up a regression tree**

Repeat the experiment described in Example 14.2 by coding up a recursively defined regression tree. You need not reproduce Figure 14.7. Instead, measure and plot the Least Squares error at each depth of your tree.

14.4   **Code up a two-class classification tree**

Repeat the first experiment described in Example 14.4 by coding up a recursively defined two-class classification tree. You need not reproduce Figure 14.11. Instead, measure and plot the number of misclassifications at each depth of your tree.

14.5   **Code up a multi-class classification tree**

Repeat the second experiment described in Example 14.4 by coding up a recursively defined multi-class classification tree. You need not reproduce Figure 14.12. Instead, measure and plot the number of misclassifications at each depth of your tree.

14.6   **Gradient boosting for regression**

Repeat the experiment described in Example 14.5 by coding up a gradient boosting algorithm employing regression stumps. Reproduce Figure 14.13, illustrating the boosted tree as well as the best stump fit to the residual at rounds one, two, and ten of boosting.

14.7   **Gradient boosting for classification**

Determine the leaf values of a stump added at the mth round of boosting to a classification tree by minimizing the Softmax costs in Equation (14.13) via taking a *single* step of Newton's method.

14.8   **Random forests**

Repeat the experiment described in Example 14.7 by coding up a random forest built from classification trees. You need not reproduce Figure 14.15. However, you can verify that your implementation is working properly by checking that the final accuracy of your random forest classifier outstrips the accuracy of many of the individual trees in the ensemble or, alternatively, you can employ a testing set by setting aside a small portion of the original data.

14.9 **Limitation of trees outside their training range**
We have seen in this chapter that trees are efficient nonlinear approximations, and do not suffer from the sort of oscillatory behavior that can adversely affect global approximators like polynomials and neural networks (see Section 14.6). However, tree-based learners – by nature – fail to work effectively outside their training range. In this exercise you will see why this is the case by training a regression tree using the student debt data first shown in Figure 1.8. Use your trained tree to predict what the total student debt will be in the year 2050. Does it make sense? Explain why.

14.10 **Naive cross-validation**
Repeat the experiment outlined in Example 14.8 whose results are shown in Figure 14.16.

Part IV

Appendices

A   Advanced First- and Second-Order Optimization Methods

A.1   Introduction

In this chapter we study advanced first- and second-order optimization techniques that are designed to ameliorate the natural weaknessess associated with gradient descent and Newton's method – as detailed previously in Sections 3.6 and 4.4, respectively.

A.2   Momentum-Accelerated Gradient Descent

In Section 3.6 we discussed a fundamental issue associated with the *direction* of the negative gradient: it can (depending on the function being minimized) oscillate rapidly, leading to zig-zagging gradient descent steps that slow down minimization. In this section we describe a popular enhancement to the standard gradient descent step, called *momentum acceleration*, that is specifically designed to ameliorate this issue. The core of this idea comes from the field of *time series analysis*, and in particular is a tool for smoothing time series data known as the *exponential average*. Here we first introduce the exponential average and then detail how it can be integrated into the standard gradient descent step in order to help ameliorate some of this undesirable zig-zagging behavior (when it occurs), and consequently speed up gradient descent.

A.2.1   The exponential average

In Figure A.1 we show an example of a time series data. This particular example shows a real snippet of the price of a financial stock measured at 450 consecutive points in time. In general time series data consists of a sequence of K ordered points $w^1, w^2, ..., w^K$, meaning that the point w^1 comes before (i.e., it is created and/or collected before) w^2, the point w^2 before w^3, and so on. For example, we generate a (potentially multi-dimensional) time series of points whenever we run a local optimization scheme with steps $\mathbf{w}^k = \mathbf{w}^{k-1} + \alpha \mathbf{d}^{k-1}$ since it produces the sequence of ordered points $\mathbf{w}^1, \mathbf{w}^2, ..., \mathbf{w}^K$.

Because the raw values of a time series often oscillate, it is common practice to *smooth* them (in order to remove these zig-zagging motions) for better

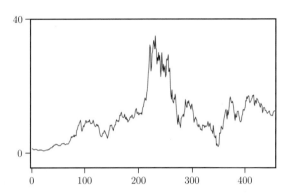

Figure A.1 An example of a time series data, representing the price of a financial stock measured at 450 consecutive points in time.

visualization or prior to further analysis. The *exponential average* is one of the most popular such smoothing techniques for time series, and is used in virtually every application area in which this sort of data arises. In Figure A.2 we show the result of smoothing the data shown in Figure A.1. Before we see how the exponential average is computed, it is first helpful to see how to compute a *cumulative average* of K input points w^1, w^2, ..., w^K, that is the average of the first two points, the average of the first three points, and so forth. Denoting the average of the first k points as h^k we can write

$$
\begin{aligned}
h^1 &= w^1 \\
h^2 &= \frac{w^1 + w^2}{2} \\
h^3 &= \frac{w^1 + w^2 + w^3}{3} \\
&\vdots \\
h^K &= \frac{w^1 + w^2 + w^3 + \cdots + w^K}{K}.
\end{aligned}
\tag{A.1}
$$

Notice how at each step h^k essentially *summarizes* the input points w^1 through w^k via the simplest statistic: their sample mean. The way the cumulative average is written in Equation (A.1), we need access to every raw point w^1 through w^k in order to compute the kth cumulative average h^k. Alternatively, we can write this cumulative average by expressing h^k for $k > 1$ in a recursive manner involving only its preceding cumulative average h^{k-1} and current time series value w^k, as

$$
h^k = \frac{k-1}{k} h^{k-1} + \frac{1}{k} w^k.
\tag{A.2}
$$

From a computational perspective, the recursive way of defining the cumulative average is far more efficient since at the kth step we only need to store and deal with two values as opposed to k of them.

The exponential average is a simple twist on the cumulative average formula.

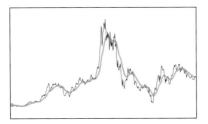 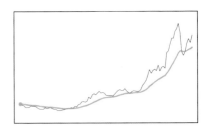

Figure A.2 (left panel) An exponential average (in pink) of the time series data shown in Figure A.1. (right panel) An exponential average of just the first 100 points of the time series, which is a smooth approximation to the underlying time series data.

Notice, at every step in Equation (A.2) that the coefficients on h^{k-1} and w^k *always* sum to 1, i.e., $\frac{k-1}{k} + \frac{1}{k} = 1$. As k grows larger both coefficients change: the coefficient on h^{k-1} gets closer to 1 while the one on w^k gets closer to 0. With the exponential average we *freeze* these coefficients. That is, we replace the coefficient on h^{k-1} with a constant value $\beta \in [0, 1]$, and the coefficient on w^k with $1 - \beta$, giving a similar recursive formula for the exponential average, as

$$h^k = \beta\, h^{k-1} + (1 - \beta)\, w^k. \tag{A.3}$$

Clearly the parameter β here controls a trade-off: the smaller we set β the more our exponential average approximates the raw (zig-zagging) time series itself, while the larger we set it the more each subsequent average looks like its predecessor (resulting in a smoother curve). Regardless of how we set β each h^k in an exponential average can still be thought of as a summary for w^k and all time series points that precede it.

Why is this slightly adjusted version of the cumulative average called an *exponential* average? Because if we roll back the update shown in Equation (A.3) to express h^k only in terms of preceding time series elements, as we did for cumulative average in Equation (A.1), an exponential (or power) pattern in the coefficients will emerge.

Note that in deriving the exponential average we assumed our time series data was *one-dimensional*, that is each raw point w^k is a scalar. However, this idea holds regardless of the input dimension. We can likewise define the exponential average of a time series of general N-dimensional points $\mathbf{w}^1, \mathbf{w}^2, ..., \mathbf{w}^K$ by initializing $\mathbf{h}^1 = \mathbf{w}^1$, and then for $k > 1$ building $\mathbf{h}^k$ as

$$\mathbf{h}^k = \beta\, \mathbf{h}^{k-1} + (1 - \beta)\, \mathbf{w}^k. \tag{A.4}$$

Here the exponential average $\mathbf{h}^k$ at step k is also N-dimensional.

A.2.2   Ameliorating the zig-zagging behavior of gradient descent

As mentioned previously, a sequence of gradient descent steps can be thought of as a *time series*. Indeed if we take K steps of a gradient descent run we do create a time series of ordered gradient descent steps $\mathbf{w}^1, \mathbf{w}^2, ..., \mathbf{w}^K$ and descent directions $-\nabla g\left(\mathbf{w}^0\right), -\nabla g\left(\mathbf{w}^1\right), ..., -\nabla g\left(\mathbf{w}^{K-1}\right)$.

To attempt to ameliorate some of the zig-zagging behavior of our gradient descent steps $\mathbf{w}^1, \mathbf{w}^2, ..., \mathbf{w}^K$ – as detailed in Section 3.6.3 – we could compute their *exponential average*. However, we do not want to smooth the gradient descent steps *after* they have been created – as the "damage is already done" in the sense that the zig-zagging has already slowed the progress of a gradient descent run. Instead what we want is to smooth the steps *as they are created*, so that our algorithm makes more progress in minimization.

How do we smooth the steps as they are created? Remember from Section 3.6.3 that the root cause of zig-zagging gradient descent is the oscillating nature of the (negative) gradient directions themselves. In other words, if the descent directions $-\nabla g\left(\mathbf{w}^0\right), -\nabla g\left(\mathbf{w}^1\right), ..., -\nabla g\left(\mathbf{w}^{K-1}\right)$ zig-zag, so will the gradient descent steps themselves. Therefore it seems reasonable to suppose that if we smooth out these directions themselves, as they are created during a run of gradient descent, we can as a consequence produce gradient descent steps that do not zig-zag as much and therefore make more progress in minimization.

To do this we first initialize $\mathbf{d}^0 = -\nabla g\left(\mathbf{w}^0\right)$ and then for $k > 1$ the exponentially averaged descent direction $\mathbf{d}^{k-1}$ (using the formula in Equation (A.4)) takes the form

$$\mathbf{d}^{k-1} = \beta\,\mathbf{d}^{k-2} + (1-\beta)\left(-\nabla g\left(\mathbf{w}^{k-1}\right)\right). \qquad (A.5)$$

We can then use this descent direction in our generic local optimization framework to take a step as

$$\mathbf{w}^k = \mathbf{w}^{k-1} + \alpha\,\mathbf{d}^{k-1}. \qquad (A.6)$$

Together this exponential averaging adds only a single extra step to our basic gradient descent scheme, forming a *momentum-accelerated* gradient descent step of the form[1]

---

[1] Sometimes this step is written slightly differently: instead of averaging the *negative* gradient directions the gradient itself is exponentially averaged, and then the *step* is taken in their *negative* direction. This means that we initialize our exponential average at the first *negative* descent direction $\mathbf{d}^0 = -\nabla g\left(\mathbf{w}^0\right)$, and for $k > 1$ the general descent direction and corresponding step is computed as

$$\mathbf{d}^{k-1} = \beta\,\mathbf{d}^{k-2} + (1-\beta)\,\nabla g\left(\mathbf{w}^{k-1}\right)$$
$$\mathbf{w}^k = \mathbf{w}^{k-1} - \alpha\,\mathbf{d}^{k-1}. \qquad (A.7)$$

$$\mathbf{d}^{k-1} = \beta\,\mathbf{d}^{k-2} + (1-\beta)\left(-\nabla g\left(\mathbf{w}^{k-1}\right)\right)$$
$$\mathbf{w}^k = \mathbf{w}^{k-1} + \alpha\,\mathbf{d}^{k-1}. \tag{A.8}$$

The term "momentum" here refers to the new exponentially averaged descent direction $\mathbf{d}^{k-1}$, that by definition is a function of *every* negative gradient which precedes it. Hence $\mathbf{d}^{k-1}$ captures the average or "momentum" of the directions preceding it.

As with any exponential average the choice of $\beta \in [0,1]$ provides a trade-off. On the one hand, the smaller β is chosen the *more* the exponential average resembles the actual sequence of negative descent directions since *more* of each negative gradient direction is used in the update, but the *less* these descent directions summarize all of the previously seen negative gradients. On the other hand, the larger β is chosen the *less* these exponentially averaged descent steps resemble the negative gradient directions, since each update will use *less* of each subsequent negative gradient direction, but the *more* they represent a summary of them. Often in practice larger values of β are used, e.g., in the range $[0.7,1]$.

Example A.1 Accelerating gradient descent on a simple quadratic
In this example we compare a run of standard gradient descent to the *momentum-accelerated* version using a quadratic function of the form

$$g(\mathbf{w}) = a + \mathbf{b}^T\mathbf{w} + \mathbf{w}^T\mathbf{C}\mathbf{w} \tag{A.9}$$

where $a = 0$, $\mathbf{b} = \begin{bmatrix} 0 \\ 0 \end{bmatrix}$, and $\mathbf{C} = \begin{bmatrix} 0.5 & 0 \\ 0 & 9.75 \end{bmatrix}$.

Here we make three runs of 25 steps: a run of gradient descent and two runs of momentum-accelerated gradient descent using two choices of the parameter $\beta \in \{0.2, 0.7\}$. All three runs are initialized at the same point $\mathbf{w}^0 = [10\ 1]^T$, and use the same steplength $\alpha = 10^{-1}$.

We show the resulting steps taken by the standard gradient descent run in the top panel of Figure A.3 (where significant zig-zagging is present), and the momentum-accelerated versions using $\beta = 0.2$ and $\beta = 0.7$ in the middle and bottom panels of this figure, respectively. Both momentum-accelerated versions clearly outperform the standard scheme, in that they reach a point closer to the true minimum of the quadratic. Also note that the overall path taken by gradient descent is smoother in the bottom panel, due to the larger value of its corresponding β.

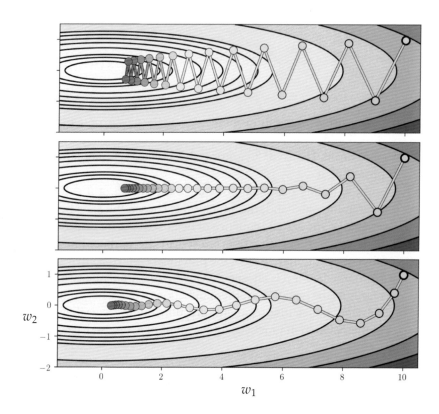

Figure A.3 Figure associated with Example A.1. The zig-zagging behavior of gradient descent can be ameliorated using the momentum-accelerated gradient descent step in Equation (A.8). See text for further details.

A.3   Normalized Gradient Descent

In Section 3.6 we discussed a fundamental issue associated with the *magnitude* of the negative gradient and the fact that it vanishes near stationary points, causing gradient descent to slowly crawl near stationary points. In particular this means – depending on the function being minimized – that it can halt near saddle points. In this section we describe a popular enhancement to the standard gradient descent scheme, called *normalized gradient descent*, that is specifically designed to ameliorate this issue. The core of this idea lies in a simple inquiry: since the (vanishing) *magnitude* of the negative gradient is what causes gradient descent to slowly crawl near stationary points or halt at saddle points, what happens if we simply ignore the magnitude at each step by *normalizing* it out?

A.3.1   Normalizing out the full gradient magnitude

In Section 3.6.4 we saw how the length of a standard gradient descent step is proportional to the magnitude of the gradient, expressed algebraically as

$\alpha \left\| \nabla g(\mathbf{w}^{k-1}) \right\|_2$. Moreover, we also saw there how this fact explains why gradient descent slowly crawls near stationary points, since near such points the *magnitude* of the gradient vanishes.

Since the magnitude of the gradient is to blame for slow crawling near stationary points, what happens if we simply ignore it by normalizing it out of the update step and just travel in the direction of negative gradient itself?

One way to normalize a (gradient) descent direction is via dividing it by its magnitude. Doing so gives a *normalized gradient descent* step of the form

$$\mathbf{w}^k = \mathbf{w}^{k-1} - \alpha \frac{\nabla g(\mathbf{w}^{k-1})}{\left\| \nabla g(\mathbf{w}^{k-1}) \right\|_2}. \tag{A.10}$$

In doing this we do indeed ignore the magnitude of the gradient, since

$$\left\| \mathbf{w}^k - \mathbf{w}^{k-1} \right\|_2 = \left\| -\alpha \frac{\nabla g(\mathbf{w}^{k-1})}{\left\| \nabla g(\mathbf{w}^{k-1}) \right\|_2} \right\|_2 = \alpha. \tag{A.11}$$

In other words, if we normalize out the magnitude of the gradient at each step of gradient descent then the length of each step is exactly equal to the value of our steplength parameter α. This is precisely what we did with the random search method in Section 2.5.2.

Notice that if we slightly rewrite the fully normalized step in Equation (A.10) as

$$\mathbf{w}^k = \mathbf{w}^{k-1} - \frac{\alpha}{\left\| \nabla g(\mathbf{w}^{k-1}) \right\|_2} \nabla g(\mathbf{w}^{k-1}) \tag{A.12}$$

we can interpret our fully magnitude-normalized step as a standard gradient descent step with a steplength value $\frac{\alpha}{\left\| \nabla g(\mathbf{w}^{k-1}) \right\|_2}$ that *adjusts itself* at each step based on the magnitude of the gradient to ensure that the length of each step is precisely α.

Also notice that in practice it is often useful to add a small constant ϵ (e.g., 10^{-7} or smaller) to the gradient magnitude to avoid potential division by zero (where the magnitude completely vanishes)

$$\mathbf{w}^k = \mathbf{w}^{k-1} - \frac{\alpha}{\left\| \nabla g(\mathbf{w}^{k-1}) \right\|_2 + \epsilon} \nabla g(\mathbf{w}^{k-1}). \tag{A.13}$$

Example A.2   Ameliorating slow-crawling near minima and saddle points
Shown in the left panel of Figure A.4 is a repeat of the run of gradient descent first detailed in Example 3.14, only here we use a fully normalized gradient descent step. We use the same number of steps and steplength value used in that example (which led to slow-crawling with the standard scheme). Here,

however, the normalized step – unaffected by the vanishing gradient magnitude – is able to pass easily through the flat region of this function and find a point very close to the minimum at the origin. Comparing this run to the original run (of standard gradient descent) in Figure 3.14 we can see that the normalized run gets considerably closer to the global minimum of the function.

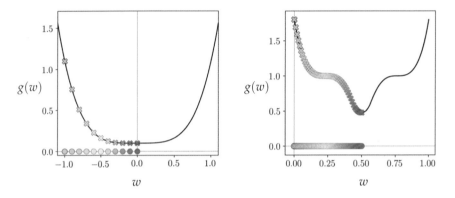

Figure A.4 Figure associated with Example A.2. By normalizing the gradient we can overcome the slow-crawling behavior of gradient descent near a function's minima (left panel) and saddle points (right panel). See text for further details.

Shown in the right panel of Figure A.4 is a repeat of the run of gradient descent first detailed in Example 3.14, only here we use a fully normalized gradient descent step. We use the same number of steps and steplength value used in that example (which led to halting at a saddle point with the standard scheme). Here, however, the normalized step – unaffected by the vanishing gradient magnitude – is able to pass easily through the flat region of the saddle point and reach a point of this function close to the minimum.

Example A.3 A trade-off when using normalized gradient descent
In Figure A.5 we show a comparison of fully normalized (left panel) and standard (right panel) gradient descent on the simple quadratic function

$$g(w) = w^2. \tag{A.14}$$

Both algorithms use the same initial point ($w^0 = -3$), steplength parameter ($\alpha = 0.1$), and maximum number of iterations (20 each). Steps are colored from green to red to indicate the starting and ending points of each run, with circles denoting the actual steps in the input space and x marks denoting their respective function evaluations.

Notice, how the standard version races to the global minimum of the function, while the normalized version – taking fixed-length steps – gets only a fraction of the way there. This behavior is indicative of how a normalized step will fail to leverage the gradient when it is large – as the standard method does – in order to take larger steps at the beginning of a run.

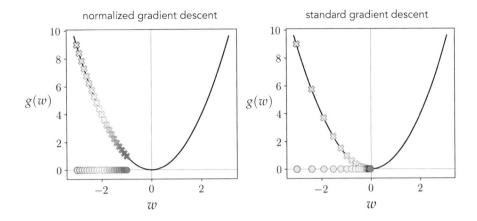

Figure A.5 Figure associated with Example A.3. While normalizing the gradient speeds up gradient descent near flat regions of a function where the magnitude of the gradient is small, it likewise fails to leverage the often large magnitude of the gradient far from a function's minima. See text for further details.

A.3.2 Normalizing out the magnitude component-wise

Remember that the gradient is a vector of N partial derivatives

$$
\nabla g(\mathbf{w}) = \begin{bmatrix} \frac{\partial}{\partial w_1} g(\mathbf{w}) \\ \frac{\partial}{\partial w_2} g(\mathbf{w}) \\ \vdots \\ \frac{\partial}{\partial w_N} g(\mathbf{w}) \end{bmatrix} \tag{A.15}
$$

with the jth partial derivative $\frac{\partial}{\partial w_j} g(\mathbf{w})$ defining how the gradient behaves along the jth coordinate axis. If we then look at what happens to the jth partial derivative of the gradient when we normalize off the *full magnitude* of the gradient

$$
\frac{\frac{\partial}{\partial w_j} g(\mathbf{w})}{\left\| \nabla g(\mathbf{w}) \right\|_2} = \frac{\frac{\partial}{\partial w_j} g(\mathbf{w})}{\sqrt{\sum_{n=1}^{N} \left(\frac{\partial}{\partial w_n} g(\mathbf{w}) \right)^2}} \tag{A.16}
$$

we can see that the jth partial derivative is normalized using a sum of the magnitudes of every partial derivative. This means that if the jth partial derivative is already small in magnitude itself, doing this will erase virtually all of its contribution to the final descent step. Therefore normalizing by the magnitude of the entire gradient can be problematic when dealing with functions containing regions that are flat with respect to only some of the partial derivative directions, as it *diminishes* the contribution of the very partial derivatives we wish to enhance by ignoring magnitude.

As an alternative we can normalize out the magnitude of the gradient *component-wise*. In other words, instead of normalizing each partial derivative by the magnitude of the entire gradient we can normalize each partial derivative with respect to only itself as

$$\frac{\frac{\partial}{\partial w_j} g(\mathbf{w})}{\sqrt{\left(\frac{\partial}{\partial w_j} g(\mathbf{w})\right)^2}} = \frac{\frac{\partial}{\partial w_j} g(\mathbf{w})}{\left|\frac{\partial}{\partial w_j} g(\mathbf{w})\right|} = \text{sign}\left(\frac{\partial}{\partial w_j} g(\mathbf{w})\right). \tag{A.17}$$

Therefore in the jth direction we can write this component-normalized gradient descent step as

$$w_j^k = w_j^{k-1} - \alpha \, \text{sign}\left(\frac{\partial}{\partial w_j} g\left(\mathbf{w}^{k-1}\right)\right). \tag{A.18}$$

We can then write the entire component-wise normalized step as

$$\mathbf{w}^k = \mathbf{w}^{k-1} - \alpha \, \text{sign}\left(\nabla g\left(\mathbf{w}^{k-1}\right)\right) \tag{A.19}$$

where here the sign function acts component-wise on the gradient vector. We can easily compute the length of a single step of this component-normalized gradient descent step (provided the partial derivatives of the gradient are all nonzero) as

$$\left\|\mathbf{w}^k - \mathbf{w}^{k-1}\right\|_2 = \left\|-\alpha \, \text{sign}\left(\nabla g\left(\mathbf{w}^{k-1}\right)\right)\right\|_2 = \sqrt{N}\,\alpha. \tag{A.20}$$

Notice, additionally, that if we slightly rewrite the jth component-normalized step in Equation (A.18) as

$$w_j^k = w_j^{k-1} - \frac{\alpha}{\sqrt{\left(\frac{\partial}{\partial w_j} g(\mathbf{w}^{k-1})\right)^2}} \frac{\partial}{\partial w_j} g\left(\mathbf{w}^{k-1}\right) \tag{A.21}$$

we can interpret our component-normalized step as a standard gradient descent step with an individual steplength value

$$\text{steplength} = \frac{\alpha}{\sqrt{\left(\frac{\partial}{\partial w_j} g(\mathbf{w}^{k-1})\right)^2}} \tag{A.22}$$

per component that all adjusts themselves individually at each step based on component-wise magnitude of the gradient to ensure that the length of each step is precisely $\sqrt{N}\,\alpha$. Indeed if we write

$$\mathbf{a}^{k-1} = \begin{bmatrix} \dfrac{\alpha}{\sqrt{\left(\frac{\partial}{\partial w_1} g(\mathbf{w}^{k-1})\right)^2}} \\ \dfrac{\alpha}{\sqrt{\left(\frac{\partial}{\partial w_2} g(\mathbf{w}^{k-1})\right)^2}} \\ \vdots \\ \dfrac{\alpha}{\sqrt{\left(\frac{\partial}{\partial w_N} g(\mathbf{w}^{k-1})\right)^2}} \end{bmatrix} \qquad (A.23)$$

then the full component-normalized descent step can also be written as

$$\mathbf{w}^k = \mathbf{w}^{k-1} - \mathbf{a}^{k-1} \circ \nabla g(\mathbf{w}^{k-1}) \qquad (A.24)$$

where the $\circ$ symbol denotes component-wise multiplication (see Appendix Section C.2.3). In practice, a small $\epsilon > 0$ is added to the denominator of each value of each entry of $\mathbf{a}^{k-1}$ to avoid division by zero.

Example A.4   Full versus component-normalized gradient descent
In this example we use the function

$$g(w_1, w_2) = \max\left(0, \tanh(4w_1 + 4w_2)\right) + |0.4w_1| + 1 \qquad (A.25)$$

to show the difference between full and component-normalized gradient descent steps on a function that has a very narrow flat region along only a single dimension of its input. Here this function – whose surface and contour plots can be seen in the left and right panels of Figure A.6, respectively – is very flat along the w_2 dimension for any fixed value of w_1, and has a very narrow valley leading towards its minima in the w_2 dimension where $w_1 = 0$. If initialized at a point where $w_2 > 2$ this function cannot be minimized very easily using standard gradient descent *or* the fully normalized version. In the latter case, the magnitude of the partial derivative in w_2 is nearly zero everywhere, and so fully normalizing makes this contribution smaller, and halts progress. In the top row of the figure we show the result of 1000 steps of fully normalized gradient descent starting at the point $\mathbf{w}^0 = [2 \ 2]^T$, colored green (at the start of the run) to red (at its finale). As can be seen, little progress is made.

In the bottom row of the figure we show the results of using component-normalized gradient descent starting at the same initialization and employing the same steplength. Here we only need 50 steps in order to make significant progress.

In summary, normalizing out the gradient magnitude – using either of the approaches detailed previously – ameliorates the "slow-crawling" problem of

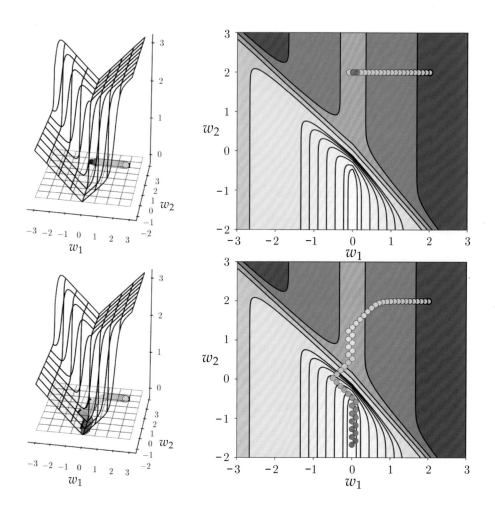

Figure A.6 Figure associated with Example A.4. See text for details.

standard gradient descent and empowers the method to push through flat regions of a function with much greater ease. This includes flat regions of a
function that may lead to a local minimum, or the region around a saddle point
of a nonconvex function where standard gradient descent can halt. However –
as highlighted in Example A.3 – in normalizing every step of standard gradient
descent we do *shorten* the first few steps of the run that are typically large (since
random initializations are often far from stationary points of a function). This is
the trade-off of the normalized step when compared with the standard gradient
descent scheme: we trade shorter initial steps for longer ones around stationary
points.

A.4   Advanced Gradient-Based Methods

In Section A.2 we described the notion of momentum-accelerated gradient descent, and how it is a natural remedy for the zig-zagging problem the standard gradient descent algorithm suffers from when run along long narrow valleys. As we saw, the momentum-accelerated descent direction $\mathbf{d}^{k-1}$ is simply an exponential average of gradient descent directions taking the form

$$\mathbf{d}^{k-1} = \beta\,\mathbf{d}^{k-2} + (1-\beta)\left(-\nabla g\left(\mathbf{w}^{k-1}\right)\right)$$
$$\mathbf{w}^{k} = \mathbf{w}^{k-1} + \alpha\,\mathbf{d}^{k-1} \tag{A.26}$$

where $\beta \in [0,1]$ is typically set at a value of $\beta = 0.7$ or higher.

Then in Section A.3.2 we saw how normalizing the gradient descent direction component-wise helps deal with the problem standard gradient descent has when traversing flat regions of a function. We saw there how a component-normalized gradient descent step takes the form (for the jth component of $\mathbf{w}$)

$$w_j^k = w_j^{k-1} - \alpha\,\frac{\frac{\partial}{\partial w_j} g\left(\mathbf{w}^{k-1}\right)}{\sqrt{\left(\frac{\partial}{\partial w_j} g\left(\mathbf{w}\right)\right)^2}} \tag{A.27}$$

where in practice a small fixed value $\epsilon > 0$ is often added to the denominator on the right-hand side to avoid division by zero.

With the knowledge that these two additions to the standard gradient descent step help solve two fundamental problems associated with the descent direction used with gradient descent, it is natural to try to *combine them* in order to leverage both enhancements.

One way of combining the two ideas would be to component-normalize the exponential average descent direction computed in momentum-accelerated gradient descent. That is, compute the exponential average direction in the top line of Equation (A.8) and then normalize *it* (instead of the raw gradient descent direction). With this idea we can write out the update for the jth component of the resulting descent direction as

$$d_j^{k-1} = \operatorname{sign}\left(\beta\,d_j^{k-2} - (1-\beta)\,\frac{\partial}{\partial w_j} g\left(\mathbf{w}^{k-1}\right)\right). \tag{A.28}$$

Many popular first-order steps used to tune machine learning models – particularly those involving deep neural networks (see Chapter 13) – combine momentum and normalized gradient descent in this sort of way. Below we list a few examples, including the popular *Adam* and *RMSProp* first-order steps.

Example A.5   **Adaptive Moment Estimation (Adam)**

Adaptive Moment Estimation (Adam) [69] is a component-wise normalized gradient step employing independently-calculated exponential averages for both the descent direction *and* its magnitude. That is, we compute the jth coordinate of the updated descent direction by first computing the exponential average of the gradient descent direction d_j^k and the squared magnitude h_j^k separately along this coordinate as

$$d_j^{k-1} = \beta_1\, d_j^{k-2} + (1 - \beta_1) \frac{\partial}{\partial w_j} g\left(\mathbf{w}^{k-1}\right)$$

$$h_j^{k-1} = \beta_2\, h_j^{k-2} + (1 - \beta_2) \left(\frac{\partial}{\partial w_j} g\left(\mathbf{w}^{k-1}\right)\right)^2$$

(A.29)

where β_1 and β_2 are exponential average parameters that lie in the range $[0, 1]$. Popular values for the parameters of this update step are $\beta_1 = 0.9$ and $\beta_2 = 0.999$. Note that, as with any exponential average, these two updates apply when $k > 1$, and should be initialized[2] at first values from the series they respectively model, i.e., $d_j^0 = \frac{\partial}{\partial w_j} g\left(\mathbf{w}^0\right)$ and $h_j^0 = \left(\frac{\partial}{\partial w_j} g\left(\mathbf{w}^0\right)\right)^2$.

The Adam step is then a component-wise normalized descent step using this exponential average descent direction and magnitude, with a step in the jth coordinate taking the form

$$w_j^k = w_j^{k-1} - \alpha \frac{d_j^{k-1}}{\sqrt{h_j^{k-1}}}.$$

(A.30)

Example A.6   **Root Mean Squared Propagation (RMSProp)**

This popular first-order step is a variant of the component-wise normalized step where – instead of normalizing each component of the gradient by its magnitude – each component is normalized by the exponential average of the component-wise magnitudes of previous gradient directions.

Denoting by h_j^k the exponential average of the squared magnitude of the jth partial derivative at step k, we have

$$h_j^k = \gamma\, h_j^{k-1} + (1 - \gamma)\left(\frac{\partial}{\partial w_j} g\left(\mathbf{w}^{k-1}\right)\right)^2.$$

(A.31)

---

[2] The authors of this particular update step proposed that each exponential average be initialized at zero – i.e., $d_j^0 = 0$ and $h_j^0 = 0$ – instead of the first step in each series they respectively model. This initialization – along with the values for β_1 and β_2 that are typically chosen to be greater than 0.9 – causes the first few update steps of these exponential averages to be "biased" towards zero as well. Because of this they also employ a "bias-correction" term to compensate for this initialization.

The Root Mean Squared Error Propagation (RMSProp) [70] step is then a component-wise normalized descent step using this exponential average, with a step in the jth coordinate taking the form

$$w_j^k = w_j^{k-1} - \alpha \frac{\frac{\partial}{\partial w_j} g\left(\mathbf{w}^{k-1}\right)}{\sqrt{h_j^{k-1}}}. \tag{A.32}$$

Popular values for the parameters of this update step are $\gamma = 0.9$ and $\alpha = 10^{-2}$.

A.5 Mini-Batch Optimization

In machine learning applications we are almost always tasked with minimizing a *sum* of P functions *of the same form*. Written algebraically such a cost takes the form

$$g(\mathbf{w}) = \sum_{p=1}^{P} g_p(\mathbf{w}) \tag{A.33}$$

where $g_1, g_2, ..., g_P$ are functions of the same type, e.g., convex quadratic functions (as with Least Squares linear regression discussed in Chapter 5), all parameterized using the same set of weights $\mathbf{w}$.

This special *summation structure* allows for a simple but very effective enhancement to virtually any local optimization scheme, called *mini-batch optimization*. Mini-batch optimization is most often used in combination with a gradient-based step.

A.5.1 A simple idea with powerful consequences

The motivation for mini-batch optimization rests on a simple inquiry: for this sort of function g shown in Equation (A.33), what would happen if instead of taking one descent step in g – that is, one descent step in the entire sum of the functions $g_1, g_2, ..., g_P$ *simultaneously* – we took a sequence of P descent steps in $g_1, g_2, ..., g_P$ *sequentially* by first descending in g_1, then in g_2, etc., until finally we descend in g_P? As we will see throughout this text, in many instances this idea can lead to considerably faster optimization of a such a function. While this finding is largely *empirical*, it can be interpreted in the framework of machine learning as we will see in Section 7.8.

The gist of this idea is drawn in Figure A.7 for the case $P = 3$, where we graphically compare the idea of taking a descent step simultaneously in $g_1, g_2, ..., g_P$ versus a sequence of P descent steps in g_1, then g_2, etc., up to g_P.

Taking the first step of a local method to minimize a cost function g of the form in

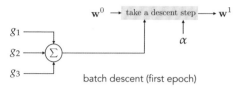

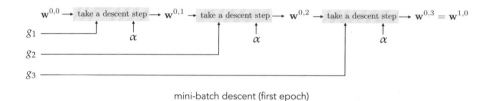

Figure A.7 An abstract illustration of the batch (top panel) and mini-batch descent approaches to local optimization. See text for further details.

Equation (A.33), we begin at some initial point $\mathbf{w}^0$, determine a descent direction $\mathbf{d}^0$, and transition to a new point $\mathbf{w}^1$ as

$$\mathbf{w}^1 = \mathbf{w}^0 + \alpha\,\mathbf{d}^0. \tag{A.34}$$

By analogy, if we were to follow the mini-batch idea detailed above this entails taking a sequence of P steps. If we call our initial point $\mathbf{w}^{0,0} = \mathbf{w}^0$, we then first determine a descent direction $\mathbf{d}^{0,1}$ in g_1 alone, the first function in the sum for g, and take a step in this direction as

$$\mathbf{w}^{0,1} = \mathbf{w}^{0,0} + \alpha\,\mathbf{d}^{0,1}. \tag{A.35}$$

Next we determine a descent direction $\mathbf{d}^{0,2}$ in g_2, the second function in the sum for g, and take a step in this direction

$$\mathbf{w}^{0,2} = \mathbf{w}^{0,1} + \alpha\,\mathbf{d}^{0,2} \tag{A.36}$$

and so forth. Continuing this pattern we take a sequence of P steps, where $\mathbf{d}^{0,p}$ is the descent direction found in g_p, that takes the following form

$$\mathbf{w}^{0,1} = \mathbf{w}^{0,0} + \alpha \, \mathbf{d}^{0,1}$$
$$\mathbf{w}^{0,2} = \mathbf{w}^{0,1} + \alpha \, \mathbf{d}^{0,2}$$
$$\vdots$$
$$\mathbf{w}^{0,p} = \mathbf{w}^{0,p-1} + \alpha \, \mathbf{d}^{0,p} \tag{A.37}$$
$$\vdots$$
$$\mathbf{w}^{0,P} = \mathbf{w}^{0,P-1} + \alpha \, \mathbf{d}^{0,P}.$$

This sequence of updates completes one sweep through the functions $g_1, g_2, ..., g_P$, and is commonly referred to as an *epoch*. If we continued this pattern and took another sweep through each of the P functions we perform a second *epoch* of steps, and so on.

A.5.2 Descending with larger mini-batch sizes

Instead of taking P sequential steps in single functions g_p, one at a time (a mini-batch of *size* 1), we can more generally take fewer steps in one epoch, but take each step with respect to *several* of the functions g_p, e.g., two functions at a time, or three functions at a time, etc. With this slight twist on the idea detailed above we take fewer steps per epoch but take each with respect to larger nonoverlapping subsets of the functions $g_1, g_2, ..., g_P$, but still sweep through each g_p exactly once per epoch.

The size/cardinality of the subsets used is called the *batch size* of the process (mini-batch optimization using a batch size of 1 is also often referred to as *stochastic optimization*). What batch size works best in practice – in terms of providing the greatest speed up in optimization – varies and is often problem dependent.

A.5.3 General performance

Is the trade-off – taking more steps per epoch with a mini-batch approach as opposed to a full descent step – worth the extra effort? Typically *yes*. Often in practice, when minimizing machine learning functions an epoch of mini-batch steps like those detailed above will drastically outperform an analogous full descent step – often referred to as a *full batch* or simply a *batch* epoch in the context of mini-batch optimization. This is particularly true when P is large, typically in the thousands or greater.

A prototypical comparison of a cost function history employing a batch and corresponding epochs of mini-batch optimization applied to the same hypothetical function g (with the same initialization) is shown in Figure A.8. Because we take far more steps with the mini-batch approach and because each g_p takes the same form, each epoch of the mini-batch approach typically outperforms its full batch analog. Even when taking into account that far more descent steps are taken during an epoch of mini-batch optimization the method often greatly outperforms its full batch analog (see, e.g., Exercise 7.11) – again, particularly when P is large.

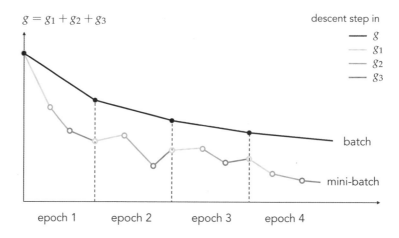

Figure A.8 A prototypical cost function history comparison of batch and mini-batch descent. Here $P = 3$. Epoch per epoch, the mini-batch approach tends to outperform the full batch step, reaching points nearer to local minima of functions of the form in Equation (A.33) faster than the full batch approach.

A.6   Conservative Steplength Rules

In Section 3.5 we described how the steplength parameter α for the gradient descent step – whether fixed for all iterations or diminishing – is very often determined by trial and error in machine learning applications. However, it is possible to derive proper steplength parameter settings mathematically that are guaranteed to produce convergence of the algorithm. These steplength choices are often quite *conservative*, specifically designed to force descent in the function at *every step*, and are therefore quite expensive computationally speaking. In this section we briefly review such steplength schemes for the sake of the interested reader.

A.6.1   Gradient descent and simple quadratic surrogates

Crucial to the analysis of theoretically convergent steplength parameter choices for gradient descent is the following quadratic function

$$h_\alpha(\mathbf{w}) = g(\mathbf{w}^0) + \nabla g(\mathbf{w}^0)^T (\mathbf{w} - \mathbf{w}^0) + \frac{1}{2\alpha} \left\| \mathbf{w} - \mathbf{w}^0 \right\|_2^2 \qquad (A.38)$$

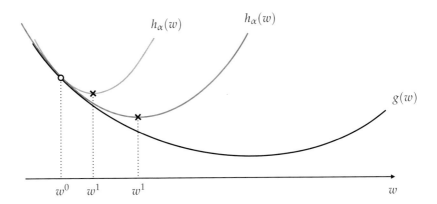

Figure A.9 Two quadratic functions approximating the function g around w^0 given by the quadratic approximation in Equation (A.38). The value of α is larger with the red quadratic than with the blue one.

where $\alpha > 0$. The first two terms on the right-hand side constitute the first-order Taylor series approximation to $g(\mathbf{w})$ at a point $\mathbf{w}^0$ or, in other words, the formula for the tangent hyperplane there. The final term on the right-hand side is the simplest quadratic component imaginable, turning the tangent hyperplane – regardless of whether or not it is tangent at a point that is locally convex or concave – into a convex and perfectly symmetric quadratic whose curvature is controlled in every dimension by the parameter α. Moreover, note that like the hyperplane, this quadratic is still tangent to $g(\mathbf{w})$ at $\mathbf{w}^0$, matching both the function and derivative values at this point.

What happens to this quadratic when we change the value of α? In Figure A.9 we illustrate the approximation for two different values of α with a generic convex function. Note the connection to α: the larger the value α, the wider the associated quadratic becomes.

One of the beautiful things about such a simple quadratic approximation as h_α is that we can easily compute a unique global minimum for it, regardless of the value of α, by checking the first-order optimality condition (see Section 3.2). Setting its gradient to zero we have

$$\nabla h_\alpha(\mathbf{w}) = \nabla g\left(\mathbf{w}^0\right) + \frac{1}{\alpha}\left(\mathbf{w} - \mathbf{w}^0\right) = \mathbf{0}. \tag{A.39}$$

Rearranging the above and solving for $\mathbf{w}$, we can find the minimizer of h_α, which we call $\mathbf{w}^1$, as

$$\mathbf{w}^1 = \mathbf{w}^0 - \alpha \nabla g\left(\mathbf{w}^0\right). \tag{A.40}$$

Thus the minimum of our simple quadratic approximation is precisely a standard gradient descent step at $\mathbf{w}^0$ with a steplength parameter α.

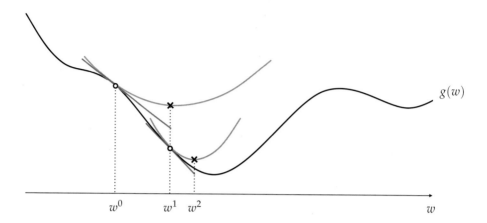

Figure A.10 Gradient descent can be viewed simultaneously as using either linear or simple quadratic surrogates to find a stationary point of g. At each step the associated steplength defines both how far along the linear surrogate we move before hopping back onto the function g, and at the same time the width of the simple quadratic surrogate which we minimize to reach the same point on g.

If we continue taking steps in this manner the kth update is found as the minimum of the simple quadratic approximation associated with the previous update $\mathbf{w}^{k-1}$, which is likewise

$$h_\alpha\left(\mathbf{w}\right) = g\left(\mathbf{w}^{k-1}\right) + \nabla g\left(\mathbf{w}^{k-1}\right)^T\left(\mathbf{w} - \mathbf{w}^{k-1}\right) + \frac{1}{2\alpha}\left\|\mathbf{w} - \mathbf{w}^{k-1}\right\|_2^2 \qquad (A.41)$$

where the minimum is once again given as the kth gradient descent step

$$\mathbf{w}^k = \mathbf{w}^{k-1} - \alpha\nabla g\left(\mathbf{w}^{k-1}\right). \qquad (A.42)$$

In sum, our exercise with the simple quadratic yields an alternative perspective on the standard gradient descent algorithm (detailed in Section 3.5): we can interpret gradient descent as an algorithm that uses linear approximation to move towards a function's minimum, or simultaneously as an algorithm that uses simple quadratic approximations to do the same. In particular this new perspective says that as we move along the direction of steepest descent of the hyperplane, moving from step to step, we are simultaneously "hopping" down the global minima of these simple quadratic approximations. These two simultaneous perspectives are illustrated prototypically in Figure A.10.

A.6.2   Backtracking line search

Since the negative gradient is a descent direction, if we are at a step $\mathbf{w}^{k-1}$ then – with a small enough α – the gradient descent step to $\mathbf{w}^k$ will decrease the value

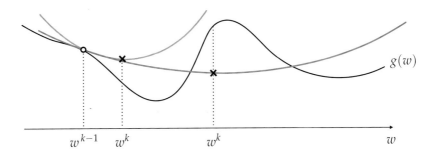

$$w^{k-1} \qquad w^k \qquad\qquad w^k \qquad\qquad\qquad\qquad w$$

Figure A.11 An illustration of our second perspective on how to choose a value for the steplength parameter α that is guaranteed to decrease the underlying function's value by taking a single gradient descent step. The value of α should be decreased until its minimum lies over the function. A step to such a point must decrease the function's value because at this point the quadratic is by definition at its lowest, and so is in particular lower than where it began tangent to g. Here the α value associated with the red quadratic is too large, while the one associated with the blue quadratic is small enough so that the quadratic lies above the function. A (gradient descent) step to this point decreases the value of the function g.

of g, i.e., $g\left(\mathbf{w}^k\right) \leq g\left(\mathbf{w}^{k-1}\right)$. Our first perspective on gradient descent (detailed in Section 3.5) tells us that this will work because as we shrink α we are traveling a shorter distance in the descent direction of the tangent hyperplane at $\mathbf{w}^{k-1}$, and if we shrink this distance enough the underlying function should also be decreasing in this direction. Our second perspective gives us a different but completely equivalent take on this issue: it tells us that in shrinking α we are *increasing* the curvature of the associated quadratic approximation shown in Equation (A.41) (whose minimum is the point we will move to), so that the minimum point on the quadratic approximation lies *above* the function g. A step to such a point must decrease the function's value because at this point the quadratic is by definition at its lowest, and is in particular lower than where it began tangent to g.

How can we find a value of α that does just this at the point $\mathbf{w}^{k-1}$? If our generic gradient descent step is $\mathbf{w}^k = \mathbf{w}^{k-1} - \alpha \nabla g\left(\mathbf{w}^{k-1}\right)$, we want to determine a value of α so that at $\mathbf{w}^k$ the function is lower than the minimum of the quadratic, i.e., $g\left(\mathbf{w}^k\right) \leq h_\alpha\left(\mathbf{w}^k\right)$, as illustrated in Figure A.11. We could select a large number of values for α and test this condition, keeping the one that provides the biggest decrease. However, this is a computationally expensive and somewhat unwieldy prospect. Instead, we test out values of α via an efficient *bisection* process by which we gradually decrease the value of α from some initial value until the inequality is satisfied. This procedure, referred to as *backtracking line search*, generally runs as follows.

1. Choose an initial value for α, e.g., $\alpha = 1$, and a scalar "dampening factor" $t \in (0, 1)$.

2. Create the candidate descent step $\mathbf{w}^k = \mathbf{w}^{k-1} - \alpha \nabla g\left(\mathbf{w}^{k-1}\right)$.

3. Test if $g\left(\mathbf{w}^k\right) \leq h_\alpha\left(\mathbf{w}^k\right)$. If yes, then choose $\mathbf{w}^k$ as the next gradient descent step; otherwise decrease the value of α as $\alpha \longleftarrow t\alpha$, and go back to step 2.

Note that the inequality $g\left(\mathbf{w}^k\right) \leq h_\alpha\left(\mathbf{w}^k\right)$ cab be written equivalently as

$$g\left(\mathbf{w}^k\right) \leq g\left(\mathbf{w}^{k-1}\right) - \frac{\alpha}{2}\left\|\nabla g\left(\mathbf{w}^{k-1}\right)\right\|_2^2 \tag{A.43}$$

by plugging in the step $\mathbf{w}^k = \mathbf{w}^{k-1} - \alpha \nabla g\left(\mathbf{w}^{k-1}\right)$ into the quadratic $h_\alpha\left(\mathbf{w}^k\right)$, completing the square, and simplifying. This equivalent version tells us that so long as we have not reached a stationary point of g the term

$$\frac{\alpha}{2}\left\|\nabla g\left(\mathbf{w}^{k-1}\right)\right\|_2^2 \tag{A.44}$$

will always be positive, hence finding a value of α that satisfies our inequality means that $g\left(\mathbf{w}^k\right)$ will be strictly smaller than $g\left(\mathbf{w}^{k-1}\right)$.

It is the logic of trying out a large value for α first and then decreasing it until we satisfy the inequality that prevents an otherwise unwieldy number of tests to be performed here. Note, however, the way the dampening factor $t \in (0, 1)$ controls how coarsely we sample the α values: in setting t closer to 1 we decrease the amount by which α is shrunk at each failure, which could mean more evaluations are required to determine an adequate α value. Conversely, setting t closer to 0 here shrinks α considerably with every failure, leading to completion more quickly but at the possible cost of outputting a small value for α (and hence a short gradient descent step).

Backtracking line search is a convenient rule for determining a steplength value at each iteration of gradient descent and works "right out of the box." However, each gradient step using backtracking line search, compared to using a fixed steplength value, typically includes higher computational cost due to the search for proper steplength.

A.6.3   Exact line search

In thinking on how one could automatically adjust the steplength value α, one might also think about trying to determine the steplength α that minimizes the function g directly along the kth gradient descent step $\mathbf{w}^k = \mathbf{w}^{k-1} - \alpha \nabla g\left(\mathbf{w}^{k-1}\right)$, that is

$$\underset{\alpha > 0}{\text{minimize}} \; g\left(\mathbf{w}^{k-1} - \alpha \nabla g\left(\mathbf{w}^{k-1}\right)\right). \tag{A.45}$$

This idea is known as *exact line search*. Practically speaking, however, this idea must be implemented via a backtracking line search approach in much the same way we saw previously – by successively examining smaller values until we find a value of α at the kth step $\mathbf{w}^k = \mathbf{w}^{k-1} - \alpha \nabla g\left(\mathbf{w}^{k-1}\right)$ such that

$$g\left(\mathbf{w}^k\right) \leq g\left(\mathbf{w}^{k-1}\right). \tag{A.46}$$

A.6.4 Conservatively optimal fixed steplength values

Suppose we construct the simple quadratic approximation of the form in Equation (A.41) and we turn up the value of α in the simple quadratic approximation so that it reflects the greatest amount of curvature or change in the function's first derivative. Setting the quadratic's parameter α to this maximum curvature, called the *Lipschitz constant*, means that the quadratic approximation will lie completely above the function everywhere except at its point of tangency with the function at $\left(\mathbf{w}^{k-1}, g\left(\mathbf{w}^{k-1}\right)\right)$.

When α is set so that the entire quadratic itself lies above the function this means that in particular the quadratic's minimum lies above the function. In other words, our gradient descent step must lead to a smaller evaluation of g since

$$g\left(\mathbf{w}^k\right) < h_\alpha\left(\mathbf{w}^k\right) \leq h_\alpha\left(\mathbf{w}^{k-1}\right) = g\left(\mathbf{w}^{k-1}\right). \tag{A.47}$$

As detailed in Section 4.1, curvature information of a function is found in its second derivative(s). More specifically, for a single-input function maximum curvature is defined as the maximum (in absolute value) taken by its second derivative, i.e.,

$$\max_w \left| \frac{d^2}{dw^2} g(w) \right|. \tag{A.48}$$

Analogously for a multi-input function $g(\mathbf{w})$ to determine its maximum curvature we must determine the largest possible eigenvalue (in magnitude) of its Hessian matrix or, written algebraically, employing the spectral norm $\|\cdot\|_2$ (see Section C.5)

$$\max_{\mathbf{w}} \left\| \nabla^2 g\left(\mathbf{w}\right) \right\|_2. \tag{A.49}$$

As daunting a task as this may seem it can in fact be done analytically for a range of common machine learning functions including linear regression, (two-class and multi-class) logistic regression, support vector machines, as well as shallow neural networks.

Once determined this maximum curvature L – or an upper bound on it – gives a fixed[3] steplength $\alpha = \frac{1}{L}$ that can be used so that the kth descent step

$$\mathbf{w}^k = \mathbf{w}^{k-1} - \frac{1}{L}\nabla g\left(\mathbf{w}^{k-1}\right) \qquad (A.51)$$

is guaranteed to always descend in the function g.[4] With this steplength we can initialize gradient descent anywhere in the input domain of a function and gradient descent will converge to a stationary point.

This conservatively optimal steplength can be a very convenient rule to use in practice. However, as its name implies, it is indeed a conservative rule by nature. Therefore, in practice, one should use it as a benchmark to search for larger convergence-forcing fixed steplength values. In other words, with the steplength $\alpha = \frac{1}{L}$ calculated one can easily test larger steplengths of the form $\alpha = \frac{t}{L}$ for any constant $t > 1$. Indeed depending on the problem values of t ranging from 1 to 100 can work well in practice.

Example A.7   Computing the Lipschitz constant of a single-input sinusoid
Let us compute the Lipschitz constant – or maximum curvature – of the sinusoid function

$$g(w) = \sin(w). \qquad (A.55)$$

---

[3] If we use *local* instead of *global* curvature to define the steplength our corresponding step will take the form

$$\mathbf{w}^k = \mathbf{w}^{k-1} - \frac{1}{\left\|\nabla^2 g\left(\mathbf{w}^{k-1}\right)\right\|_2}\nabla g\left(\mathbf{w}^{k-1}\right) \qquad (A.50)$$

which we can interpret as a gradient descent step with self-adjusting steplength (that adjusts itself based on the local curvature of the function g). Newton's method – as discussed in Section 4.3 – can be thought of as an extension of this idea.

[4] It is fairly easy to rigorously show that the simple quadratic surrogate tangent to g at the point $\left(\mathbf{w}^{k-1}, g\left(\mathbf{w}^{k-1}\right)\right)$ with $\alpha = \frac{1}{L}$

$$h_{\frac{1}{L}}(\mathbf{w}) = g\left(\mathbf{w}^{k-1}\right) + \nabla g\left(\mathbf{w}^{k-1}\right)^T\left(\mathbf{w} - \mathbf{w}^{k-1}\right) + \frac{L}{2}\left\|\mathbf{w} - \mathbf{w}^{k-1}\right\|_2^2 \qquad (A.52)$$

indeed lies completely above the function g at all points. Writing out the first-order Taylor's formula for g centered at $\mathbf{w}^{k-1}$, we have

$$g(\mathbf{w}) = g\left(\mathbf{w}^{k-1}\right) + \nabla g\left(\mathbf{w}^{k-1}\right)^T\left(\mathbf{w} - \mathbf{w}^{k-1}\right) + \frac{1}{2}\left(\mathbf{w} - \mathbf{w}^{k-1}\right)^T\nabla^2 g\left(\mathbf{c}\right)\left(\mathbf{w} - \mathbf{w}^{k-1}\right) \qquad (A.53)$$

where $\mathbf{c}$ is a point on the line segment connecting $\mathbf{w}$ and $\mathbf{w}^{k-1}$. Since $\nabla^2 g \leq L\mathbf{I}_{N\times N}$ we have

$$\mathbf{a}^T\nabla^2 g\left(\mathbf{c}\right)\mathbf{a} \leq L\|\mathbf{a}\|_2^2 \qquad (A.54)$$

for all $\mathbf{a}$, and in particular for $\mathbf{a} = \mathbf{w} - \mathbf{w}^{k-1}$, which implies $g(\mathbf{w}) \leq h_{\frac{1}{L}}(\mathbf{w})$.

We can easily compute the second derivative of this function as

$$\frac{\mathrm{d}^2}{\mathrm{d}w^2} g(w) = -\sin(w). \tag{A.56}$$

The maximum value this (second derivative) function can take is 1, hence $L = 1$, and therefore $\alpha = \frac{1}{L} = 1$ guarantees descent at every step.

Example A.8 Computing the Lipschitz constant of a multi-input quadratic
In this example we look at computing the Lipschitz constant of the quadratic function

$$g(\mathbf{w}) = a + \mathbf{b}^T \mathbf{w} + \mathbf{w}^T \mathbf{C} \mathbf{w} \tag{A.57}$$

where $a = 1$, $\mathbf{b} = \begin{bmatrix} 1 \\ 1 \end{bmatrix}$, and $\mathbf{C} = \begin{bmatrix} 2 & 0 \\ 0 & 1 \end{bmatrix}$.

Here the Hessian is simply $\nabla^2 g(\mathbf{w}) = \mathbf{C} + \mathbf{C}^T = 2\mathbf{C}$ for all input $\mathbf{w}$, and since the eigenvalues of a diagonal matrix are precisely its diagonal elements, the maximum (in magnitude) eigenvalue is clearly 4. Thus we can set $L = 4$, giving a conservative optimal steplength value of $\alpha = \frac{1}{4}$.

A.6.5 Convergence proofs

To set the stage for the material of this section, it will be helpful to briefly point out the specific set of mild conditions satisfied by all of the cost functions we aim to minimize in this book, as these conditions are relied upon explicitly in the upcoming convergence proofs. These three basic conditions are listed below.

1. They have piecewise-differentiable first derivative.

2. They are bounded from below.

3. They have bounded curvature.

Gradient descent with fixed Lipschitz steplength

With the gradient of g being Lipschitz continuous with constant L, from Section A.6.4 we know that at the kth iteration of gradient descent we have a corresponding quadratic upper bound on g of the form

$$g(\mathbf{w}) \leq g\left(\mathbf{w}^{k-1}\right) + \nabla g\left(\mathbf{w}^{k-1}\right)^T \left(\mathbf{w} - \mathbf{w}^{k-1}\right) + \frac{L}{2}\|\mathbf{w} - \mathbf{w}^{k-1}\|_2^2 \tag{A.58}$$

for all $\mathbf{w}$ in the domain of g. Now plugging in the form of the gradient step $\mathbf{w}^k = \mathbf{w}^{k-1} - \frac{1}{L}\nabla g\left(\mathbf{w}^{k-1}\right)$ into the above and simplifying gives

$$g\left(\mathbf{w}^k\right) \le g\left(\mathbf{w}^{k-1}\right) - \frac{1}{2L}\|\nabla g\left(\mathbf{w}^{k-1}\right)\|_2^2 \tag{A.59}$$

which, since $\|\nabla g\left(\mathbf{w}^{k-1}\right)\|_2^2 \ge 0$, indeed shows that the sequence of gradient steps is decreasing. To show that it converges to a stationary point where the gradient vanishes we subtract $g\left(\mathbf{w}^{k-1}\right)$ from both sides of Equation (A.59), and sum the result over $1 \le k \le K$, giving

$$\sum_{k=1}^{K}\left[g\left(\mathbf{w}^k\right) - g\left(\mathbf{w}^{k-1}\right)\right] = g\left(\mathbf{w}^K\right) - g\left(\mathbf{w}^0\right) \le -\frac{1}{2L}\sum_{k=1}^{K}\left\|\nabla g\left(\mathbf{w}^{k-1}\right)\right\|_2^2. \tag{A.60}$$

Note importantly here that since g is bounded from below, taking $K \longrightarrow \infty$, we *must* have that

$$\sum_{k=1}^{\infty}\left\|\nabla g\left(\mathbf{w}^{k-1}\right)\right\|_2^2 < \infty. \tag{A.61}$$

Hence the fact that the infinite sum above must be finite implies that as $k \longrightarrow \infty$ we have that

$$\left\|\nabla g\left(\mathbf{w}^{k-1}\right)\right\|_2^2 \longrightarrow 0 \tag{A.62}$$

meaning that the sequence of gradient descent steps with steplength determined by the Lipschitz constant of the gradient of g produces a sequence with vanishing gradient that converges to a stationary point of g. Note that we could have made the same argument above using any fixed steplength smaller than $\frac{1}{L}$ as well.

Gradient descent with backtracking line search

With the assumption that g has a maximum bounded curvature of L, it follows that with any fixed choice of initial steplength $\alpha > 0$ and $t \in (0,1)$ we can always find an integer n_0 such that

$$t^{n_0}\alpha \le \frac{1}{L}. \tag{A.63}$$

Thus the backtracking-found steplength at the kth gradient descent step will always be larger than this lower bound, i.e.,

$$\alpha_k \ge t^{n_0}\alpha > 0 \tag{A.64}$$

for all k.

Recall from Equation (A.43) that by running the backtracking procedure at the kth gradient step we have

$$g\left(\mathbf{w}^k\right) \leq g\left(\mathbf{w}^{k-1}\right) - \frac{\alpha_k}{2}\left\|\nabla g\left(\mathbf{w}^{k-1}\right)\right\|_2^2. \tag{A.65}$$

To show that the sequence of gradient steps converges to a stationary point of g we first subtract $g\left(\mathbf{w}^{k-1}\right)$ from both sides of Equation (A.65), and sum the result over $1 \leq k \leq K$, which gives

$$\sum_{k=1}^{K}\left[g\left(\mathbf{w}^k\right) - g\left(\mathbf{w}^{k-1}\right)\right] = g\left(\mathbf{w}^K\right) - g\left(\mathbf{w}^0\right) \leq -\frac{1}{2}\sum_{k=1}^{K}\alpha_k\left\|\nabla g\left(\mathbf{w}^{k-1}\right)\right\|_2^2. \tag{A.66}$$

Since g is bounded from below, taking $K \longrightarrow \infty$, we must have

$$\sum_{k=1}^{\infty}\alpha_k\left\|\nabla g\left(\mathbf{w}^{k-1}\right)\right\|_2^2 < \infty. \tag{A.67}$$

Now, we know from Equation (A.64) that

$$\sum_{k=1}^{K}\alpha_k \geq K\, t^{n_0}\alpha \tag{A.68}$$

implying

$$\sum_{k=1}^{\infty}\alpha_k = \infty. \tag{A.69}$$

In order for Equations (A.67) and (A.69) to hold simultaneously we *must* have that

$$\left\|\nabla g\left(\mathbf{w}^{k-1}\right)\right\|_2^2 \longrightarrow 0 \tag{A.70}$$

as $k \longrightarrow \infty$. This shows that the sequence of gradient steps determined by backtracking line search converges to a stationary point of g.

A.7 Newton's Method, Regularization, and Nonconvex Functions

As we saw in Section 4.3, Newton's method is naturally incapable of properly minimizing generic nonconvex functions. In this section we describe the regularized Newton step as a common approach to ameliorate this particular issue (which we have essentially seen before albeit without the more in-depth context we provide here).

A.7.1   Turning up ϵ

In Section 4.3.3 we saw how adding a very small positive value ϵ to the second derivative of a single-input function, or analogously a weighted identity matrix of the form $\epsilon \mathbf{I}_{N \times N}$ to the Hessian in the multi-input case, helps Newton's method avoid numerical problems in flat regions of a convex function. This adjusted Newton step, which takes the form

$$\mathbf{w}^k = \mathbf{w}^{k-1} - \left(\nabla^2 g(\mathbf{w}^{k-1}) + \epsilon \mathbf{I}_{N \times N} \right)^{-1} \nabla g(\mathbf{w}^{k-1}), \qquad (A.71)$$

can be interpreted as the stationary point of a slightly adjusted second-order Taylor series approximation centered at $\mathbf{w}^{k-1}$

$$h(\mathbf{w}) = g(\mathbf{w}^{k-1}) + \nabla g(\mathbf{w}^{k-1})^T (\mathbf{w} - \mathbf{w}^{k-1}) + \frac{1}{2}(\mathbf{w} - \mathbf{w}^{k-1})^T \nabla^2 g\left(\mathbf{w}^{k-1}\right)(\mathbf{w} - \mathbf{w}^{k-1})$$
$$+ \frac{\epsilon}{2} \left\| \mathbf{w} - \mathbf{w}^{k-1} \right\|_2^2 . \qquad (A.72)$$

The first three terms (on the right-hand side of the equality) still represent the second-order Taylor series at $\mathbf{w}^{k-1}$, and to it we have added $\frac{\epsilon}{2} \left\| \mathbf{w} - \mathbf{w}^{k-1} \right\|_2^2$, a *convex* and perfectly symmetric quadratic centered at $\mathbf{w}^{k-1}$ with N positive eigenvalues (each equal to $\frac{\epsilon}{2}$). In other words, we have a sum of two quadratic functions. When $\mathbf{w}^{k-1}$ is at a nonconvex or flat portion of a function the first quadratic is likewise nonconvex or flat. However, the second one is *always* convex, and the larger ϵ is set the greater its (convex) curvature. This means that if we set ϵ larger we can *convexify* the entire approximation, forcing the stationary point we solve for to be a minimum and the direction in which we travel is one of guaranteed descent.

Example A.9   The effect of regularization

In Figure A.12 we illustrate the regularization of a nonconvex function

$$h_1(w_1, w_2) = w_1^2 - w_2^2 \qquad (A.73)$$

using the convex regularizer

$$h_2(w_1, w_2) = w_1^2 + w_2^2. \qquad (A.74)$$

In particular we show what the resulting sum $h_1 + \epsilon h_2$ looks like over four progressively increasing values of ϵ, from $\epsilon = 0$ (leftmost panel) to $\epsilon = 2$ (rightmost panel).

Since h_1 is nonconvex, and has a single stationary point that is a saddle point at the origin, the addition of h_2 pulls up its downward-facing dimension. Not surprisingly as ϵ is increased, the shape of the sum is dictated more and more by h_2. Eventually, turning up ϵ sufficiently, the sum becomes convex.

$\epsilon = 0$ $\epsilon = 0.6$ $\epsilon = 1$ $\epsilon = 2$

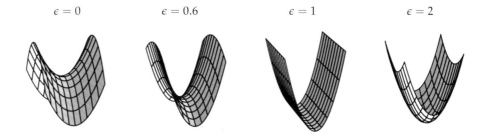

Figure A.12 Figure associated with Example A.9. From left to right, a nonconvex quadratic function is slowly turned into a convex function via the weighted addition of a convex quadratic. See text for further details.

Example A.10   Minimization of a nonconvex function
In Figure A.13 we illustrate five regularized Newton steps (using the update step shown in Equation (A.71)) to minimize the nonconvex function

$$g(w) = 2 - e^{-w^2}. \tag{A.75}$$

We initialize the algorithm at a point of local nonconvexity for this function, and gradually increase ϵ from $\epsilon = 0$ (top panel) where Newton's method diverges, to $\epsilon = 4$ (bottom panel) where it has sufficiently *convexified* the Newton step, leading the algorithm to take downhill steps towards the minimum of the function. In the figure the steps from each run are colored from green (the first step) to red (the final step), with the regularized second-order approximation at each step colored accordingly.

   To determine how high we need to turn up ϵ in order to make the regularized second-order Taylor series approximation convex, recall from Section 4.2 that a quadratic function is convex *if and only if* it has all nonnegative eigenvalues. Thus ϵ must be made larger than the magnitude of the smallest eigenvalue of the Hessian $\nabla^2 g\left(\mathbf{w}^{k-1}\right)$ in order for the regularized second-order quadratic to be convex. For a single-input function this reduces to ϵ being larger in magnitude than value of the function's second derivative at w^{k-1} (if it is negative there).

   While ϵ in the regularized Newton step in Equation (A.71) is typically set to a relatively small value, it is interesting to note that as we increase ϵ, the direction we travel in tilts toward the gradient descent direction at $\mathbf{w}^{k-1}$. In other words, when ϵ is large the direction we travel when taking the regularized Newton step becomes the gradient descent direction (albeit with a very small magnitude)

Figure A.13 Figure associated with Example A.10. See text for details.

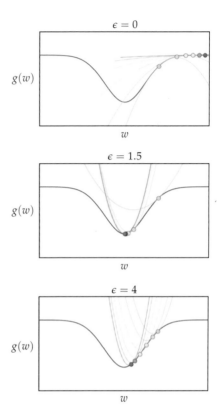

$$-\left(\nabla^2 g(\mathbf{w}^{k-1}) + \epsilon \mathbf{I}_{N \times N}\right)^{-1} \nabla g(\mathbf{w}^{k-1}) \approx -\left(\epsilon \mathbf{I}_{N \times N}\right)^{-1} \nabla g(\mathbf{w}^{k-1}) = -\frac{1}{\epsilon} \nabla g(\mathbf{w}^{k-1}).$$

$$\text{(A.76)}$$

A.8 Hessian-Free Methods

While Newton's method is a powerful technique that exhibits rapid convergence due to its employment of second-order information, it is naturally constrained by the input dimension N of a general function $g(\mathbf{w})$. More specifically, the $N \times N$ Hessian matrix $\nabla^2 g(\mathbf{w})$, with its N^2 entries, naturally limits Newton's method's use to cases where N is (roughly speaking) in the thousands, since it is difficult to even store such a matrix when N is larger (let alone compute with it).

In this section we discuss a two variations on Newton's method, collectively referred to as *Hessian-free* optimization methods, that ameliorate this issue by replacing the Hessian (in each Newton's method step) with a close approximation that does not suffer from the same scaling issue. Because of this, both approaches naturally trade the precision of each Newton's method step with the ability to scale the basic algorithm to high-dimensional input. The first of these approaches is the simplest conceptually speaking, and involves *subsampling* the Hessian, using only a fraction of its entries. The latter method, often referred to as *quasi-Newton*, involves replacing the Hessian with a *low-rank* approximation that can be computed effectively.

A.8.1   Subsampling the Hessian

The simplest way to deal with the scaling issue inherent with Newton's method, that of the massive number of entries in the $N \times N$ Hessian matrix $\nabla^2 g(\mathbf{w})$ as N increases, is to simply *subsample* the Hessian. That is, instead of using the entire Hessian matrix we use only a fraction of its entries, setting the remainder of the entries to zero. This of course dilutes the power of complete second-order information leveraged at each Newton's step, and thus the corresponding efficacy of each corresponding Newton's step, but salvages the otherwise untenable method when N is too large. There are a variety of ways one can consider subsampling the Hessian that trade-off between maintaining second-order information and the allowance of graceful scaling with input dimension (see, e.g., [15, 71]).

One popular subsampling scheme involves simply retaining only the diagonal entries of the Hessian matrix, i.e., only the N pure partial second derivatives of the form $\frac{\partial^2}{\partial w_n^2} g(w)$ for $n = 1, 2, ..., N$. This drastically reduces the number of entries of the Hessian, and greatly simplifies the Newton's step from the solution to a linear system of equations

$$\mathbf{w}^k = \mathbf{w}^{k-1} - \left(\nabla^2 g(\mathbf{w}^{k-1})\right)^{-1} \nabla g(\mathbf{w}^{k-1}) \tag{A.77}$$

to the straightforward component-wise update for each $n = 1, 2, ..., N$

$$w_n^k = w_n^{k-1} - \frac{\frac{\partial}{\partial w_n} g(\mathbf{w}^{k-1})}{\frac{\partial^2}{\partial w_n^2} g(\mathbf{w}^{k-1})}. \tag{A.78}$$

In other words, keeping only the diagonal entries of the Hessian we decouple each coordinate and no longer have a system of equations to solve. The downside, of course, is that we ignore all cross-partial derivatives retaining only the second-order information corresponding to the curvature along each input dimension independently. Nonetheless, this subsampled Newton's step can be quite effective in practice when used to minimize machine learning cost functions (see, e.g., Exercise 9.8) and can scale as gracefully as a first-order method like gradient descent.

A.8.2   Secant methods

In studying Newton's method as a zero-finding algorithm, at the kth step in finding a zero of the first-order equation $\frac{d}{dw}g(w) = 0$ we form the first-order Taylor series

$$h(w) = \frac{d}{dw}g\left(w^{k-1}\right) + \frac{d^2}{dw^2}g\left(w^{k-1}\right)\left(w - w^{k-1}\right) \tag{A.79}$$

and find where this linear function equals zero, which is given by the corresponding Newton update

$$w^k = w^{k-1} - \frac{\frac{d}{dw}g\left(w^{k-1}\right)}{\frac{d^2}{dw^2}g\left(w^{k-1}\right)}. \tag{A.80}$$

If we replace the slope of the tangent line – here the second derivative evaluation $\frac{d^2}{dw^2}g\left(w^{k-1}\right)$ – with the slope provided by a closely related[5] *secant line*

$$\frac{d^2}{dw^2}g\left(w^{k-1}\right) \approx \frac{\frac{d}{dw}g\left(w^{k-1}\right) - \frac{d}{dw}g\left(w^{k-2}\right)}{w^{k-1} - w^{k-2}} \tag{A.81}$$

we will produce an algorithm highly related to Newton's method (but with no need to employ the second derivative). Replacing the second derivative with this approximation in our Newton's step we have a *secant method* update

$$w^k = w^{k-1} - \frac{\frac{d}{dw}g\left(w^{k-1}\right)}{\frac{\frac{d}{dw}g(w^{k-1}) - \frac{d}{dw}g(w^{k-2})}{w^{k-1} - w^{k-2}}} \tag{A.82}$$

which we can write in a less cumbersome manner as

$$w^k = w^{k-1} - \frac{\frac{d}{dw}g\left(w^{k-1}\right)}{s^{k-1}} \tag{A.83}$$

where s^{k-1} has been used to denote the slope of the secant line

$$s^{k-1} = \frac{\frac{d}{dw}g\left(w^{k-1}\right) - \frac{d}{dw}g\left(w^{k-2}\right)}{w^{k-1} - w^{k-2}}. \tag{A.84}$$

While less accurate than Newton's method this approach, which does not rely directly on the second derivative, can still be generally used to solve the first-order equation and find stationary points of the function $g(w)$. This fact, while fairly inconsequential for a single-input function with $N = 1$, gains significantly

---

[5] Remember that the derivative of a single-input function defines the slope of the tangent line to the function at the point of tangency. This slope can be roughly approximated as the slope of a nearby *secant* line, that is a line that passes through the same point as well as another point nearby on the function (see Section B.2.1).

more value when this secant method is generalized to multi-input functions. This is because, as we have already discussed, it is the very size of the Hessian matrix that prohibits Newton's method's use for functions with large values of N.

Everything we have discussed for the generic single-input case tracks to the multi-input instance as well. Notice, looking back at Equation (A.84), that by multiplying both sides by $w^{k-1} - w^{k-2}$ we can write it equivalently as

$$s^{k-1}\left(w^{k-1} - w^{k-2}\right) = \frac{d}{dw}g\left(w^{k-1}\right) - \frac{d}{dw}g\left(w^{k-2}\right). \tag{A.85}$$

This is often referred to as the single-input *secant condition*.
Replacing each component of Equation (A.85) with its multi-input analog gives the multi-input secant condition

$$\mathbf{S}^{k-1}\left(\mathbf{w}^{k-1} - \mathbf{w}^{k-2}\right) = \nabla g\left(\mathbf{w}^{k-1}\right) - \nabla g\left(\mathbf{w}^{k-2}\right). \tag{A.86}$$

Here we have replaced the scalar s^{k-1} with its analog, an $N \times N$ matrix $\mathbf{S}^{k-1}$, and the one-dimensional terms $w^{k-1}, w^{k-2}, \frac{d}{dw}g\left(w^{k-1}\right)$, and $\frac{d}{dw}g\left(w^{k-2}\right)$ with their respective N-dimensional analogs: $\mathbf{w}^{k-1}, \mathbf{w}^{k-2}, \nabla g\left(\mathbf{w}^{k-1}\right)$, and $\nabla g\left(\mathbf{w}^{k-2}\right)$. Assuming for a moment that $\mathbf{S}^{k-1}$ is invertible, we can also express the secant condition as

$$\mathbf{w}^{k-1} - \mathbf{w}^{k-2} = \left(\mathbf{S}^{k-1}\right)^{-1}\left(\nabla g\left(\mathbf{w}^{k-1}\right) - \nabla g\left(\mathbf{w}^{k-2}\right)\right). \tag{A.87}$$

In either instance, we can see that the secant method requires we *solve* for the matrix $\mathbf{S}^{k-1}$ or its inverse $\left(\mathbf{S}^{k-1}\right)^{-1}$. Unlike the one-dimensional instance of the secant condition in Equation (A.85) where each update has a unique solution, with the N-dimensional case in Equation (A.86) we must solve a system of equations which will generally have infinitely many solutions, since there are only N equations but N^2 entries to solve for in the matrix $\mathbf{S}^{k-1}$.

A.8.3   Quasi-Newton methods

As discussed in Section 4.3.2, the standard Newton step

$$\mathbf{w}^k = \mathbf{w}^{k-1} - \left(\nabla^2 g\left(\mathbf{w}^{k-1}\right)\right)^{-1}\nabla g\left(\mathbf{w}^{k-1}\right) \tag{A.88}$$

is an example of a local optimization step of the generic form $\mathbf{w}^k = \mathbf{w}^{k-1} + \alpha\mathbf{d}^k$, with the descent direction given by

$$\mathbf{d}^k = -\left(\nabla^2 g\left(\mathbf{w}^{k-1}\right)\right)^{-1}\nabla g\left(\mathbf{w}^{k-1}\right). \tag{A.89}$$

The term "quasi-Newton" method is the jargon phrase used for any descent step of the form

$$\mathbf{d}^k = - \left(\mathbf{S}^{k-1}\right)^{-1} \nabla g \left(\mathbf{w}^{k-1}\right) \tag{A.90}$$

where the true Hessian matrix $\nabla^2 g \left(\mathbf{w}^{k-1}\right)$ is replaced with a secant approximation $\mathbf{S}^{k-1}$. As with the single-input case this kind of update – while less accurate at each step than the true Newton's method – can still define an effective local optimization method depending on how $\mathbf{S}^{k-1}$ is constructed (all while avoiding the need for direct second-order information).

In other words, in taking quasi-Newton steps we are no longer required to compute a sequence of Hessian matrices

$$\nabla^2 g \left(\mathbf{w}^0\right), \ \nabla^2 g \left(\mathbf{w}^1\right), \ \nabla^2 g \left(\mathbf{w}^2\right), \ \text{etc.,} \tag{A.91}$$

and instead we construct a sequence of corresponding secant matrices

$$\mathbf{S}^1, \ \mathbf{S}^2, \ \mathbf{S}^3, \ \text{etc.,} \tag{A.92}$$

as an approximation to the Hessian sequence. To construct this secant sequence, note the following.

- **$\mathbf{S}^{k-1}$ should be a solution to the secant condition.** Defining for notational convenience $\mathbf{a}^k = \mathbf{w}^{k-1} - \mathbf{w}^{k-2}$ and $\mathbf{b}^k = \nabla g \left(\mathbf{w}^{k-1}\right) - \nabla g \left(\mathbf{w}^{k-2}\right)$, the secant condition in Equation (A.86) dictates we must have that $\mathbf{S}^{k-1}\mathbf{a}^k = \mathbf{b}^k$. Letting $\mathbf{F}^k = \left(\mathbf{S}^k\right)^{-1}$ we can write this equivalently as $\mathbf{a}^k = \mathbf{F}^k \mathbf{b}^k$.

- **$\mathbf{S}^{k-1}$ should be symmetric.** Since the Hessian $\nabla^2 g \left(\mathbf{w}^{k-1}\right)$ is always symmetric and ideally we want $\mathbf{S}^{k-1}$ to mimic the Hessian closely, it is reasonable to expect $\mathbf{S}^{k-1}$ to be symmetric as well.

- **The secant sequence should converge.** As the quasi-Newton method progresses, the sequence of steps $\mathbf{w}^{k-1}$ should converge (to a minimum of g), so too should the secant sequence $\mathbf{S}^{k-1}$ (to the Hessian evaluated at that minimum).

We now explore, by example, a number of ways to construct secant sequences that satisfy these conditions. These constructions are *recursive* in nature and take the general form

$$\mathbf{F}^k = \mathbf{F}^{k-1} + \mathbf{D}^{k-1} \tag{A.93}$$

where the $N \times N$ *difference* matrix $\mathbf{D}^{k-1}$ is designed to be *symmetric*, of a particular *low rank*, and *diminishing* in magnitude.

Symmetry of $\mathbf{D}^{k-1}$ guarantees if we initialize the very first $\mathbf{F}^0$ to a symmetric matrix (most commonly the identity matrix), this recursive update will retain our desired symmetry property (since the sum of two symmetric matrices is always symmetric). Similarly, if $\mathbf{D}^{k-1}$ is designed to be positive-definite and the initialization $\mathbf{F}^0$ is also positive-definite (as the identity matrix is) then this property is inherited by all matrices $\mathbf{F}^k$. Constraining $\mathbf{D}^{k-1}$ to be of low rank makes it structurally simple and allows to compute it in closed form at each step. Finally, the magnitude/norm of $\mathbf{D}^{k-1}$ should shrink as k gets larger, otherwise $\mathbf{F}^k$ simply will not converge.

Example A.11   Rank-1 difference matrix

In this example we describe one of the simplest recursive formula for $\mathbf{S}^k$ (or more precisely its inverse $\mathbf{F}^k$) where the difference matrix $\mathbf{D}^{k-1}$ in Equation (A.93) is a rank-1 outer-product matrix

$$\mathbf{D}^{k-1} = \mathbf{u}\,\mathbf{u}^T. \tag{A.94}$$

By first assuming this form of the difference matrix does indeed satisfy the secant condition, we can then work *backwards* from it to determine the proper value for $\mathbf{u}$.

Substituting $\mathbf{F}^{k-1} + \mathbf{u}\,\mathbf{u}^T$ for $\mathbf{F}^k$ into the secant condition gives

$$\left(\mathbf{F}^{k-1} + \mathbf{u}\,\mathbf{u}^T\right)\mathbf{b}^k = \mathbf{a}^k \tag{A.95}$$

or rearranging equivalently

$$\mathbf{u}\,\mathbf{u}^T\mathbf{b}^k = \mathbf{a}^k - \mathbf{F}^{k-1}\mathbf{b}^k. \tag{A.96}$$

Multiplying both sides by $\left(\mathbf{b}^k\right)^T$

$$\left(\mathbf{b}^k\right)^T \mathbf{u}\,\mathbf{u}^T\mathbf{b}^k = \left(\mathbf{b}^k\right)^T \mathbf{a}^k - \left(\mathbf{b}^k\right)^T \mathbf{F}^{k-1}\mathbf{b}^k \tag{A.97}$$

and taking the square root of both sides gives

$$\mathbf{u}^T\mathbf{b}^k = \left(\left(\mathbf{b}^k\right)^T \mathbf{a}^k - \left(\mathbf{b}^k\right)^T \mathbf{F}^{k-1}\mathbf{b}^k\right)^{\frac{1}{2}}. \tag{A.98}$$

Substituting the value for $\mathbf{u}^T\mathbf{b}^k$ from Equation (A.98) into Equation (A.96), we arrive at the desired form for the vector $\mathbf{u}$

$$\mathbf{u} = \frac{\mathbf{a}^k - \mathbf{F}^{k-1}\mathbf{b}^k}{\left(\left(\mathbf{b}^k\right)^T \mathbf{a}^k - \left(\mathbf{b}^k\right)^T \mathbf{F}^{k-1}\mathbf{b}^k\right)^{\frac{1}{2}}} \tag{A.99}$$

with the corresponding recursive formula for $\mathbf{F}^k$ given as

$$\mathbf{F}^k = \mathbf{F}^{k-1} + \frac{\left(\mathbf{a}^k - \mathbf{F}^{k-1}\mathbf{b}^k\right)\left(\mathbf{a}^k - \mathbf{F}^{k-1}\mathbf{b}^k\right)^T}{\left(\mathbf{b}^k\right)^T \mathbf{a}^k - \left(\mathbf{b}^k\right)^T \mathbf{F}^{k-1}\mathbf{b}^k}. \tag{A.100}$$

Example A.12   The Davidon–Fletcher–Powell (DFP) method

We can use a slightly more complex structure for the difference matrix by constructing it as a sum of two rank-1 matrices

$$\mathbf{D}^{k-1} = \mathbf{u}\,\mathbf{u}^T + \mathbf{v}\,\mathbf{v}^T. \tag{A.101}$$

By allowing for a rank-2 difference between the subsequent matrices (as opposed to a rank-1 difference) we encode an additional level of complexity into our approximation to subsequent inverse Hessian evaluations.

In order to determine the proper values for $\mathbf{u}$ and $\mathbf{v}$, we substitute $\mathbf{F}^{k-1} + \mathbf{D}^{k-1}$ for $\mathbf{F}^k$ into the secant condition (as we did in Example A.11), giving

$$\left(\mathbf{F}^{k-1} + \mathbf{u}\,\mathbf{u}^T + \mathbf{v}\,\mathbf{v}^T\right)\mathbf{b}^k = \mathbf{a}^k, \tag{A.102}$$

or rearranging equivalently

$$\mathbf{u}\,\mathbf{u}^T\mathbf{b}^k + \mathbf{v}\,\mathbf{v}^T\mathbf{b}^k = \mathbf{a}^k - \mathbf{F}^{k-1}\mathbf{b}^k. \tag{A.103}$$

Note that here we have only a *single* equation, and so there are infinitely many choices for our *two* unknown vectors $\mathbf{u}$ and $\mathbf{v}$. A very simple yet common way of determining a single set of values for $\mathbf{u}$ and $\mathbf{v}$ is to suppose the first/second term on the left-hand side of Equation (A.103) equals the corresponding term on the right-hand side, that is

$$\mathbf{u}\,\mathbf{u}^T\mathbf{b}^k = \mathbf{a}^k \qquad \text{and} \qquad \mathbf{v}\,\mathbf{v}^T\mathbf{b}^k = -\mathbf{F}^{k-1}\mathbf{b}^k. \tag{A.104}$$

This added assumption allows us to solve for a valid pair of $\mathbf{u}$ and $\mathbf{v}$ in a way that closely mirrors the solution method from Example A.11. First, we multiply each by $\left(\mathbf{b}^k\right)^T$, giving

$$\left(\mathbf{b}^k\right)^T \mathbf{u}\,\mathbf{u}^T\mathbf{b}^k = \left(\mathbf{b}^k\right)^T \mathbf{a}^k \qquad \text{and} \qquad \left(\mathbf{b}^k\right)^T \mathbf{v}\,\mathbf{v}^T\mathbf{b}^k = -\left(\mathbf{b}^k\right)^T \mathbf{F}^{k-1}\mathbf{b}^k. \tag{A.105}$$

Taking the square root of both sides in both equations, we then have the set of equations

$$\mathbf{u}^T\mathbf{b}^k = \left(\left(\mathbf{b}^k\right)^T \mathbf{a}^k\right)^{\frac{1}{2}} \qquad \text{and} \qquad \mathbf{v}^T\mathbf{b}^k = \left(-\left(\mathbf{b}^k\right)^T \mathbf{F}^{k-1}\mathbf{b}^k\right)^{\frac{1}{2}}. \tag{A.106}$$

Substituting these value for $\mathbf{u}^T\mathbf{b}^k$ and $\mathbf{v}^T\mathbf{b}^k$ from Equation (A.106) into Equation (A.104), we have

$$\mathbf{u} = \frac{\mathbf{a}^k}{\left(\left(\mathbf{b}^k\right)^T \mathbf{a}^k\right)^{\frac{1}{2}}} \quad \text{and} \quad \mathbf{v} = \frac{-\mathbf{F}^{k-1}\mathbf{b}^k}{\left(-\left(\mathbf{b}^k\right)^T \mathbf{F}^{k-1}\mathbf{b}^k\right)^{\frac{1}{2}}}. \tag{A.107}$$

with the corresponding recursive formula for $\mathbf{F}^k$ given as

$$\mathbf{F}^k = \mathbf{F}^{k-1} + \frac{\mathbf{a}^k \left(\mathbf{a}^k\right)^T}{\left(\mathbf{b}^k\right)^T \mathbf{a}^k} - \frac{\left(\mathbf{F}^{k-1}\mathbf{b}^k\right)\left(\mathbf{F}^{k-1}\mathbf{b}^k\right)^T}{\left(\mathbf{b}^k\right)^T \mathbf{F}^{k-1}\mathbf{b}^k}. \tag{A.108}$$

This is called the Davidon–Fletcher–Powell (DFP) method based on the authors who first put forth this solution [14, 72].

While the update derived here is for the inverse matrix $\mathbf{F}^k = \left(\mathbf{S}^k\right)^{-1}$, an entirely similar recursive expression can be formulated for $\mathbf{S}^k$ itself, leading to the Broyden–Fletcher–Goldfarb–Shanno (BFGS) update, named after its original authors [14, 72, 73] (see Exercise 4.10).

A.8.4 Low-memory quasi-Newton methods

In the previous examples we saw how to construct a sequence of (inverse) secant matrices via the recursion

$$\left(\mathbf{S}^k\right)^{-1} = \left(\mathbf{S}^{k-1}\right)^{-1} + \mathbf{D}^{k-1} \tag{A.109}$$

to replace the true Hessian sequence. The *descent direction* for the kth step of a quasi-Newton method then takes the form

$$\mathbf{d}^k = -\left(\mathbf{S}^{k-1}\right)^{-1} \nabla g(\mathbf{w}^{k-1}) \tag{A.110}$$

where $\mathbf{S}^{k-1}$ is our approximation to $\nabla^2 g\left(\mathbf{w}^{k-1}\right)$. Notice, however, that as written in Equation (A.110), computing the descent direction still involves the explicit form of an $N \times N$ matrix: the inverse of $\mathbf{S}^{k-1}$. But recall, it was precisely the presence of the $N \times N$ Hessian matrix (with its N^2 values) that drove us to examine quasi-Newton methods to begin with. So, at first glance, it appears that we have not avoided the serious scaling issues associated with employing an $N \times N$ matrix: originally a Hessian matrix, now a secant matrix.

Luckily we can avoid explicit construction of the (inverse) secant matrix by focusing on how it acts on the gradient vector $\nabla g(\mathbf{w}^{k-1})$ in Equation (A.110). For example, denoting for notational convenience $\mathbf{z}^{k-1} = \nabla g(\mathbf{w}^{k-1})$ and $\mathbf{F}^k = \left(\mathbf{S}^k\right)^{-1}$, the descent direction $\mathbf{d}^k$ using the recursive update derived in Example A.11 can be written as

$$\mathbf{d}^k = -\mathbf{F}^k \mathbf{z}^{k-1} = -\left(\mathbf{F}^{k-1} + \frac{\left(\mathbf{a}^k - \mathbf{F}^{k-1}\mathbf{b}^k \right)\left(\mathbf{a}^k - \mathbf{F}^{k-1}\mathbf{b}^k \right)^T}{\left(\mathbf{b}^k \right)^T \mathbf{a}^k - \left(\mathbf{b}^k \right)^T \mathbf{F}^{k-1}\mathbf{b}^k} \right) \mathbf{z}^{k-1}$$

$$= -\mathbf{F}^{k-1}\mathbf{z}^{k-1} - \frac{\left(\mathbf{a}^k - \mathbf{F}^{k-1}\mathbf{b}^k \right)^T \mathbf{z}^{k-1}}{\left(\mathbf{b}^k \right)^T \mathbf{a}^k - \left(\mathbf{b}^k \right)^T \mathbf{F}^{k-1}\mathbf{b}^k} \left(\mathbf{a}^k - \mathbf{F}^{k-1}\mathbf{b}^k \right).$$

(A.111)

Notice wherever the matrix $\mathbf{F}^{k-1}$ appears in the final line of Equation (A.111), it is always attached to some other vector, in this case $\mathbf{b}^k$ and $\mathbf{z}^{k-1}$. Thus if we somehow manage to compute $\mathbf{F}^{k-1}\mathbf{b}^k$ and $\mathbf{F}^{k-1}\mathbf{z}^{k-1}$ directly as vectors, we can avoid explicitly forming $\mathbf{F}^{k-1}$.

This is where our particular recursive construction of $\mathbf{F}^k$ as

$$\mathbf{F}^k = \mathbf{F}^{k-1} + \mathbf{u}^{k-1}\left(\mathbf{u}^{k-1} \right)^T$$

(A.112)

comes to rescue. Rolling this recursion all the way back to $\mathbf{F}^0$ and initializing $\mathbf{F}^0 = \mathbf{I}_{N\times N}$, we can write it equivalently as

$$\mathbf{F}^k = \mathbf{I}_{N\times N} + \left(\mathbf{u}^0 \left(\mathbf{u}^0 \right)^T + \cdots + \mathbf{u}^{k-1}\left(\mathbf{u}^{k-1} \right)^T \right).$$

(A.113)

Multiplication of any vector $\mathbf{t}$ with $\mathbf{F}^k$ can then be written as

$$\mathbf{F}^k\mathbf{t} = \mathbf{I}_{N\times N}\mathbf{t} + \left(\mathbf{u}^0 \left(\mathbf{u}^0 \right)^T + \cdots + \mathbf{u}^{k-1}\left(\mathbf{u}^{k-1} \right)^T \right)\mathbf{t}$$

(A.114)

and simplified as

$$\mathbf{F}^k\mathbf{t} = \mathbf{t} + \mathbf{u}^0 \left(\left(\mathbf{u}^0 \right)^T \mathbf{t} \right) + \cdots + \mathbf{u}^{k-1}\left(\left(\mathbf{u}^{k-1} \right)^T \mathbf{t} \right).$$

(A.115)

To compute $\mathbf{F}^k\mathbf{t}$, as written in Equation (A.115), we only need access to vectors $\mathbf{t}$ as well as $\mathbf{u}^0$ through $\mathbf{u}^{k-1}$, and perform simple vector inner-product.

We can make very similar kinds of arguments with any rank-1 or rank-2 quasi-Newton update formula, meaning that any of the update formulae given in the preceding examples can be implemented in a way that we never need explicitly construct a secant or inverse secant matrix (see, e.g., [14, 72]).

B Derivatives and Automatic Differentiation

B.1 Introduction

The concept of the *derivative* from calculus is fundamental to forming an intuitive and practical understanding of local optimization techniques, which are the workhorse of machine learning. Because of its critical importance this appendix chapter provides a self-contained overview of derivatives, practical computation of derivatives via automatic differentiation (including an outline of the *backpropagation algorithm*), and Taylor series approximations.

B.2 The Derivative

The derivative is a simple tool for understanding a (continuous) mathematical function *locally*, meaning at and around a single point. In this section we review the notion of the derivative at a point, and how it naturally defines the best *linear approximation* (i.e., a line in two dimensions, a hyperplane in higher dimensions) that matches the given function at that point as well as a line/hyperplane can.

B.2.1 The secant and tangent lines

In Figure B.1 we illustrate a single-input function $g(w)$ along with the tangent line to this function at various points in its input domain. At any input point the *slope* of the tangent line to this function is referred to as its *derivative*. Notice at different input points how this quantity (i.e., the slope of the tangent line) seems to match the general local steepness of the underlying function itself. The derivative naturally encodes this information.

To formally define the derivative, we start by examining a *secant line* formed by taking any two points on a (single-input) function and connecting them with a straight line. The equation of any secant line constructed in this way is easy to find since its slope is given by the "rise over run" difference between the two points, and the equation of any line can be formed using its slope and any point on said line. For a generic single-input function g and the input points w^0 and w^1, the secant line h passes through the points $\left(w^0,\, g\left(w^0\right)\right)$ and $\left(w^1,\, g\left(w^1\right)\right)$, with its slope defined as

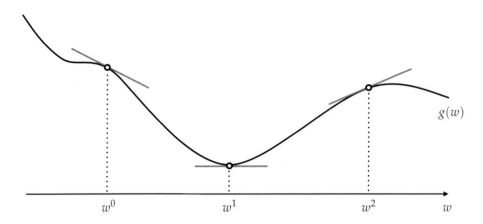

Figure B.1 A generic function with the tangent line drawn at various points of its input domain. See text for further details.

$$\frac{g(w^1) - g(w^0)}{w^1 - w^0} \tag{B.1}$$

and its equation defined as

$$h(w) = g(w^0) + \frac{g(w^1) - g(w^0)}{w^1 - w^0}(w - w^0). \tag{B.2}$$

Now imagine we fix the point w^0 and begin slowly pushing the point w^1 towards it. As w^1 gets closer and closer to w^0 the secant line resembles the tangent line at w^0 more and more. In other words, if we denote the difference between w^1 and w^0 by ϵ

$$w^1 = w^0 + \epsilon \tag{B.3}$$

then the slope of the secant line connecting $\left(w^0, g\left(w^0\right)\right)$ to $\left(w^1, g\left(w^1\right)\right)$, given as

$$\frac{g(w^1) - g(w^0)}{w^1 - w^0} = \frac{g(w^0 + \epsilon) - g(w^0)}{w^0 + \epsilon - w^0} = \frac{g(w^0 + \epsilon) - g(w^0)}{\epsilon} \tag{B.4}$$

matches the slope of the tangent line at w^0 (i.e., the derivative at w^0) as $|\epsilon|$ gets infinitesimally small. Notice that ϵ may be positive or negative here, depending on whether w^1 lies to the right or left of w^0. For the derivative to be defined at w^0 the quantity in Equation (B.4) needs to converge to the same value as ϵ approaches zero, regardless of the mathematical sign of ϵ (or equivalently, the direction in which w^1 approaches w^0). For example, the derivative at $w^0 = 0$ *is* defined for the function $g(w) = \max(0, w^2)$ since

$$\frac{g(w^0 + \epsilon) - g(w^0)}{\epsilon} = \frac{\max(0, \epsilon^2) - \max(0, 0)}{\epsilon} = \frac{\epsilon^2}{\epsilon} = \epsilon \tag{B.5}$$

which converges to zero as ϵ approaches zero. On the other hand, the derivative at $w^0 = 0$ is *not* defined for the function $g(w) = \max(0, w)$ since

$$\frac{g(w^0 + \epsilon) - g(w^0)}{\epsilon} = \frac{\max(0, \epsilon)}{\epsilon} \tag{B.6}$$

equals zero when a negative ϵ approaches zero, and equals one when a positive ϵ approaches zero.

One common notation used to denote the derivative of g at the point w^0 is

$$\frac{dg(w^0)}{dw} \tag{B.7}$$

where the symbol d means "infinitely small change in the value of." Notice this is precisely what the fraction in Equation (B.4) expresses when $|\epsilon|$ is infinitesimally small. A common variation on this notation puts $g(w^0)$ out front, as

$$\frac{d}{dw} g(w^0). \tag{B.8}$$

There are also other notations commonly used in practice to denote the derivative of g at w^0, e.g., $g'(w^0)$. Finally, note that with this notation the equation of the tangent line to g at the point w^0 can be written as

$$h(w) = g(w^0) + \frac{d}{dw} g(w^0)(w - w^0). \tag{B.9}$$

B.2.2 Numerical differentiation

To avoid having to find the derivative expression in Equation (B.4) analytically, especially when the function g is rather complicated, we can create a derivative calculator that *estimates* its derivative at a user-defined point w^0 by simply setting ϵ to some small number (in absolute value) like $\epsilon = 0.0001$. To be more robust in our approximation of the derivative we can replace the "one-way" secant slope in Equation (B.8) with the average slope of the right and left secant lines, as

$$\frac{d}{dw} g(w^0) \approx \frac{g(w^0 + \epsilon) - g(w^0 - \epsilon)}{2\epsilon}. \tag{B.10}$$

In either case the smaller we set ϵ the better our estimation of the actual derivative value becomes. However, setting ϵ too small can create *round-off errors* due to the fact that numerical values (whether or not they are produced from a mathematical function) can be represented only up to a certain precision on a computer, and that both the numerator and denominator in Equations (B.4) and (B.10) shrink to zero rapidly. This numerical stability issue does not completely invalidate numerical differentiation, but is worth being aware of.

B.3   Derivative Rules for Elementary Functions and Operations

For certain elementary functions and operations we need not resort to numerical differentiation since finding their exact derivative value is quite straightforward. We organize derivative formulae for popular *elementary functions* and *operations* in Tables B.1 and B.2, respectively. We provide formal proof for one elementary function from Table B.1 (monomial of degree d) and one elementary operation from Table B.2 (multiplication) in Examples B.1 and B.2, respectively. One can easily confirm the remainder of the derivative rules by following a similar argument, or consulting any standard calculus reference.

Table B.1 Derivative formulae for elementary functions.

Elementary function	Equation	Derivative
Sine	$\sin(w)$	$\cos(w)$
Cosine	$\cos(w)$	$-\sin(w)$
Exponential	e^w	e^w
Logarithm	$\log(w)$	$\frac{1}{w}$
Hyperbolic tangent	$\tanh(w)$	$1 - \tanh^2(w)$
Rectified Linear Unit (ReLU)	$\max(0, w)$	$\begin{cases} 0 & \text{if } w < 0 \\ 1 & \text{if } w > 0 \end{cases}$

Table B.2 Derivative formulae for elementary operations.

Elementary operation	Equation	Derivative
Addition of a constant	$c + g(w)$	$\frac{d}{dw}g(w)$
Multiplication by a constant	$c\,g(w)$	$c\frac{d}{dw}g(w)$
Addition of functions	$f(w) + g(w)$	$\frac{d}{dw}f(w) + \frac{d}{dw}g(w)$
Multiplication of functions	$f(w)\,g(w)$	$\left[\frac{d}{dw}f(w)\right]g(w) + f(w)\left[\frac{d}{dw}g(w)\right]$
Composition of functions	$f(g(w))$	$\frac{d}{dg}f(g)\frac{d}{dw}g(w)$

Example B.1   Derivative of general monomial terms

Starting with a degree-two monomial $g(w) = w^2$, we can write for general w and small ϵ

$$\frac{g(w + \epsilon) - g(w)}{\epsilon} = \frac{(w + \epsilon)^2 - w^2}{\epsilon} = \frac{(w^2 + 2w\epsilon + \epsilon^2) - w^2}{\epsilon} = \frac{2w\epsilon + \epsilon^2}{\epsilon} = 2w + \epsilon.$$

(B.11)

Sending ϵ to zero we get

$$\frac{d}{dw} g(w) = 2w.$$

(B.12)

Now let us examine a general degree-d monomial $g(w) = w^d$. All we need to do here is expand $(w + \epsilon)^d$ and rearrange its terms appropriately, as

$$(w + \epsilon)^d = \sum_{j=0}^{d} \binom{d}{j} \epsilon^j w^{d-j} = w^d + d \epsilon w^{d-1} + \epsilon^2 \sum_{j=2}^{d} \binom{d}{j} \epsilon^{j-2} w^{d-j}$$

(B.13)

where

$$\binom{d}{j} = \frac{d!}{j!(d-j)!}.$$

(B.14)

Plugging this expansion into the definition of the derivative

$$\frac{(w + \epsilon)^d - w^d}{\epsilon} = dw^{d-1} + \epsilon \sum_{j=2}^{d} \binom{d}{j} \epsilon^{j-2} w^{d-j}$$

(B.15)

we can see that the second term on the right-hand side vanishes as $\epsilon \longrightarrow 0$.

Example B.2   The product rule

With two functions $f(w)$ and $g(w)$ we want to evaluate

$$\frac{f(w + \epsilon)g(w + \epsilon) - f(w)g(w)}{\epsilon}$$

(B.16)

as ϵ approaches zero. Adding and subtracting $f(w + \epsilon)g(w)$ in the numerator gives

$$\frac{f(w + \epsilon)g(w + \epsilon) - f(w + \epsilon)g(w) + f(w + \epsilon)g(w) - f(w)g(w)}{\epsilon}$$

(B.17)

which then simplifies to

$$\frac{f(w + \epsilon) - f(w)}{\epsilon} g(w) + f(w + \epsilon) \frac{g(w + \epsilon) - g(w)}{\epsilon}.$$

(B.18)

Notice as $\epsilon \longrightarrow 0$, the first and second term in Equation (B.18) goes to $\left[\frac{d}{dw}f(w)\right]g(w)$ and $f(w)\left[\frac{d}{dw}g(w)\right]$, respectively, together giving

$$\frac{d}{dw}\left[f(w)\,g(w)\right] = \left[\frac{d}{dw}f(w)\right]g(w) + f(w)\left[\frac{d}{dw}g(w)\right]. \tag{B.19}$$

B.4   The Gradient

The *gradient* is a straightforward generalization of the notion of derivative for a multi-input function $g(w_1, w_2, \ldots, w_N)$. Treating all inputs but the first one (i.e., w_1) as fixed values (not *variables*), the function g momentarily reduces to a single-input function for which we have seen how to define the derivative (with respect to its only input variable w_1). This *partial* derivative, written as $\frac{\partial}{\partial w_1}g(w_1, w_2, \ldots, w_N)$, determines the slope of the hyperplane tangent to g at a given point, along the first input dimension. Repeating this for all inputs to g gives N partial derivatives (one along each input dimension) that collectively define the set of slopes of the *tangent hyperplane*. This is completely analogous to the single-input case where the derivative provides the slope of the *tangent line*.

For notational convenience these partial derivatives are typically collected into a vector, called the *gradient* and denoted by $\nabla g(w_1, w_2, \ldots, w_N)$, as

$$\nabla g(w_1, w_2, \ldots, w_N) = \begin{bmatrix} \frac{\partial}{\partial w_1}g(w_1, w_2, \ldots, w_N) \\ \frac{\partial}{\partial w_2}g(w_1, w_2, \ldots, w_N) \\ \vdots \\ \frac{\partial}{\partial w_N}g(w_1, w_2, \ldots, w_N) \end{bmatrix}. \tag{B.20}$$

Stacking all N inputs (w_1 through w_N) similarly into a column vector

$$\mathbf{w} = \begin{bmatrix} w_1 \\ w_2 \\ \vdots \\ w_N \end{bmatrix}, \tag{B.21}$$

the hyperplane tangent to the function $g(\mathbf{w})$ at the point $\left(\mathbf{w}^0, g\left(\mathbf{w}^0\right)\right)$ can be compactly characterized as

$$h(\mathbf{w}) = g(\mathbf{w}^0) + \nabla g(\mathbf{w}^0)^T(\mathbf{w} - \mathbf{w}^0), \tag{B.22}$$

which is the direct generalization in higher dimensions of the formula for a tangent line defined by the derivative of a single-input function given in Equation (B.9), as illustrated in Figure B.2.

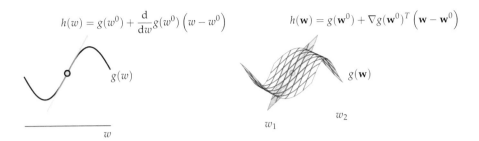

Figure B.2 (left panel) The derivative of a single-input function defines the slope of the tangent line at a point. (right panel) The gradient of a multi-input function analogously defines the set of slopes of the tangent hyperplane at a point. Here $N = 2$. In each panel the point of tangency is highlighted on the function as a green circle.

B.5   The Computation Graph

Virtually any function g expressed via an algebraic formula can be broken down (akin to the way physical substances may be broken down into their atomic parts) into a combination of elementary functions (e.g., $\sin(\cdot)$, $e^{(\cdot)}$, $\log(\cdot)$, etc.) and operations (e.g., addition, multiplication, composition, etc.). One very useful way of organizing the elementary decomposition of a generic function is via a so-called *computation graph*. The computation graph of a function g not only allows us to more easily understand its anatomy, as a combination of elementary functions and operations, but it also allows us to evaluate a function in a programmatic way. Here we describe the computation graph by studying two simple examples employing a single-input and a multi-input function.

Example B.3   The computation graph of a single-input function
Take the single-input function

$$g(w) = \tanh(w)\cos(w) + \log(w). \qquad (B.23)$$

We can represent this function by decomposing it into its simplest parts, as shown in the top panel of Figure B.3. This graphical depiction is like a blueprint, showing us precisely how g is constructed from elementary functions *and* operations. We read this computation graph from left to right starting with the input node representing w, and ending with the full computation of $g(w)$ on the right. Each yellow node in the graph (aside from the input node that is colored differently in gray) represents a single elementary function or operation, and is marked as such. The directed arrows or *edges* connecting pairs of nodes then show how computation flows when evaluating $g(w)$.

The terms *parent* and *child* are often used to describe the local topology of computation graphs for any pair of nodes in the graph connected via a directed edge. The parent node is where the edge/arrow originates from and the child node is where it points to. Because these parent–child relationships are defined locally, a particular node can be both a parent and a child with respect to other nodes in the graph. For instance, in the computation graph shown in the top panel of Figure B.3, the input node w (colored gray) is parent to the nodes a, b, and c, while a and b themselves are parents to the node d.

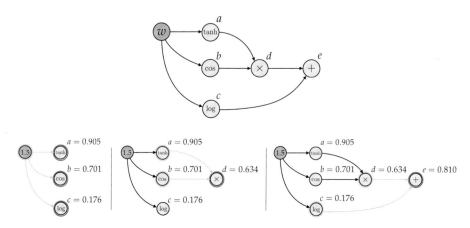

Figure B.3 Figure associated with Example B.3. (top panel) The computation graph for the single-input function defined in Equation (B.23). (bottom panels) Visualizing the flow of computation through this graph. See text for further details.

If we were to write out the formula of each child node in terms of its parent(s) we would have the following list of formulae

$$
\begin{aligned}
a &= \tanh(w) \\
b &= \cos(w) \\
c &= \log(w) \\
d &= a \times b \\
e &= c + d
\end{aligned}
\qquad\text{(B.24)}
$$

where the final node evaluates the whole function, i.e., $e = g(w)$. Note, once again, how each child node is a function of its parent(s). Therefore we can, for example, write a as $a(w)$ since w is the (only) parent node of a, or likewise write d as $d(a, b)$ since both a and b are parents to d. At the same time, if we unravel the definition of each node, every node in the end is really a function of the input w alone. In other words, we can also write each node as a function of their common *ancestor* w, e.g., $d(w) = \tanh(w) \times \cos(w)$.

The computation flows *forward* through the graph (from left to right) in sets of parent–child nodes. In the bottom panels of Figure B.3 we illustrate how $g(w)$ is evaluated for the particular input value $w = 1.5$ using the computation graph shown in the top panel of the figure. Beginning on the left we first substitute the value $w = 1.5$ in the input node, and evaluate each of the input node's children, here computing $a(1.5) = \tanh(1.5) = 0.905$, $b(1.5) = \cos(1.5) = 0.701$, and $c(1.5) = \log(1.5) = 0.176$, as illustrated in the bottom left panel with the parent node highlighted in blue and the children in red. Computation next flows to any child whose parents have all been evaluated, here the node d, as illustrated in the middle panel of the figure where we have used the same coloring to denote the parent–child relationship. Note how in computing $d(1.5) = a(1.5) \times b(1.5) = 0.634$ we only need access to its evaluated parents, i.e., $a(1.5)$ and $b(1.5)$, which we have indeed already computed. We then evaluate the final child node in the graph, e, at our desired input value. Once again, to compute $e(1.5) = c(1.5) + d(1.5) = 0.810$ we only need access to the evaluations made at its parents, here $c(1.5)$ and $d(1.5)$, which have already been computed.

Example B.4 **The computation graph of a multi-input function**
Computation graphs can similarly be constructed for multi-input functions as well. For example, in the top row of Figure B.4 we show the computation graph for the simple multi-input quadratic

$$g(w_1, w_2) = w_1^2 + w_2^2. \tag{B.25}$$

The two inputs w_1 and w_2 here are each represented by a distinct node. Just as with the single-input case in Example B.3, the computation flows from left to right or *forward* through the graph. Also, as with single-input case, one forward sweep through the graph is sufficient to calculate any value $g(w_1, w_2)$.

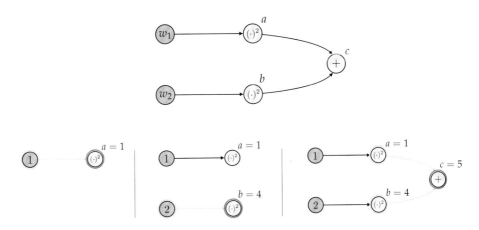

Figure B.4 Figure associated with Example B.4. (top panel) The computation graph for the multi-input quadratic function defined in Equation (B.25). (bottom panels) Visualizing the flow of computation through this graph. See text for further details.

In the bottom panels of Figure B.4 we illustrate how $g(w_1, w_2)$ is evaluated for the particular input values of $w_1 = 1$ and $w_2 = 2$. Beginning on the left we first substitute in the value $w_1 = 1$ and evaluate its only child $a(1) = 1^2$, as illustrated in the bottom-left panel of the figure. Next, we do the same for the second input node, substituting in the value $w_2 = 2$ and evaluating its only child $b(2) = 2^2$, as illustrated in the bottom-middle panel of the figure. Finally, we move to the last child node, computing it as $c = a(1) + b(2) = 5$ where the evaluations $a(1)$ and $b(2)$ have already been computed in the previous steps of the process.

The notion of a computation graph is quite flexible, as functions can be decomposed in various ways. Here we have broken two example functions down into their simplest, most elementary parts. However, it is more useful to decompose more sophisticated functions (e.g., fully connected neural networks, as detailed in Section 13.2) into computation graphs consisting of more sophisticated elementary building blocks such as matrix multiplication, vector-wise functions, etc.

B.6   The Forward Mode of Automatic Differentiation

In the previous section we saw how representing a function via its computation graph allows us to evaluate it at any input point by traversing the graph in a forward direction, from left to right, recursively evaluating each node in the graph with respect to the function's original input. The computation graph of a function can similarly be used to form and evaluate a function's *gradient* by similarly sweeping forward through the function's computation graph from left to right, forming and evaluating the gradient of each node with respect to the function's original input. In doing this we also naturally evaluate the original function at each node along with the gradient evaluation. This recursive algorithm, called the *forward mode of automatic differentiation*, is easily programmable and allows for the transfer of the tedious chore of gradient computation to a computer program, which makes gradient computation faster and more reliable (than when performed manually and then hard-coded into a computer program). Moreover, unlike numerical differentiation (see Section B.2.2), automatic differentiation provides the *exact* derivative or gradient evaluation, not just an approximation. Here we describe the forward mode of automatic differentiation by studying two examples, employing the algorithm to differentiate a single-input and a multi-input function.

Example B.5   Forward-mode differentiation of a single-input function
Take the function $g(w) = \tanh(w)\cos(w) + \log(w)$ whose computation graph we previously illustrated in Figure B.3. To evaluate the derivative $\frac{d}{dw}g(w)$ using the forward mode of automatic differentiation, we traverse this computation graph

much in the same way we did to compute its evaluation $g(w)$ in Example B.3. We begin at the input node all the way to the left where w and its derivative $\frac{d}{dw}w$ both have known values: the value of w is chosen by the user and $\frac{d}{dw}w$ is always, trivially, equal to 1. With the form of both the node and its derivative in hand we then move to the children of the input node, i.e., nodes a, b, and c. For each child node, we form both the node and its derivative with respect to the input, i.e., a and $\frac{d}{dw}a$ for the first child node, b and $\frac{d}{dw}b$ for the second child node, and c and $\frac{d}{dw}c$ for the third. These first steps are illustrated on the computation graph of the function in the top panel of Figure B.5, where the parent node (the input) is highlighted in blue and the children (nodes a, b, and c) are colored in red.

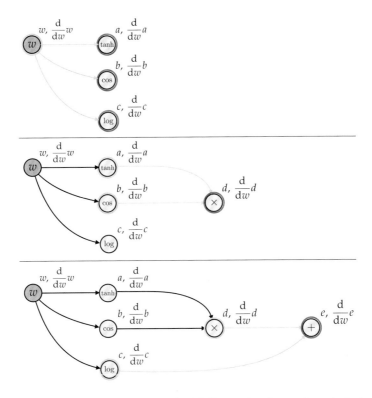

Figure B.5 Figure associated with Example B.5 illustrating forward-mode derivative computation of the single-input function defined in Equation (B.23). At each step of the process a child node and its derivative are both formed with respect to w, which are recursively constructed using the node/derivative evaluations of its parent(s). See text for further details.

Forming each child node and its derivative requires only the values computed at its parent(s), along with the derivative rules for elementary functions and operations described in Section B.3, which tell us how to combine the derivatives computed at the parent(s) to compute the child derivative. For each child node here we can form their derivatives with respect to the input w as

$$\frac{\mathrm{d}}{\mathrm{d}w}a(w) = 1 - \tanh^2(w)$$

$$\frac{\mathrm{d}}{\mathrm{d}w}b(w) = -\sin(w) \tag{B.26}$$

$$\frac{\mathrm{d}}{\mathrm{d}w}c(w) = \frac{1}{w}$$

using Table B.1 as a look-up table.

With the current function values/derivatives computed with respect to w we move forward to the next child nodes in the graph where we will see the same pattern emerge, seeking to form the nodes and their derivatives with respect to w. Examining the computation graph in the top panel of Figure B.5 we can see that we have already formed all parents of d (i.e., nodes a and b) as well as their derivatives, and thus we move to node d next and form $d(w) = a(w) \times b(w)$ using the values of $a(w)$ and $b(w)$ just computed. To compute the derivative $\frac{\mathrm{d}}{\mathrm{d}w}d(a,b)$ we employ the *chain rule* and write it, in terms of the derivatives of its parents/inputs, as

$$\frac{\mathrm{d}}{\mathrm{d}w}d(a,b) = \frac{\partial}{\partial a}d(a,b) \times \frac{\mathrm{d}}{\mathrm{d}w}a(w) + \frac{\partial}{\partial b}d(a,b) \times \frac{\mathrm{d}}{\mathrm{d}w}b(w). \tag{B.27}$$

Notice, because we have already formed the derivatives of a and b with respect to w, we need only compute the parent–child derivatives $\frac{\partial}{\partial a}d(a,b)$ and $\frac{\partial}{\partial b}d(a,b)$. Since the parent–child relationship here is multiplicative both of these derivatives can be found, using the *product rule* from Table B.2, as

$$\frac{\partial}{\partial a}d(a,b) = b$$

$$\frac{\partial}{\partial b}d(a,b) = a. \tag{B.28}$$

All together we have the entire form of the derivative at node d as

$$\frac{\mathrm{d}}{\mathrm{d}w}d(a,b) = (1 - \tanh^2(w))\cos(w) - \tanh(w)\sin(w), \tag{B.29}$$

which we illustrate pictorially in the middle panel of Figure B.5.

Now that we have resolved d and its derivative, we can work on the final node e, which is a child of nodes d and c, defined in terms of them as $e(d,c) = d + c$. Once again, using the chain rule the derivative of e with respect to w is written as

$$\frac{\mathrm{d}}{\mathrm{d}w}e(d,c) = \frac{\partial}{\partial d}e(d,c) \times \frac{\mathrm{d}}{\mathrm{d}w}d(w) + \frac{\partial}{\partial c}e(d,c) \times \frac{\mathrm{d}}{\mathrm{d}w}c(w). \tag{B.30}$$

We have already computed the derivatives of d and c with respect to w, so plugging 1 for both $\frac{\partial}{\partial d}e(d,c)$ and $\frac{\partial}{\partial c}e(d,c)$ we have our desired derivative

$$\frac{\mathrm{d}}{\mathrm{d}w}g(w) = \frac{\mathrm{d}}{\mathrm{d}w}e(d,c) = (1 - \tanh^2(w))\cos(w) - \tanh(w)\sin(w) + \frac{1}{w}. \quad \text{(B.31)}$$

With the form of the derivative computed at each node of the graph we can imagine appending each derivative to its respective node, leaving us with an upgraded computation graph for our function $g(w)$ that can now evaluate both function and derivative values. This is done by plugging in a value for w at the start of the graph, and propagating the function and derivative forward through the graph from left to right. This is why the method is called the *forward mode* of automatic differentiation, as all computation is done moving forward through the computation graph. For example, in Figure B.6 we illustrate how $g(1.5)$ and $\frac{\mathrm{d}}{\mathrm{d}w}g(1.5)$ are computed together traversing forward through the computation graph. Note that in evaluating the derivative in this manner we naturally evaluate the function as well.

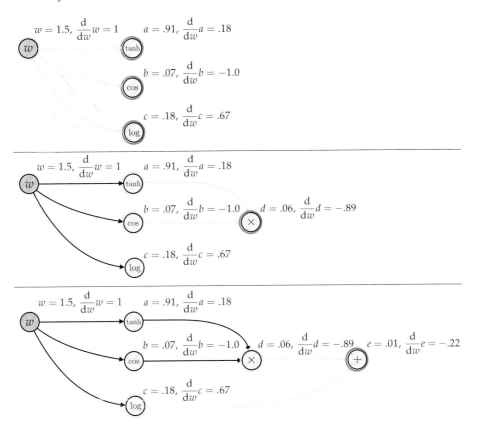

Figure B.6 Figure associated with Example B.5 illustrating evaluation of the function $g(w)$ given in Equation (B.23), as well as its derivative, at the input point $w = 1.5$ using the forward mode of automatic differentiation. See text for further details.

Example B.6 **Forward-mode differentiation of a multi-input function**
The forward mode of automatic differentiation works similarly for multi-input functions as well, only now we must compute the form of the gradient at each node in the graph (instead of a single derivative). Here we illustrate how this is done using the multi-input quadratic $g(w_1, w_2) = w_1^2 + w_2^2$ whose computation graph was first shown in Figure B.4.

Following the pattern set forth with the single-input function in Example B.5, we begin by computing the gradient at each input node, which are trivially

$$\nabla w_1 = \begin{bmatrix} 1 \\ 0 \end{bmatrix} \quad \text{and} \quad \nabla w_2 = \begin{bmatrix} 0 \\ 1 \end{bmatrix}. \quad (B.32)$$

We then move to the children of our input nodes beginning at node a, where we compute the *gradient* of a or, in other words, the partial derivatives of $a(w_1) = w_1^2$ with respect to both w_1 and w_2, as

$$\frac{\partial}{\partial w_1} a = 2w_1 \quad \text{and} \quad \frac{\partial}{\partial w_2} a = 0. \quad (B.33)$$

Similarly, we can compute the partial derivatives of $b(w_2) = w_2^2$ with respect to w_1 and w_2, as

$$\frac{\partial}{\partial w_1} b = 0 \quad \text{and} \quad \frac{\partial}{\partial w_2} b = 2w_2. \quad (B.34)$$

These two steps are illustrated in the left panel of Figure B.7. With the form of the gradient computed at nodes a and b we can finally compute the gradient at their common child node, c. Employing the chain rule we have

$$\frac{\partial}{\partial w_1} c = \frac{\partial}{\partial a} c \frac{\partial}{\partial w_1} a + \frac{\partial}{\partial b} c \frac{\partial}{\partial w_1} b = 1 \times 2w_1 + 1 \times 0 = 2w_1$$

$$\frac{\partial}{\partial w_2} c = \frac{\partial}{\partial a} c \frac{\partial}{\partial w_2} a + \frac{\partial}{\partial b} c \frac{\partial}{\partial w_2} b = 1 \times 0 + 1 \times 2w_2 = 2w_2. \quad (B.35)$$

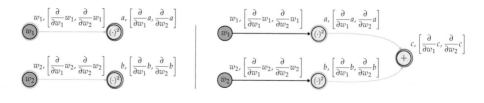

Figure B.7 Figure associated with Example B.6 illustrating forward-mode gradient computation of the multi-input function $g(w_1, w_2)$ defined in Equation (B.25). See text for further details.

The forward-mode differentiator discussed here does not provide an algebraic description of a function's derivative, but a programmatic function (a computation graph) that can be used to evaluate the function and its derivative at any set of input points. Conversely, one can build an algorithm that employs the basic derivative rules to provide an algebraic derivative, but this requires the implementation of a computer algebra system. Such a derivative calculator that deals with derivatives using symbolic computation (i.e., algebra on the computer) is called a *symbolic differentiator*. However, expressing equations algebraically can be quite unwieldy. For example, the rather complicated-looking function

$$g(w) = \frac{w^2 \sin(w^2 + w) \cos(w^2 + 1)}{\log(w + 1)} \tag{B.36}$$

has an expansive algebraic derivative

$$\begin{aligned} \frac{\mathrm{d}}{\mathrm{d}w} g(w) &= \frac{(2w + 1)\, w^2 \cos(w^2 + 1) \cos(w^2 + w)}{\log(w + 1)} \\ &\quad - \frac{w^2 \sin(w^2 + w) \cos(w^2 + 1)}{(w + 1) \log^2(w + 1)} + \frac{2w \sin(w^2 + w) \cos(w^2 + 1)}{\log(w + 1)} \\ &\quad - \frac{2w^3 \sin(w^2 + 1) \sin(w^2 + w)}{\log(w + 1)}. \end{aligned} \tag{B.37}$$

This problem is exponentially worse, to the point of being a considerable computational burden, when dealing with multi-input functions (which we commonly deal with in machine learning). The (forward-mode) automatic differentiator, which produces a computation graph of the derivative instead of an algebraic form, does not have this problem.

Finally, note that because calculations involved in the forward mode of automatic differentiation are made using the computation graph of a given function g, one engineering choice to be made is to decide how the graph will be constructed and manipulated. Essentially we have two choices: we can either construct the computation graph *implicitly* by carefully implementing the elementary derivative rules, or we can parse the input function g and construct its computation graph *explicitly* as we did in explaining the method in this section. The advantage of implicitly constructing the graph is that the corresponding calculator is light-weight (as there is no need to store a graph) and easy to construct. On the other hand, implementing a calculator that explicitly constructs computation graphs requires additional tools (e.g., a parser), but allows for easier computation of higher-order derivatives. This is because in the latter case the differentiator takes in a function to differentiate and treats it as a computation graph, and often outputs a computation graph of its derivative which can then be plugged back into the same algorithm to create second-order derivatives, and so forth.

B.7   The Reverse Mode of Automatic Differentiation

While the forward mode of automatic differentiation introduced in the previous section provides a wonderful programmatic way of computing derivatives, it can be inefficient for many kinds multi-input functions (particularly those in machine learning involving *fully connected networks*). This is because while most of the nodes in the computation graph of a multi-input function may only take in just a few inputs, we compute the *complete gradient* with respect to *all the inputs* at each and every node. This leads to considerable computation waste since we know that the partial derivative of any node with respect to an original input of the function that it does *not* take in will always be equal to zero.

This obvious waste is the motivation for what is called the *reverse mode of automatic differentiation* or, in the jargon of machine learning, the *backpropagation algorithm*. With the reverse mode we begin by traversing a function's computation graph in the forward direction (starting with the input nodes and moving from left to right) computing *only* the form of the partial derivatives needed at each node of the computation graph (ignoring partial derivatives that will always be zero). Once this forward sweep is complete we then (starting with the final node(s) in the graph) move in the *reverse* direction and sweep *backwards* through the graph, collecting and combining the previously computed partial derivatives to appropriately construct the gradient. While this means that we must construct the computation graph explicitly (and store it), the trade-off with wasted computation is often well worth it when dealing with machine learning functions. This makes the *reverse mode* a more popular choice (than the forward mode detailed in the prior section) in machine learning applications, and is in particular the brand of automatic differentiator implemented in `autograd` (the `Python`-based automatic differentiator we recommend one uses with this text).

Example B.7   Reverse-mode differentiation of a multi-input function

In Example B.6 of the previous section we described how to compute the gradient of the function $g(w_1, w_2) = w_1^2 + w_2^2$ using forward-mode automatic differentiation. Notice how, in calculating the full gradient at each node of this function's computation graph, we performed several wasteful computations: whenever partial derivatives were taken with respect to input not taken in by a node (through its complex web of parent–child relations leading back to the original input of the function) we know by default this partial derivative will always equal zero. For example, the partial derivative $\frac{\partial}{\partial w_2} a = 0$ since a is not a function of the original input w_2.

In Figure B.8 we redraw the computation graph of this quadratic with the gradient expressed in terms of the partial derivatives at each node and with all zero partials marked accordingly. Examining this graph we can see that a good deal of the partials trivially equal zero. Trivial zeros such as these waste computation as we form them traversing *forward* through the graph.

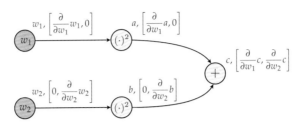

Figure B.8 Figure associated with Example B.7. The computation graph of the quadratic function described in the text, with zero partials marked. See text for further details.

This issue becomes much more severe with multi-input functions that take in larger numbers of input variables. For example, in Figure B.9 we illustrate the computational graph of the analogous quadratic function taking four inputs $g(w_1, w_2, w_3, w_4) = w_1^2 + w_2^2 + w_3^2 + w_4^2$. In this case more than *half* of all the gradient entries at the nodes in this graph are zero due to the fact that certain nodes are not functions of certain inputs, and hence their partial derivatives are always zero.

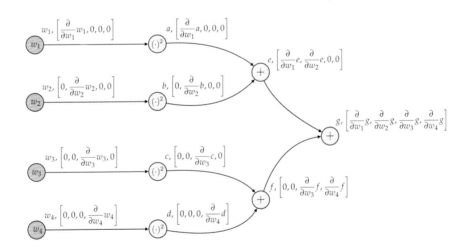

Figure B.9 Figure associated with Example B.7. The computation graph of a simple four-input quadratic with the gradient expressed at each node in terms of partial derivatives. Here over half of the partial derivatives computed are trivial zeros. See text for further details.

To remedy this inefficiency, automatic differentiation can also be performed in a *reverse mode*, which consists of a forward and reverse (or backward) sweep through the computation graph of a function. In the forward sweep of the reverse mode we traverse the computation graph in forward direction (from left to right) recursively just as with the forward mode, only at each node we compute the partial derivatives of each child node with respect to its parent(s) *only*, and *not* the full gradient with respect to the function input.

This is illustrated for the quadratic function $g(w_1, w_2) = w_1^2 + w_2^2$ in the top panels of Figure B.10. In the top-left panel we show the computing of partial derivatives for child nodes a and b (colored red), which are taken only with respect to their parents (which here are w_1 and w_2, respectively, colored blue). In the top-right panel we illustrate the next computation in the forward sweep, the partial derivatives computed at the child node c (colored red) with respect to its parents a and b (colored blue).

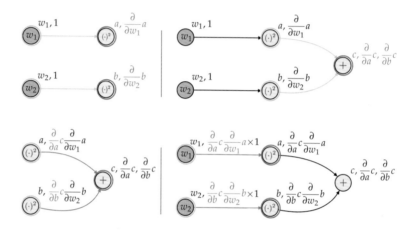

Figure B.10 Figure associated with Example B.7. Forward (top panels) and backward sweep (bottom panels) illustrated for a simple two-input quadratic function. See text for further details.

Once the forward sweep is complete, we change course and traverse the computation graph in the *reverse* direction, starting at the end node and traversing backwards recursively from right to left until we reach the input nodes. At every step of the process we update the partial derivative of each parent by multiplying it by the partial derivative of its children node with respect to that parent. When the backward sweep is completed we will have recursively constructed the gradient of the function with respect to all of its inputs.

The backward sweep is illustrated for the two-input quadratic function in the bottom panels of Figure B.10. Starting from end node in our forward sweep, i.e., node c, we observe that c has two parents: a and b. Therefore we update the derivative at a by (left) multiplying it by $\frac{\partial}{\partial a}c$ giving $\frac{\partial}{\partial a}c\frac{\partial}{\partial w_1}a$, and similarly update the derivative at b by (left) multiplying it by $\frac{\partial}{\partial b}c$ giving $\frac{\partial}{\partial b}c\frac{\partial}{\partial w_2}b$. We then repeat this procedure recursively with the children of a and b, ending with the partial derivative $\frac{\partial}{\partial a}c\frac{\partial}{\partial w_1}a\frac{\partial}{\partial w_1}w_1 = \frac{\partial}{\partial a}c\frac{\partial}{\partial w_1}a$ and $\frac{\partial}{\partial b}c\frac{\partial}{\partial w_2}b\frac{\partial}{\partial w_2}w_2 = \frac{\partial}{\partial b}c\frac{\partial}{\partial w_2}b$. These are precisely the two partial derivatives of the complete gradient of the quadratic with respect to its input w_1 and w_2.

B.8 Higher-Order Derivatives

In previous sections we have seen how we can efficiently compute the derivative of functions composed of elementary building blocks, and that these derivatives are themselves functions built from elementary building blocks. Because of this we can likewise compute derivatives of derivatives, commonly referred to as *higher-order derivatives*, which are the subject of this section.

B.8.1 Higher-order derivatives of single-input functions

Here we explore the concept of higher-order derivatives of single-input functions by looking at a simple example.

Example B.8 Higher-order derivatives

To compute the second-order derivative of the function

$$g(w) = w^4 \tag{B.38}$$

we first find its first-order derivative as

$$\frac{d}{dw} g(w) = 4w^3 \tag{B.39}$$

and then differentiate it one more time to get

$$\frac{d}{dw}\left(\frac{d}{dw} g(w)\right) = 12w^2. \tag{B.40}$$

Taking the derivative of the resulting function one more time gives the third-order derivative of g as

$$\frac{d}{dw}\left(\frac{d}{dw}\left(\frac{d}{dw} g(w)\right)\right) = 24w. \tag{B.41}$$

Similarly, we can compute the first three derivatives of the function

$$g(w) = \cos(3w) + w^2 + w^3 \tag{B.42}$$

explicitly as

$$\frac{d}{dw} g(w) = -3\sin(3w) + 2w + 3w^2$$

$$\frac{d}{dw}\left(\frac{d}{dw} g(w)\right) = -9\cos(3w) + 2 + 6w \tag{B.43}$$

$$\frac{d}{dw}\left(\frac{d}{dw}\left(\frac{d}{dw} g(w)\right)\right) = 27\sin(3w) + 6.$$

Higher-order derivatives are also often expressed using more compact notation than given in Example B.8. For instance, the second derivative is very often denoted more compactly using the following notation

$$\frac{d^2}{dw^2}g(w) = \frac{d}{dw}\left(\frac{d}{dw}g(w)\right). \tag{B.44}$$

Likewise, the third derivative is often denoted more compactly as

$$\frac{d^3}{dw^3}g(w) = \frac{d}{dw}\left(\frac{d}{dw}\left(\frac{d}{dw}g(w)\right)\right) \tag{B.45}$$

and, in general, the nth derivative is written as

$$\frac{d^n}{dw^n}g(w). \tag{B.46}$$

B.8.2 Higher-order derivatives of multi-input functions

We have seen how the gradient of a multi-input function is a collection of partial derivatives

$$\nabla g(w_1, w_2, \ldots, w_N) = \begin{bmatrix} \frac{\partial}{\partial w_1}g(w_1, w_2, \ldots, w_N) \\ \frac{\partial}{\partial w_2}g(w_1, w_2, \ldots, w_N) \\ \vdots \\ \frac{\partial}{\partial w_N}g(w_1, w_2, \ldots, w_N) \end{bmatrix} \tag{B.47}$$

where the gradient contains the nth partial derivative $\frac{\partial}{\partial w_n}g(w_1, w_2, \ldots, w_N)$ as its nth entry. This partial derivative (like the original function itself) is a function, taking in the N inputs abbreviated $\mathbf{w}$, which we can differentiate along each input axis. For instance, we can take the mth partial derivative of $\frac{\partial}{\partial w_n}g(w_1, w_2, .., w_N)$ as

$$\frac{\partial}{\partial w_m}\frac{\partial}{\partial w_n}g(w_1, w_2, \ldots, w_N). \tag{B.48}$$

This is a *second-order* derivative. How many of these does the function g have? Since every one of g's N first-order derivatives (each being a function of N inputs) has N partial derivatives, $g(\mathbf{w})$ has a total of N^2 second-order derivatives.

As with the notion of the gradient, this large set of second-order derivatives are typically organized in a particular way so that they can be more easily communicated and computed with. The *Hessian* – which is written notationally as $\nabla^2 g(\mathbf{w})$ – is the $N \times N$ matrix of second-order derivatives whose (m,n)th element is $\frac{\partial}{\partial w_m}\frac{\partial}{\partial w_n}g(\mathbf{w})$, or $\frac{\partial}{\partial w_m}\frac{\partial}{\partial w_n}g$ for short. The full Hessian matrix is written as

$$\nabla^2 g(\mathbf{w}) = \begin{bmatrix} \frac{\partial}{\partial w_1} \frac{\partial}{\partial w_1} g & \frac{\partial}{\partial w_1} \frac{\partial}{\partial w_2} g & \cdots & \frac{\partial}{\partial w_1} \frac{\partial}{\partial w_N} g \\ \frac{\partial}{\partial w_2} \frac{\partial}{\partial w_1} g & \frac{\partial}{\partial w_2} \frac{\partial}{\partial w_2} g & \cdots & \frac{\partial}{\partial w_2} \frac{\partial}{\partial w_N} g \\ \vdots & \vdots & \ddots & \vdots \\ \frac{\partial}{\partial w_N} \frac{\partial}{\partial w_1} g & \frac{\partial}{\partial w_N} \frac{\partial}{\partial w_2} g & \cdots & \frac{\partial}{\partial w_N} \frac{\partial}{\partial w_N} g \end{bmatrix}. \tag{B.49}$$

Moreover, since it is virtually always the case that $\frac{\partial}{\partial w_m} \frac{\partial}{\partial w_n} g = \frac{\partial}{\partial w_n} \frac{\partial}{\partial w_m} g$, particularly with the sort of functions used in machine learning, the Hessian is always a *symmetric* matrix.

The number of partial derivatives of a multi-input function grows *exponentially* with the order. We have just seen that a function taking in N inputs has N^2 second-order derivatives. In general such a function has N^D partial derivatives of order D.

B.9   Taylor Series

In this section we describe the *Taylor series* of a function, a fundamental tool of calculus that is critically important for first- and second-order local optimization (as detailed in Chapters 3 and 4 of this text). We begin by deriving this crucial concept for single-input functions, and follow by generalizing to the case of multi-input functions.

B.9.1   Linear approximation is only the beginning

We began our discussion of derivatives in Section B.2 by defining the derivative at a point as the slope of the tangent line to a given input function. For a function $g(w)$ we then formally described the tangent line at a point w^0 as

$$h(w) = g(w^0) + \frac{\mathrm{d}}{\mathrm{d}w} g(w^0)(w - w^0) \tag{B.50}$$

with the slope here given by the derivative $\frac{\mathrm{d}}{\mathrm{d}w} g(w^0)$. The justification for examining the tangent line to begin with is fairly straightforward: locally (close to the point w^0) the tangent line looks awfully similar to the function, and so if we want to better understand g near w^0 we can just as well look at the tangent line there. This makes our lives a lot easier because a line (compared to a generic g) is always a comparatively simple object, and therefore understanding the tangent line is always a simple affair.

If we study the form of our tangent line $h(w)$ closely, we can define in precise mathematical terms how it matches the function g. Notice, first of all, that the tangent line takes on the same value as the function g at the point w^0, i.e.,

$$h(w^0) = g(w^0) + \frac{\mathrm{d}}{\mathrm{d}w} g(w^0)(w^0 - w^0) = g(w^0). \tag{B.51}$$

Next, notice that the first derivative value of these two functions match as well. That is, if we take the first derivative of h with respect to w we can see that

$$\frac{d}{dw}h(w^0) = \frac{d}{dw}g(w^0). \tag{B.52}$$

In short, when the tangent line h matches g so that at w^0 both the function *and* derivative values are equal, we can write

$$h(w^0) = g(w^0)$$
$$\frac{d}{dw}h(w^0) = \frac{d}{dw}g(w^0). \tag{B.53}$$

What if we turned this idea around and tried to find a line that satisfies these two properties. In other words, we start with a general line

$$h(w) = a_0 + a_1(w - w^0) \tag{B.54}$$

with unknown coefficients a_0 and a_1, and we want to determine the right value for these coefficients so that the line satisfies the two criteria in Equation (B.53). These two criteria constitute a *system of equations* we can solve for the correct values of a_0 and a_1. Computing the left-hand side of each, where h is our general line in Equation (B.54), we end up with a trivial system of equations to solve for both unknowns simultaneously

$$h(w^0) = a_0 = g(w^0)$$
$$\frac{d}{dw}h(w^0) = a_1 = \frac{d}{dw}g(w^0) \tag{B.55}$$

and behold, the coefficients are precisely those of the tangent line.

B.9.2   From tangent line to tangent quadratic

Given that the function and derivative values of the tangent line match those of its underlying function, can we do better? Can we find a simple function that matches the function value, first derivative, *and* the second derivative value at a point w_0? In other words, is it possible to determine a simple function h that satisfies the following *three* conditions?

$$h(w^0) = g(w^0)$$
$$\frac{d}{dw}h(w^0) = \frac{d}{dw}g(w^0) \tag{B.56}$$
$$\frac{d^2}{dw^2}h(w^0) = \frac{d^2}{dw^2}g(w^0)$$

Notice how a (tangent) line h can only satisfy the first two of these properties and never the third, since it is a degree-one polynomial and $\frac{d^2}{dw^2}h(w) = 0$ for all w. This fact implies that we need *at least* a degree-two polynomial to satisfy all three criteria. Starting with a general degree-two polynomial

$$h(w) = a_0 + a_1(w - w^0) + a_2(w - w^0)^2 \tag{B.57}$$

with unknown coefficients a_0, a_1, and a_2, we can evaluate the left-hand side of each criterion in Equation (B.56), forming a system of three equations, and solve for these coefficients.

$$h(w^0) = a_0 = g(w^0)$$
$$\frac{d}{dw}h(w^0) = a_1 = \frac{d}{dw}g(w^0) \tag{B.58}$$
$$\frac{d^2}{dw^2}h(w^0) = 2a_2 = \frac{d^2}{dw^2}g(w^0)$$

With all of our coefficients determined (in terms of the derivatives of g at w^0) we have a degree-two polynomial that satisfies the three desired criteria

$$h(w) = g(w^0) + \frac{d}{dw}g(w^0)(w - w^0) + \frac{1}{2}\frac{d^2}{dw^2}g(w^0)(w - w^0)^2. \tag{B.59}$$

This is one step beyond the tangent line (an approximating quadratic function) but note that the first two terms are indeed the tangent line itself. Such a quadratic approximation matches a generic function g near the point w_0 far more closely than the tangent line, as illustrated in Figure B.11.

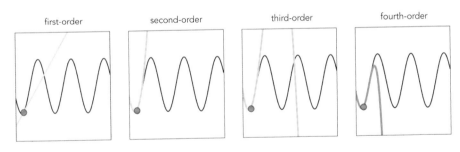

Figure B.11 From left to right, the first-, second-, third-, and fourth-order Taylor series approximation to the function $g(w) = \sin(w)$, drawn in different colors, evaluated at the same input point.

B.9.3   Building better and better local approximations

Having derived this quadratic based on our reflection on the tangent line, one can naturally think of going one step further. That is, finding a simple function h that satisfies even one more condition than the quadratic.

$$h(w^0) = g(w^0)$$

$$\frac{d}{dw}h(w^0) = \frac{d}{dw}g(w^0)$$

$$\frac{d^2}{dw^2}h(w^0) = \frac{d^2}{dw^2}g(w^0) \tag{B.60}$$

$$\frac{d^3}{dw^3}h(w^0) = \frac{d^3}{dw^3}g(w^0)$$

Noting that no degree-two polynomial could satisfy this last condition, since its third derivative is always equal to zero, we could seek out a degree-three polynomial. Using the same analysis as we performed previously, setting up the corresponding system of equations based on a generic degree-three polynomial leads to the conclusion that the following does indeed satisfy all of the criteria in Equation (B.60)

$$h(w) = g(w^0) + \frac{d}{dw}g(w^0)(w - w^0) + \frac{1}{2}\frac{d^2}{dw^2}g(w^0)(w - w^0)^2 + \frac{1}{6}\frac{d^3}{dw^3}g(w^0)(w - w^0)^3. \tag{B.61}$$

This is an even better approximation of g near the point w^0 than the quadratic, as illustrated for a particular example in Figure B.11. Examining this figure we can clearly see that the approximation becomes better and better as we increase the order of the approximation. This makes sense, as each polynomial contains more of the underlying function's derivative information as we increase the degree. However, we can never expect it to match the entire function everywhere: we build each polynomial to match g at only a single point, so regardless of degree we can expect it only to match the underlying function near the point w_0.

Setting up a set of $N + 1$ criteria, the first demanding that $h(w^0) = g(w^0)$ and the remaining N demanding that the first N derivatives of h match those of g at w^0, leads to construction of the degree-N polynomial

$$h(w) = g(w^0) + \sum_{n=1}^{N} \frac{1}{n!}\frac{d^n}{dw^n}g(w^0)(w - w^0)^n. \tag{B.62}$$

This general degree-N polynomial is called the *Taylor series* of g at the point w^0.

B.9.4   Multi-input Taylor series

We have now seen, with single-input functions, how the general Taylor series approximation can be thought of as a natural extension of the tangent line for higher-degree polynomial approximations. The story with multi-input functions is precisely analogous.

If we asked what sort of degree-one polynomial $h(\mathbf{w})$ matched a function $g(\mathbf{w})$ at a point $\mathbf{w}^0$ in terms of both its function and gradient value there, i.e.,

Figure B.12 The first-order (colored in lime green) and second-order (colored in turquoise) Taylor series approximations to the function $g(w_1, w_2) = \sin(w_1)$ at a point near the origin.

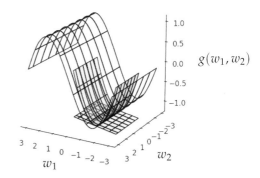

$$
\begin{aligned}
h\left(\mathbf{w}^0\right) &= g\left(\mathbf{w}^0\right) \\
\nabla h\left(\mathbf{w}^0\right) &= \nabla g\left(\mathbf{w}^0\right)
\end{aligned}
\tag{B.63}
$$

we could set up a system of equations (mirroring the one we set up when asking the analogous question for single-input functions) and recover the tangent hyperplane (our first-order Taylor series approximation) we saw back in Section B.4

$$
h(\mathbf{w}^0) = g(\mathbf{w}^0) + \nabla g(\mathbf{w}^0)^T (\mathbf{w} - \mathbf{w}^0).
\tag{B.64}
$$

Notice how this is the exact analog of the first-order approximation for single-input functions, and reduces to it (a tangent line) when $N = 1$.

Likewise, inquiring about what sort of degree-two (quadratic) function h could match g at a point $\mathbf{w}^0$ in terms of the value it takes, as well as the values its first and second derivatives take, i.e.,

$$
\begin{aligned}
h\left(\mathbf{w}^0\right) &= g\left(\mathbf{w}^0\right) \\
\nabla h\left(\mathbf{w}^0\right) &= \nabla g\left(\mathbf{w}^0\right) \\
\nabla^2 h\left(\mathbf{w}^0\right) &= \nabla^2 g\left(\mathbf{w}^0\right)
\end{aligned}
\tag{B.65}
$$

we would likewise derive (as we did explicitly with the single-input case) the second-order Taylor series approximation

$$
h(\mathbf{w}) = g(\mathbf{w}^0) + \nabla g(\mathbf{w}^0)^T (\mathbf{w} - \mathbf{w}^0) + \frac{1}{2}(\mathbf{w} - \mathbf{w}^0)^T \nabla^2 g\left(\mathbf{w}^0\right)(\mathbf{w} - \mathbf{w}^0).
\tag{B.66}
$$

Notice once again how this is the exact analog of the second-order approximation for single-input functions, and reduces to it when $N = 1$.

In Figure B.12 we plot the first- and second-order Taylor series approximation (shown in lime green and turquoise, respectively) of the function $g(w_1, w_2) =$

$\sin(w_1)$ at a point near the origin. As was the case with single-input functions the second-order approximation is a much better local approximator than the first, as it contains more derivative information there.

Higher-order Taylor series approximations can be defined precisely as in the single-input case. The main difference with multi-input functions is that higher-order approximations, starting with the third-order derivative, require serious manipulation of tensors of partial derivatives.

While this can be readily defined, only approximations up to the second order are ever used in practice. This is because (as we saw in Section B.8.2) the number of partial derivatives grows *exponentially* in the order of a derivative. Thus even though we get a better (local) approximator as we increase the order of a Taylor series, the serious pitfall of calculating/storing exponentially many partial derivatives nullifies the benefit.

B.10   Using the autograd Library

In this section we demonstrate how to use a simple yet powerful *automatic differentiator* written in Python, called autograd [10, 11] – a tool we make extensive use of throughout the text.

B.10.1   Installing autograd

autograd is an open-source professional-grade gradient calculator, or *automatic differentiator*, whose default is the *reverse mode* outlined in the previous section. It allows you to automatically compute the derivative of arbitrarily complex functions built using basic Python and NumPy functions.

It is also very easy to install: simply open a terminal and type

```
pip install autograd
```

to install the program. You can also visit the github repository for autograd at

```
https://github.com/HIPS/autograd
```

to download the same set of files to your machine. Another tool called JAX – built by the same community as an extension of autograd for use on GPUs and TPUs – can be used in a very similar manner as described here, and can be downloaded by visiting

```
https://github.com/google/jax
```

Along with autograd we also highly recommend the Anaconda Python 3 distribution, which can be installed by visiting

$$\texttt{https://anaconda.com}$$

This standard Python distribution includes a number of useful libraries, including NumPy, Matplotlib, and Jupyter notebooks.

B.10.2   Using autograd

Here we show a number of examples highlighting the basic usage of the autograd automatic differentiator. With simple modules we can easily compute derivatives of single-input functions, as well as partial derivatives and complete gradients of multi-input functions implemented in Python and NumPy.

Example B.9   Computing derivatives of single-input functions

Since autograd is specially designed to automatically compute the derivative(s) of NumPy code, it comes with its own wrapper on the basic NumPy library. This is where the differentiation rules (applied specifically to NumPy functionality) are defined. You can use autograd's version of NumPy exactly like you would the standard version, as virtually nothing about the user interface has been changed. To import this autograd wrapped version of NumPy, type

```
1  # import statement for autograd wrapped NumPy
2  import autograd.numpy as np
```

We begin by demonstrating the use of autograd with the simple function

$$g((w) = \tanh(w) \tag{B.67}$$

whose derivative, written algebraically, is

$$\frac{d}{dw} g(w) = 1 - \tanh^2(w). \tag{B.68}$$

There are two common ways of defining functions in Python. First is the standard Python function declaration (as shown below).

```
1  # a named Python function
2  def g(w):
3      return np.tanh(w)
```

You can also create *anonymous* functions in Python (functions you can define in a single line of code) by using the lambda command.

```
1  # a function defined via lambda
2  g = lambda w: np.tanh(w)
```

Regardless of how we define our function in Python it still amounts to the same thing mathematically/computationally.

To compute the derivative of our function we must first import the gradient calculator, conveniently called grad.

```
1  # import autograd's basic automatic differentiator
2  from autograd import grad
```

To use the grad function we simply pass in the function we wish to differentiate. grad works by *explicitly* computing the computation graph of our input, returning its derivative that we can then evaluate wherever we want. It does not provide an *algebraic function*, but a Python *function*. Here we call the derivative function of our input dgdw.

```
1  # create the gradient of g
2  dgdw = grad(g)
```

We can compute higher-order derivatives of our input function by using the same autograd functionality recursively, i.e., by plugging in the derivative function dgdw into autograd's grad function. Doing this once gives us the second derivative Python function, which we call d2gdw2.

```
1  # compute the second derivative of g
2  d2gdw2 = grad(dgdw)
```

We plot the input function as well as its first and second derivatives in Figure B.13.

Figure B.13 Figure associated with Example B.9. See text for details.

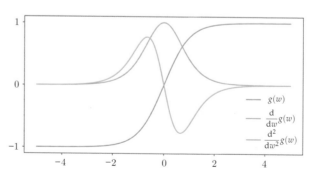

Example B.10   Function and gradient evaluations

As mentioned in the previous section, when we use an automatic differentiator to evaluate the gradient of a function we evaluate *the function itself* in the process. In other words, whenever we evaluate the gradient, we get the function evaluation "for free."

However, the grad function in the previous example returned only a single value: the evaluation of the derivative. The function evaluation is indeed being computed "under the hood" of grad, and is simply not returned for ease of use.

There is another autograd method called value_and_grad that returns everything computed "under the hood" including both the derivative(s) and function evaluation(s). Below we use this autograd functionality to reproduce the previous example's first derivative calculations.

```
1  # import autograd's automatic differentiator
2  from autograd import value_and_grad
3
4  # create the gradient of g
5  dgdw = value_and_grad(g)
6
7  # evaluate g and its gradient at w=0
8  w = 0
9  g_val, grad_val = dgdw(w)
```

Example B.11   Computing Taylor series approximations

Using autograd we can easily compute Taylor series approximations (see Section B.9) of any single-input function. Take, for instance, the function $g(w) = \tanh(w)$ along with its first-order Taylor series approximation

$$h(w) = g(w^0) + \frac{d}{dw}g(w^0)(w - w^0) \tag{B.69}$$

centered at the point $w^0 = 1$. First, we produce this function and its first-order Taylor series approximation in Python as follows.

```
1  # create the function g and its first derivative
2  g = lambda w: np.tanh(w)
3  dgdw = grad(g)
4
5  # create first-order Taylor series approximation
6  first_order = lambda w0, w: g(w0) + dgdw(w0)*(w - w0)
```

Next, we evaluate and plot the function (in black) and its first-order approximation (in green) in Figure B.14. It is just as easy to compute the second-order Taylor series approximation as well, whose formula is given as

$$q(w) = g(w^0) + \frac{d}{dw}g(w^0)(w - w^0) + \frac{1}{2}\frac{d^2}{dw^2}g(w^0)(w - w^0)^2. \qquad (\text{B.70})$$

```
1  # create the second derivative of g
2  d2gdw2 = grad(dgdw)
3
4  # create second-order Taylor series approximation
5  second_order = lambda w0, w: g(w0) + dgdw(w0)*(w - w0) + 0.5*d2gdw2(w0
      )*(w - w0)**2
```

The second-order Taylor series approximation is shown (in blue) in Figure B.14, with the point of expansion/tangency shown as a red circle.

Figure B.14 Figure associated with Example B.11. See text for details.

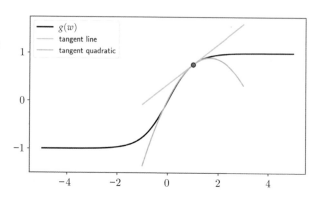

Example B.12 Computing individual partial derivatives

There are a number of ways we can go about using autograd to compute the partial derivatives of a multi-input function. First, let us look at how to use autograd to compute partial derivatives individually or, one at a time, beginning with the function

$$g(w_1, w_2) = \tanh(w_1 w_2). \qquad (\text{B.71})$$

We translate this function into Python below.

```
1  # a simple multi-input function
2  def g(w_1, w_2):
3      return np.tanh(w_1*w_2)
```

Taking in two inputs, this function lives in three dimensions, as plotted in the left panel of Figure B.15.

If we use the exact same call we used in the previous examples, and write

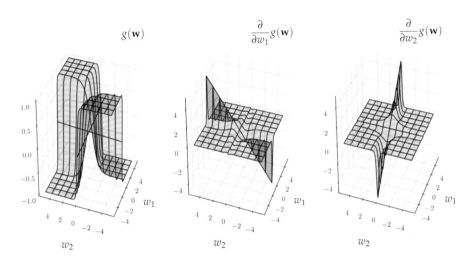

Figure B.15 Figure associated with Example B.12. A multi-input function (left panel) and its partial derivative in the first (middle panel) and second input (right panel). See text for further details.

$$\texttt{grad(g)}$$

since our function takes in two inputs, this will return the *first* partial derivative $\frac{\partial}{\partial w_1} g\,(w_1, w_2)$. This is the *default* setting of each automatic differentiation method in autograd.

Alternatively to compute this partial derivative function we can explicitly pass a second argument to grad, or any of the other autograd methods, which is a simple index denoting which partial derivative we want. To create the same (first) partial derivative this way, we pass in the index 0 (since Python indexing starts with 0) as follows.

$$\texttt{grad(g, 0)}$$

Similarly, to construct the second partial derivative we pass in the index 1.

$$\texttt{grad(g, 1)}$$

More generally, if g takes in N inputs (w_1 through w_N) we can construct its nth partial derivative using the same pattern as follows.

$$\texttt{grad(g, n-1)}$$

We plot these two partial derivative functions in the middle and right panels of Figure B.15.

Example B.13　Computing several derivatives or the full gradient

Building on the previous example, here we look at how to use autograd to construct several partial derivative functions at once or the entire gradient of a multi-input function. We do this via example, using the same function employed in the previous example.

There are two ways to do this using autograd. The first way is to simply index all the partial derivatives desired using the same sort of notation introduced previously. For instance, if we wish to construct the full gradient of the two-input function employed in the previous example, we tell autograd of this desire by feeding in the two indices (0,1) as shown below.

$$\text{grad}(g, (0,1))$$

More generally, for a function taking in N inputs, to construct any subset of partial derivatives at once we use the same sort of indexing notation. Note that this usage applies to all methods in the autograd automatic differentiation library.

The second way to construct several derivatives at once using autograd is by writing a function in NumPy where the desired variables we wish to differentiate with respect to are all input into the function as a single argument. For example, instead of writing out our function in Python as

```
1  # a simple multi-input function defined in Python
2  def g(w_1, w_2):
3      return np.tanh(w_1*w_2)
```

where both w_1 and w_2 are fed in one at a time, if we write it equivalently using vector notation as

```
1  def g(w):
2      return np.tanh(w[0]*w[1])
```

then the call

$$\text{grad}(g)$$

or equivalently

$$\text{grad}(g, 0)$$

will produce derivatives of g with respect to its first argument, which here will give us the complete gradient of g.

This indexing format holds more generally as well, that is, the statement

$$\text{grad(g, n-1)}$$

computes the derivatives of the function with respect to the nth input (whether it is a single variable or multiple variables).

B.10.3 Flattening mathematical functions using autograd

Mathematical functions come in all shapes and sizes, and moreover, we can often express individual functions algebraically in a variety of different ways. This short section discusses *function flattening*, which is a convenient way to *standardize* functions implemented in Python so that, for example, we can more quickly apply (in code) local optimization without the need to loop over weights of a particularly complicated function.

Example B.14 **Flattening a multi-input function**
Consider the function

$$g\left(a, \mathbf{b}, \mathbf{C}\right) = \left(a + \mathbf{r}^T \mathbf{b} + \mathbf{z}^T \mathbf{C} \mathbf{z}\right)^2 \tag{B.72}$$

where the input variable a is a scalar, $\mathbf{b}$ is a 2×1 vector, $\mathbf{C}$ is a 2×2 matrix, and the nonvariable vectors $\mathbf{r}$ and $\mathbf{z}$ are fixed at $\mathbf{r} = \begin{bmatrix} 1 & 2 \end{bmatrix}^T$ and $\mathbf{z} = \begin{bmatrix} 1 & 3 \end{bmatrix}^T$, respectively. This function is not written in the standard form $g(\mathbf{w})$ in which we discuss local optimization in Chapters 2 through 4 of this text. While, of course, all of the principles and algorithms described there still apply to this function, the implementation of any of those methods will be naturally more complicated as each step will need to be explicitly written in all three inputs a, $\mathbf{b}$, and $\mathbf{C}$, which is more cumbersome to write (and implement). This annoyance is greatly amplified when dealing with functions of many more explicit inputs variables, which we regularly encounter in machine learning. For such functions, in order to take a single descent step using some local method we must *loop* over their many different input variables.

Thankfully, every mathematical function can be expressed so that *all* of its input variables are represented as a single contiguous vector $\mathbf{w}$, which alleviates this irritation. For example, by defining a single vector

$$\mathbf{w} = \begin{bmatrix} w_1 \\ w_2 \\ w_3 \\ w_4 \\ w_5 \\ w_6 \\ w_7 \end{bmatrix} = \begin{bmatrix} a \\ b_1 \\ b_2 \\ c_{11} \\ c_{12} \\ c_{21} \\ c_{22} \end{bmatrix} \tag{B.73}$$

the original function in Equation (B.72) can then be equivalently written as

$$g(\mathbf{w}) = \left(\mathbf{s}^T \mathbf{w}\right)^2 \tag{B.74}$$

where

$$\mathbf{s} = \begin{bmatrix} 1 \\ 1 \\ 2 \\ 1 \\ 3 \\ 3 \\ 9 \end{bmatrix}. \tag{B.75}$$

Again note that all we have really done here is *reindexed* the entries of the input vectors in a contiguous manner so that we can implement local optimization schemes in a less cumbersome way in a single line of algebra or autograd code, instead of requiring a loop over each input variable. This variable reindexing scheme is called *function flattening*, and is depicted visually in Figure B.16.

Figure B.16 Figure associated with Example B.14. A figurative illustration of flattening of the function given in Equation (B.72). See text for further details.

While performing the reindexing required to flatten a function is important, it (like derivative computation itself) is a repetitive and time-consuming operation for humans to perform themselves. Thankfully, the autograd library has a built-in module that flattens functions, which can be imported as shown below.

```
1  # import function flattening module from autograd
2  from autograd.misc.flatten import flatten_func
```

Below we define a Python version of the function in Equation (B.72).

```
1  # Python implementation of g
2  r = np.array([[1],[2]])
3  z = np.array([[1],[3]])
4  def g(input_weights):
5      a = input_weights[0]
6      b = input_weights[1]
7      C = input_weights[2]
8      return (((a + np.dot(r.T, b) + np.dot(np.dot(z.T, C), z)))**2)
              [0][0]
```

To flatten g we then simply call

```
1  # flatten an input function g
2  g_flat, unflatten_func, w = flatten_func(g, input_weights)
```

Here on the right-hand side input_weights is a list of initializations for input variables to the function g. The outputs g_flat, unflatten_func, and w are the flattened version of g, a module to unflatten the input weights, and a flattened version of the initial weights, respectively.

C Linear Algebra

C.1 Introduction

In this appendix chapter we briefly review basic ideas from linear algebra that are fundamental to understanding machine learning. These include vector and matrix arithmetic, vector and matrix norms, and eigenvalue decomposition. The reader is strongly encouraged to ensure familiarity with all concepts mentioned in this chapter before proceeding with the rest of the text.

C.2 Vectors and Vector Operations

We begin by reviewing the fundamental notion of a vector, as well as vector arithmetic.

C.2.1 The vector

A *vector* is another word for an ordered listing of numbers. For example,

$$[-3 \ 4 \ 1] \tag{C.1}$$

is a vector of three *elements* or *entries*, also referred to as a vector of *size* or *dimension* three. In general, a vector can have an arbitrary number of elements, and can contain numbers, variables, or both. For example,

$$[x_1 \ x_2 \ x_3 \ x_4] \tag{C.2}$$

is a vector of four variables. When numbers or variables are listed out horizontally (or in a row) we call the resulting vector a *row* vector. However, we can list them vertically (or in a column) just as well, in which case we refer to the resulting vector as a *column* vector. For instance,

$$\begin{bmatrix} -3 \\ 4 \\ 1 \end{bmatrix} \tag{C.3}$$

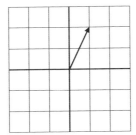 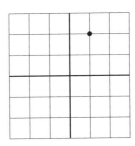

Figure C.1 A two-dimensional vector visualized as an *arrow* stemming from the origin (left panel), or equivalently as a single *point* in a two-dimensional plane (right panel).

is now a column vector of size three. We can swap back and forth between a row and column version of a vector by *transposing* each. Transposition is usually denoted by a superscript T placed just to the right and above a vector, and simply turns a row vector into an equivalent column vector and vice versa. For example, we have

$$\begin{bmatrix} -3 \\ 4 \\ 1 \end{bmatrix}^T = [-3 \; 4 \; 1] \quad \text{and} \quad [-3 \; 4 \; 1]^T = \begin{bmatrix} -3 \\ 4 \\ 1 \end{bmatrix}. \tag{C.4}$$

To discuss vectors more generally we use algebraic notation, typically a bold lowercase (often Roman) letter, e.g., $\mathbf{x}$. The transpose of $\mathbf{x}$ is then denoted as $\mathbf{x}^T$. This notation does not denote whether or not the vector is a row or column, or how many elements it contains. Such information must therefore be given explicitly. Throughout the text, unless stated otherwise, we assume all vectors are column vectors by default.

Vectors of length two (or three) are easy to intuit since they live in two- (or three-) dimensional spaces that are familiar to our human senses. For example, the two-dimensional vector

$$\mathbf{x} = \begin{bmatrix} 1 \\ 2 \end{bmatrix} \tag{C.5}$$

can be drawn in a two-dimensional plane as an *arrow* stemming from the origin and ending at the point whose horizontal and vertical coordinates are 1 and 2, respectively, as illustrated in the left panel of Figure C.1. However, as shown in the right panel of the figure, $\mathbf{x}$ can alternatively be drawn (and thought of) as a single *point*, i.e., the arrow's endpoint. When plotting a low-dimensional machine learning dataset (that is simply a collection of vectors) we often employ the latter visual style.

C.2.2   Vector addition

We add (and subtract) two vectors element-wise, noting that, in order to be able to do so, the two vectors must have the same number of elements (or dimension), and both must be row or column vectors. For example, vectors

$$\mathbf{x} = \begin{bmatrix} x_1 \\ x_2 \\ \vdots \\ x_N \end{bmatrix} \quad \text{and} \quad \mathbf{y} = \begin{bmatrix} y_1 \\ y_2 \\ \vdots \\ y_N \end{bmatrix} \tag{C.6}$$

are added element-wise to form

$$\mathbf{x} + \mathbf{y} = \begin{bmatrix} x_1 + y_1 \\ x_2 + y_2 \\ \vdots \\ x_N + y_N \end{bmatrix}. \tag{C.7}$$

Subtraction of $\mathbf{y}$ from $\mathbf{x}$ is defined similarly as

$$\mathbf{x} - \mathbf{y} = \begin{bmatrix} x_1 - y_1 \\ x_2 - y_2 \\ \vdots \\ x_N - y_N \end{bmatrix}. \tag{C.8}$$

Thinking of vectors as arrows stemming from the origin, the addition of two vectors is equal to the vector representing the far corner of the parallelogram formed by the two vectors in the sum. This is typically called the *parallelogram law*, and is illustrated in Figure C.2 for two input vectors colored black, with their sum shown in red. The dashed lines here are merely visual guides helping to outline the parallelogram underlying the sum.

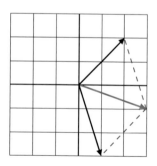

Figure C.2 The parallelogram law illustrated.

C.2.3   Vector multiplication

Unlike addition, there is more than one way to define vector multiplication. In what follows we review multiplication of a vector by a scalar, element-wise multiplication of two vectors, as well as inner- and outer-product of two vectors.

Multiplication of a vector by a scalar

We can multiply any vector $\mathbf{x}$ by a scalar c, by treating the multiplication element-wise as

$$c\mathbf{x} = \begin{bmatrix} c\,x_1 \\ c\,x_2 \\ \vdots \\ c\,x_N \end{bmatrix}. \tag{C.9}$$

Element-wise product of two vectors

The element-wise product, sometimes called the Hadamard product, works precisely how it sounds: we multiply two vectors element by element. Note that, just like addition, we need both vectors to have the same dimension in order to make this work. Notationally, the element-wise product of two vectors $\mathbf{x}$ and $\mathbf{y}$ is written as

$$\mathbf{x} \circ \mathbf{y} = \begin{bmatrix} x_1 y_1 \\ x_2 y_2 \\ \vdots \\ x_N y_N \end{bmatrix}. \tag{C.10}$$

Inner-product of two vectors

The inner-product (also referred to as the dot product) is another way to multiply two vectors of the same dimension. Unlike the element-wise product, the inner-product of two vectors produces a *scalar* output. To take the inner-product of two vectors we first multiply them together element-wise, and then simply add up the elements in the resulting vector. The inner-product of vectors $\mathbf{x}$ and $\mathbf{y}$ is written as

$$\mathbf{x}^T \mathbf{y} = x_1 y_1 + x_2 y_2 + \cdots + x_N y_N = \sum_{n=1}^{N} x_n y_n. \tag{C.11}$$

Vector length or magnitude

The well-known Pythagorean theorem provides a useful way to measure the length of a vector in two dimensions. Using the Pythagorean theorem we can treat the general two-dimensional vector

$$\mathbf{x} = \begin{bmatrix} x_1 \\ x_2 \end{bmatrix} \tag{C.12}$$

as the hypotenuse of a right triangle, and write

$$\text{length of } \mathbf{x} = \sqrt{x_1^2 + x_2^2}. \tag{C.13}$$

Notice, we can also express the length of $\mathbf{x}$ in terms of the inner-product of $\mathbf{x}$ with itself, as

$$\text{length of } \mathbf{x} = \sqrt{\mathbf{x}^T \mathbf{x}}, \tag{C.14}$$

and this generalizes to vectors of any dimension. Using the notation $\|\mathbf{x}\|_2$ to denote the length of an N-dimensional vector $\mathbf{x}$, we have

$$\|\mathbf{x}\|_2 = \sqrt{\mathbf{x}^T \mathbf{x}} = \sqrt{\sum_{n=1}^{N} x_n^2}. \tag{C.15}$$

Geometric interpretation of the inner-product

The inner-product of two vectors $\mathbf{x}$ and $\mathbf{y}$

$$\mathbf{x}^T \mathbf{y} = \sum_{n=1}^{N} x_n y_n \tag{C.16}$$

can be expressed in terms of the lengths of $\mathbf{x}$ and $\mathbf{y}$, via the so-called *inner-product rule*, as

$$\mathbf{x}^T \mathbf{y} = \|\mathbf{x}\|_2 \|\mathbf{y}\|_2 \cos(\theta) \tag{C.17}$$

where θ is the angle between $\mathbf{x}$ and $\mathbf{y}$. This rule is perhaps best intuited after a slight rearrangement of its terms, as

$$\left(\frac{\mathbf{x}}{\|\mathbf{x}\|_2} \right)^T \left(\frac{\mathbf{y}}{\|\mathbf{y}\|_2} \right) = \cos(\theta) \tag{C.18}$$

where vectors $\frac{\mathbf{x}}{\|\mathbf{x}\|_2}$ and $\frac{\mathbf{y}}{\|\mathbf{y}\|_2}$ still point in the same direction as $\mathbf{x}$ and $\mathbf{y}$, respectively, but both have been normalized to have unit length, since

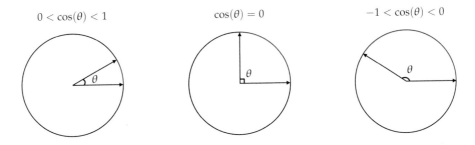

Figure C.3 The inner-product of two unit-length vectors is equal to the cosine of the angle θ created between them.

$$\left\|\frac{\mathbf{x}}{\|\mathbf{x}\|_2}\right\|_2 = \left\|\frac{\mathbf{y}}{\|\mathbf{y}\|_2}\right\|_2 = 1. \tag{C.19}$$

Note that because cosine always lies between -1 and 1, so too does the inner product of any two unit-length vectors. When they point in the exact same direction, $\theta = 0$, and their inner-product is maximal (i.e., 1). As the two vector start to point away from each other, θ increases, and the inner-product starts to shrink. When the two vectors are *perpendicular* to each other, their inner-product is equal to zero. The inner-product reaches its minimal value (i.e., -1) when $\theta = \pi$, and the two vectors point in completely opposite directions (see Figure C.3).

Outer-product of two vectors
The outer-product is another way to define multiplication between two vectors. With two column vectors (of not necessarily the same dimension)

$$\mathbf{x} = \begin{bmatrix} x_1 \\ x_2 \\ \vdots \\ x_N \end{bmatrix} \quad \text{and} \quad \mathbf{y} = \begin{bmatrix} y_1 \\ y_2 \\ \vdots \\ y_M \end{bmatrix} \tag{C.20}$$

their outer-product is written as $\mathbf{xy}^T$, and defined as

$$\mathbf{xy}^T = \begin{bmatrix} x_1 \\ x_2 \\ \vdots \\ x_N \end{bmatrix} \begin{bmatrix} y_1 & y_2 & \cdots & y_M \end{bmatrix} = \begin{bmatrix} x_1 y_1 & x_1 y_2 & \cdots & x_1 y_M \\ x_2 y_1 & x_2 y_2 & \cdots & x_2 y_M \\ \vdots & \vdots & \ddots & \vdots \\ x_N y_1 & x_N y_2 & \cdots & x_N y_M \end{bmatrix}. \tag{C.21}$$

The result is an $N \times M$ matrix, which can be thought of as a collection of M column vectors of length N stacked side by side (or likewise, as a collection of N

row vectors of length M stacked on top of each other). We will return to matrices and discuss them further in the next section.

C.2.4   Linear combination of vectors

A linear combination is an operation that generalizes simple addition of two vectors by combining addition and scalar multiplication. Given two vectors x_1 and x_2 of the same dimension, their linear combination is formed by multiplying each with a scalar first and then adding up the result, as

$$\alpha_1 x_1 + \alpha_2 x_2 \tag{C.22}$$

where α_1 and α_2 are real numbers. Notice that for a given pair of values (α_1, α_2) the linear combination is a vector itself with the same dimension as x_1 and x_2. In Figure C.4 we show the linear combination of vectors

$$x_1 = \begin{bmatrix} 2 \\ 1 \end{bmatrix} \quad \text{and} \quad x_2 = \begin{bmatrix} -1 \\ 1 \end{bmatrix} \tag{C.23}$$

for three distinct settings of (α_1, α_2). The set of all such vectors created by taking a linear combination of vectors x_1 and x_2 is referred to as the *span* of x_1 and x_2, and written as

$$\text{span of } x_1 \text{ and } x_2 = \left\{ \alpha_1 x_1 + \alpha_2 x_2 \mid (\alpha_1, \alpha_2) \in \mathbb{R}^2 \right\}. \tag{C.24}$$

$(\alpha_1, \alpha_2) = (1, 1)$ $(\alpha_1, \alpha_2) = (2, 3)$ $(\alpha_1, \alpha_2) = (-1, -2)$

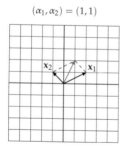 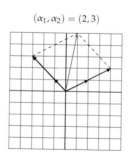 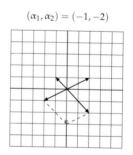

Figure C.4 The linear combination (in red) of vectors x_1 and x_2 defined in Equation (C.23) for three different settings of (α_1, α_2). As you can see by changing the values of α_1 and α_2 in $\alpha_1 x_1 + \alpha_2 x_2$ we get a new vector each time. The set of all such vectors is referred to as the span of x_1 and x_2, which in this case is the entire two-dimensional plane.

For the vectors x_1 and x_2 in Equation (C.23) the span is the entire two-dimensional plane. But this is not necessarily always the case for any pair of vectors x_1 and x_2. Take

$$\mathbf{x}_1 = \begin{bmatrix} 1 \\ 1 \end{bmatrix} \quad \text{and} \quad \mathbf{x}_2 = \begin{bmatrix} 3 \\ 3 \end{bmatrix} \tag{C.25}$$

for instance. Because these two vectors point at the same direction (one is a scalar multiple of the other), any linear combination of the two will have the same direction. In this case the span of $\mathbf{x}_1$ and $\mathbf{x}_2$ is no longer the entire two-dimensional plane, but a one-dimensional line that can be traced out using scalar multiples of any of the two vectors. In other words, given either one of $\mathbf{x}_1$ or $\mathbf{x}_2$ the other one becomes redundant (in terms of finding their span). In linear algebra terms such vectors are called *linearly dependent*.

The notion of linear combination of vectors can be extended in general to a set of k vectors $\{\mathbf{x}_1, \mathbf{x}_2, \ldots, \mathbf{x}_k\}$ (all of the same dimension), taking the form

$$\sum_{i=1}^{k} \alpha_i \mathbf{x}_i = \alpha_1 \mathbf{x}_1 + \alpha_2 \mathbf{x}_2 + \cdots + \alpha_k \mathbf{x}_k. \tag{C.26}$$

If these vectors span a k-dimensional space they are called *linearly independent*. Otherwise, there is at least one vector in the set that can be written as a linear combination of the rest.

C.3   Matrices and Matrix Operations

In this section we review the concept of a matrix as well as the basic operations one can perform on a single matrix or pairs of matrices. These completely mirror those of the vector in the previous section, including the transpose operation, addition/subtraction, and several multiplication operations. Because of the close similarity to vectors this section is much more terse than the previous section.

C.3.1   The matrix

If we take a set of N row vectors, each of dimension M

$$\mathbf{x}_1 = \begin{bmatrix} x_{11} & x_{12} & \cdots & x_{1M} \end{bmatrix}$$

$$\mathbf{x}_2 = \begin{bmatrix} x_{21} & x_{22} & \cdots & x_{2M} \end{bmatrix}$$

$$\vdots$$

$$\mathbf{x}_N = \begin{bmatrix} x_{N1} & x_{N2} & \cdots & x_{NM} \end{bmatrix}$$

and stack them one by one on top of each other we form an object called a *matrix*

$$
\mathbf{X} =
\begin{bmatrix}
x_{11} & x_{12} & \cdots & x_{1M} \\
x_{21} & x_{22} & \cdots & x_{2M} \\
\vdots & \vdots & \ddots & \vdots \\
x_{N1} & x_{N2} & \cdots & x_{NM}
\end{bmatrix}
\tag{C.27}
$$

of dimension $N \times M$, where the first number N is the number of rows in the matrix, with the second number M denoting the number of columns. The notation we use to describe a matrix in the text is a bold uppercase letter, e.g., $\mathbf{X}$. Like the vector notation nothing about the dimensions of the matrix is detailed by its notation and they must be explicitly stated.

The transpose operation we originally saw for vectors is defined by extension for matrices. When performed on a matrix, the transpose operation flips the entire matrix around: every column is turned into a row, and then these rows are stacked one on top of the other, forming an $M \times N$ matrix

$$
\mathbf{X}^T =
\begin{bmatrix}
x_{11} & x_{21} & \cdots & x_{N1} \\
x_{12} & x_{22} & \cdots & x_{N2} \\
\vdots & \vdots & \ddots & \vdots \\
x_{1M} & x_{2M} & \cdots & x_{NM}
\end{bmatrix}.
\tag{C.28}
$$

C.3.2   Matrix addition

As with vectors, addition (and subtraction) is performed element-wise on matrices of the same dimensions. For example, with two $N \times M$ matrices

$$
\mathbf{X} =
\begin{bmatrix}
x_{11} & x_{12} & \cdots & x_{1M} \\
x_{21} & x_{22} & \cdots & x_{2M} \\
\vdots & \vdots & \ddots & \vdots \\
x_{N1} & x_{N2} & \cdots & x_{NM}
\end{bmatrix}
\quad \text{and} \quad
\mathbf{Y} =
\begin{bmatrix}
y_{11} & y_{12} & \cdots & y_{1M} \\
y_{21} & y_{22} & \cdots & y_{2M} \\
\vdots & \vdots & \ddots & \vdots \\
y_{N1} & y_{N2} & \cdots & y_{NM}
\end{bmatrix}
\tag{C.29}
$$

their sum is defined as

$$
\mathbf{X} + \mathbf{Y} =
\begin{bmatrix}
x_{11} + y_{11} & x_{12} + y_{12} & \cdots & x_{1M} + y_{1M} \\
x_{21} + y_{21} & x_{22} + y_{22} & \cdots & x_{2M} + y_{2M} \\
\vdots & \vdots & \ddots & \vdots \\
x_{N1} + y_{N1} & x_{N2} + y_{N2} & \cdots & x_{NM} + y_{NM}
\end{bmatrix}.
\tag{C.30}
$$

C.3.3   Matrix multiplication

As with vectors, there are a variety of ways to define matrix multiplication which we review here.

Multiplication of a matrix by a scalar

We can multiply any matrix $\mathbf{X}$ by a scalar c, and this operation is performed element by element as

$$c\,\mathbf{X} = \begin{bmatrix} c\,x_{11} & c\,x_{12} & \cdots & c\,x_{1M} \\ c\,x_{21} & c\,x_{22} & \cdots & c\,x_{2M} \\ \vdots & \vdots & \ddots & \vdots \\ c\,x_{N1} & c\,x_{N2} & \cdots & c\,x_{NM} \end{bmatrix}. \tag{C.31}$$

Multiplication of a matrix by a vector

Generally speaking, there are two ways to multiply an $N \times M$ matrix $\mathbf{X}$ by a vector $\mathbf{y}$. The first, referred to as *left multiplication*, involves multiplication by an N-dimensional row vector $\mathbf{y}$. This operation, written as $\mathbf{y}\mathbf{X}$, results in a row vector of dimension M whose mth element is the inner-product of $\mathbf{y}$ with the mth column of $\mathbf{X}$

$$\mathbf{y}\mathbf{X} = \left[\sum_{n=1}^{N} y_n x_{n1} \quad \sum_{n=1}^{N} y_n x_{n2} \quad \cdots \quad \sum_{n=1}^{N} y_n x_{nM} \right]. \tag{C.32}$$

Likewise, *right multiplication* is defined by multiplying $\mathbf{X}$ on the right by an M-dimensional column vector $\mathbf{y}$. Written as $\mathbf{X}\mathbf{y}$, right multiplication results in an N-dimensional column vector whose nth element is the inner-product of $\mathbf{y}$ with the nth row of $\mathbf{X}$

$$\mathbf{X}\mathbf{y} = \begin{bmatrix} \sum_{m=1}^{M} y_m x_{1m} \\ \sum_{m=1}^{M} y_m x_{2m} \\ \vdots \\ \sum_{m=1}^{M} y_m x_{Nm} \end{bmatrix}. \tag{C.33}$$

Element-wise multiplication of two matrices

As with vectors, we can define element-wise multiplication on two matrices of the same dimensions. The element-wise product of two $N \times M$ matrices $\mathbf{X}$ and $\mathbf{Y}$ is written as

$$\mathbf{X} \circ \mathbf{Y} = \begin{bmatrix} x_{11}\,y_{11} & x_{12}\,y_{12} & \cdots & x_{1M}\,y_{1M} \\ x_{21}\,y_{21} & x_{22}\,y_{22} & \cdots & x_{2M}\,y_{2M} \\ \vdots & \vdots & \ddots & \vdots \\ x_{N1}\,y_{N1} & x_{N2}\,y_{N2} & \cdots & x_{NM}\,y_{NM} \end{bmatrix}. \tag{C.34}$$

General multiplication of two matrices

The general product (or simply product) of two matrices $\mathbf{X}$ and $\mathbf{Y}$ can be defined based on the vector outer-product operation, provided that the number of columns in $\mathbf{X}$ matches the number of rows in $\mathbf{Y}$. That is, we must have $\mathbf{X}$ and $\mathbf{Y}$ of dimensions $N \times M$ and $M \times P$ respectively, for the matrix product to be defined as

$$\mathbf{XY} = \sum_{m=1}^{M} \mathbf{x}_m \mathbf{y}_m^T \tag{C.35}$$

where $\mathbf{x}_m$ is the mth column of $\mathbf{X}$, and $\mathbf{y}_m^T$ is the transpose of the mth column of $\mathbf{Y}^T$ (or equivalently, the mth row of $\mathbf{Y}$). Note that each summand in Equation (C.35) is itself a matrix of dimension $N \times P$, and so too is the final matrix $\mathbf{XY}$.

General matrix multiplication can also be defined element-wise, using vector inner-products, where the entry in the nth row and pth column of $\mathbf{XY}$ is found as the inner-product of (transpose of) the nth row in $\mathbf{X}$ and the pth column in $\mathbf{Y}$.

C.4   Eigenvalues and Eigenvectors

In this section we review general linear functions and their relationship to matrices. We particularly focus on the special case of the square matrix, for which we discuss the important topics of eigenvectors and eigenvalues.

C.4.1   Linear functions and matrix multiplication

As we discussed in the previous section, the product of an $N \times M$ matrix

$$\mathbf{X} = \begin{bmatrix} x_{11} & x_{12} & \cdots & x_{1M} \\ x_{21} & x_{22} & \cdots & x_{2M} \\ \vdots & \vdots & \ddots & \vdots \\ x_{N1} & x_{N2} & \cdots & x_{NM} \end{bmatrix} \tag{C.36}$$

by an M-dimensional column vector

$$\mathbf{w} = \begin{bmatrix} w_1 \\ w_2 \\ \vdots \\ w_M \end{bmatrix} \tag{C.37}$$

is an N-dimensional column vector written as $\mathbf{Xw}$. Treating the vector $\mathbf{w}$ as input, $\mathbf{Xw}$ defines a function g written formally as

$$g(\mathbf{w}) = \mathbf{Xw}. \tag{C.38}$$

Writing $g(\mathbf{w})$ explicitly as

$$g(\mathbf{w}) = \begin{bmatrix} x_{11}w_1 + x_{12}w_2 + \cdots + x_{1M}w_M \\ x_{21}w_1 + x_{22}w_2 + \cdots + x_{2M}w_M \\ \vdots \\ x_{N1}w_1 + x_{N2}w_2 + \cdots + x_{NM}w_M \end{bmatrix} \tag{C.39}$$

it is clear that each of its elements is a linear function in variables w_1 through w_M, and hence g itself is called a *linear* function.

C.4.2 Linear functions and square matrices

When the number of rows in a matrix $\mathbf{X}$ is identical to the number of columns in it, i.e., $N = M$, the matrix is called a *square* matrix. When $N = M = 2$ we can visually examine the effect of a linear function $g(\mathbf{w}) = \mathbf{X}\mathbf{w}$ by viewing the way two-dimensional points $\mathbf{w}$ are transformed via g.

In Figure C.5 we provide just such a visualization using the 2×2 matrix $\mathbf{X}$ whose entries were set at random as

$$\mathbf{X} = \begin{bmatrix} 0.726 & -1.059 \\ -0.200 & -0.947 \end{bmatrix}. \tag{C.40}$$

In the left panel of the figure we show a coarse set of grid lines. The point of this grid is to help visualize how each point constituting the grid lines (and thus the entire space itself) is transformed using the matrix $\mathbf{X}$ in Equation (C.40). For visualization purposes, a circle of radius 2 is drawn on top of the grid, and is transformed along with it. In the right panel of the figure we illustrate how the space shown in the left panel is warped by the function $g(\mathbf{w}) = \mathbf{X}\mathbf{w}$.

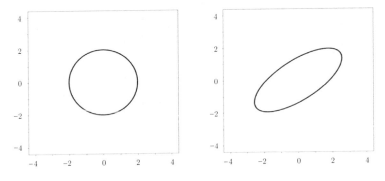

Figure C.5 The input (left panel) and output (right panel) spaces of the linear function $g(\mathbf{w}) = \mathbf{X}\mathbf{w}$ with the matrix $\mathbf{X}$ defined in Equation (C.40).

C.4.3   Eigenvalues and eigenvectors

The previous visualization in Figure C.5 is interesting in its own right, but if examined closely can also be used to provoke the notion of what are called *eigenvectors*. These are the handful of directions that, unlike most others that are warped and twisted by multiplication with the given matrix, are only *scaled* by the function. In other words, eigenvectors are those special vectors in the input space that retain their direction, after having gone through the linear transformation g. In Figure C.6 we again show the transformation provided by the random matrix **X** in Equation (C.40). This time, however, we also highlight two such eigenvectors as black arrows. Comparing the left and right panels of the figure, notice how neither direction gets twisted or warped by the transformation: they are only scaled.

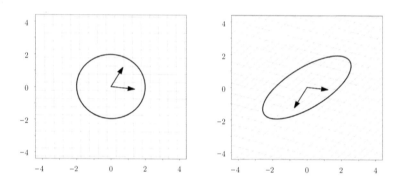

Figure C.6 A redrawing of Figure C.5 with the two eigenvectors of **X** added as two black arrows in both input and output spaces.

What we saw with linear functions based on 2×2 square matrices holds more generally for higher dimensions as well: a linear function based on an $N \times N$ matrix affects *at most* N linearly independent directions by simply scaling them. For an $N \times N$ matrix **X** each such direction $\mathbf{v} \neq \mathbf{0}_{N \times 1}$ satisfying

$$\mathbf{X}\mathbf{v} = \lambda \mathbf{v} \tag{C.41}$$

is called an *eigenvector*. Here the value λ is precisely the amount by which **X** scales $\mathbf{v}$, and is called an *eigenvalue*. In general, λ can take on real or complex values.

C.4.4   The special case of the symmetric matrix

A *symmetric* matrix, that is a square matrix **X** where $\mathbf{X} = \mathbf{X}^T$, is an important special case of a square matrix that arises in a wide range of contexts (e.g., Hessian matrices, covariance matrices, etc.). One of the main advantages such matrices have over merely square ones is the following: their eigenvectors are *always* perpendicular to each other, and their eigenvalues are *always* real numbers [74, 75, 76]. This fact has significant repercussions in the analysis of such matrices as we can *diagonalize* them as follows.

Stacking all of the eigenvectors of $\mathbf{X}$ column-wise into a matrix $\mathbf{V}$, and placing the corresponding eigenvalues along the diagonal of a matrix $\mathbf{D}$, we can write the Equation (C.41) simultaneously for all eigenvectors/values, as

$$\mathbf{XV} = \mathbf{VD}. \tag{C.42}$$

When the eigenvectors are all perpendicular to each other, $\mathbf{V}$ is an orthonormal matrix[1] and we have $\mathbf{VV}^T = \mathbf{I}$. Thus multiplying both sides of Equation (C.42) by $\mathbf{V}^T$ (on the right) we can express $\mathbf{X}$ completely in terms of its eigenvectors/values as

$$\mathbf{X} = \mathbf{VDV}^T. \tag{C.43}$$

C.5   Vector and Matrix Norms

In this section we discuss popular vector and matrix norms that will arise frequently in our study of machine learning, particularly when discussing regularization. A norm is a kind of function that measures the length of real vectors and matrices. The notion of length is extremely useful as it enables us to define distance (or similarity) between any two vectors (or matrices) living in the same space.

C.5.1   Vector norms

The ℓ_2 norm
We begin with the most widely used vector norm in machine learning, the ℓ_2 norm, defined for an N-dimensional vector $\mathbf{x}$ as

$$\|\mathbf{x}\|_2 = \sqrt{\sum_{n=1}^{N} x_n^2}. \tag{C.44}$$

Using the ℓ_2 norm we can measure the distance between any two points $\mathbf{x}$ and $\mathbf{y}$ via $\|\mathbf{x} - \mathbf{y}\|_2$, which is simply the length of the vector connecting $\mathbf{x}$ and $\mathbf{y}$. For example, the distance between

$$\mathbf{x} = \begin{bmatrix} 1 \\ 2 \end{bmatrix} \quad \text{and} \quad \mathbf{y} = \begin{bmatrix} 9 \\ 8 \end{bmatrix} \tag{C.45}$$

---

[1] Here we have assumed every eigenvector $\mathbf{v}$ that satisfies Equation (C.41) has unit length, i.e., $\|\mathbf{v}\|_2 = 1$. If not, we can always replace $\mathbf{v}$ with $\frac{\mathbf{v}}{\|\mathbf{v}\|_2}$ and Equation (C.41) will still hold.

is calculated as $\sqrt{(1-9)^2 + (2-8)^2} = 10$, as shown pictorially (in red) in Figure C.7.

Figure C.7 The ℓ_1 (blue), ℓ_2 (red), and ℓ_∞ (green) based distances between the points **x** and **y** defined in Equation (C.45).

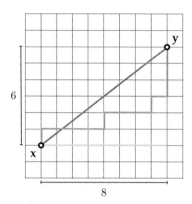

The ℓ_1 norm

The ℓ_1 norm of a vector **x** is another way to measure its length, defined as the sum of the absolute values of its entries

$$\|\mathbf{x}\|_1 = \sum_{n=1}^{N} |x_n|. \tag{C.46}$$

In terms of the ℓ_1 norm the distance between **x** and **y** is given by $\|\mathbf{x} - \mathbf{y}\|_1$, which provides a measurement of distance different from the ℓ_2 norm. As illustrated in Figure C.7 the distance defined by the ℓ_1 norm is the length of a path consisting of perpendicular pieces (shown in blue). Because these paths are somewhat akin to how an automobile might travel from **x** to **y** if they were two locations in a gridded city, having to traverse perpendicular city blocks one after the other, the ℓ_1 norm is sometimes referred to as the *taxicab norm*, and the distance measured via the ℓ_1 norm, the *Manhattan distance*. For **x** and **y** in Equation (C.45) the Manhattan distance is calculated as $|1-9| + |2-8| = 14$.

The ℓ_∞ norm

The ℓ_∞ norm of a vector **x** is equal to its largest entry (in terms of absolute value), defined mathematically as

$$\|\mathbf{x}\|_\infty = \max_n |x_n|. \tag{C.47}$$

For example, the distance between **x** and **y** in Equation (C.45) in terms of the ℓ_∞ norm is found as $\max(|1-9|, |2-8|) = 8$, as illustrated in Figure C.7 (in green).

C.5.2 Common properties of vector norms

The ℓ_2, ℓ_1, and ℓ_∞ norms share a number of useful properties that we detail below. Since these properties hold in general for *any* vector norm, we momentarily drop the subscript and represent the generic norm of **x** simply by $\|\mathbf{x}\|$.

1. Norms are always nonnegative, that is, $\|\mathbf{x}\| \geq 0$ for any **x**. Furthermore, the equality holds if and only if $\mathbf{x} = \mathbf{0}$, implying that the norm of any nonzero vector is always greater than zero.

2. The norm of $\alpha\mathbf{x}$, that is a scalar multiple of **x**, can be written in terms of the norm of **x** as $\|\alpha\mathbf{x}\| = |\alpha| \|\mathbf{x}\|$. With $\alpha = -1$ for example, we have that $\| - \mathbf{x}\| = \|\mathbf{x}\|$.

3. Norms also satisfy the so-called *triangle inequality* where for any three vectors **x**, **y**, and **z** we have $\|\mathbf{x} - \mathbf{z}\| + \|\mathbf{z} - \mathbf{y}\| \geq \|\mathbf{x} - \mathbf{y}\|$. As illustrated in Figure C.8 for the ℓ_2 norm (left panel), the ℓ_1 norm (middle panel), and the ℓ_∞ norm (right panel), the triangle inequality simply states that the distance between **x** and **y** is always smaller than (or equal to) the distance between **x** and **z**, and the distance between **z** and **y**, combined. In other words, if one wanted to travel from a given point **x** to a given point **y**, it would be always better to travel directly from **x** to **y** than to travel first to a third point z, and then to y. With the change of variables $\mathbf{u} = \mathbf{x} - \mathbf{z}$ and $\mathbf{v} = \mathbf{z} - \mathbf{y}$, the triangle inequality is sometimes written in the simpler form of $\|\mathbf{u}\| + \|\mathbf{v}\| \geq \|\mathbf{u} + \mathbf{v}\|$ for all vectors **u** and **v**.

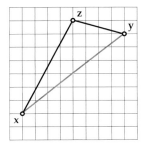 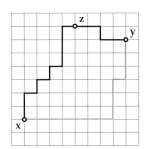 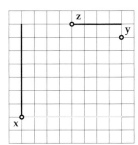

Figure C.8 The triangle inequality illustrated for the ℓ_2 norm (left panel), ℓ_1 norm (middle panel), and ℓ_∞ norm (right panel).

In addition to the general properties mentioned above and held by any norm, the ℓ_2, ℓ_1, and ℓ_∞ norms share a stronger bond that ties them together: they are all members of the ℓ_p norm family. The ℓ_p norm is generally defined as

$$\|\mathbf{x}\|_p = \left(\sum_{n=1}^{N} |x_n|^p \right)^{\frac{1}{p}} \tag{C.48}$$

Figure C.9 Illustration of the ℓ_1 (blue), ℓ_2 (red), and ℓ_∞ (green) unit balls.

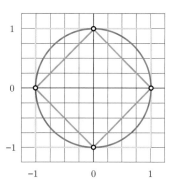

for $p \geq 1$. One can easily verify that with $p = 1$, $p = 2$, and as $p \longrightarrow \infty$, the ℓ_p norm reduces to the ℓ_1, ℓ_2, and ℓ_∞ norm, respectively.

The ℓ_p norm balls

A norm ball is the set of all vectors $\mathbf{x}$ with same norm value, that is, all $\mathbf{x}$ such that $\|\mathbf{x}\| = c$ for some constant $c > 0$. When $c = 1$, this set is called the unit norm ball, or simply the unit ball. The ℓ_1, ℓ_2, and ℓ_∞ unit balls are plotted in Figure C.9.

The ℓ_0 norm

The ℓ_0 norm is yet another way of defining a vector's length as

$$\|\mathbf{x}\|_0 = \text{number of nonzero entries of } \mathbf{x}. \tag{C.49}$$

Calling the ℓ_0 norm a *norm* is technically a misnomer as it does not hold the scalability property held by all vector norms. That is, $\|\alpha\mathbf{x}\|_0$ is generally not equal to $|\alpha| \|\mathbf{x}\|_0$. Nevertheless, the ℓ_0 norm arises frequently when modeling vectors with a large number of zeros (also called *sparse* vectors).

C.5.3 Matrix norms

The Frobenius norm

Recall that the ℓ_2 norm of a vector is defined as the square root of the sum of the squares of its elements. The Frobenius norm is the intuitive extension of the ℓ_2 norm for vectors to matrices, defined similarly as the square root of the sum of the squares of all the elements in the matrix, and written for an $N \times M$ matrix $\mathbf{X}$ as

$$\|\mathbf{X}\|_F = \sqrt{\sum_{n=1}^{N}\sum_{m=1}^{M} x_{nm}^2}. \tag{C.50}$$

For example, the Frobenius norm of the matrix $\mathbf{X} = \begin{bmatrix} -1 & 2 \\ 0 & 5 \end{bmatrix}$ is calculated as

$\sqrt{(-1)^2 + 2^2 + 0^2 + 5^2} = \sqrt{30}.$

The connection between the ℓ_2 norm and the Frobenius norm goes further: collecting all singular values of $\mathbf{X}$ in the vector $\mathbf{s}$ we have

$$\|\mathbf{X}\|_F = \|\mathbf{s}\|_2 . \tag{C.51}$$

The spectral and nuclear norms

The observation that the ℓ_2 norm of the vector of singular values of a matrix is identical to its Frobenius norm motivates the use of other ℓ_p norms on the vector $\mathbf{s}$. In particular, the ℓ_1 norm of $\mathbf{s}$ defines the nuclear norm of $\mathbf{X}$ denoted by $\|\mathbf{X}\|_*$

$$\|\mathbf{X}\|_* = \|\mathbf{s}\|_1 \tag{C.52}$$

and the ℓ_∞ norm of $\mathbf{s}$ defines the spectral norm of $\mathbf{X}$ denoted by $\|\mathbf{X}\|_2$

$$\|\mathbf{X}\|_2 = \|\mathbf{s}\|_\infty . \tag{C.53}$$

Because the singular values of real matrices are always nonnegative, the spectral norm and the nuclear norm of such a matrix are simply its largest and the sum of all its singular values, respectively.

References

[1] J. Elson, J. R. Douceur, J. Howell, and J. Saul, "Asirra: a CAPTCHA that exploits interest-aligned manual image categorization," *Proceedings of ACM Conference on Computer and Communications Security*, pp. 366–374, 2007.

[2] D. Lee, W. Van der Klaauw, A. Haughwout, M. Brown, and J. Scally, "Measuring student debt and its performance," *FRB of New York Staff Report*, no. 668, 2014.

[3] R. Panaligan and A. Chen, "Quantifying movie magic with google search," *Google Whitepaper*, 2013.

[4] S. Asur and B. A. Huberman, "Predicting the future with social media," *Proceedings of IEEE/WIC/ACM International Conference on Web Intelligence and Intelligent Agent Technology (WI-IAT)*, vol. 1, pp. 492–499, 2010.

[5] N. Dalal and B. Triggs, "Histograms of oriented gradients for human detection," *Proceedings of IEEE Computer Society Conference on Computer Vision and Pattern Recognition*, vol. 1, pp. 886–893, 2005.

[6] M. Enzweiler and D. M. Gavrila, "Monocular pedestrian detection: survey and experiments," *IEEE Transactions on Pattern Analysis and Machine Intelligence*, vol. 31, no. 12, pp. 2179–2195, 2009.

[7] S. Maldonado-Bascon, S. Lafuente-Arroyo, P. Gil-Jimenez, H. Gomez-Moreno, and F. López-Ferreras, "Road-sign detection and recognition based on support vector machines," *IEEE Transactions on Intelligent Transportation Systems*, vol. 8, no. 2, pp. 264–278, 2007.

[8] B. Pang, L. Lee, and S. Vaithyanathan, "Thumbs up?: sentiment classification using machine learning techniques," *Proceedings of the ACL-02 Conference on Empirical Methods in Natural Language Processing*, vol. 10, pp. 79–86, 2002.

[9] R. Hammer, J. R. Booth, R. Borhani, and A. K. Katsaggelos, "Pattern analysis based on fMRI data collected while subjects perform working memory tasks allowing high-precision diagnosis of ADHD," *US Patent App. 15317724*, 2017.

[10] D. Maclaurin, D. Duvenaud, and R. P. Adams, "Autograd: reverse-mode differentiation of native Python," *ICML Workshop on Automatic Machine Learning*, 2015.

[11] M. Johnson, R. Frostig, and C. Leary, "Compiling machine learning programs via high-level tracing," *Systems and Machine Learning (SysML)*, 2018.

[12] A. G. Baydin, B. A. Pearlmutter, A. A. Radul, and J. M. Siskind, "Automatic differentiation in machine learning: a survey," *Journal of Marchine Learning Research*, vol. 18, pp. 1–43, 2018.

[13] R. D. Neidinger, "Introduction to automatic differentiation and MATLAB object-oriented programming," *SIAM Review*, vol. 52, no. 3, pp. 545–563, 2010.

[14] D. G. Luenberger, *Linear and Nonlinear Programming*. Springer, 2003.

[15] S. P. Boyd and L. Vandenberghe, *Convex Optimization*. Cambridge University Press, 2004.

[16] D. Harrison Jr and D. L. Rubinfeld, "Hedonic housing prices and the demand for clean air," *Journal of Environmental Economics and Management*, vol. 5, no. 1, pp. 81–102, 1978.

[17] D. Dua and C. Graff, *Auto MPG dataset*. UCI Machine Learning Repository available at https://archive.ics.uci.edu/ml/datasets/auto+mpg, 2017.

[18] C. Cortes and V. Vapnik, "Support-vector networks," *Machine Learning*, vol. 20, no. 3, pp. 273–297, 1995.

[19] S. Boyd, N. Parikh, E. Chu, B. Peleato, and J. Eckstein, "Distributed optimization and statistical learning via the alternating direction method of multipliers," *Foundations and Trends® in Machine Learning*, vol. 3, no. 1, pp. 1–122, 2011.

[20] B. Schölkopf and A. J. Smola, *Learning with Kernels: Support Vector Machines, Regularization, Optimization, and Beyond*. MIT Press, 2002.

[21] L. Bottou, "Large-scale machine learning with stochastic gradient descent," pp. 177–186, 2010.

[22] O. Chapelle, "Training a support vector machine in the primal," *Neural Computation*, vol. 19, no. 5, pp. 1155–1178, 2007.

[23] D. Dua and C. Graff, *Spambase dataset*. UCI Machine Learning Repository available at https://archive.ics.uci.edu/ml/datasets/spambase, 2017.

[24] D. Dua and C. Graff, *Statlog dataset*. UCI Machine Learning Repository available at https://archive.ics.uci.edu/ml/datasets/statlog+(german+credit+data), 2017.

[25] R. Rifkin and A. Klautau, "In defense of one-vs-all classification," *Journal of Machine Learning Research*, vol. 5, pp. 101–141, 2004.

[26] Y. Tang, "Deep learning using support vector machines," *CoRR, abs/1306.0239*, 2013.

[27] D. Dua and C. Graff, *Iris dataset*. UCI Machine Learning Repository available at https://archive.ics.uci.edu/ml/datasets/iris, 2017.

[28] D. P. Bertsekas, "Incremental gradient, subgradient, and proximal methods for convex optimization: a survey," *Optimization for Machine Learning*, vol. 2010, pp. 1–38, 2011.

[29] Y. LeCun and C. Cortes, *MNIST handwritten digit database*. Available at http://yann.lecun.com/exdb/mnist/, 2010.

[30] B. A. Olshausen and D. J. Field, "Sparse coding with an overcomplete basis set: a strategy employed by V1?" *Vision Research*, vol. 37, no. 23, pp. 3311–3325, 1997.

[31] D. D. Lee and H. S. Seung, "Algorithms for nonnegative matrix factorization," *Advances in Neural Information Processing Systems*, pp. 556–562, 2001.

[32] C. D. Manning and H. Schütze, *Foundations of Statistical Natural Language Processing*. MIT Press, 1999.

[33] H. Barlow, "The coding of sensory messages," *Current Problems in Animal Behaviour*, pp. 331–360, 1961.

[34] H. Barlow, "Redundancy reduction revisited," *Network: Computation in Neural Systems*, vol. 12, no. 3, pp. 241–253, 2001.

[35] S. J. Prince, *Computer Vision: Models, Learning, and Inference*. Cambridge University Press, 2012.

[36] Y. LeCun, K. Kavukcuoglu, and C. Farabet, "Convolutional networks and applications in vision," *Proceedings of IEEE International Symposium on Circuits and Systems (ISCAS)*, pp. 253–256, 2010.

[37] Y. LeCun and Y. Bengio, "Convolutional networks for images, speech, and time series," *The Handbook of Brain Theory and Neural Networks*, vol. 3361, no. 10, 1995.

[38] A. Krizhevsky, I. Sutskever, and G. E. Hinton, "Imagenet classification with deep convolutional neural networks," *Advances in Neural Information Processing Systems*, pp. 1097–1105, 2012.

[39] S. Marčelja, "Mathematical description of the responses of simple cortical cells," *JOSA*, vol. 70, no. 11, pp. 1297–1300, 1980.

[40] J. P. Jones and L. A. Palmer, "An evaluation of the two-dimensional gabor filter model of simple receptive fields in cat striate cortex," *Journal of neurophysiology*, vol. 58, no. 6, pp. 1233–1258, 1987.

[41] X. Huang, A. Acero, and H. W. Hon, *Spoken Language Processing: A Guide to Theory, Algorithm and System Development.*   Prentice Hall, 2001.

[42] L. R. Rabiner and B. H. Juang, *Fundamentals of Speech Recognition.*   Prentice Hall, 1993.

[43] D. Dua and C. Graff, *Breast cancer Wisconsin (diagnostic) dataset.*   The University of California, Irvine (UCI) Machine Learning Repository available at https://archive.ics.uci.edu/ml/datasets/breast+cancer+wisconsin+(diagnostic), 2017.

[44] G. Galilei, *Dialogues Concerning Two New Sciences.*   Dover, 1914.

[45] S. Straulino, "Reconstruction of Galileo Galilei's experiment: the inclined plane," *Physics Education*, vol. 43, no. 3, p. 316, 2008.

[46] J. Lin, S. M. Lee, H. J. Lee, and Y. M. Koo, "Modeling of typical microbial cell growth in batch culture," *Biotechnology and Bioprocess Engineering*, vol. 5, no. 5, pp. 382–385, 2000.

[47] G. E. Moore, "Cramming more components onto integrated circuits," *Proceedings of the IEEE*, vol. 86, no. 1, pp. 82–85, 1998.

[48] V. Mayer and E. Varaksina, "Modern analogue of Ohm's historical experiment," *Physics Education*, vol. 49, no. 6, p. 689, 2014.

[49] W. Rudin, *Principles of Mathematical Analysis.*   McGraw-Hill New York, 1964.

[50] G. Cybenko, "Approximation by superpositions of a sigmoidal function," *Mathematics of Control, Signals and Systems*, vol. 2, no. 4, pp. 303–314, 1989.

[51] J. Park and I. W. Sandberg, "Universal approximation using radial-basis-function networks," *Neural Computation*, vol. 3, no. 2, pp. 246–257, 1991.

[52] K. Hornik, M. Stinchcombe, and H. White, "Multilayer feedforward networks are universal approximators," *Neural Networks*, vol. 2, no. 5, pp. 359–366, 1989.

[53] A. Rahimi and B. Recht, "Uniform approximation of functions with random bases," *Proceedings of the 46th Annual Allerton Conference on Communication, Control, and Computing*, pp. 555–561, 2008.

[54] A. Rahimi and B. Recht, "Random features for large-scale kernel machines," *Advances in Neural Information Processing Systems*, pp. 1177–1184, 2008.

[55] H. Buhrman and R. De Wolf, "Complexity measures and decision tree complexity: a survey," *Theoretical Computer Science*, vol. 288, no. 1, pp. 21–43, 2002.

[56] B. Osgood, *Lectures on the Fourier Transform and Its Applications.*   American Mathematical Society, 2019.

[57] D. J. MacKay, "Introduction to Gaussian processes," *NATO ASI Series F Computer and Systems Sciences*, vol. 168, pp. 133–166, 1998.

[58] C. M. Bishop, *Pattern Recognition and Machine Learning.*   Springer, 2011.

[59] F. Rosenblatt, *The Perceptron, A Perceiving and Recognizing Automaton.* Cornell Aeronautical Laboratory, 1957.

[60] X. Glorot, A. Bordes, and Y. Bengio, "Deep sparse rectifier networks," *Proceedings of the 14th International Conference on Artificial Intelligence and Statistics*, vol. 15, pp. 315–323, 2011.

[61] I. J. Goodfellow, D. Warde-Farley, M. Mirza, A. Courville, and Y. Bengio, "Maxout networks," *arXiv preprint arXiv:1302.4389*, 2013.

[62] S. Ioffe and C. Szegedy, "Batch normalization: accelerating deep network training by reducing internal covariate shift," *arXiv preprint arXiv:1502.03167*, 2015.

[63] G. Huang, Y. Li, G. Pleiss, Z. Liu, J. E. Hopcroft, and K. Q. Weinberger, "Snapshot ensembles: train 1, get M for free," *arXiv preprint arXiv:1704.00109*, 2017.

[64] L. N. Smith, "Cyclical learning rates for training neural networks," *Proceedings of 2017 IEEE Winter Conference on Applications of Computer Vision (WACV)*, pp. 464–472, 2017.

[65] L. Breiman, J. Friedman, C. J. Stone, and R. A. Olshen, *Classification and Regression Trees.* Chapman and Hall CRC, 1984.

[66] J. H. Friedman, "Greedy function approximation: a gradient boosting machine," *Annals of Statistics*, pp. 1189–1232, 2001.

[67] T. Chen and C. Guestrin, "XGBoost: a scalable tree boosting system," *Proceedings of the 22nd ACM SIGKDD International Conference on Knowledge Discovery and Data Mining*, pp. 785–794, 2016.

[68] A. Liaw and M. Wiener, "Classification and regression by randomForest," *R News*, vol. 2, no. 3, pp. 18–22, 2002.

[69] D. P. Kingma and J. Ba, "Adam: a method for stochastic optimization," *arXiv preprint arXiv:1412.6980*, 2014.

[70] T. Tieleman and G. Hinton, *Lecture 6a: Overview of Mini?Batch Gradient Descent.* Neural Networks for Machine Learning (Coursera Lecture Slides), 2012.

[71] C. M. Bishop, *Neural Networks for Pattern Recognition.* Oxford University Press, 1995.

[72] R. Fletcher, *Practical Methods of Optimization.* John Wiley & Sons, 2013.

[73] C. G. Broyden, "The convergence of a class of double-rank minimization algorithms," *Journal of the Institute of Mathematics and Its Applications*, vol. 6, no. 3-4, pp. 76–90, 1970.

[74] S. Boyd and L. Vandenberghe, *Introduction to Applied Linear Algebra: Vectors, Matrices, and Least Squares.* Cambridge University Press, 2018.

[75] M. Lobo, L. Vandenberghe, S. Boyd, and H. Lebret, "Applications of second-order cone programming," *Linear Algebra and Its Applications*, vol. 284, no. 1, pp. 193–228, 1998.

[76] L. N. Trefethen and D. Bau III, *Numerical Linear Algebra.* SIAM: Society for Industrial and Applied Mathematics, 1997.

Index

Z
zero-order optimization, 21